Phylogeny of the Primates

A Multidisciplinary Approach

Phylogeny
of the Primates
A Multidisciplinary Approach

EDITED BY

W. Patrick Luckett
Creighton University
Omaha, Nebraska

AND

Frederick S. Szalay
Hunter College, C.U.N.Y.
and the American Museum of Natural History
New York, New York

PLENUM PRESS • NEW YORK AND LONDON

Library of Congress Cataloging in Publication Data

Main entry under title:

Phylogeny of the primates.

"Proceedings of the Wenner Gren Symposium No. 61 in Burg Wartenstein, Austria, July 6-14, 1974."
   Includes bibliographies and index.
   1. Primates — Evolution — Congresses. 2. Primates — Classification — Congresses. I. Luckett, Winter Patrick. II. Szalay, Frederick S.
QL737.P945   1975   599'.8'0438   75-30714
ISBN 0-306-30852-5

Proceedings of the Wenner Gren Symposium No. 61
in Burg Wartenstein, Austria, July 6-14, 1974

© 1975 Plenum Press, New York
A Division of Plenum Publishing Corporation
227 West 17th Street, New York, N.Y. 10011

United Kingdom edition published by Plenum Press, London
A Division of Plenum Publishing Company, Ltd.
Davis House (4th Floor), 8 Scrubs Lane, Harlesden, London, NW10 6SE, England

Printed in the United States of America

Preface

The past decade has witnessed a tremendous surge of interest in varied aspects of primate biology, encompassing virtually all disciplines of the biological sciences. Regardless of whether these studies have been approached from a paleontological, morphological, developmental, biochemical, neuroanatomical, or behavioral point of view, one underlying theme has been a common interest in the possible phylogenetic relationships suggested by the results of such studies. In some cases, sound taxonomic principles have not been followed in the interpretation of these data, and this has led to skepticism among many taxonomists with regard to the validity of some of the genealogical relationships and conclusions suggested by comparative studies of living primates.

It is generally agreed that the fossil record alone provides the essential *time dimension* for directly observing changes in characteristics, but unfortunately this record is limited both in the number of genera represented and particularly in the incomplete nature of the available preserved material. On the other hand, extensive comparative analyses of numerous characteristics in living primates have provided additional insight into possible phylogenetic relationships, despite the lack of a time dimension. Such studies of both fossil and living primates are enhanced considerably by a cladistic analysis of the probable primitive (ancestral) or advanced (derived) condition of each character state discussed, based upon their distribution (and ontogeny, wherever possible) in a wide variety of primate and nonprimate taxa, including other eutherian mammals, marsupials, monotremes, and reptiles.

It is virtually impossible for any single student of primate evolution to understand and evaluate critically all of the available comparative data that now bear on the phylogenetic relationships among Primates. It is primarily for this reason that we organized an international conference at which a number of major questions pertaining to the phylogenetic relationships among Primates were discussed from a multidisciplinary point of view.

v

This volume is the result of the proceedings of that conference, Burg Wartenstein Symposium No. 61 on the Phylogeny of the Primates, held during July 6–14, 1974, at the Burg Wartenstein European Conference Center of the Wenner–Gren Foundation for Anthropological Research, near Gloggnitz, Austria. Participants were drawn from the following areas of research which at present appear to be among the most useful in assessing phylogenetic relationships:

1. The dental, cranial, and musculoskeletal systems of fossil and living groups, including ontogenetic aspects where feasible.
2. The anatomy and physiology of the nervous system.
3. The morphogenesis of the fetal membranes, placenta, and other aspects of reproductive biology.
4. Biochemical analyses of tissue and serum proteins and of DNA.
5. Evolutionary appraisal of behavior (ethology).

Participants were asked to discuss the methods of phylogenetic inference used in assessing the primitive or derived nature of the characters studied (the polarity of character clines) and to evaluate the reliability of judgments on homology, parallelism, or convergence of character states in their fields of research.

Throughout the conference sporadic, but often lively and stimulating, discussion focused on the methods of phylogenetic inference. There was a general endorsement of Hennig's method of cladistic analysis for the determination of branching sequences, and the subsequent use of derived character states for the establishment of recency of relationships. Participants were encouraged to construct phylogenetic trees or cladograms based solely on an analysis of their particular data, rather than to propose new classifications based on these data. We have not attempted an overall synthesis of the presently available data relating to primate phylogeny; however, fundamentally similar conclusions concerning phylogenetic relationships were reached independently by many of the contributors, when an adequate sampling of data was available and sound methods of phylogenetic inference were employed. Some of the relationships evaluated from a multidisciplinary point of view included the possible affinities of Tupaiidae and other Eutheria to Primates, the haplorhine–strepsirhine dichotomy, relationships between Lemuriformes and Lorisiformes, cladistic relationships of the Tarsiiformes, the monophyly of the Anthropoidea, and the relationships within the Catarrhini and the Hominoidea.

We thank the Wenner–Gren Foundation for Anthropological Research for their generous financial support which made the conference possible. On behalf of the conference participants, we wish to express our appreciation to Mrs. Lita Osmundsen, Director of Research, the Wenner–Gren Foundation, and to Dr. Karl Frey, Charlotte Frey, Maria Brunet, Jane Barden, Steve Montayne, and Laura Montayne, for their warmth and hospitality which made our stay at the Burg Wartenstein Conference Center both enjoyable and intellectually profitable.

W. PATRICK LUCKETT
FREDERICK S. SZALAY

Contributors

PETER ANDREWS · Sub-Department of Anthropology, British Museum (Natural History), London, England.

C. B. G. CAMPBELL · Departments of Anatomy and Radiological Sciences, University of California, Irvine, California College of Medicine, Irvine, California.

MATT CARTMILL · Departments of Anatomy and Anthropology, Duke University, Durham, North Carolina.

PIERRE CHARLES-DOMINIQUE · Laboratoire de Primatologie et d'Écologie des Forêts Equatoriales, C.N.R.S., and Museum National d'Histoire Naturelle, Laboratoire d'Écologie Générale, Brunoy, France.

ERIC DELSON · Department of Anthropology, Lehman College, City University of New York, Bronx, New York.

JOHN F. EISENBERG · National Zoological Park, Smithsonian Institution, Washington, D.C.

MORRIS GOODMAN · Department of Anatomy, Wayne State University School of Medicine, Detroit, Michigan.

DAVID E. KOHNE · Department of Experimental Pathology, Scripps Clinic and Research Foundation, La Jolla, California.

W. PATRICK LUCKETT · Department of Anatomy, Creighton University School of Medicine, Omaha, Nebraska.

R. D. MARTIN · Senior Research Fellow, Wellcome Institute of Comparative Physiology, The Zoological Society of London, London, England.

MALCOLM C. MCKENNA · Department of Vertebrate Paleontology, The American Museum of Natural History and Department of Geology, Columbia University, New York, New York.

CHARLES R. NOBACK · Department of Anatomy, College of Physicians and Surgeons, Columbia University, New York, New York.

JEFFREY H. SCHWARTZ · Department of Anthropology, University of Pittsburgh, Pittsburgh, Pennsylvania.

GEORGE G. SIMPSON · University of Arizona and Simroe Foundation, Tuscon, Arizona.

DIETRICH STARCK · Anatomisches Institut der Universität, Frankfurt am Main, West Germany.

FREDERICK S. SZALAY · Department of Anthropology, Hunter College of the City University of New York, New York, and Department of Vertebrate Paleontology, The American Museum of Natural History, New York.

IAN TATTERSALL · Department of Anthropology, The American Museum of Natural History, New York, New York.

RUSSELL TUTTLE · Department of Anthropology and Committee on Evolutionary Biology, University of Chicago, and Yerkes Regional Primate Research Center, Emory University.

Contents

PHYLOGENY OF PRIMATE HIGHER TAXA

BACKGROUND TO PRIMATE PHYLOGENY

I

Recent Advances in Methods of Phylogenetic Inference*

G‍EORGE G‍AYLORD S‍IMPSON

I. Introduction

It would be possible to form an impression that zoologists are not currently much in‑
terested in the phylogeny of mammals. For example among 111 papers scheduled for
presentation at the 1973 meeting of the American Society of Mammalogists, none were
directly phylogenetic and only 9 (8%) dealt with systematics. Among 100 recent papers
sampled at random in the journals *Systematic Zoology* and *Evolution*, 9 were on mam‑
malian systematics in a rather narrow sense, but of these only 4 involved phylogeny.
Nevertheless, the existence of the present conference and examination of a broader
spectrum of the scattered literature do show that an interest in phylogeny does exist
and that indeed this is now quite an active field of study and debate.

Much of the recent work is theoretical in nature, and it is difficult to avoid a certain
distaste for dogmatic statements of principles by authors who seem to have a limited

---

* This paper was originally prepared early in 1974 as a preprint to be circulated at the conference. Although
Dr. Simpson planned extensive rewriting following the conference, circumstances entirely beyond his control
prevented him from making any of the desired changes. For this reason Dr. Simpson planned to withdraw his
contribution, but at the request of the editors he graciously consented to have the original preconference paper
published. For this and for his illuminating presence and participation throughout the conference, we the
editors are deeply grateful to him.

G‍EORGE G‍AYLORD S‍IMPSON · University of Arizona and the Simroe Foundation, Tucson, Arizona.

3

acquaintance with the broad application of those principles to actual cases. There are, however, many exceptions and, while the exceptions apply to many groups of animals and of plants as well, it is interesting that many occur in three specialties: entomology, primatology, and molecular biology. The background of these studies is almost always taxonomic, and the foreground is usually taxonomic as well. Almost all of the taxonomic studies that deal with principles or methodology have contributions, potential, at least, to the theories and methods of phylogeny. It is perhaps curious that this is true even of work by taxonomists who maintain that in principle there should be no relationship between phylogeny and taxonomy. That it is difficult, in the present state of knowledge virtually impossible, to keep taxonomy and phylogeny quite separate is illustrated by a recent symposium (Hull, 1973). This was explicitly "to concern problems in *classification*, not *phylogeny reconstruction*" (italics in the original), but all the authors wrote about reconstruction of phylogeny, indeed two of them wrote about little else.

It is, then, appropriate and should be useful to start a background discussion about principles and methods in phylogeny with some consideration of current schools of taxonomy.

II. Taxonomy

The most frequent or at least the most discussed approaches to taxonomy today are those that I shall designate as essentialist, phenetic, evolutionary, cladistic, and non-selectionist although, as I shall note in passing, all often go by other names.

In their extreme forms, these schools of taxonomy are incompatible. I am a frank partisan of evolutionary taxonomy, and I believe that the other four schools specified have weaknesses that make them less useful, and indeed less valid, in those extreme forms. However, in their less extreme development they tend to approach, even to merge with, evolutionary taxonomy. Moreover, although two of them (essentialist and phenetic) in principle are nonphylogenetic, all have made and can continue to make definite contributions to the study of phylogeny.

I will state the elements of their extreme forms as briefly as possible. Such summaries do not do them justice, and I will cite more extended, authoritative treatments of each. I will state, also as briefly as possible, what I consider weaknesses in them, and will also cite some other criticisms. Subsequently I will discuss their methodological bearings on the study of taxonomy.

I will take them approximately in historical order, and here essentialism has right of place for its philosophy derives from the ancient Greeks and then the medieval scholastics, although it still has adherents today. This now ancient approach has also been called "classical" and "omnispective." Its most complete recent exposition is a book by Blackwelder (1967), who gives many background references. While modern essentialists recognize evolution and phylogeny as facts, they reject them as bearing on the practice of taxonomy, which is held to be concerned only with the results of those facts. Classification is merely the grouping of organisms according to resemblances and differences in their essential beings, their "essences" in the scholastic sense. Although Blackwelder avoids that term in his book, it is implicit there and explicit in previous publications. The greatest weakness is just that point: the implicit basis is undefined and is left on a completely subjective and intuitive level. For more extended criticism see, for example, Hull (1965) and Mayr (1969).

Phenetic taxonomy has flourished mainly within the last 15 years, but it is consciously a revival of a theoretical approach advanced by Adanson (1763) more than two centuries ago. In both its early and its modern form it was and is an attempt to reduce (some extreme enthusiasts believe to eliminate) the subjective and intuitive elements in the still older essentialist taxonomy. It has a large literature, much of it in the journal *Systematic Zoology*, and its principal full statement in an extreme form was by Sokal and Sneath (1963). In that form, its basic principles are that classification should be based solely on overall similarity in phenotypic characters and that every character should be given equal weight. "Affinity" is defined as "overall similarity,"and any phylogenetic or other evolutionary considerations are ruled out. In order to reduce observations of a large number of characters to a single measure of "affinity," they are put in numerical form and run through programmed computers, hence the commonly used misnomer "numerical taxonomy."

The results were claimed to be strictly empirical, objective, and consistently repeatable. Subsequent criticisms include the fact that none of those claims is strictly true. The selection of characters does in effect weigh them, and followers of other philosophies insist that indeed characters should be consciously weighed. The numerical expression of characters not inherently numerical is an artifice and makes different measures not truly comparable. Correlation of characters within organisms and their parallelism and convergence between groups of organisms are ignored, although they are highly relevant to a rational concept of affinity. Results are fully repeatable only if procedures are exactly copied, which is true of anything. The principles are not in fact strictly empirical but have a somewhat veiled theoretical basis while needlessly and confusingly discarding the theoretical basis, that of evolution, that is most relevant of all. (See especially Hull, 1970, and many references there.)

In historical sequence evolutionary taxonomy should follow here. It can be dated from Darwin and publication of *The Origin of Species* in 1859, although Ghiselin (1969) has shown that Darwin was in fact beginning to practice evolutionary taxonomy, especially in his study of barnacles, before he revealed his views on evolution. (Lamarck was an earlier distinguished taxonomist and a kind of evolutionist, but his taxonomy was not truly evolutionary.) Its fairly current status has been discussed at book length by Mayr (1969) and by me (Simpson, 1961). It has been cogently compared with the phenetic and cladistic systems in Hull's (1970) interesting review. Its main tenet is that although classification is achieved mainly on the basis of resemblances and differences among groups of organisms, these must be interpreted in the light of their evolutionary significance. The groups classified, taxa, are populations, and they vary in practically all their characters. Characters closely correlated phenetically and connected genetically should be given lower individual weighting or not counted separately. Inferences as to phylogeny should be based on as many different characters as possible and derived by the clearest available and most significant methods. Phylogeny includes not only cladistic relationships but also patristic, not only splitting but also phyletic divergence and advance. Such evolutionary phenomena as parallelism and convergence must be recognized and resemblances due solely to them given appropriate phylogenetic interpretations. A strict interpretation of monophyly is impractical in classification (although not necessarily in a phylogeny as such). Phylogeny as a whole, in all its relevant aspects, cannot be expressed in classification. Classification should, however, be consistent with an inferred phylogeny. Different classifications can be consistent with the same phylogeny, and preference for one looks to the need for clear communication and requires individual judgment. The term "phylogenetic taxonomy" sometimes applied to this system is thus a misnomer, and it also causes confusion with cladistic taxonomy, most of whose adherents call it "phylogenetic."

The principal objections to evolutionary taxonomy are, first, that phylogeny cannot be observed and is often difficult or even impossible to infer from available observations, and, second, that the system is said to be insufficiently objective, allowing some measure of judgment and art rather than being stringently scientific. I believe those objections to be invalid and will discuss some of the reasons later in more direct relationship to phylogenetic studies.

Cladistic taxonomy in its extreme or perhaps one should say purest form was largely the construction of one man, Willi Hennig, although he had forerunners. He has provided the canonical version in two books (Hennig, 1950, 1966; the latter version, put into English from a manuscript probably written about 1960, is not a direct translation of the earlier one). Brundin (1966) should also be mentioned as further exposition and extended application of the strict theory. It is not a coincidence that both Hennig and Brundin are students of Recent insects. The essence of this system as applied to classification is: that phylogeny consists entirely of a series of dichotomies, each representing the splitting of an ancestral species into two sister species; that the ancestral species ceases to exist at the time of the dichotomy; that taxa including the descendants of sister species must be given the same hierarchical rank; that the relative ranks of all taxa therefore depend on the relative time of origin of the first species belonging to them; and that all taxa must be monophyletic in the strictest·sense of descent from one stem species that belongs to the taxon in question and not to any contemporaneous taxon of equal rank.

As Hull (1970) has said in a discussion that is nevertheless critical of Hennig and should be read on that account, "Hennig's principles of classification are extremely straightforward." If followed precisely, they enable the author's beliefs as to the sequence of phylogenetic dichotomies to be inferred from the classification itself. Nevertheless Hennig's system is not practical as a general methodology for classification, and while (as later noted) it has great merit as an aid to inferences about the strict cladistic aspects of phylogeny, it is not adequate for taxonomy or for the study of phylogeny in general. Among its defects are these: it is not necessarily or probably true that dichotomy is the sole pattern of cladistic evolution; ancestral species commonly persist with little or no genetic, hence phylogenetic change, although daughter species have split off from them; the extreme requirement of monophyly leads to such practical impossibilities as having to classify two hardly or not distinguishable sister species (products of dichotomy from one parent species) in different classes or phyla; the ranking of taxa by antiquity leads to further absurdities in practice (even if not in inapplicable principle) such as having to classify all Precambian species extinct without progeny as separate phyla; the system takes no account of evolutionary factors that nevertheless are of great importance for genetic affinities and hence for phylogeny, notably the degree of genetic divergence in different phyletic sequences or of genetic change within single phyletic lines; and attainment of a truly objective classification by this system would require knowledge of all past dichotomies, knowledge not available and certainly not obtainable. Besides Hull (1970), before cited, see especially Bock (1968) and Darlington (1970) for further critical discussion.

The nonselectionist school, most recent of those here considered differs in that its aim is primarily phylogenetic, not taxonomic. It becomes involved in taxonomy because a phylogeny inferred according to its principles and by its methods may then be involved in taxonomy according to the further principles and methods of any of the schools of taxonomy, notably the evolutionary or the cladistic. The basis is the composition of large biomolecules, usually amino acid sequences in polypeptides and proteins and also corresponding codon or base sequences in nucleic acids. No one questions the validity of such data or their relevance, in some way, to both phylogeny and, either directly or through phylogeny, to taxonomy. Questions arise only as regards the direct transfer of those data and, usually, no others into phylogeny on the basis of three principal axioms or postulates:

1. The genetic relationships of organisms are directly proportional to the minimum number of changes (mutations) in amino acid (codon, base) sequences required to produce the differences between them.
2. These changes occur at an approximately (or exactly) constant rate during phylogeny.
3. They are not affected by natural selection, that is, their effects are selectively neutral.

Because of the first postulate, this system is occasionally called allozymic. Because of the third, it is usually called non–Darwinian, in the incorrect belief that natural selection is the only Darwinian factor in evolution.

Interpretation of the molecular data on the basis of these postulates leads to conclusions as to both the relative and the absolute places of dichotomies in a phylogeny. All three postulates are open to serious question. For discussion on both sides, see for instance Le Cam *et al.* (1972). Phylogenetic problems are further considered later in the present paper.

III. The Data of Phylogeny

Taxonomy and its application in classification long antedated any idea of phylogeny. When the Darwinian revolution introduced phylogeny as a major concern of biologists, the data involved were inevitably the same as those earlier used in classification. That continues to be true, and I have sufficiently emphasized that studies of taxonomy and phylogeny are related and sometimes confused disciplines. The classical data of taxonomy were anatomical and that continues to be largely, but now far from solely, true in the study of phylogeny. Recent changes involve for the most part the use of different kinds of characters and methods for handling larger numbers of characters.

The data for both taxonomy and phylogeny potentially but not practically include anything that is known about the organisms under study. I know of no complete roster of the kinds of data obtainable, and I will not attempt one in detail here. Sokal and Sneath (1963) give as categories: (1) morphological, (2) physiological, (3) behavioral, and (4) ecological and distributional characters. Extensive as the categories are, they are still not complete. Mayr (1969) gives a more extended formal list, with 18 categories under these main headings: (1) morphological, (2) physiological, (3) ecological, (4) ethological, and (5) geographical characters. Even his minor subdivisions are incomplete, as some of his own subsequent examples do not fall clearly into one of them.

Morphological characters still are the most used data in phylogeny. They have been extended greatly by the invention of electron microscopes and perhaps still more by electron scanning microscopes. Even the older but recently improved light microscopes provide relatively new kinds of data, especially now in comparative karyology. It is not necessary or possible to review here the classical and still usual morphological data for taxonomy and phylogeny. Even for the relatively recent development of karyology, the general nature and many examples of such data are familiar to everyone. It should never-theless be noted that the diversity of these general kinds of data raises problems of measurement or expression that are still far from adequate solution. For example it could be reasonably maintained that the *forms* of organisms and their parts are the most fundamental characteristics relevant to their taxonomic and phylogenetic inferences. There is an enormous literature on this subject; for a rather elementary introduction see, for example, the delightful but rather anachronistic classic by D'Arcy Thompson (1917, 1942) and such subsequent symposial volumes as those edited by Clark and Medawar (1945), by Zuckerman (1950), and by L. L. Whyte (1968), the latter extending its scope beyond biology. Despite much attention and labor, except in relatively extremely simple cases no method of observation has been developed superior to simple comparison with the human eye and no method of record and communication superior to straightforward pictorial reproduction and verbal description.

That fact reduces the value of entirely numerical approaches to comparisons of organisms. It should, however, be noted that there are now many ingenious methods for basing indices of resemblance on multidimensional data. These data can include essential determinants of total morphology and thus to some extent reduce the need for subjective comparison alone. In general they are simple dimensions, often linear, which cannot practically correspond with all characteristics of a complex form, or they are observed morphological differences reduced to "character states," itself a subjective procedure. Although the numerical treatment of form is not adequate at present, some progress has been made. In addition to references previously given, see Andrews (1972) and Oxnard (1973), who, although his study is functional rather than phylogenetic, interestingly illustrates the use of such data for primates.

Most distinctive of modern data are those molecular or biochemical in nature. Among the earliest approaches in this general field is that of systematic serology or immunology. A pioneer in this subject has reviewed much of it as a central part of a somewhat idiosyncratic book (Boyden, 1973). An earlier review was edited by Leone (1964). Numerous applications, generally using other but allied procedures, are known to anthropologists because of special concentration on primates, for example Goodman (1963), Sarich and Wilson (1966), and Williams (1964). In all the various procedures the data quantify resemblances in proteins of different organisms. These may be made specific for certain homologous protein molecules, as by the microcomplement fixation method of Wasserman and Levine (1960). In usual serological methods, however, the data are from the totality of a mixture of individually unidentified proteins. There is some advantage from thus automatically combining in one sample a considerable part of the genetic code, but a disadvantage in not knowing what concentrations of particular proteins may affect and bias the result.

A next step is identification of molecules, almost always again proteins, that vary within and between species. Of several available methods electrophoresis and various forms of chromatography are most widely known and used. The separations achieved by these methods are assumed to represent alleles of genes, hence codon differences in DNA; see especially Lewontin (1974). It is believed that such procedures nevertheless miss a considerable number of alleles (or genetic variants per locus), but refinement of techniques increases their selectivity (for instance: Bernstein et al., 1973).

A next step is the identification of amino acid sequences in proteins. Although the procedures involved are onerous and data still refer to relatively few proteins (or polypeptides) in a small minority of species of organisms, the body of data already achieved is impressive and highly significant for taxonomy and phylogeny. See, for instance, the compilations of Bryson and Vogel (1965), Dayhoff (1969), and Lewontin (1974), the brief review of comparison techniques by Jukes (1972), and the popular exposition of one of the best-known and most interesting examples (cytochrome *c*) by Dickerson (1972).

Although there has also been much discussion of DNA evolution, most of this has been based on knowledge of amino acid sequences and inferences from them to base sequences in DNA codons. Sequences in DNA itself do not necessarily correspond exactly or perhaps closely with such inferences, not only because of the "degeneracy" (apparent redundancy) of the code and the fact that some codons do not code for amino acids but also because of the distinct possibility that "the great majority of DNA . . . does not code for proteins" (Crow, 1972). If that should prove to be true, DNA sequences in themselves alone might prove to have little bearing on evolution, a shocking idea. Nevertheless it is of interest that techniques have been developed for direct comparison of base sequences in DNA from different, even widely different, species (Hoyer et al., 1964; Kohne et al., 1972—the latter paper has special reference to primates).

So far reference has been to physical parts of the organisms themselves, and such data are overwhelmingly most used in studies of taxonomy and phylogeny. There are nevertheless many relevant data of other sorts, mostly behavioral and distributional. An early review of the application of behavioral characteristics to systematics was given by Mayr (1958), and the symposium in which it appeared (Roe and Simpson, 1958) was a pioneering and, as it turned out, highly heuristic rapprochement of behavioral and evolutionary, including phylogenetic, studies. Anthropologists do not need to be reminded of the proliferation of behavioral data on primates, such as the already classic collection edited by DeVore (1965), and their incorporation into general studies of primate and especially human evolution and phylogeny, such as that by Campbell (1974).

Indeed it is not surprising that Darwin was a founder in that respect, as in practically everything that has to do with evolution above the molecular level. Behavioral aspects of evolution, or evolutionary aspects of behavior, figured in *The Origin* (1859), more so in *The Descent of Man* (1871), and a special aspect of behavior and evolution is the whole subject of *The Expression of the Emotions* (1872). It was also Darwin who first clearly connected geographic distribution with phylogeny. Indeed principal clues that set him on the trail of evolution were the discovery that fossils in Argentina were related to characteristic living South American groups and that the Galápagos biota could best be understood as descended with modification from, for the most part, continental South American animals and plants (see Barlow, 1958; also any of the many biographies of Darwin). The most recent general compilation known to me is that by Cox *et al.* (1973). The subject is now in a particularly active state of flux, especially because of the new background involved in recent general acceptance of the theory of continental drift. I have in hand but have not completed a review of the impact of this theory on biogeography.

Connected by its author with continental drift (unnecessarily, since he admits that contact between South America and Africa was not involved) is the resuscitation of an old hypothesis that the South American primates were of African ancestry (Hoffstetter, 1972), an idea of special interest to primatologists.

Last, as far as this brief outline is concerned, but certainly not least, are data as to the relative ages or succession of relevant fossils. It used to be commonly claimed that it is impossible to reach valid conclusions about phylogeny except on the basis of fossil evidence. That position is still held by some phenetic extremists. For example Sokal and Sheath (1963) wrote that, "The classification could not then [i.e., without fossils] be based on the evolution, or at most it could only be based on guesses about the evolution; the interpretation of the evolution would be based, rather, on the classification." (I will indicate later that many pheneticists, including Sokal, no longer hold quite such an extreme view.) If strictly true, that disability would make much or most study of phylogeny futile. Pertinent fossil evidence is available for only a minority of organisms, and *fully* adequate fossil evidence at low taxonomic levels over a geologically long time for almost none. However, there certainly is now, and I believe has always been since Darwin, a strong consensus that reasonable approximations of phylogeny at various taxonomic levels can be made from extinct organisms alone. Indeed the authors of a recent study have swung quite to the opposite extreme from that of the pheneticists and have concluded, with strong reliance on Hennig's cladistics, that even if fossils are known for a group under study their succession should not be taken into account in making inferences as to its phylogeny. They do admit that succession has other significant uses and may even have some veto power over phylogenies otherwise derived. It is remarkable that the authors are paleontologists (Schaeffer *et al.*, 1972).

Despite the antiphylogenetic claims of extreme pheneticists and the antipaleontological views of extreme cladists, I still consider it self-evident that when relevant fossils are available and are well interpreted, their characteristics and their succession provide both the most direct and the most important data bearing on phylogeny. (See, for example, Simpson, 1961; Raup and Stanley, 1971; or almost any able text on paleontology, systematics, and evolution.) That various phenetic and cladistic concepts and methods are applicable to the phylogenetic interpretation of paleontological data is a different point. It can also be shown that paleontological data, when available, bear in crucial ways on operational aspects even of those theoretical taxonomic systems that seek to minimize or to reject paleontological evidence.

IV. Some Methods

From what has already been said about data and taxonomic principles, it is clear that recent advances in phylogenetic inference involve especially a proliferation in kinds and bulk of data and in variety of interpretive methods both logical and mathematical. On the first point, it has sometimes been held that phylogenetic inferences should ideally be based on all available or possible data. From a practical point of view that is not realistic. In principle, the number of possible linear measurements, for example, is infinite; the distance between measurements of the same anatomical part can be successively reduced to the infinitesimal. As the number of molecules in an organism is finite, data on their composition is also finite, but the number is so enormous and their composition is so varied that determination of all of them for even one species is not conceivable. The original pheneticist doctrine that all characters should be equally weighted to begin with, at least, is reduced to nullity because "all characters" cannot be defined and could not be measured if defined. The doctrine of "minimal attributes" occasionally advanced, is equally undefined, and impractical as well as illogical (e.g., Hull, 1970).

It is clear that these recent possibilities do permit advances because the use of more data and of more kinds of data certainly does increase the probability of phylogenetic inferences. There is, however, an effect of diminishing returns not only because even with computers the number of usable data is always limited (computers do not make the observations that they manipulate), but also because a point is reached where additional data do not appreciably affect the inferences drawn. The latter fact indicates also the appropriate cutoff point for addition of new data. However, one simply cannot go through the process of adding data one by one for each two-by-two comparison involved in a phylogeny until no appreciable change results. As a matter of experience with related groups, a student of phylogeny achieves an appreciation of what data and how many data are likely to be adequate, and transfers this judgment to further examples. He learns to select data that discriminate populations at the level involved and that also are adequate for measuring or estimating the degree and nature of their differences and similarities. No one capable of effective phylogenetic study starts with a *tabula rasa*. The early extreme pheneticists did in some cases postulate that as a principle, but none of them really followed it in practice and many have abandoned it in theory.

The addition of new kinds of data has made possible great advances when these are brought into conjunction. That is, however, dependent on the combined use of such data, and there is a tendency for the users of special techniques to draw taxonomic or phylogenetic inferences based on them alone. The inferences drawn from different techniques, hence different kinds of data, are often in conflict, in which case some, at least, must be wrong, and sometimes they are downright absurd. For example, the molecular composition of hemoglobin is certainly a relevant datum, but the conclusion that "therefore from the point of view of hemoglobin structure . . . the two species [gorillas and humans] form actually one continuous population," although properly conditional, is certainly open to misunderstanding (Zuckerkandl, 1963). Although the value of numerical estimation of similarity by computer methods is not questioned, and although hard-core pheneticists use the term "affinities" in a nonphylogenetic sense, it surely is absurd to conclude by that method that female gorillas have closer affinities with male *Homo sapiens* than with male gorillas (Boyle, 1964). And although no reasonable person doubts the importance of immunological evidence, it certainly is not reasonable to reject all other evidence and declare (on the further basis of a hotly disputed interpretation of the immunological evidence) that no fossil over about 8 million years in age can possibly be hominid "no matter what it looks like" (Sarich, 1971). These examples show that the availability of new data and new methods can be actually counterproductive if these are not used in conjunction among themselves and to add to, rather than to replace, others. Examples could readily be drawn from other groups as well, but I think it is no coincidence that those given all bear on human phylogeny.

It is an encouraging sign that disputes, often polemic and sometimes ill-natured, have tended to abate somewhat and that the cooler-headed disputants do seem to be approaching some common ground. There is thus some chance that an eventual synthesis will be reached by contributions from all the varied approaches. In my opinion that is likely to be closest to a modified and improved version of evolutionary taxonomy and of allied methods of inference as to phylogeny. I next propose to mention, very summarily, some of the things that I believe may reasonably be adopted (and what rejected) as contributions from the various schools of thought.

Nothing in essentialism can be considered a recent advance, even if "recent" be interpreted as broadly as it is here. Moreover essentialists disavow any involvement of phylogeny in their methods, and no further contribution to phylogenetic principles can be expected from this quarter. It may be remembered, however, that the strictly observational methods still most commonly used in phylogenetic research were developed, in at least elementary form, by early essentialists.

The earlier, purist modern pheneticists also eschewed phylogenetic principles, or thought that they did although that was not strictly true. The results of numerical phenetic analysis are commonly presented as tree diagrams, dendrograms, which resemble phylogenetic trees but which, as their makers correctly emphasize, are not meant to represent phylogenies. In Mayr's (1969) terms they are phenograms, not phylograms. It was, however, noted even by the most extreme pheneticists that phenograms tend to approximate phylograms (e.g., by Sokal and Sneath, 1963, using other terms). A or the major cause of resemblance, or "affinity" in pheneticist terms, is phylogenetic relationship as the sum of cladistic and patristic genetic similarity. If that were the only cause of resemblance, the phenograms would be acceptable as approximations of phylograms (not necessarily as cladograms), but it is not. A phenogram could be converted to an approximate phylogram by sorting out a sufficient number of characters reasonably ascribable to common ancestry of any two (or eventually more) of what the pheneticists call OTUs (operational taxonomic units). That can conceivably be done, especially by applying Hennig's criteria for cladistic relationships and adjusting patristic relationships for homology and phenetic affinity. In fact since 1960, pheneticists (including one of the founding fathers, Sokal) have been working on the relationship of phenograms and phylograms but have followed different approaches. They usually take the phenogram as given, without sorting or weighting for phylogenetic significance, and then calculate hypothetical or approximate phylograms usually either on a minimum principle, for the least number of steps or the shortest connections between the given organisms and a common ancestor, or on a maximum likelihood principle according to a postulated probabilistic model of evolution. Some of the key papers, which give other references, are: Edwards and Cavalli-Sforza (1964); Camin and Sokal (1965); Farris (1973); Felsenstein (1973).

It is often considered a matter of principle by pheneticists that the data used have not been screened for probable phylogenetic significance, but to many students of phylogeny that will seem a severe weakness. However, Kluge and Farris (1969) did select and weigh characters on phylogenetic criteria in a relatively early application of phenetics to construction of a phylogram of a real group (anurans). Some of the methods treated in theory only seem rigid to the point of artificiality or conversely based on flexible models without clear criteria of choice. Here, too, as in phenetics generally, the problem of polythetic taxa (or OTUs) arises. Pheneticists recognize the existence of this problem, but in my opinion they have not successfully coped with it; they often cope with polytypy just by ignoring it.

In any case, all phylogenetic inquiry has to be based on observation of resemblances and differences among organisms. Pheneticists have developed numerous and ingenious methods of dealing with multidimensional data. Their further phylogenetic interpretation by computer methods is still largely in a theoretical and debatable stage.

Evolutionist taxonomy has generally been associated with the synthetic theory of evolution, now espoused by a clear majority of organismic biologists. As Hull (1970) has remarked, criticisms of that taxonomy, coming mostly from pheneticists and recently from cladists and nonselectionists, have often been directed more at the synthetic theory than at the associated principles of taxonomy. Such criticism is hardly relevant in application to phylogeny as distinct from classification. The synthetic theory is itself evolving and is accretionary, absorbing methods and data, although more rarely principles, from the other taxonomic schools. As almost all agree that approximate phylogenies can be obtained for many groups of organisms, there is now little serious argument on that score. The evolutionist view that a number of different classifications could be erected in conformity with the same phylogeny annoys taxonomists who wish to be given "objective," unique answers (a futile hope), but that is not relevant to the problems of phylogeny as distinct from those of classification.

In this connection the status of Hennig's "phylogenetic," that is, cladistic, systematics is interesting and peculiar. For reasons that are outside the scope of a brief paper on phylogeny but that have been suggested in quite inadequate summary above, I believe that Hennig's system is impractical and unacceptable for its ostensible purpose, application to classification. However, I believe that it includes a major contribution to the methodology of phylogenetic research, apart from questions as to how phylogeny is to relate to classification. Greatly oversimplified, the main point is this: characters of the ancestry of a group of organisms tend to change in the course of time; as the group diversifies, the ancestral condition will tend to be retained in some but not all of the descendants; derived conditions shared by some but not all members of the larger group indicate origin from a later common ancestry. The essential is discrimination between primitive and derived characters at different levels, and Hennig's extensive discussion clarifies many dubious points. Successful analysis reveals the probable sequence of dichotomies in the phylogeny under study. A cladogram can be based on the results. [A cladogram on Hennig's principles shows only the succession of dichotomies; it has no time scale and it takes no account at all of morphological change; Mayr (1969) is mistaken in saying that a cladogram according to Hennig has time on the ordinate and degree of difference on the abscissa.]

In fact students of phylogeny have been following procedures similar to Hennig's for years, but they have rarely if ever done it quite so carefully and consistently or so clearly demonstrated the theoretical basis in a hierarchic sequence of derived characters as the sole (or at worst most reliable) criterion for cladistic succession.

A trivial but real disadvantage is that Hennig invented a complex, esoteric, and for the most part unnecessarily idiosyncratic terminology to expound his views. A serious point is that he not only ignores but, as a matter of principle, rejects any consideration of patristic affinity, which nevertheless is a real and extremely important aspect of phylogeny. Since the whole system depends on distinguishing between primitive and derived conditions, it is a curious further defect that Hennig seriously undervalues paleontological evidence of primitive conditions. It is still more curious that the paleontologists previously cited (Schaeffer et al., 1972) virtually reject this approach, while stressing the need to make the distinction. The most important point of their argument is that "primitiveness and apparent ancientness are not necessarily correlated." That is true, but they usually are correlated, and for any group with even a fair fossil record there is seldom any doubt that characters usual or shared by older members are almost always more primitive than those of later members. It is merely eccentric to claim that time is not a desirable parameter in working out phylogenies, even though what is desirable is not invariably available or clear in its significance.

It has been noted that the inferences of the nonselectionists, or so-called non-Darwinian, school are more phylogenetic than taxonomic. The basic and, as far as it goes, evidently valid phylogenetic application depends mainly on amino acid sequences in selected proteins or polypeptides. When different amino acids occur in homologous positions, it is possible to determine the minimum number of base changes (minimal mutations) in DNA codons involved. It is then inferred that in comparing any two species there is likely to be some positive correlation between that minimum number and the remoteness of their common ancestry. At best the correlation must be less than complete because of "degeneracy" (redundancy) in the code, back-mutation, and the possibility of multiple-step changes with the same end result as fewer-step or single-step. It is also true that not nearly the whole DNA code is known for even one organism, so that while the results may be interestingly indicative, they should never be accepted as conclusive. It is possible to argue that significant total genetic evolutionary, hence phylogenetic, change is more adequately represented by multidimensional phenetic comparisons above the molecular level than by any presently available data on amino acid sequences and DNA codons.

The nonselectionist theory goes well beyond that point. It is claimed that a majority of base substitutions are neutral as regards natural selection. It is impossible here to summarize all the arguments pro and con, but among the pro arguments are the great polymorphism at gene loci commonly found in wild populations, the existence of "synonymous" codons, the approximate correlation of codon abundance with their numbers of "synonyms," and distribution of codons according to probability of random fixation (King and Jukes, 1969; recent reviews in several chapters of Le Cam *et al.*, 1972). Con arguments include the conclusion that these facts, so far as confirmed, are themselves the results of natural selection, that the existence and distribution of "synonymous" codons have a buffering effect in populations, that even though "synonyms" may make no difference in individuals they do so in populations because they have different probabilities of mutation, that supposedly neutral molecular differences (as between hemoglobins or blood groups) commonly prove to be selective on further study, and that the molecular biologists have fallen into the common fallacy that if they do not know the cause of a phenomenon it has none, i.e., is random (Richmond, 1970; and also recent reviews in Le Cam *et al.*, 1972). I would add that it is inconceivable to me that any considerable number of mutations could be completely neutral to selection if they have definite effects above the molecular level. If a majority of fixed mutations have no such effects, then molecular evolution is a phenomenon so distinct from total phenetic evolution as to have relatively little significance for organismic phylogeny. It seems most likely that some mutations are fixed at random if their selective effects are slight (not necessarily zero), but that the DNA code as a whole is the result of and subject to the directive influence of natural selection.

Probably of most interest as regards phylogeny, however, is the further claim that amino acid or DNA codon substitutions have occurred at a steady average rate. If true, this would tend to fix the cladistic sequence of splittings of lineages, subject always to the previously mentioned uncertainties as to how many substitutions have in fact occurred. Moreover, if these changes can be calibrated against the fossil record, it is claimed that they can thereafter provide a time scale for the significant cladistic events of a phylogeny (e.g., Sarich and Wilson, 1967). The most impressive evidence for this view was a supposed fit to a straight line when numbers of substitutions (or sometimes immunological distances) were graphed against year dates for last common ancestor as indicated by fossils. An early effort, for cytochrome *c*, made so bad a fit that it could only be taken as contrary evidence: later modification improved the fit, but the *ad hoc* adjustment of evidence or procedure to fit a hypothesis is not acceptable in proof (Margoliash and Fitch, 1968, and cited earlier work by Margoliash from 1961).

Some positive correlation between degree of biochemical differentiation and remoteness of last common ancestor would of course be expected under any view of evolution. In my opinion the actually demonstrated correlation for relatively few molecules in relatively few groups is not such either in degree or kind as to make the hypothesis of uniform rate acceptable at present. (With reference to the supposed clock for primates, and the hypothesis in general, see especially Pilbeam, 1972.) To an organismic biologist there is something almost charmingly naive when nonselectionist molecular biologists discover that molecular evolutionary rates (like all others) can be species specific (Jukes and Holmquist, 1972) or when another, having calibrated the molecular rate from the fossil record, insists that the fossil record must be wrong if it does not agree with his ensuing projection of the molecular rate (Sarich, 1971).

What, then, does remain from studies of amino acid substitutions and DNA codon changes that is of special value for phylogeny? They are another kind of resemblance and difference among organisms to be considered along with all the others, and one of special interest because, contrary to some of its students, it probably does usually have a basic importance for supramolecular phenetics and because its changes occur in definite (one might say quantum) steps the minimum possible number of which can be counted. Another point as regards immunological and molecular data in general is the fact that they are applicable over remarkably wide phylogenetic distances. Some morphological data, such as raw linear measurements of contemporaneous organisms, have little phylogenetic application beyond the confines of a genus. Some molecular data, at the other extreme, are readily applicable between orders, classes, and sometimes even phyla.

V. Homology and Polarity

As these topics were especially recommended to all conferees and space here is becoming short, I shall be particularly brief and trust others to elaborate.

There is, I believe, little to say about homology under the title of "recent advances." Everyone agrees that comparisons used for taxonomic and phylogenetic inferences must be of homologous entities, but opinions differ as to how to define and recognize homology. The concept of homology was long preevolutionary and the term, adopted (not, in fact, coined) by Richard Owen, was also pre- and in Owen's hands became even anti-evolutionary. [It is an example of the slipperiness of history, that a recent reviewer, Adler (1974) credits Owen, who did not believe in phylogeny, with making homology the basis of phylogeny.] Darwin demonstrated that homology is explained and may be defined by community of ancestry. Ever since, however, there has been disagreement between those who insist on a purely descriptive or (misleadingly) so-called "objective" definition of homology and those who insist on the evolutionary definition by community of descent.

Some 30 years ago there was a flurry of discussion on that point, e.g., by Boyden (1947, and earlier papers there cited), nonevolutionary, and Haas and Simpson (1946), evolutionary. [Boyden has returned to the attack in his recent book (1973).] The evolutionary view is at present held by most biologists, but not by all as Margoliash (1969), for example, believes. Modern pheneticists have accepted the view that homology by phylogenetic definition has no function in taxonomy (Sokal and Sneath, 1963). Phylogeny was equated with sameness; characters were said to be "operational homologs" if they could not be distinguished, "numerical homologs" if some not-defined level of pairing of morphological correspondences could be established.

The principal argument against phylogenetic homology is that it cannot be directly observed and sometimes cannot be identified with complete certainty. But if homology is to mean only our inability to distinguish characters, the whole subject becomes merely trivial. If it is to be established by pairing correspondences, the selection of characters to be paired and judgment of the nature and degree of correspondence become just as subjective and uncertain as phylogenetic inference and definitely inferior in having only a covert and inadequate theoretical base in place of the clear and explicatory phylogenetic theory. The pheneticist definitions of homology do *not* depend only on observation (see again Hull, 1970). In fact pure observation is completely sterile and has no place in science. Scientific observation must have some philosophical and theoretical basis, and the need is for conscious selection of the best established and most meaningful basis, which in this case is clearly evolutionary.

In fact much of the trouble in this respect, as in many others such as the age-old problem of the definition of species, is simply the failure to distinguish between definition and the evidence available and used to determine whether the definition can be taken as applicable. That observation of characters is used as evidence of homology does not require that homology be defined solely, or at all, in terms of the evidence for it.

Adequate discussion of the whole subject of lineation, or specifically of clines, would require at least a whole book, which no one has written but someone should. Again recent discussions are certainly not adequate and can hardly be hailed as advances in a subject that has been discussed since Aristotle, at least, without reaching full clarity. To speak of the polarity of a cline is new language but not a new subject.

As a small step toward clarification, I suggest that a cline should be a seriation either of characters or of taxa in orderly sequence along a dimension either of space (chorocline) or of time (chronocline). A chronocline is an element in a phylogeny or is itself a phylogeny. There is no guarantee and in most instances not much likelihood that a chorocline will approximate a chronocline or that either end or any member of it will represent an ancestral condition. Many seriations that are not clines, not sequential in a spatial or temporal dimension, are possible. It is improbable that they will cast any useful light on phylogeny. It was a seriation of that sort to which the term "orthogenesis" was first applied, and the subsequent transfer of that concept to chronoclines led to great confusion and falsification of phylogeny and evolutionary theory.

VI. References

ADANSON, M. 1763. *Familles des plantes*. Vincent, Paris.
ADLER, N. T. 1974. Essays in ethology. *Science* **183**:191–192.
ANDREWS, D. F. 1972. Plots of high-dimensional data. *Biometrics* **28**:125–136.
BARLOW, N., ed. 1958. *The Autobiography of Charles Darwin*. 1809–1882. Collins, London.
BERNSTEIN, S. C., L. H. THROCKMORTON, and J. L. HUBBY. 1973. Still more genetic variability in natural populations. *Proc. Natl. Acad. Sci. U.S.A.* **70**:3928–3931.
BLACKWELDER, R. E. 1967. *Taxonomy, A Text and Reference Book*. John Wiley and Sons, New York.
BOCK, W. J. 1968. Phylogenetic systematics, cladistics, and evolution. *Evolution* **22**:646–648.

BOYDEN, A. 1947. Homology and analogy, a critical review of the meanings and implications of these concepts in biology. *Am. Midl. Nat.* **37**:648–669.

BOYDEN, A. 1973. *Perspectives in Biology*. Pergamon Press, Oxford.

BOYLE, A. J. 1964. The value of some methods of numerical taxonomy with reference to hominoid classification. *In* V. H. Heywood and J. McNeill ed. *Phenetic and Phylogenetic Classification*. Systematics Association, London.

BRUNDIN, L. 1966. Transantarctic relationships and their significance, as evidenced by chironomid midges with a monograph of the subfamilies. Podonominae and Aphroteniinae and the austral Heptagyinae. *K. Sven. Vetensk. apsakad. Handl.*, Ser. 4, **11**(1):1–472.

BRYSON, V., and H. J. VOGEL ed. 1965. *Evolving Genes and Proteins*. Academic Press, New York and London.

CAMIN, J. H., and R. R. SOKAL. 1965. A method for producing branching sequences in phylogeny. *Evolution* **19**:311–326.

CAMPBELL, B. 1974. *Human Evolution. An Introduction to Man's Adaptations*. Aldine, Chicago.

CLARK, W. E. LE GROS, and P. B. MEDAWAR, ed. 1945. *Essays on Growth and Form Presented to D'Arcy Wentworth Thompson*. Clarendon Press, Oxford.

COX, C. B., I. N. HEALEY, and P. D. MOORE. 1973. *Biogeography. An Ecological and Evolutionary Approach*. Halsted (Wiley), New York.

CROW, J. F. 1972. Darwinian and non-Darwinian evolution. *In* L. M. Le Cam, J. Neyman, and E. L. Scott, eds. *Darwinian, Neo-Darwinian, and Non-Darwinian Evolution*, University of California Press, Berkeley and Los Angeles.

DARLINGTON, P. J., Jr. 1970. A practical criticism of Hennig–Brundin "phylogenetic systematics" and Antarctic biogeography. *Syst. Zool.* **19**:1–18.

DARWIN, C. 1859. *On the Origin of Species by Means of Natural Selection, or the Preservation of Favoured Races in the Struggle for Life*. Murray, London.

DARWIN, C. 1871. *The Descent of Man and Selection in Relation to Sex*. Murray, London.

DARWIN, C. 1872. *The Expression of the Emotions in Man and Animals*. Murray, London.

DAYHOFF, M. O. 1969. *Atlas of Protein Sequence and Structure*, Vol. 4. National Biochemical Research Foundation, Silver Spring, Maryland.

DEVORE, I., ed. 1965. *Primate Behavior. Field Studies of Monkeys and Apes*. Holt, Rinehart and Winston, New York.

DICKERSON, R. E. 1972. The structure and history of an ancient protein. *Sci. Am.* **226**(4):58–72.

EDWARDS, A. W. F., and L. L. CAVALLI-SFORZA. 1964. Reconstruction of evolutionary trees. *In* V. H. Heywood and J. McNeill, eds., *Phenetic and Phylogenetic Classification*, Systematics Association, London.

FARRIS, J. S. 1973. On the use of the parsimony criterion for inferring evolutionary trees. *Syst. Zool.* **22**:250–256.

FELSENSTEIN, J. 1973. Maximum likelihood and minimum-steps methods for estimating evolutionary trees from data on discrete characters. *Syst. Zool.* **22**:240–249.

GHISELIN, M. T. 1969. *The Triumph of the Darwinian Method*. University of California Press, Berkeley.

GOODMAN, M. 1963. Man's place in the phylogeny of the Primates as reflected in serum proteins. *In* S. L. Washburn, ed. *Classification and Human Evolution*, Aldine, Chicago.

HAAS, O., and G. G. SIMPSON. 1946. Analysis of some phylogenetic terms, with attempts at redefinition. *Proc. Am. Phil. Soc.* **90**:319–349.

HENNIG, W. 1950. *Grundzüge einer Theorie der Phylogenetischen Systematik*. Deutscher Zentralverlag, Berlin.

HENNIG, W. 1966. *Phylogenetic Systematics*. University of Illinois Press, Urbana.

HOFFSTETTER, R. 1972. Relationships, origins, and history of the ceboid monkeys and caviomorph rodents: a modern reinterpretation. *Evol. Biol.* **6**:323–347.

HOYER, B. H., B. J. MCCARTHY, and E. T. BOLTON. 1964. A molecular approach in the systematics of higher organisms. *Science* **1944**:959–967.

HULL, D. L. 1965. The effect of essentialism on taxonomy. *Br. J. Phil. Sci.* **15**:314–326; **16**:1–18.

HULL, D. L. 1970. Contemporary systematic philosophies. *Ann. Rev. Ecol. Syst.* **1**:19–54.

HULL, D. L., ed. 1973. Contemporary systematic philosophies. *Syst. Zool.* **22**:337–400. (A symposium by five authors, with discussion by others.)

JUKES, T. H. 1972. Comparison of polypeptide sequences. *In* L. M. Le Cam, J. Neyman, and E. L. Scott, eds., *Darwinian, Neo-Darwinian, and Non-Darwinian Evolution*, University of California Press, Berkeley and Los Angeles.

JUKES, T. H., and R. Holmquist. 1972. Evolutionary clock: non-constancy of rate in different species. *Science* **177**:530–532.

KING, J. L., and T. H. JUKES. 1969. Non-Darwinian evolution. *Science* **164**:788–798.

KLUGE, A. G., and J. S. FARRIS. 1969. Quantitative phyletics and the evolution of anurans. *Syst. Zool.* **18**:1–32.

KOHNE, D. E., J. A. CHISCON, and B. H. HOYER. 1972. Evolution of mammalian DNA. *In* L. M. Le Cam, J. Neyman, and E. L. Scott, eds., *Darwinian, Neo-Darwinian, and Non-Darwinian Evolution.* University of California Press, Berkeley and Los Angeles.

LE CAM, L. M., J. NEYMAN, and E. L. SCOTT, eds. 1972. Proceedings of the sixth Berkeley symposium on mathematical statistics and probability, Vol. 5. *Darwinian, Neo-Darwinian, and Non-Darwinian Evolution.* University of California Press, Berkeley and Los Angeles.

LEONE, C. A., ed. 1964. *Taxonomic Biochemistry and Serology.* Ronald Press, New York.

LEWONTIN, R. C. 1974. *The Genetic Basis of Evolutionary Change.* Columbia University Press, New York.

MARGOLIASH, E. 1969. Homology: a definition. *Science* **163**:127.

MARGOLIASH, E., and W. M. FITCH. 1968. Evolutionary variability of cytochrome c primary structures. *Ann. N.Y. Acad. Sci.* **151**:359–381.

MAYR, E. 1958. Behavior and systematics. *In* A. Roe and G. G. Simpson, eds., *Behavior and Evolution*, Yale University Press, New Haven.

MAYR, E. 1969. *Principles of Systematic Zoology.* McGraw-Hill, New York.

OXNARD, C. E. 1973. Functional inferences for morphometrics: Problems posed by uniqueness and diversity among primates. *Syst. Zool.* **22**:409–424.

PILBEAM, D. 1972. Adaptive response of hominids to their environment as ascertained by fossil evidence. *Soc. Biol.* **19**:115–127.

RAUP, D. M., and S. M. STANLEY. 1971. *Principles of Paleontology.* W. H. Freeman, San Francisco.

RICHMOND, R. C. 1970. Non-Darwinian evolution: a critique. *Nature* **225**:1025–1028.

ROE, A., and G. G. SIMPSON, eds. 1958. *Behavior and Evolution.* Yale University Press, New Haven.

SARICH, V. M. 1971. A molecular approach to the question of human origins. *In* P. Dolhinow and V. M. Sarich, eds., *Background for Man*, Little, Brown, Boston.

SARICH, V. M., and A. C. WILSON. 1966. Quantitative immunochemistry and the evolution of primate albumins: Micro-complement fixation. *Science* **154**:1563–1566.

SARICH, V. M., and A. C. WILSON. 1967. Immunological time scale for hominid evolution. *Science* **158**:1200–1203.

SCHAEFFER, B., M. K. HECHT, and N. ELDREDGE. 1972. Phylogeny and paleontology. *Evol. Biol.* **6**:31–46.

SIMPSON, G. G. 1961. *Principles of Animal Taxonomy.* Columbia University Press, New York.

SOKAL, R. R., and P. H. A. SNEATH. 1963. *Principles of Numerical Taxonomy.* W. H. Freeman, San Francisco and London. (A revised edition, 1973, relaxes much of the dogmatism but retains the premises of the original statement.)

THOMPSON, D'ARCY W. 1917. *On Growth and Form.* Cambridge University Press, Cambridge. (Revised edition, 1942.)

WAHLERT, J. H. 1968. Variability of rodent incisor enamel as viewed in thin section and the microstructure of the enamel in fossil and recent rodent groups. *Breviora Mus. Comp. Zool.* **309**:1–18.

WASSERMAN, E., and L. LEVINE. 1960. Quantitative micro-complement fixation and its use in the study of antigenic structure by specific antigen–antibody fixation. *J. Immunol.* **87**:290–295.

WHYTE, L. L., ed. 1968. *Aspects of Form.* American Elsevier, New York.

WILLIAMS, C. A., Jr. 1964. Immunochemical analysis of serum proteins of the primates. *In* J. Buettner-Janusch, ed., *Evolutionary and Genetic Biology of the Primates*, Academic Press, New York.

ZUCKERKANDL, E. 1963. Perspectives in molecular anthropology. *In* S. L. Washburn, ed., *Classification and Human Evolution*, Aldine, Chicago.

ZUCKERMAN, S., ed. 1950. *A Discussion on the Measurement of Growth and Form.* Royal Society, London.

2

Toward a Phylogenetic Classification of the Mammalia

MALCOLM C. MCKENNA

I. Introduction

Periodically it is worthwhile to assess our knowledge and understanding of mammalian phylogeny and one of its expressions, classification. This short paper is yet another attempt to do so, taking into account the results of recently published paleontological research and drawing heavily on work in progress by many researchers in many fields and in various parts of the world. Concepts of mammalian phylogeny and classification have changed markedly during the last few years. A good many of the ideas expressed here are frankly speculative, but they are presented anyway in order to determine how well they will stand scrutiny, especially by nonpaleontologists. A few years ago I prepared a paper with a similar aim (McKenna, 1969), but that paper is now outdated. In the present offering I attempt to update certain aspects of my previous review by taking into account research published since 1969, as well as work being incorporated into a new classification of the Mammalia now being prepared which will deal with all taxonomic levels down to the subgeneric level in essentially the same style as Simpson's (1945) classification.

Obviously, in a paper of the present length the subject of the whole mammalian phylogeny and its taxonomic expression can only be touched upon; for this reason I have

MALCOLM C. MCKENNA · Department of Vertebrate Paleontology, The American Museum of Natural History, and Department of Geology, Columbia University, New York, New York.

21

concentrated on what seem to be some of the critical issues. I have accepted the invitation to prepare this review partly in order to discuss new developments in phylogenetic analysis and conclusions and partly with the aim of finding out what would become of the classification of mammals if its higher categories were to be based upon kinship and relative recency of common ancestry rather than upon "morphological distance." Wherever possible I have tried to separate synapomorphy, autapomorphy, and symplesiomorphy. Readers unfamiliar with these cladistic terms and their meanings are strongly advised to consult the somewhat contradictory works of Hennig (1965, 1966), Brundin (1966), Griffiths (1972), Kavanaugh (1972), Schaeffer *et al.* (1972), and Ashlock (1974) and to become acquainted with numerous articles about cladistic procedures in recent issues of the journal *Systematic Zoology* before reading further here. Identification of these three conditions is not always easy, particularly if a series of major phylogenetic branchings followed close upon one another as seems to have been the case in the late Cretaceous and again in the early Eocene; nevertheless cladistic procedures offer what I take to be new rigor in the determination of phylogeny and one of its expressions, genealogical classification. Although there are many ways to classify organisms, a cladistic classification has the highest genealogical information content, a nonarbitrary rationale for formulation, vulnerability, or falsifiability in the sense of Popper (1968), and is maximally heuristic.

On the debit side of the ledger lies the apparent bugaboo of instability. Because of its maximized genealogical information content there is always more to take back in a cladistic classification if the phylogeny upon which it is based proves to be wrong. The more ancient the site of phylogenetic repairs, the more a cladistic classification will crumble at higher and higher categories. Some workers would regard this instability as bad because it represents confusion; others would rejoice because it *identifies* unsuspected confusion and points to areas of research where more work is needed. There has been only one actual sequence of genealogical branching, however, so that it seems to me that after the initial shock of conversion to a cladistic classification, one can expect successive iterations toward a stable phylogenetically based system, successively diminishing adjustments being made on the basis of ever smaller increases in knowledge rather than on authority, arbitrary judgment, consensus, and fiat.

Another apparent disability is the lack of reflection of "morphologic distance" in a cladistic classification, a point made much of by Mayr (1974 and elsewhere). But once a cladistic general reference system is established, "morphologic distance" can then be discussed on the logical basis of a minimally arbitrary genealogical framework. There is no law that says that classifications must directly reflect "morphologic distance" nor any mandate for regarding Mayr's or Simpson's syncretistic philosophy of classification as more "evolutionary" than cladistic systematics.

I have taken mammalian phylogeny and classification as far as the present cladistic analysis will permit. The subdivisions of certain higher taxa (e.g., marsupials, rodents, artiodactyls, and carnivores) are not treated here and, in other cases, multichotomies are left untouched by the cladistic method; that work can be accomplished in the future. Nor have I given reasons for every decision made; this paper is merely an abstract of the present state of a classification which is itself evolving. For reasons of economy it is necessary to be cheerfully dogmatic on many points if the whole subject is to be covered in the space available: the reasons behind various changes from Simpson's (1945) classification of mammals being entombed in an extensive file system accompanying a more complete and detailed mammalian classification now being prepared for publication by Susan Bell, Karl Koopman, Guy Musser, Richard Tedford, and myself. I am wholly responsible for the present arrangement of higher categories, however.

The basis of most of my knowledge of mammalian phylogeny is osteological and odontological morphology, available from paleontological excavation as well as from living animals. Needless to say, the speculations reached here primarily from these hard parts can and should be tested by comparison with evidence from other organ systems and from embryology.

II. Dental Evidence and Phylogeny

A. Prototheria–Theria Dichotomy

The most basic division of the Mammalia as now classified is the dichotomy into subclasses Prototheria and Theria. MacIntyre (1967, 1972) would restrict the term Mammalia to the latter, which solves nothing. In terms of derived character states, both subclasses share the basic mammalian ear ossicle apparatus, which many workers would agree has arisen only once, even if dentary contacts with the squamosal bone were achieved several times independently by various early mammal-like animals. The Prototheria, however, independently evolved a detrahens muscle for depressing the jaw. Dental occlusion in prototherians possessing teeth is not based upon the interlocking of reversed triangles as in therians. In contrast, the Theria developed a digastric muscle for depressing the jaw and adopted the reversed triangle method of dental occlusion, which was ultimately transformed into the tribosphenic type (Simpson, 1936, p. 8) at or even before the beginning of the Cretaceous, when a new cusp with the confusing name *proto*cone came to occupy the lingual side of upper cheek teeth and occlude with basined talonids of the lower cheek teeth. In the ear region, therian petrosal bones are trisulcate (MacIntyre, 1972) and their cochleas acquired at least one full coil. The scapula developed a supraspinous fossa (McKenna, 1961*b*). In the urogenital system, therians are apomorphous as well, in that the ureters empty into the bladder rather than into a cloaca, urine began to pass through the external genitalia, and of course living therians are viviparous. In therians the possibly composite alisphenoid bone is an important part of the braincase wall, whereas in prototherians in which the facts are known, a large anterior lamina of membrane bone is fused to the petrosal bone as part of the braincase and the epipterygoid component of the wall is less important, at least in ornithodelphians.

Prototherians retain many plesiomorph character states as well, especially in the urogenital system, but members of one important taxonomic subdivision of the prototherians that is known thus far only from Laurasia, the multituberculates, developed many autapomorphous features, especially in the dentition. Multituberculates are very well represented in collections of Holarctic Mesozoic and earliest Tertiary fossil mammals, and they evidently were highly successful small herbivores in their day. But except for the two extant monotreme suborders, subsuming just three genera in the Australian fragment of Gondwanaland, all prototherian mammals are now extinct; thus the subject of multituberculate soft anatomy will remain conjectural. The prototherian classification utilized here is based essentially on that of Hopson (1970) but takes into account the conclusions reached by Kielan-Jaworowska (1971) resulting from her studies of recently discovered and well-preserved cranial and postcranial material from the Cretaceous of Mongolia. The reader is referred to Kielan-Jaworowska's excellent but noncladistic discussion for additional details. Cladistic analysis of the information marshalled by Kielan-Jaworowska is not attempted here.

Early prototherians have recently become much better known as descriptions of Triassic and Jurassic finds have appeared. Recent papers that provide a good introduction to the subject have been provided by Parrington (1971), Hahn (1973), Kermack *et al.* (1973), and Crompton (1974).

The subclass Theria, possessors of dentitions in which the upper and lower post-canine teeth occlude in a reversed triangular pattern, were divided originally by early students of the subject into two infraclasses, the extant and tribosphenic Eutheria and Marsupialia (each here given supercohort rank), but a third and basal protribosphenic infraclass, Pantotheria, was added by Simpson (1931*a*, p. 261) to include the Mesozoic groups here called symmetrodonta, Dryolestoidea, and Peramurida. Simpson's para-phylectic infraclass Pantotheria had not yet developed the protocone, this important cusp being the innermost cusp of both eutherian and marsupial tribosphenic posterior upper cheek teeth that are not secondarily simplified. Conversion of this previously accepted tripartite system of mammalian infraclasses to a cladistic nested dichotomous scheme is attempted below.

In recent years there has been a renewal in the study of Rhaetic and Jurassic therians. This has followed a long period during which little or no new material came to light. But now extensive work in the Triassic–Jurassic fissure fillings of Wales and in various Jurassic deposits has been accomplished in Europe. For example, Kühne has done much by the use of ingenious collecting techniques in the Jurassic of Portugal (Kühne, 1968). Small but important new collections have been obtained recently by others from the Jurassic of both Great Britain and Wyoming, and several important Jurassic specimens long hidden in matrix have now been carefully prepared after lying for many decades in museum cabinets. A flurry of restudy has occurred, and opinion now appears to have crystallized that advanced symmetrodonts did not give rise to any higher mammals and that advanced dryolestoids are also a dead-end group without known issue. The early as well as plesiomorphous therian genus *Kuehneotherium*, now fairly well known thanks especially to the work of Kermack *et al.* (1968) and Parrington (1971), has been shuttled back and forth by various authors to comply with different usages of the words " panto-there," " symmetrodont," " eupantothere," etc., but all would agree that this plesio-morphous Rhaetic (or early Liassic?) genus is close to the morphology to be expected in a sister group of the more apomorphous members of the Theria.

According to Crompton (1974, p. 421), *Kuehneotherium* has the beginnings of the reversed triangle system of dental occlusion, but only the anterior wall of the trigon of upper molars wore against the posterior wall of the trigonid of the dental counterpart in the lower dentition. In life, *Kuehneotherium* did not yet wear the posterior wall of its upper molar trigons against the anterior wall of succeeding lower molar trigonids. Principally for this reason, the kuehneotheriids are considered here to be the sister group of all other known therians. Kuehneotheriids are neither symmetrodonts nor dryolestoids in the cladistic taxonomy adopted here. Within the remaining therians, here dubbed Trechnotheria, two sister groups again are seen: Symmetrodonta and all other trechnotheres, here dubbed Cladotheria. Within the Cladotheria we have the Dryolestoidea and all other cladotheres, here dubbed Zatheria. And within the Zatheria we have *Peramus* and its sister group the Tribosphenida, the latter in turn being divisible into Marsupialia and Eutheria (Fig. 1).

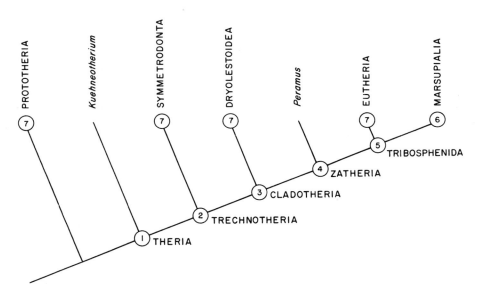

FIG. 1. Cladogram of mammals, using dental characters. Dentally advanced forms progress from left to right: (1) reversed triangle dental pattern results from labial shift of stylocone and protoconid; anterior face of trigon wears against posterior face of trigonid; (2) posterior face of trigon begins to wear against anterior face of trigonid of succeeding lower tooth; (3) talonid is expanded; (4) molars posterior to M_3^3 not present (see text); (5) protocone added; (6) reduction of premolar loci from 5 to 3 in number, of which only P_4^4 is replaced; this results in 4 molars, $dP_5^5 \ M_1^1 \ M_2^2 \ M_3^3$; (7) various apomorphous character states.

Dryolestoids as here defined (i.e., minus *Peramus*) had generally a large number of cheek teeth (Dryolestidae), but in the medial Jurassic Stonesfield genus *Amphitherium* similarities to zatherian mammals can be detected (Mills, 1964) even though *Amphitherium* has some synapomorphies with the dryolestids and is classified with them and the paurodontids as a poorly defined member of the Dryolestoidea in this paper.

It also seems possible to me that *Amphitherium* may not yet have reduced its premolar locus count to 4. A premolar locus count of 4 or less occurs in all other dryolestoids in which the dentition is known well enough to decide. This may be seen by the break in morphology between premolars and molars or, more reliably, by demonstrated replacement such as that published by Butler (1973). Although it has become traditional through repetition to regard the dental formula of the lower teeth of *Amphitherium* as consisting of 4 incisors, 1 canine, 4 premolars, and 7 or more molars, nevertheless I suspect on the basis of study of models prepared many years ago by Pycraft that 4 is perhaps merely a minimum number of premolar loci and that 5 premolar loci is a more likely number, as even Pycraft's specimen label admits. Unfortunately, I have not yet had the opportunity to study the original material, but perhaps someone will reinvestigate the dental formula of *Amphitherium* in view of the interpretation made later in this paper concerning the cheek tooth formula of the late Jurassic genus *Peramus*, which I regard as the closest protribosphenic relative (sister group) of the tribosphenic marsupials and eutherians (Tribosphenida). I suspect that not only *Peramus* but possibly also *Amphitherium* had 5 rather than 4 premolar loci. A premolar locus count of 5, rather than 4, is judged here to be primitive for the Theria,[*] reductions to smaller numbers having occurred later in various therian lineages as well as in nontherians. Tendencies to increase the number of molars,[†] to reduce the length of the talonids thereon, and to decrease the number of premolars were characteristic of advanced dryolestoids and, as P. M. Butler (personal communication) believes, *Amphitherium* is similar to dryolestids in that its lower molar roots are unequal.

I propose here a cladistic reclassification of the therian groups Marsupialia, Eutheria, and Simpson's infraclass Pantotheria, emphasizing kinship and recency of common ancestry. Based upon the assumption that in the upper cheek teeth the acquisition of a protocone is synapomorphous for marsupials and eutherians, these last two groups can be linked cladistically, but Simpson's infraclass Pantotheria is merely a paraphyletic and arbitrary linkage of Symmetrodonta, Dryolestoidea, and *Peramus* and is therefore abandoned. Reference to the cladogram in Fig. 1 will make this immediately clear in terms of cladistic principles.

Osborn (1893, p. 320) informally called certain late Cretaceous marsupials and eutherians primitive trituberculates, but his casual term has no legal status in taxonomy. He did not regard any Jurassic mammal as a trituberculate (i.e., informally a mammal whose upper molar pattern was primarily based upon three cusps: protocone, paracone, and metacone). Unfortunately, however, Zittel's (1893, p. 96) formal taxonomic term Trituberculata, which Zittel wrongly (1893, p. 100) attributed to Osborn, was meant to include only what are now known as dryolestoids and certain polyprotodont marsupials. The Eutheria were omitted by Zittel and the dryolestoids lack a protocone (Patterson, 1956). Therefore, unless we wish to return to Gill's (1872, pp. v, vi) first but confusing usage of the term Eutheria, which originally included both the marsupials and the placentals and which was later changed by Huxley (1880, p. 657) to its now almost

---

[*] *Kuehneotherium* may possess 6 premolars, but the matter is not certain (Kermack *et al.*, 1968). If so, 6 is the plesiomorphous number from which other therians departed, assuming that *Kuehneotherium* did not increase its complement of premolars. Five premolars would thus characterize the therians that in this paper are dubbed trechnotheres, i.e., all known therians except *Kuehneotherium*. A possible alternative to this scheme would be to exclude *Kuehneotherium* from the Theria, thus synonymizing trechnotheres with the therians, but such action would not reduce the number of names required for mammalian higher taxa and therefore is not taken.

[†] The methods are not known. Was it by addition of posterior molars alone? In part by retention of posterior milk premolars? Always by the same method?

universally accepted meaning, or we are able to surmount the technicalities surrounding the useful but informal Osbornian term "trituberculates," I would prefer to employ a new formal taxonomic term, Tribosphenida, to encompass eutherians and marsupials. This reflects the view that their acquisition of a protocone is synapomorphous and is meant to be the cladistic taxonomic equivalent of Simpson's (1936, p. 8) descriptive term tribosphenic. According to the proposed theory of relationships, bearers of tribosphenic dentitions and their descendants, whether tribosphenic or secondarily nontribosphenic, thus would be members of the Tribosphenida. Tribosphenida is an equivalent of Osborn's informal term "trituberculates" but not an equivalent of Zittel's taxonomic term Trituberculata. For these reasons both "trituberculates" and Trituberculata should be allowed to lapse further into obscurity. Thus the major subdivisions of the subclass Theria would be as follows:

Subclass THERIA Parker and Haswell, 1897
    Superlegion KUEHNEOTHERIA, new
    Superlegion TRECHNOTHERIA, new
      Legion SYMMETRODONTA Simpson, 1925, new rank
      Legion CLADOTHERIA, new
        Sublegion DRYOLESTOIDEA Butler, 1939, new rank
        Sublegion ZATHERIA, new
          Infraclass PERAMURA, new
          Infraclass TRIBOSPHENIDA, new
            Supercohort MARSUPIALIA Illiger, 1811, new rank
            Supercohort EUTHERIA Gill, 1872 (as modified by Huxley, 1880), new rank

To the critics who would ask, "Where will all this proliferation of names and ranks end?," I suggest that if the terms are not found useful to convey exact genealogical meanings dictated by phylogeny, then they can be ignored by those who so choose. They are, however, based upon cladistic principles, not on "art" or caprice.

In the course of preparation of this paper I have considered four possible ways to avoid the problem of proliferation of ranks: (1) by use of cladograms alone, without a written classification, (2) by omission of the rank name of a higher category, rank being indicated only by position in the hierarchy of names, (3) by taking Hennig at his word and indicating rank by a number representing millions of years, inasmuch as rank is proportional to relative recency of common ancestry in cladistic systems, and (4) by assigning relative code numbers to ranks. I have not found either of the first two schemes compatible with attempts to communicate via prose, however, and the third scheme requires knowledge of the time of actual branching, which in fact is never known but only estimated, in many cases not even approximately. The fourth solution merely substitutes numbers for words, and both the third and fourth schemes would be difficult to remember. I therefore reluctantly continue to proliferate ranks and rank names. The theoretical number of ranks necessary to classify n organisms would lie between $\log_2 n$ and $n - 1$, probably near the former if one assumes essentially random cladogenesis. Thus even a million kinds of organisms would not appear to require an outrageously unwieldy number of ranks, the theoretical absolute minimum for that number being 20.

B. Marsupialia–Eutheria Dichotomy

We now proceed to the Eutheria, with a brief consideration of the marsupials.

The extant Eutheria differ from the present-day Marsupialia not only in such matters as the dental formula and mode of dental replacement, which can be deduced from fossils as well as from the living animals, but also in such basic features as brain organization, trophoblast development in early blastocyst formation to prevent immunological rejection and therefore permit long gestation (Lillegraven, 1975), position of the ureters with regard to the Müllerian ducts, and in many other aspects of reproductive anatomy and function, which cannot be deduced from fossils. Lillegraven (1969) has provided some strong arguments for considering the chorioallantoic placenta of living eutherians to be apomorphous, the yolk sac placenta (except the independently derived chorioallantoic placenta of perameloids) of extant marsupials being considered plesiomorphous for both marsupials and eutherians. Living eutherians autapomorphously pass the ureters ventral to, instead of dorsal to, the Müllerian ducts,* thus permitting eventual complete fusion of the latter in eutherians in which this occurs. Complete fusion without incorporation of the ureters is impossible in Recent marsupials and in them has been supplemented by the development of a medial pseudovagina present in both American and Australian living marsupials. The abandonment of replacement in all but the most posterior remaining premolar, interpreted in this paper to be P^4_4, is another synapomorphous feature shared by Australian and American marsupials. These and other features are taken here to mean that the marsupial–eutherian dichotomy was extremely ancient, during the Cretaceous or even earlier, presumably occurring in response to vicariance (Hennig, 1966) and geographic isolation on a major scale, e.g., tectonic separation of major land areas, spread of epicontinental seas, or other associated barriers. Essentially this view has recently been stated well by Hoffstetter (1973, p. 21, addendum). In Hoffstetter's view, marsupials became isolated in the early Cretaceous on a chain of land masses extending from North America to Australia via South America and Antarctica, but they managed to escape to Asia via the Bering route in the late Cretaceous (he regards deltatheridiids as marsupials) and also to Europe around the north end of the Atlantic Ocean via Greenland (McKenna, 1972) in the early Tertiary. The various subdivisions of the marsupials will not be discussed in this paper.

C. Edentata–Epitheria Dichotomy

The first major fragmentation of the Eutheria into subgroups appears to have taken place well back in the Cretaceous, when cohort Edentata (Xenarthra) separated from all other eutherians, which are here named cohort Epitheria. If one accepts Lillegraven's (1969) arguments, the time of the edentate–epithere split was necessarily after the marsupial–eutherian dichotomy and some time before, probably well before, the late Paleocene when edentates first made their appearance in the known fossil record. It is at

---

* Godet (1949, Fig. 1; repeated by Grassé, 1955b, Fig. 1626) demonstrated that the ureters pass dorsal to the uterine horns in adult females of the genus *Talpa*. I do not know how widespread this condition is in other talpids, but I assume the condition to be a convergence with that of marsupials.

least possible that the edentate–epithere dichotomy took place very early, during the Cretaceous, but the exact details of the genealogy and relative phylogenetic age of the Edentata are unknown.

1. Edentata. A reading of accounts of edentate anatomy, such as those of Forbes (1882), Flower (1882), Grassé (1955*a*), or Hoffstetter (1958), yields a wealth of truly archaic features long since modified by other eutherians but retained by the edentates in combination with various autapomorphous features unique to members of that cohort. Presence in various edentates of septomaxillary bones, ossified ribs reaching the sternum, and low body temperature with poor thermoregulation are plesiomorphous character states modified by most other eutherians. Significant palaeotelic (Gregory, 1910, p. 111) autapomorphous features of edentates include the presence of xenarthrous vertebrae in the trunk, reduction and loss of enamel on the teeth, (primitive) presence of ossifications in the skin, development of a somewhat birdlike synsacrum as the result of ischial fusion to anterior caudal vertebrae, and independent acquisition of a simplex uterus that is nevertheless little differentiated into uterine and vaginal regions.

Unfortunately, the origin of the edentate dental formula and replacement pattern in primitive forms of 7 of the 8 cheek teeth is not known yet, although perhaps the South American Miocene genus *Necrolestes* will prove to have a bearing on the problem. Intermediate stages between edentates and other mammals are greatly needed.

At least as long ago as late Paleocene time the Edentata are known to have already been in existence, for their remains are known to occur in sedimentary rocks in South America reaching back to that time. They were apparently already living in isolation there. Until recently the early Tertiary North American palaeanodonts were believed by most authors to be allied in some way with the edentates, but recent work by Emry (1970) does not support that view. The only remaining evidence of edentate-like animals in North America before the late Tertiary consists of two odd specimens from the medial Eocene of Wyoming (Robinson, 1963) that represent an animal with several fused cervical vertebrae similar to those of the armadillo *Priodontes* and, in a lesser degree, to those of certain other armadillos. Such a condition has not yet been found in palaeano-donts or in other North American Eocene mammals, so it remains to be seen whether the possessors of these fused cervical vertebrae were synapomorphous Eocene North American members of one branch of the Edentata or were simply nonedentates con-vergent in that regard with *Priodontes*. Although it cannot be certainly ruled out that armadillos with fused cervical vertebrae similar to those of *Priodontes* might have been present in South America as early as the Eocene or Paleocene and that they might have sent emissaries northwards, as the Notoungulata appear to have done in the late Paleo-cene, nevertheless *Utaetus*, an early Eocene South American armadillo, lacks the fusion and probably was representative of armadillos of the time. Edentate affinities of the bearers of the North American medial Eocene fused cervicals thus seem doubtful, and the vertebrae are at present best regarded as poorly known examples of convergence.

Regardless of whether edentates escaped from South America or not in the early Tertiary, their main known area of evolution has been in South America for as far back in time as an abundant fossil record there reaches. With the possible but improbable exception noted above, almost their entire spectacular evolutionary flowering took place on that continent, supplemented at the end of the Tertiary by a northward expansion that reached as far as Alaska. It is reasonable to speculate, as Hoffstetter (1970) has done, that edentate ancestors were in South America long before their known appearance there, bringing up the possibility or even probability, but by no means proving, a Gondwanaland origin for the Edentata. The simplest and most appealing hypothesis is that the Edentata arose in South America as the result of vicariance caused by Cretaceous separation of the South American land mass from Africa, but until the Mesozoic mammalian record of the southern continents, and even northern ones, is made clearer by use of the pick and shovel, rather than by more speculation, it is still possible to suppose that edentates might have arisen from some sort of unknown Cretaceous eutherians, crossing an oceanic water gap from North to South America by some unspecified means. Such an improbable scenario, however, is a relic of stabilist tectonic and paleogeographic views and is now unnecessary to postulate if South America was still connected to Africa in the early part of the Cretaceous.

   2. Dental homologies of Peramus and Tribosphenida. Returning from the edentates to their sister group the epitheres, the main stem of the Eutheria, it is necessary first to discuss the dentition of the long-known but poorly understood late Jurassic protribosphenic genus. *Peramus*, whose upper dentition was recently described for the first time and whose lower dentition has been made better known by Clemens and Mills (1971). *Peramus* is very similar in many dental features to epitheres but still lacked a protocone on its upper molars and is thus by definition not a member of the Tribosphenida, although clearly the genus has a phylogenetic position at or very close to the origin of the Tribosphenida. *Peramus* had 8 postcanine teeth[*] in each row, identified as P^1_1 P^2_2 P^3_3 P^4_4 M^1_1 M^2_2 M^3_3 M^4_4 by Clemens and Mills (1971, p. 92) for traditional reasons, but which might with equal or greater logic be termed dP^1_1 P^2_2 P^3_3 P^4_4 P^5_5 M^1_1 M^2_2 M^3_3 if the partially molariform teeth regarded as M^1_1 by Clemens and Mills are instead simply regarded as P^5_5.[†] If I have correctly identified the essentials of the dental formula (Fig. 2), the enlarged upper premolar just anterior to P^5 is homologous with the tooth that is usually called P^3 in marsupials and certain Mesozoic and later mammals such as *Kennalestes* and *Zalambdalestes* (see Kielan-Jaworowska, 1969) and their allies (Ernotheria of this chapter). The same tooth is suggested here to be homologous with the enlarged last upper premolar of various other epitheres, P^4. This seeming impossibility can be explained if originally the postcanine dental allotment of early members of the Tribosphenida was 5 premolar loci and 3 molar loci, the number of teeth at premolar loci later having been reduced to 4 in several ways in adult eutherians but to 2 nonreplaced or nonreplacing premolar loci, a replacing pre-

---

[*] I accept the conventional view that dP^1_1, dP^2_2, dP^3_3, dP^4_4, dP^5_5, M^1_1, M^2_2, and M^3_3 are members of a single zahnreihe. P^2_2, P^3_3, P^4_4, and P^5_5 are members of a second such sequence. Some authors logically refer to dP^x_x as dM^x_x. Another logical solution would be to regard M^x_x as better denoted by dM^x_x or dP^x_x, but such a scheme would be cumbersome and for that reason probably would not be generally adopted. The essential point is that molars are unreplaced posterior members of the "milk tooth" zahnreihe, no matter how one writes dental formulae. The reasons for including 5 premolar loci are developed in this paper. If there is no evidence that replacement occurs at the first postcanine locus, I identify that tooth as dP^1_1. This may be an incorrect assumption in some cases, if in the earliest mammals all premolar loci were originally subject to replacement.

[†] *Erythrotherium*, *Megazostrodon*, and *Morganucodon* also appear to have the same formula in adults, acquired convergently, although no one else has suggested this. I suspect that both P^5_5 and dP^5_5 have been identified as M^1_1 in various individuals assigned to these early nontherians.

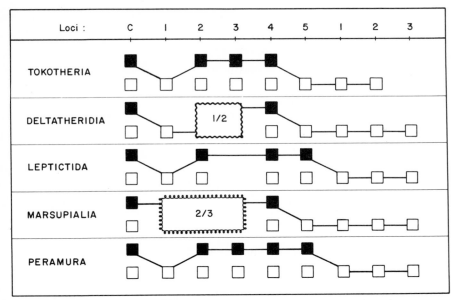

FIG. 2. Schematic representation of postulated normal tooth family homologies of canine and postcanine teeth of selected taxa. Dark squares: replacing canine and premolars; empty squares: milk teeth or molars; zig-zag outline: uncertainty as to which one of two loci are represented; coiled outline: uncertainty as to which two of three loci are represented; connected squares: adult dentition, exclusive of incisors.

molar (P_4^4), and a retained last milk premolar in primitive marsupials. The retained dP_5^5 of adult marsupials has always been known as M_1^1 and would in any case be part of the same series as the 3 more posterior molars, which in a sense are nonreplaced milk teeth. Thus the traditionally identified M_1^1 of marsupials might be the nonreplaced milk predecessors of the teeth identified as M_1^1 by Clemens and Mills in *Peramus* but here called P_5^5. The marsupial "M_2^2 M_3^3 M_4^4" would thus really be M_1^1 M_2^2 M_3^3. Another, more traditional view is that marsupials had 4 molars to begin with, which would mean that *Peramus* could not have been a plesiomorphous sister group of marsupials without evolutionary reversal because *Peramus* as here interpreted had already lost M_4^4. Separate and convergent origins of marsupial and eutherian protocones would also be implied. A more parsimonious solution is that the marsupial "M_1^1" is dP_5^5.

D. Relationships within Ernotheria

We still lack an adequate record of the evolution of dental formulae between the late Jurassic and the late Cretaceous,[*] but at the end of that long gap in the fossil record Lillegraven (1969, p. 57) and Clemens (1973, p. 6) have recently demonstrated that the late Cretaceous leptictid-like genus *Gypsonictops* still retained 5 lower premolar loci in various individuals, reduction to 4 in number in older individuals by loss of the reduced tooth at the P_3 locus (P_c of Clemens' notation) occurring in at least one known specimen.

[*] Slaughter (1971, Plate 9) has figured a jaw bearing 4 teeth that he referred to *Pappotherium pattersoni*, an Albian eutherian. I interpret its teeth to be P_3 P_4 P_5 M_1.

A young specimen of the Mongolian Cretaceous genus *Kennalestes*, MgM-I/1, also displays 5 premolar loci, but in adult specimens the number is 4, P_c^c having been lost. Significantly, replacement occurs at the fifth locus. MgM-I/1 has been studied through the courtesy of Dr. Zofia Kielan-Jaworowska, who will publish the details of its immature morphology elsewhere. *Kennalestes* is not classified as a leptictid here, but it is obviously quite close to them. I consider it quite unlikely that these occurrences of P_c are fortuitous. Rather, I suspect that we have been fortunate enough to observe in *Kennalestes* and *Gypsonictops* that the plesiomorphous premolar locus count is 5 in the Zatheria, in keeping with my interpretation of *Peramus*. Five or more lower premolars were present in the even more primitive genus *Kuehneotherium*. Presumably in *Kennalestes* and *Gypsonictops* replacement had been suppressed at the P_3^3 locus, so that P_c^c are probably dP_3^3. In Tertiary leptictids even dP_3^3 are no longer present. Their premolars are interpreted here to be dP_1^1 P_2^2 P_4^4 P_5^5. Unequivocal confirmation that 5 lower premolar loci existed in the adults of primitive *Kennalestes*-like animals in the Albian Cretaceous Khobor (Khovboor) fauna of Mongolia (see Belyaeva *et al.*, 1974, p. 20) has recently been obtained by the joint Soviet–Mongolian Paleontological Expeditions and presumably will be discussed in detail by Trofimov in a forthcoming paper. Through the courtesy of Dr. Trofimov, I have had the opportunity to study these specimens briefly during June, 1974, and have been generously permitted to mention here the lower premolar locus count.

The loci usually identified as P_3^3 and P_4^4 in adults of *Kennalestes* and its allies which have lost the actual P_3^3 locus should therefore be known as the loci of P_4^4 and P_5^5, respectively. In members of this epithere radiation (Fig. 3), here called Magnorder Ernotheria (offshoot therians), the premolars of adults are dP_1^1 P_2^2 P_4^4 P_5^5, or are further reduced anteriorly. I refer "*Asioryctes*" Kielan-Jaworowska (1975*b*)★ and palaeoryctines, *Kennalestes*, *Gypsonictops*, leptictids, didymoconids, macroscelidids, anagalids, pseudictopids, zalambdalestids, eurymylids, ochotonids, and leporids to the Ernotheria.

Uniting the ernotheres is a highly unorthodox departure, in part suggested by published work of Evans (1942), Van Valen (1964), Sulimski (1969), Kielan-Jaworowska (1969), and Szalay and McKenna (1971). I have also been influenced by unpublished work of S. B. McDowell, Jr., and by Z. Kielan-Jaworowska, although the interpretation given here is not necessarily endorsed by either of them.

In the past the ernotheres have been studied in much the same way as the famous research on the elephant was conducted by the blind men. No comprehensive study has appeared in which palaeoryctines,† leptictids, lagomorphs, macroscelidids, and various families assigned to Szalay and McKenna's Anagalida have been cross-compared fully.

★ "*Asioryctes*" Kielan-Jaworowska (1975*b*) is mentioned here by permission. Kielan-Jaworowska concludes, and I concur, that "*Asioryctes*" is an early palaeoryctine. Palaeoryctines would thus seem to be ernotheres in that "*Asioryctes*" shares the derived character state of P_3^3 locus loss. In other dental features the genus is rather similar to "*Prokennalestes*" from the Mongolian Albian.

† *Palaeoryctes* Matthew, 1913, has been discussed by McDowell (1958, pp. 176–180), who mistook its expanded tympanic ring for an entotympanic ossification. Van Valen (1966, pp. 52–55) has noted some additional errors in McDowell's otherwise excellent interpretation. In agreement with McDowell, however, both Van Valen and I concur that a medial branch of the carotid artery appears to have been present medial to the petrosal bone in *Palaeoryctes*. The postcanine dental formula of the genus is interpreted here to be dP_1^1 P_2^2 P_4^4 P_5^5 M_1^1 M_2^2 M_3^3. Matthew's (1913, Figs. 2 and 3) dental reconstruction was not well founded, nor is there reason to believe that Van Valen's (1966, p. 62) interpretation was yet fully correct.

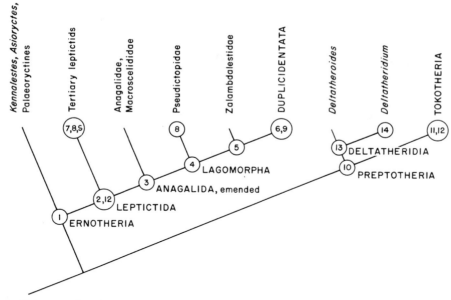

FIG. 3. Cladogram of Epitheria, exclusive of rodents, edentates, and pangolins. Advanced forms progress from left to right: (1) loss of P^3_3; (2) loss of dP^3_3, astragalus and calcaneum become lagomorph-like; (3) cheek teeth become somewhat prismatic; (4) jaw deepens; (5) enlargement of I_1 and I^2; (6) loss of I_1 I_2 I_3 C^1_1 dP^1_1 P_2; (7) inflated sternum; (8) multicuspate incisors; (9) fibula fuses to tibia; (10) retain dP^5_5; (11) lose M^3_3; (12) reduce incisor formula to $\frac{3}{3}$; (13) lose dP^2_2 and P^2_2 or dP^3_3 and P^3_3, shorten snout; (14) lose M^3, reduce M_3, lose 2 or even 3 lower incisors.

Simpson (1931*b*) made comparisons among macroscelidids, tupaiids, and *Anagale*, concluding that the last was most closely related to the Tupaiidae but merited its own family, Anagalidae. Evans (1942) made extensive comparisons between *Anagale* and macroscelidids, but merely concluded that anagalids were intermediate between tupaiids and macroscelidids. It is unfortunate that his study did not separate plesiomorph from apomorph character states. Bohlin (1951) and McKenna (1963) brought out new details of anagalid anatomy, but simply denied tupaiid special relationships without going further, leaving the anagalids as Mammalia, *incertae sedis*. None of these authors considered lagomorphs. Van Valen (1964) studied mostly dental features of the Zalambdalestidae, then still very inadequately known, and Pseudictopidae, concluding that lagomorphs originated from closely related stock. In his study he ignored the macroscelidids. Szalay and McKenna (1971) created an order Anagalida for the Anagalidae, Pseudictopidae, Eurymylidae, Zalambdalestidae, and possibly the Didymoconidae as well, but made no comparisons with palaeoryctines, leptictids, macroscelidids, or lagomorphs. The task cannot be accomplished adequately in these few pages, but it is to be hoped that detailed work can begin in the near future.

In addition to their common heritage from animals which reduced their premolar locus count from 5 to 4 by the loss of the P^3_3 locus, what seems to me to unite the ernotheres is a combination, which I take to have been acquired independently of any of the remaining epitheres (Preptotheria of this paper), of the following features: high condyle of jaw, often well above plane of teeth; coronoid process hooklike; elaboration of the primitive palatal fenestration in several groups; frontals tending to develop supraorbital processes; nasals tending to become narrow and arched; P^4_4 P^5_5 M^1_1 M^2_2 M^3_3 quintet enhanced at expense of anterior premolars (M^3_3 lost, however, in most macroscelidids); anterior lower incisor function increased; tendency toward unilaterally hypsodont or prismatic teeth with reduced stylar shelves and loss of anterior and posterior cingula on worn upper teeth of high-crowned forms; possession of somewhat lagomorph-like chewing action, in which the wear surfaces of the lower molar trigonids continue posterodorsally the wear surfaces of the talonids of the teeth in front of them; and lagomorph-like foot structure. These character states are of course admixed with plesiomorph features and with a variety of autapomorphous character states individually acquired by the various ernothere branches.

The most primitive known fossil ernotheres are "*Asioryctes,*" *Kennalestes*, and various members of the families Leptictidae, Zalambdalestidae, Anagalidae, and Pseudictopidae. "*Asioryctes*" and *Kennalestes* are the most plesiomorphous of them all and are known only from the Mongolian late Cretaceous. The zalambdalestids are known from the Asian Cretaceous and late Paleocene, the anagalids from the Asian late Paleocene to Oligocene, and the pseudictopids only from the Asian late Paleocene. Leptictids (minus *Gypsonictops*) are known from the Paleocene to the Oligocene in North America, and also from the European early Tertiary. Advanced leptictids greatly reduced the medial branch of the internal carotid artery, acquired an ossified but incomplete entotympanic bulla, completely lost the P^3_3 locus, and by Eocene time were beginning to fuse the tibia with the fibula (Matthew, 1918). They possessed relatively long hind limbs and short front ones (Guth, 1962) and developed a peculiar type of sternum.

Other than those of "*Asioryctes*" and *Kennalestes*, the generally most plesiomorph ernothere cheek-tooth dental cusp pattern is possessed by the zalambdalestids, which generally have been classified as leptictid-like insectivores because of numerous phenetic resemblances to those animals. The primitive zalambdalestid dental formula is interpreted here to have been I^1_1 I^2_2 I^3_3 C^1_1 dP^1_1 P^2_2 P^4_4 P^5_5 M^1_1 M^2_2 M^3_3. In the upper dentition of old individuals of *Zalambdalestes* dP^1 and P^2 are shed (Szalay and McKenna, 1971, p. 307). Zalambdalestids, however, possess enlarged anterior incisors, identified as I_1 and I^2, but had not yet lost I^1. P^4_4 P^5_5 M^1_1 M^2_2 M^3_3 form a quintet of cheek teeth which contrast sharply with more anteriorly placed teeth. P^4_4 are simpler than P^5_5–M^3_3, but are large teeth, especially P^4. The lower cheek teeth compress the trigonids; as M_3 is approached the paraconids are progressively reduced. Both the trigonids and the talonids are slightly prismatic, wear being best described as lagomorph-like. The talonids and neighboring trigonids to the rear of them form integrated posterodorsally elevated wear surfaces. Although the stylar shelves of the upper quintet of principal cheek teeth are reduced, the prismatic condition of advanced ernothere cheek teeth had barely been initiated in zalambdalestids. Anterior and posterior cingula were lacking on the upper teeth. Additional Cretaceous zalambdalestid material, including a new genus that is represented by postcranial as well as cranial material, together with newly collected postcranial material of *Zalambdalestes*, will soon be described by Kielan-Jaworowska. I believe *Zalambdalestes* is cladistically a Cretaceous lagomorph. A *Zalambdalestes*-like animal, possibly still with the tibia and fibula unfused, could have given rise to later lagomorphs by loss of I^1, I_2, I_3, C_1^1, dP_1^1, and P_2 to yield the dental formula I_1 I^2 I^3 P^2 P_4^4 P_5^5 M_1^1 M_2^2 M_3^3, from which both ochotonids and leporids could be derived.

The dental formula in the Pseudictopidae is interpreted here to be I_1^1 I_2^2 I_3^3 C_1^1 dP_1^1 P_2^2 P_4^4 P_5^5 M_1^1 M_2^2 M_3^3. Although the anterior teeth are unreduced (indeed, they are uniquely elaborated), the cheek-tooth pattern is lagomorph-like as Van Valen (1964) and Sych (1971) have noted.

Lagomorph cheek teeth are not zalambdodont as various authors have claimed. If they were, the ectoflexids of lower cheek teeth would be enlarged and the talonid basins would be medially displaced, which is not the case.

Another major ernothere branch is represented by the fossil and living Macroscelididae and the fossil Anagalidae. The Anagalidae have been described extensively by Simpson (1931b), Evans (1942), Bohlin (1951), McKenna (1963), and Szalay and McKenna (1971). Information about Chinese Paleocene anagalids may be expected soon as the result of ongoing research at the Paleontological Institute in Peking (Chow *et al.*, 1973; Chow, 1974). Further preparation of the skeleton of the type specimen of *Anagale gobiensis* is also desirable, but has not yet been attempted. On the basis of what is already known, however, anagalids now appear to me to be relatively plesiomorphous early Asian macroscelideans that had not yet fused the tibia with the fibula or yet strongly reduced or lost M_3^3. Otherwise, they appear to be most closely similar to African macroscelideans, as Evans (1942) documented but curiously did not conclude. The Macroscelididae can be regarded as having descended from early Tertiary stock close to the Anagalidae if not from early anagalids themselves. The time of origin of the Macroscelididae was at least as long ago as the early Oligocene but more likely much earlier, after which it appears that the family was restricted to Africa. The time of origin of the Macroscelidea as a whole appears to have been at least as long ago as the late Cretaceous. Possibly early macroscelideans first dispersed between southwest Asia and Africa in company with the ancestors of phiomyid rodents, early anthracotheres, and several other groups, but they may have been in both areas earlier. Details of macroscelidid evolutionary history after their isolation in Africa are gradually emerging and may be found in papers by Evans (1942), Patterson (1965), and Corbet and Hanks (1968).

The original dental formula of the Macroscelididae is interpreted here to have been I_1^1 I_2^2 I_3^3 C_1^1 dP_1^1 P_2^2 P_4^4 P_5^5 M_1^1 M_2^2 M_3^3, but this was reduced by the loss of M_3^3 in all living genera except *Nasilio*, which retains a tiny trigonid remnant of M_3. Among fossil macroscelidids the macroscelidine genus *Palaeothentoides* retains a trigonid remnant of M_3 and the two myohyracine genera *Myohyrax* and *Protypotheroides* retained remnants of both M^3 and M_3. *Rhynchocyon* has lost its anterior upper incisors, retaining only the tiny and presumably almost, if not entirely, functionless I^3.

E. Relationships within Preptotheria

After the phylogenetic departure of the ernotheres, the remaining epitherian magnorder, the Preptotheria, differentiated into a plethora of groups that are still rather arbitrarily classified.

The original dental formula from which the preptotheres gradually departed appears to have consisted of a plesiomorphous 5/4 incisor count, C_1^1, and a postcanine formula of $dP_1^1 P_2^2 P_3^3 P_4^4 P_5^5 M_1^1 M_2^2 M_3^3$ inherited from a *Peramus*-like form in the Jurassic. The incisor count was then reduced to $I_1^1 I_2^2 I_3^3$ except in *Deltatheridium* (McKenna *et al.*, 1971).

It is necessary at this point to remark on the recent exciting proposal by Butler and Kielan-Jaworowska (1973) that members of the Cretaceous family Deltatheridiidae, long considered eutherian mammals, are perhaps really marsupials instead or at least have the marsupial adult postcanine dental formula. Van Valen (1974), in a comment on their paper, suggests marsupial affinities as well, although for somewhat different reasons. If marsupials, the Deltatheridiidae are the only ones thus far found on the Asian continent. I concur with Butler and Kielan-Jaworowska that the genus *Deltatheridium* possesses 3 premolar loci and 4 lower (only 3 upper) molars, contrary to the arguments of McKenna *et al.* (1971) and Szalay and McKenna (1971), which were based on an extrapolation from the Mongolian Tertiary family Didymoconidae. We proposed, on the basis of the very poorly preserved material which was at that time all that was known, that the adult postcanine formula of *Deltatheridium* was $P_1^1 P_2^2 P_3^3 P_4^4 M_1^1 M_2^2 ?M_3$. Butler and Kielan-Jaworowska (1973) suggest on the basis of recently collected, well-preserved, presumably adult specimens of *Deltatheridium* and *Deltatheroides* that the deltatheridiid formula is effectively $P_1^1 P_2^2 P_3^3 M_1^1 M_2^2 M_3^3 M_4^4$ or M_4, a conclusion I accept in terms of numbers of molar and presumed replacement premolar loci but which I believe could equally well be written $dP_1^1 P_2^2$ or $\frac{3}{3} P_4^4 dP_5^5 M_1^1 M_2^2 M_3^3$ or M_3. If the adult deltatheridiid postcanine dental formula and a presumed reduction of premolar replacement are derived character states which can be shown beyond reasonable doubt to be shared synapomorphously with the marsupials, then well and good, deltatheridiids would have to be marsupials; if not, and adult retention of dP_5^5 as an anterior molar* is merely a parallel event in this branch of the Theria, then the Deltatheridiidae can be regarded as a side branch of small, carnivorous eutherian preptotheres that had retained a molariform dP_5^5 in adults while also reducing the remaining number of premolars from 4 to 3 by the loss of an anterior premolar and reducing the number of molars in at least one genus, *Deltatheridium*, by the loss of M^3 and reduction of M_3. Deltatheridiids possessed, or had in one known case recently departed from, the gross marsupial molar and premolar locus count, but they are not necessarily marsupials because of this nor are tooth-for-tooth homologies between marsupials and deltatheridiids necessarily established. No one has observed actual premolar replacement yet in deltatheridiids. The number of known specimens is still quite inadequate in spite of recent successes. Whether deltatheridiids had an Asian or North American origin cannot presently be determined; the family has now been reported from the North American late Cretaceous by Fox (1974), although this is disputed by Kielan-Jaworowska (1975*a*). Similarities in molar morphology to the American Albian Pappotheriidae may also prove significant. I utilize Van Valen's term Deltatheridia in drastically restricted form at superordinal rank for the deltatheridiids.

A speculation perhaps still worthy of study when more evidence is available is that the dental formula of modern marsupials did not evolve so early as previously believed

---

* It follows that if 4 molars can occur in this manner in Cretaceous eutherians, then we should take a second look at various fossil "marsupials" identified as such on the basis of inflected angles (plesiomorphy) and possession of 4 molars (convergence).

and that it actually arose from the epithere formula via a stage represented by *Delta-theroides*. Studies of other aspects of marsupial and ernothere morphology make such a derivation appear unlikely at present, however, and I attribute the marsupial-like features of deltatheridiids to both plesiomorphy and convergence.

F. Relationships within Tokotheria

Unfortunately, early nondeltatheridian preptothere history from a *Peramus*-like creature in the Jurassic up through the Cretaceous is still almost wholly a matter of conjecture, but, by the late Cretaceous, members of all four nondeltatheridian prepto-there grandorders recognized here are known from the fossil record in North America, where the genera *Cimolestes*, *Batodon*, *Purgatorius*, and *Protungulatum* represent the Ferae, Insectivora, Archonta, and Ungulata, respectively. I have not yet found a way to treat this apparent grandordinal tetrachotomy cladistically, however, and so will pass over this major problem here. All four groups together, here dubbed Tokotheria, represent the sister group of the Deltatheridia. All tokotheres share or further modify a postcanine dental formula consisting of $dP_1^1 \, P_2^2 \, P_3^3 \, P_4^4 \, dP_5^5 \, M_1^1 \, M_2^2$. dP_5^5 is the tooth usually called M_1^1; the last premolar is primitively a nonmolariform P_4^4, the same tooth as the penulti-mate premolar in ernotheres. These homologies are fundamental.

1. Ferae. The first members of the Ferae occur in the late Cretaceous. From various species currently placed in a single genus, *Cimolestes*, or at least from close allies, the didelphodontines, Pantodonta (including *Deltatherium*), pantolestoids, apatemyoids, taeniodonts, creodonts (oxyaenoids and hyaenodontoids), and Carnivora arose at about the beginning of the Tertiary. Rodents may have arisen from this same stock (Szalay and Decker, 1974, Fig. 2), although in agreement with Wood (1962), I have previously (McKenna, 1961a) suggested that rodents might in some way be related to primates. At present I prefer to place rodents as Epitheria, *incertae sedis*. The subdivisions of these various groups will not be discussed here, although it should be obvious that these animals are a ripe subject for cladistic analysis.

2. Insectivora. The Insectivora as restricted in this paper are regarded as synonymous with Haeckel's Lipotyphla. Insectivores first appear in the fossil record in the late Cre-taceous, but they are not well known until the mid-Tertiary. Eocene and Paleocene fossils are generally fragmentary and usually consist of dental material only. Although Insectivora is commonly regarded as a sort of taxonomic wastebasket, most animals usually placed among the insectivores share certain derived features; for this reason the lipotyphlan insectivores are regarded as a natural group here. Shared derived features of the Insectivora (= Lipotyphla) include loss of the caecum of the gut, reduced jugal bone, expansion of the maxillary bone in orbital wall, loss of the medial branch of carotid artery, and reduction of the pubic symphysis. McDowell (1958) has done much to elucidate the character states of insectivores and also of the main insectivore divisions, without, however, fully separating plesiomorphous from apomorphous features. Butler (1972) has recently reviewed the subject, although not from a cladistic viewpoint, and Schmidt-Kittler (1973) has made an important improvement in Butler's classification by demonstrating the true affinities of the Dimylidae.

The true insectivores, or lipotyphlans, can be divided into two sister groups, the Erinaceidae *sensu lato* (i.e., including Adapisoricidae) and a second group which includes all other insectivores. These groups can be called Erinaceomorpha and Soricomorpha, respectively, and these names are used here in emended form different from Gregory's (1910) or Saban's (1954) definitions, which involved somewhat different content although the Erinaceidae and Soricidae were central to the definitions. The Erinaceomorpha are monophyletic, generally plesiomorphous lipotyphlan insectivores, but the Soricomorpha have reduced and then lost the jugal bone, reduced the ectopterygoid laminae, and modified the entoglenoid process of the skull into a postglenoid process, as McDowell (1958) has shown. Although talpids and chrysochlorids actually have a zygomatic arch, it lacks a jugal bone and in each case is presumed here [in agreement with M. Novacek and H. Hutchison (personal communication)] to be a secondary reconnection. Unlike Butler (1972) or Schmidt-Kittler (1973), but in essential agreement with McDowell (1958), I include the tenrecs and chrysochlorids in the Soricomorpha. The soricomorph families are Dimylidae (see Schmidt-Kittler, 1973), Talpidae, Nyctitheriidae, Plesiosoricidae, Micropternodontidae, Nesophontidae, Soricidae, Geolabididae, Solenodontidae, Apternodontidae,* Tenrecidae and Chrysochloridae. Detailed cladistic analysis of the Soricomorpha is an attractive project that should be undertaken in the near future.

3. Archonta. Of the remaining tokotheres, another major branch appears to have led to modern tupaiids, bats, dermopterans, and primates. This branch, Gregory's (1910) Archonta (minus the macroscelidids), is not very well known from fossils except for the order Primates; thus although the tupaiids, dermopterans, and bats went their separate ways quite early, the details are still obscure and useful fossils are still rare. Tupaiids are regarded here as primitive archontans because of the possession of shared derived character states (Gregory, 1910) even though they retain many plesiomorphous character states in their morphology. These plesiomorphous "insectivore-like" character states, especially those seen in *Ptilocercus* (Clark, 1926), are of course clues concerning the morphology of the Cretaceous tokothere stem group leading to such morphologically "distinct" groups as other archontans, insectivores, and ungulates, each of which developed quite different autapomorphous morphologies later in their separate histories. When the Archonta split off from other tokotheres they were still probably partly terrestrial animals with arboreal capability, as behavioral studies of tupaiids attest (Jenkins, 1974), but most of them soon took up a mainly arboreal mode of existence and from that, in dermopterans and bats, gliding and flight.

---

* McDowell (1958, pp. 164–174, etc.) excluded *Apternodus* not only from the Soricomorpha but also from the entire Lipotyphla (= Insectivora). His work unfortunately was based on poorly preserved material, especially of *Apternodus brevirostris*, and many of his conclusions are not supported by excellent specimens now at hand representing various species of several apternodont genera. For instance, he mislocated the cribriform plate, inferred from damaged pyriform fenestrae that there were no fenestrae, mistook the promontory artery for the medial branch of the carotid, and was unable to see in the material then available that the maxillary bone is extensively exposed in the orbit. Several other discrepancies exist, notably concerning the position of faint sutures. *Apternodus* is clearly a zalambdodont soricomorph; it even has pigmented teeth as in shrews and *Solenodon*. In fairness to McDowell, it should be emphasized that the specimens available to him were not typically preserved or well prepared. He was also unaware of the morphology of the snout of *Oligoryctes*. An account of the morphology of *Apternodus* and its allies is in preparation and will appear elsewhere.

A postorbital bar began to form in many archontans, but apparently it did so independently. Certain early primates either lacked it completely (Plesiadapidae) or were just beginning to develop one (*Palaechthon*, Microsyopidae). Megachiropteran fruit bats and emballonurid microchiropterans possess the beginnings of a postorbital bar or complete it, but other bats do not possess a postorbital bar. Recent dermopterans have a large postorbital process, but the bar is not complete.

In one important osteological feature, the presence of an ossified entotympanic bulla that only fuses to the petrosal late in ontogeny, tupaiids depart significantly from primates, bats, and dermopterans. Bats expand the tympanic ring somewhat to form a partial bulla, but dermopterans developed from the tympanic bone a bulla that fused extensively with the skull. Early primates took a different route, evolving a bulla as an outgrowth of the petrosal bone (McKenna, 1966), to which a tympanic component has been added in some forms.

Tupaiids plesiomorphously retained both branches of the stapedial artery, the ramus superior, and the ramus inferior, but all modern tupaiids except *Ptilocercus* have lost the ramus inferior. The medial branch of the internal carotid artery was lost in tupaiids, as it was in most primates (see Szalay, Chapter 5, this volume).

4. Ungulata. From Cretaceous tokotheres similar to the adapisoricine erinaceids and (at least dentally) the earliest known primates arose early ungulates such as the latest Cretaceous genus *Protungulatum* (Sloan and Van Valen; 1965). The ungulates in turn divided into five great groups at the end of the Cretaceous, here called mirorders Eparctocyona, Cete, Meridiungulata, Phenacodonta, and Tethytheria. I have not been able to classify these dichotomously.

The Eparctocyona broke up into six major subdivisions before the Eocene (timing vague for embrithopods and tubulidentates). Early Tertiary North American and European arctocyonids represent the plesiomorphous sister group. From an early, didolodont-like ancestor present in South America before the end of the Cretaceous, the six South American meridiungulate orders fragmented. In the rest of the world the Phenacodonta and Tethytheria split into five orders during the Eocene or earlier.

In keeping with cladistic principles, I have placed the whales among the ungulates in this paper and I have not hesitated to place their terrestrial sister group, the Acreodi (Mesonychidae) with them there, in the mirorder Cete, carrying to a cladistic conclusion arguments presented by Van Valen (1966) and Szalay (1969a, b). Perhaps these actions will be too much for those who believe that "morphological distance" should be directly reflected in a classification, but the taxonomic equivalent of Hennig's phylogenetic method demands it if Van Valen and Szalay have correctly identified mesonychids as the sister group of cetaceans. There are different ways of looking at organisms!

III. A Higher Category Classification of Mammals

The higher category classification of mammals given below is an attempt to express in words what is perhaps better expressed as a sequence of cladograms, but for various reasons I prefer to use words if their original meaning is not done too much damage.

This classification is an abstract of a presently tentative, but much more detailed, classification in preparation. Insofar as possible I have tried to use familiar terms at ranks not too outrageously removed from their original or at least their widely accepted ranks, but in a number of cases use of new ranks and terms has seemed to offer promise as a way to preserve familiar terms in their accustomed meanings. In other cases, new terms have been invented when I could find no previous applicable name.*

Class MAMMALIA Linnaeus, 1758
Subclass PROTOTHERIA Gill, 1872
†Infraclass EOTHERIA Kermack and Mussett, 1958
†Order TRICONODONTA Osborn, 1888
†Suborder MORGANUCODONTA Kermack, 1973
†Suborder EUTRICONODONTA Kermack, 1973
†Order DOCODONTA Kretzoi, 1946
Infraclass ORNITHODELPHIA de Blainville, 1834
Order MONOTREMATA Bonaparte, 1838
Suborder TACHYGLOSSA Gill, 1872
Suborder PLATYPODA Gill, 1872
†Order MULTITUBERCULATA Cope, 1884
†Suborder HARAMIYOIDEA Hahn, 1973
†Suborder PLAGIAULACOIDEA Simpson 1925
†Infraorder PLAGIAULACIDA McKenna, 1971, new rank
†Infraorder CIMOLODONTA, new
†Parvorder PTILODONTOIDEA Sloan and Van Valen, 1965, new rank
†Parvorder TAENIOLABIDOIDEA Sloan and Van Valen, 1965, new rank
Subclass THERIA Parker and Haswell, 1897
†Superlegion KUEHNEOTHERIA, new
Superlegion TRECHNOTHERIA, new
†Legion SYMMETRODONTA Simpson, 1925, new rank
Legion CLADOTHERIA, new
†Sublegion DRYOLESTOIDEA Butler, 1939, new rank
Sublegion ZATHERIA, new
†Infraclass PERAMURA, new
Infraclass TRIBOSPHENIDA, new
Supercohort MARSUPIALIA Illiger, 1811, new rank (contents not subdivided here)
Supercohort EUTHERIA Gill, 1872 (as modified by Huxley, 1880), new rank
Cohort EDENTATA Cuvier, 1798
Order CINGULATA Illiger, 1811, new rank
Order PILOSA Flower, 1883, new rank

* A † indicates an extinct taxon.

Cohort EPITHERIA, new
  Magnorder ERNOTHERIA, new
    †Superorder KENNALESTIDA, new
    Superorder LEPTICTIDA, new
    †Grandorder ICTOPSIA, new
      Grandorder ANAGALIDA Szalay and McKenna, 1971, emended, new rank
        Order MACROSCELIDEA Butler, 1956
        Order LAGOMORPHA Brandt, 1855
  Magnorder PREPTOTHERIA, new
    †Superorder DELTATHERIDIA Van Valen, 1965, emended, new rank
    Superorder TOKOTHERIA, new
      Grandorder FERAE Linnaeus, 1758, new rank
        †Order CIMOLESTA, new
          †Suborder DIDELPHODONTA, new
          †Suborder PANTODONTA Cope, 1873, new rank
          †Suborder PANTOLESTA, new
          †Suborder APATOTHERIA Scott and Jepsen, 1936, new rank
          †Suborder TAENIODONTA Cope, 1876, new rank
        †Order CREODONTA cope, 1875 (= †HYAENODONTIA Romer, 1966)
        Order CARNIVORA Bowdich, 1821 (contents not subdivided here)
      Grandorder INSECTIVORA Illiger, 1811, new rank
        Order ERINACEOMORPHA Gregory, 1910, emended, new rank
        Order SORICOMORPHA Gregory, 1910, emended, new rank
      Grandorder ARCHONTA Gregory, 1910
        Order SCANDENTIA Wagner, 1855
        Order DERMOPTERA Illiger, 1811
        Order CHIROPTERA Blumenbach, 1779
          Suborder MEGACHIROPTERA Dobson, 1875
          Suborder MICROCHIROPTERA Dobson, 1875
        Order PRIMATES Linnaeus, 1758
          †Suborder PLESIADAPIFORMES Simons and Tattersall, 1972, new rank
          Suborder STREPSIRHINI E. Geoffroy, 1812
          Suborder HAPLORHINI Pocock, 1918
            Infraorder TARSIIFORMES Gregory, 1915
            Infraorder PLATYRRHINI E. Geoffroy, 1812
            Infraorder CATARRHINI E. Geoffroy, 1812
      Grandorder UNGULATA Linnaeus, 1766, new rank
        Mirorder EPARCTOCYONA, new

†Order ARCTOCYONIA Van Valen, 1969
†Order TILLODONTIA Marsh, 1785
Order TUBULIDENTATA Huxley, 1872
†Order DINOCERATA Marsh, 1873
†Order EMBRITHOPODA Andrews, 1906
Order ARTIODACTYLA Owen, 1848 (contents not sub-divided here)
Mirorder CETE Linnaeus, 1758, new rank
†Order ACREODI Matthew, 1909, new rank
Order CETACEA Brisson, 1762
†Suborder ARCHAEOCETI Flower, 1883
Suborder ODONTOCETI Flower, 1867
Suborder MYSTICETI Flower, 1864
†Mirorder MERIDIUNGULATA, new
†Order LITOPTERNA Ameghino, 1889
†Order NOTOUNGULATA Roth, 1903
†Suborder NOTIOPROGONIA Simpson, 1934
†Suborder TOXODONTIA Owen, 1858
†Suborder TYPOTHERIA Zittel, 1893
†Suborder HEGETOTHERIA Simpson, 1945
†Order ASTRAPOTHERIA Lydekker, 1894
†Order TRIGONOSTYLOPOIDEA Simpson, 1967
†Order XENUNGULATA Paula Couto, 1952
†Order PYROTHERIA Ameghino, 1895
Mirorder PHENACODONTA, new
†Order CONDYLARTHRA Cope, 1881, emended
Order PERISSODACTYLA Owen, 1848
Suborder HIPPOMORPHA Wood, 1937
Suborder ANCYLOPODA Cope, 1889
Suborder CERATOMORPHA Wood, 1937
Order HYRACOIDEA Huxley, 1869
Mirorder TETHYTHERIA, new
Order PROBOSCIDEA Illiger, 1911
†Suborder MOERITHERIOIDEA Osborn, 1921
†Suborder BARYTHERIOIDEA Simpson, 1945
†Suborder DEINOTHERIOIDEA Osborn, 1921
Suborder ELEPHANTOIDEA Osborn, 1921
Order SIRENIA Illiger, 1811
†Order DESMOSTYLIA Reinhart, 1953
Magnorder ?PREPTOTHERIA, *incertae sedis*:
Order PHOLIDOTA Weber, 1904
Cohort EPITHERIA, *incertae sedis*:
Order RODENTIA Bowdich, 1821 (contents not subdivided here)

A genealogical reclassification of many mammalian higher taxa is presented, expressing the topology of mammalian genealogy as worked out anew on the basis of fossil and living mammals. Various new higher taxa are proposed as logical consequences of the cladistic taxonomic treatment attempted as well as of the cladistic phylogenetic philosophy applied, but as many familiar names as possible are saved at familiar ranks by interposing new hierarchical ranks in the system. The phylogeny and classification are put forth to see how well they will withstand scrutiny. To be rejected, the phylogeny must be internally inconsistent or conflict with additional evidence available from embryology, soft anatomy, biochemistry, and other character states not studied here. The taxonomic expression of the phylogeny has not yet achieved stability, especially at lower levels because of the enormous amount of analysis still to be undertaken, but the route to stability is held to be through the adoption of a cladistic and potentially falsifiable system based upon kinship as the general reference system, as Hennig has proposed.

The genealogy of the Eutheria proposed here is in part based on a new interpretation of therian dental homologies that flows from the new concept that primitive eutherians possessed 5 premolars and had reduced the number of molars to 3. One major branch of the Eutheria, represented today by elephant shrews and rabbits, retains these 3 molars primitively and has reduced the number of premolars by the loss of P^3_3 and other teeth. A second major branch, containing the majority of eutherian orders, has lost M^3_3 but has retained dP^5_5 in adults to produce what are usually called M^1_1 in those animals. The Cretaceous Deltatheridiidae retained dP^5_5 but were just losing M^3_3. Still other major eutherian branches are represented by the Edentata and Rodentia.

ACKNOWLEDGMENTS

This paper has been through a number of drafts and has been circulated rather freely at various stages of its evolution. To the Wenner–Gren Foundation I express my gratitude for the opportunity to present my initial results to a group of colleagues at Burg Wartenstein, Austria, in July, 1974. Various other colleagues throughout the world have been most helpful as well, notably Percy Butler, William Clemens, Daryl Domning, George Engelmann, Zofia Kielan-Jaworowska, Leonard Krishtalka, Samuel McDowell, Jr., Earl Manning, Michael Novacek, Donn Rosen, Jeffrey Schwartz, and Boris Trofimov. My errors are not endorsed by any of them, of course, but they have countered my disorders with friendly diatribes and I thank them warmly for their interest and constructive contributions.

V. References

ASHLOCK, P. D. 1974. The uses of cladistics. *Ann. Rev. Ecol. Syst.* **5**:81–99.

BELYAEVA, E. I., B. A. TROFIMOV, and V. Y. RESHETOV. 1974 [Translation]. Fundamental stages in the evolution of mammals of the late Mesozoic-Paleogene of central Asia. *In* N. N. Kramerenko, V. Luvsanvandan, Y. I. Voronin, R. Barsbold, A. K. Rozhdestvensky, B. A. Trofimov, and V. Y. Reshetov, eds., *Mesozoic and*

Cenozoic Faunas and Biostratigraphy of Mongolia. The Joint Soviet–Mongolian Paleontological Expedition, Trans., *I*, Science Press, Moscow.

BOHLIN, B. 1951. Some mammalian remains from Shih-ehr-ma-ch'eng, Hui-hui-p'u area, western Kansu, pp. 1–47. *In* S. Hedin. *Reports from the Scientific Expedition to the North-Western Provinces of China*, Vol. 6. Vertebrate Paleontology 5, Stockholm.

BRUNDIN, L. 1966. Transantarctic relationships and their significance, as evidenced by chironomid midges with a monograph of the subfamilies Podonominae and Aphroteniinae and the austral Heptagyiae. *K. Sven. Vetenskakad. Handl.*, Ser. 4 **11**:1–472 (especially pp. 11–45).

BUTLER, P. M. 1972. The problem of insectivore classification, pp. 253–265. *In* K. A. Joysey and T. R. Kemp, eds., *Studies in Vertebrate Evolution*. Oliver and Boyd, Edinburgh.

BUTLER, P. M. 1973. A pantotherian milk dentition. *Paläeontol. Z.* **47**(3/4):256–258.

BUTLER, P. M., and Z. KIELAN-JAWOROWSKA. 1973. Is *Deltatheridium* a marsupial? *Nature* **245**:105–106.

CHOW, M. M. 1974. Light on fossil animals and ancient geography. *China Reconstructs* **22**(12):47–48.

CHOW, M. M., Y. CHANG, B. WANG, and S. TING. 1973. New mammalian genera and species from the Paleocene of Nanhsiung Basin, N. Kwangtung. *Vert. PalAsiatica* **11**(1):31–35.

CLARK, W. E. LE GROS. 1926. On the anatomy of the pen-tailed tree-shrew (*Ptilocercus lowii*). *Proc. Zool. Soc. London* **1926**:1179–1309.

CLEMENS, W. A. 1973. Fossil mammals of the type Lance Formation, Wyoming. Part III. Eutheria and summary. *Univ. Calif. Publ. Geol. Sci.* **94**:i–vi, 1–102.

CLEMENS, W. A., and J. R. E. MILLS. 1971. Review of *Peramus tenuirostris* Owen (Eupantotheria, Mammalia). *Bull. Br. Mus. (Nat. Hist.) Geol.* **20**(3):89–113.

CORBET, G. B., and J. HANKS. 1968. A revision of the elephant-shrews, Family Macroscelididae. *Bull. Br. Mus. (Nat. Hist.) Zool.* **16**(2):5–111.

CROMPTON, A. W. 1974. The dentitions and relationships of the southern African Triassic mammals, *Erythrotherium parringtoni* and *Megazostrodon rudnerae*. *Bull. Br. Mus. (Nat. Hist.)* **24**(7):397–437.

EMRY, R. J. 1970. A North American Oligocene pangolin and other additions to the Pholidota. *Bull. Am. Mus. Nat. Hist.* **142**(6):455–510.

EVANS, F. G. 1942. The osteology and relationships of the elephant shrews (Macroscelididae). *Bull. Am. Mus. Nat. Hist.* **80**(4):85–125.

FLOWER, W. H. 1882. On the mutual affinities of the animals composing the order Edentata. *Proc. Zool. Soc. London* **1882**:358–367.

FORBES, W. A. 1882. On some points in the anatomy of the great anteater (*Myrmecophaga jubata*). *Proc. Zool. Soc. London* **1882**:287–302.

FOX, R. C. 1974. *Deltatheroides*-like mammals from the upper Cretaceous of North America. *Nature* **249**(5455):392.

GILL, T. 1872. Arrangement of the families of mammals; with analytical tables. *Smithson. Misc. Collect.* **11**(1):v, vi, 1–98.

GODET, R. 1949. Recherches d'anatomie, d'embryologie normale et expérimentale sur l'appareil génital de la taupe. *Bull. Biol. Fr. Belg.* **83**:25–111.

GRASSÉ, P.-P. 1955a. Ordre des Édentés, pp. 1182–1246, 1262–1266. *In* P.-P. Grassé, ed., *Traité de Zoologie*, Vol. 17, Masson, Paris.

GRASSÉ, P.P. 1955b. Ordre des Insectivores, pp. 1574–1653, 1705–1712. *In* P.-P. Grassé, ed., *Traité de Zoologie*, Vol. 17, Masson, Paris.

GREGORY, W. K. 1910. The orders of mammals. *Bull. Am. Mus. Nat. Hist.* **27**:1–524.

GRIFFITHS, G. C. D. 1972. *The Phylogenetic Classification of Diptera Cyclorrhapha, with Special Reference to the Male Postabdomen.* 2. Principles and procedures of classification, pp. 5–28, W. Junk, The Hague.

GUTH, C. 1962. Un insectivore de Menat. *Ann. Paléont.* **48**:1–10.

HAHN, G. 1973. Neue Zähne von Haramiyiden aus der Deutchen Ober-Trias und ihre Beziehungen zu den Multituberculaten. *Paleontographica* **142**(A):1–15.

HENNIG, W. 1965. Phylogenetic systematics. *Ann. Rev. Entomol.* **10**:97–116.

HENNIG, W. 1966. *Phylogenetic Systematics*, University of Illinois Press, Urbana. 263 pp.

HOFFSTETTER, R. 1958. Xenarthra, pp. 535–636. *In* J. Piveteau, ed., *Traite de Paléontologie*, Vol. 6(2).

HOFFSTETTER, R. 1970. Radiation initiale des mammifères placentaires et biogéographie. *C.R. Acad. Sci. Paris*, Ser. D **270**:3027–3030.

HOFFSTETTER, R. 1973. Origine, compréhension et signification des taxons de rang supérieur: quelques enseignements tirés de l'histoire des mammifères. *Ann. Paléont.* **59**(2):137–169.

HOPSON, J. 1970. The classification of non-therian mammals. *J. Mammal.* **51**(1):1–9.

Huxley, T. H. 1880. On the application of the laws of evolution to the arrangement of the Vertebrata and more particularly of the Mammalia. *Proc. Zool. Soc. London* **1880**:649–662.

Jenkins, F. A., Jr. 1974. Tree Shrew locomotion and the origins of primate arborealism, pp. 85–115. *In* F. A. Jenkins, Jr., ed., *Primate Locomotion*. Academic Press, New York.

Kavanaugh, D. H. 1972. Hennig's principles and methods of phylogenetic systematics. *The Biologist*, **54**(3):115–127.

Kermack, D. M., K. A. Kermack, and F. Mussett. 1968. The Welsh pantothere, *Kuehneotherium praecursoris*. *J. Linn. Soc. (Zool.)* **47**:407–423.

Kermack, K. A., F. Mussett, and H. W. Rigney. 1973. The lower jaw of *Morganucodon*. *J. Linn. Soc. (Zool.)* **53**(2):87–175.

Kielan-Jaworowska, Z. 1969. Preliminary data on the Upper Cretaceous eutherian mammals from Bayn Dzak, Gobi Desert. *Palaeontol. Pol.* **1969**(19–1968):171–191.

Kielan-Jaworowska, Z. 1971. Skull structure and affinities of the Multituberculata. *Palaeontol. Pol.* **1971**(25):5–41.

Kielan-Jaworowska, Z. 1975*a* (in press). Evolution of the therian mammals in the late Cretaceous of Asia. Part I. Deltatheridiidae. Results of the Polish–Mongolian Palaeontological Expeditions, Part VI. *Palaeontol. Pol.*

Kielan-Jaworowska, Z. 1975*b* (in press). Preliminary description of two new eutherian genera from the late Cretaceous of Mongolia. Results of the Polish–Mongolian Palaeontological Expeditions, Part VI. *Palaeontol. Pol.*

Kühne, W. G. 1968. Kimeridge mammals and their bearing on the phylogeny of the Mammalia, pp. 109–123. *In* E. T. Drake, ed., *Evolution and Environment*, Yale University Press, New Haven, Connecticut.

Lillegraven, J. A. 1969. Latest Cretaceous mammals of upper part of Edmonton Formation of Alberta, Canada, and review of marsupial–placental dichotomy of mammalian evolution. *Univ. Kansas Paleont. Contrib.* art. **50**(vert. 12):1–122.

Lillegraven, J. A. 1975 (in press). Biological considerations of the marsupial–placental dichotomy. *Evolution.*

MacIntyre, G. T. 1967. Foramen pseudovale and quasi-mammals. *Evolution* **21**(4):834–841.

MacIntyre, G. T. 1972. The trisulcate petrosal pattern of mammals, pp. 275–303. *In* T. Dobzhansky, M. K. Hecht, and W. C. Steere, eds., *Evolutionary Biology*, Vol. 6, Appleton-Century-Crofts, New York.

Matthew, W. D. 1913. A zalambdodont insectivore from the basal Eocene. *Bull. Am. Mus. Nat. Hist.* **32**(17):307–314.

Matthew, W. D. 1918. Insectivora (continued), Glires, Edentata, pp. 565–657. *In* W. D. Matthew and W. Granger, A revision of the lower Eocene Wasatch and Wind River faunas, Part 5. *Bull. Am. Mus. Nat. Hist.* **38**(16).

Mayr, E. 1974. Cladistic analysis or cladistic classification? *Z. Zool. Syst. Evolutionforsch.* **12**(2):98–128.

McDowell, S. B., Jr. 1958. The greater Antillean insectivores. *Bull. Am. Mus. Nat. Hist.* **115**(3):113–214.

McKenna, M. C. 1961*a*. A note on the origin of rodents. *Am. Mus. Novit.* **2037**:1–5.

McKenna, M. C. 1961*b*. On the shoulder girdle of the mammalian subclass Allotheria. *Am. Mus. Novit.* **2066**:1–27.

McKenna, M. C. 1963. New evidence against tupaioid affinities of the mammalian family Anagalidae. *Am. Mus. Novit.* **2158**:1–16.

McKenna, M. C. 1966. Paleontology and the origin of the Primates. *Folia Primatol.* **4**(1):1–25.

McKenna, M. C. 1969. The origin and early differentiation of therian mammals. *Ann. N.Y. Acad. Sci.* **167**(1):217–240.

McKenna, M. C. 1972. Was Europe connected directly to North America prior to the middle Eocene? pp. 179–188. *In* T. Dobzhansky, M. K. Hecht, and W. C. Steere, eds., *Evolutionary Biology*, Vol. 6, Appleton-Century-Crofts, New York.

McKenna, M. C., J. S. Mellett, and F. S. Szalay. 1971. Relationships of the Cretaceous mammal *Deltatheridium*. *J. Paleontal.* **45**(3):441–442.

Mills, J. R. E. 1964. The dentitions of *Peramus* and *Amphitherium*. *Proc. Linn. Soc. London* **175**(2):117–133.

Osborn, H. F. 1893. Fossil mammals of the upper Cretaceous beds. *Bull. Am. Mus. Nat. Hist.* **5**(17):311–330.

Parrington, F. R. 1971. On the upper Triassic mammals. *Phil. Trans. Roy. Soc.*, Ser. B **261**:231–272.

Patterson, B. 1956. Early Cretaceous mammals and the evolution of mammalian molar teeth. *Fieldiana: Geol.* **13**(1):1–105.

Patterson, B. 1965. The fossil elephant shrews (family Macroscelididae). *Bull. Mus. Comp. Zool.* **133**(6):297–335.

Popper, K. R. 1968. *The Logic of Scientific Discovery*, 2nd ed., Harper and Row, Torchbook, New York. (Translation by the author of the original German edition, Vienna, 1934.)

ROBINSON, P. 1963. Fused cervical vertebrae from the Bridger Formation (Eocene) of Wyoming. *Univ. Colo. Stud., Ser. Geol.* **1**:6–9.

SABAN, R. 1954. Phylogénie des Insectivores. *Bull. Mus. Nat. Hist. Nat., Paris, Ser. 2* **26**:419–432.

SCHAEFFER, B., M. K. HECHT, and N. ELDREDGE. 1972. Phylogeny and paleontology, pp. 31–46. *In* T. Dobzhansky, M. K. Hecht, and W. C. Steere, eds., *Evolutionary Biology*, Vol. 6, Appleton-Century-Crofts, New York.

SCHMIDT-KITTLER, N. 1973. Dimyloides-Neufunde aus der oberoligozänen Spalterfüllung "Ehrenstein 4" (Süddeutschland) und die systematische stellung der Dimyliden (Insectivora, Mammalia). *Mitt. Bayer. Staatssaml. Palaeontol. Hist. Geol.* **13**:115–139.

SIMPSON, G. G. 1931*a*. A new classification of mammals. *Bull. Am. Mus. Nat. Hist.* **59**(5):259–293.

SIMPSON, G. G. 1931*b*. A new insectivore from the Oligocene, Ulan Gochu Horizon, of Mongolia. *Am. Mus. Novit.* **505**:1–22.

SIMPSON, G. G. 1936. Studies of the earliest mammalian dentitions. *Dent. Cosmos* **1936**:1–24.

SIMPSON, G. G. 1945. The principles of classification and a classification of mammals. *Bull. Am. Mus. Nat. Hist.* **85**:i–xvi, 1–350.

SLAUGHTER, B. H. 1971. Mid-Cretaceous (Albian) therians of the Butler Farm local fauna, Texas, pp. 131–143. *In* D. M. Kermack and K. A. Kermack, eds., *Early Mammals. Supp. 1, J. Linn. Soc. (Zool.)* 50.

SLOAN, R. E., and L. VAN VALEN. 1965. Cretaceous mammals from Montana. *Science* **148**(3667):220–227.

SULIMSKI, A. 1969. Paleocene genus *Pseudictops* Matthew, Granger, and Simpson, 1929, (Mammalia) and its revision. *Palaeontol. Pol.* **1969**(19–1968):101–129.

SYCH, L. 1971. Mixodontia, a new order of mammals from the Paleocene of Mongolia. *Palaeontol. Pol.* **1971**(25):147–158.

SZALAY, F. S. 1969*a*. The Hapalodectinae and a phylogeny of the Mesonychidae (Mammalia, Condylarthra). *Am. Mus. Novit.* **2361**:1–26.

SZALAY, F. S. 1969*b*. Origin and evolution of function of the mesonychid condylarth feeding mechanism. *Evolution* **23**(4):703–720.

SZALAY, F. S. 1972. Cranial morphology of the early Tertiary *Phenacolemur* and its bearing on primate phylogeny. *Am. J. Phys. Anthropol.* **36**(1):59–76.

SZALAY, F. S., and R. L. DECKER. 1974. Origins, evolution, and function of the tarsus in Late Cretaceous Eutheria and Paleocene primates, pp. 223–259. *In* F. A. Jenkins, Jr., ed., *Primate Locomotion*, Academic Press, New York.

SZALAY, F. S., and M. C. McKENNA. 1971. Beginning of the Age of Mammals in Asia; the late Paleocene Gashato fauna, Mongolia. *Bull. Am. Mus. Nat. Hist.* **144**:269–318.

VAN VALEN, L. 1964. A possible origin for rabbits. *Evolution* **18**(3):484–491.

VAN VALEN, L. 1966. Deltatheridia, a new order of mammals. *Bull. Am. Mus. Nat. Hist.* **132**(1):1–126.

VAN VALEN, L. 1974. *Deltatheridium* and marsupials. *Nature* **248**(5444):165–166.

WOOD, A. E. 1962. Early Tertiary rodents of the family Paramyidae. *Trans. Am. Phil. Soc.* **52**(1):1–261.

ZITTEL, K. A. 1893. *Handbuch der Palaeontologie*. I. Abt. Palaeozoologie, Bd. IV, Mammalia, pp. i–xi, 1–799, R. Oldenbourg, Munich and Leipzig.

3

Phylogeny, Behavior, and Ecology in the Mammalia

John F. Eisenberg

I. Introduction

In a review written some 11 and published 9 years ago, I attempted to summarize the literature concerning mammalian social behavior and then proceeded to discuss two major issues: (1) the relationship of social structure to the species' habitat and economy, and (2) the influence of evolutionary history on the form of social organization displayed (Eisenberg, 1966). The almost exponential increase of information during the last decade concerning mammalian social behavior and ecology, as well as the founding of social ecology as a subdiscipline (Crook, 1970), have rendered my earlier review out of date. My co-workers and I have recently attempted two reviews, one for primates, the other for selected carnivores (Eisenberg *et al.*, 1972; Kleiman and Eisenberg, 1973). The problems of correlation and reconstruction remain as challenging as ever.

In the fields of animal behavior studies and the ecological aspects of behavior, there are at least three types of questions which one may pose regarding phenomena observed in nature. One may seek an historical answer. Such a question involves a fusion between formism and organicism (see Pepper, 1961). The question may be stated: "From whence was such a behavioral pattern derived?" And in order to explore this, one needs to

---

John F. Eisenberg · National Zoological Park, Smithsonian Institution, Washington, D.C.

employ the comparative approach and to attempt to trace the historical steps in the appearance of the pattern.

One may seek a casual answer which is more strictly mechanistic. The question may be posed: "What are the casual events antecedent to the appearance of a given phenomenon?" In this case it is generally the procedure to set up experiments to isolate the variables giving rise to the expressed pattern.

Finally, one may ask the question: "How does this particular behavior pattern function in nature?" This is an ecological or adaptive question and falls under the category of contextualism (see Pepper, 1961). We must be careful to distinguish between the anthropomorphic answer to such a question and an objective answer. An anthropomorphic answer would imply a final cause in an Aristotelean sense. What we wish to imply is that goal-directed behavior results from neurophysiological mechanisms that are patterned in a particular way because the animal in question is the progeny of forms which have survived in competition with other forms in the same context or environment. Survival of a genotype is the outcome of reproductive success, and reproductive success is the result of a successful preceding adjustment on the part of the parental stock to environmental variables and success will vary as the genotype varies.

In essence, behavior patterns must be viewed as sets of muscular movements utilized by the organism to ensure its survival and the maintenance of its genotype. The study of behavior is in fact the study of sense organ, nerve, muscle, and endocrine relationships as reflected in movements and postures. Behavior patterns involved in intraspecific interaction are generally referred to as social behavior patterns, while a subset of these may be termed signals and as such an aspect of animal communication (Marler, 1961).

Since behavior patterns are the dynamic reflection of neuromuscular relationships, it seems logical that many relatively stereotyped behavior patterns may be treated as "structures" with a temporal dimension. Rules for comparison and determination of homology were developed by Wickler (1961) who drew heavily from Remane. One of the problems in applying Wickler's criteria to mammalian behavior patterns has been the lack of a framework defining a basic or fundamental mammalian repertoire. Another difficulty centers on the fact that Wickler's criteria grew out of a consideration of the displays of fishes and birds. In these taxa such displays are often rather stereotyped, but comparable displays are rare in many but not all groups of mammals. Andrew (1963, 1964) rather successfully provided an evolutionary perspective to the displays of primates, and he creatively departed from classical motivational models and developed a new theory of signal genesis, i.e., "stimulus contrast."

While some discrete patterns of behavior may be treated as structures and compared from one taxon to another, whole complexes of behavior are often not comparable. Although the units of behavior within, for example, a courtship bout may be compared from one species to another, whole adaptive complexes do not permit themselves to be treated as unitary phenomena. The comparison of social structures at best may involve the comparison of analogs. Working from a rather limited data base and without attempting to clearly delimit homologous behavior patterns, a recent series of facile comparisons has been published by various authors. Recently I attempted to point out the existence of behavioral analogs and convergences in the evolution of social structures

(Eisenberg, 1973), and I will not attempt to repeat the same line of reasoning here. Rather I wish to explore several questions of behavioral evolution by examining four radiations, the Tenrecidae, Edentata, Marsupialia, and Menotyphla,⋆ and then proceed to a brief scheme for interpreting the phylogenetic sequences in the evolution of mammalian behavior. Obviously this essay is speculative and does not pretend to be a comprehensive review. I do hope to highlight some aspects of the interrelationship between mammalian behavior and ecology.

II. Lessons from the Tenrecidae

The tenrecid insectivores represent a unique eutherian radiation. Isolated on Madagascar since the Eocene or earlier (Simpson, 1940), this taxon has co-evolved with lemuriform primates, viverrid carnivores, and nesomyine rodents which apparently colonized Madagascar at somewhat later dates (Petter, 1972). In terms of trophic specializations, the family has not advanced beyond the level of insectivore/omnivore. In general the species are terrestrial or semiarboreal with one exception, *Limnogale mergulus*, which is aquatic. When the various genera are compared, some of the more profound differences in morphology can be related to either feeding or antipredator mechanisms (Eisenberg and Gould, 1970). Great similarities in reproductive behavior, vocalizations, and maintenance behaviors reflect the morphological and phylogenetic unity of this group. Given their morphological conservatism and relatively isolated status on Madagascar, it may be reasonable to assume that this adaptive radiation expresses the current limits of variation within a conservative morphological framework (Fig. 1).

Specializations are manifest. The eye is relatively reduced in size, as is the tail in the Tenrecinae. Facultative hypothermia may be viewed as a special adaptation in the Tenrecinae, but of rare occurrence in the Oryzorictinae. The complex communication mechanism of the genus *Hemicentetes* involving a stridulating organ is surely a very specialized adaptation (Eisenberg and Gould, 1970).

As outlined in previous publications, this family probably shows some enduring, conservative behavioral traits. Echolocation of a simple nature has been demonstrated in the Tenrecidae by Gould (1965). This ability probably evolved early in mammalian history in conjunction with nocturnal adaptations. The auditory, tactile, and chemical senses are uniquely developed in the Mammalia. If the early mammals did initiate their evolutionary history as nocturnal forms, then the three previously mentioned sensory systems undoubtedly achieved great emphasis. Indeed, Jerison (1973) argues convincingly that the initial selective pressure favoring large brains in mammals (relative to reptiles) may have derived from the need to store, retrieve, and compare the information received from the cochlea.

The possession of a cloaca and the absence of a scrotum in the male tenrec probably represent conservative characters. The prolonged courtship attention of the male which

⋆ Use of this taxonomic category in this paper does not imply a belief by the author in the close affinity of the tree shrews and elephant shrews, but rather the term serves as a convenient reference category.

induces final ovulation by the female is probably also a conservative feature (Poduschka, 1974). Copulation is prolonged with intromission lasting 7.5 min (*Microgale dobsoni*) to 90 min (*Setifer setosus*). Spontaneous ovulation and brief mount durations are considered specialized attributes (Asdell, 1965). Gestation is rather long for the size class of this family (55–66+ days). If the time from birth to eye opening is added to the gestation, the range for the family is 65–80 days. The extended gestation nevertheless results in the production of altricial young and, although the eyes open at an earlier age postpartum than the eyes of many soricoids (Gould and Eisenberg, 1966), their overall developmental time is longer than that of soricoids of equivalent weight classes with comparable litter sizes. The longer gestation of the tenrecids does appear to shorten the postpartum development time for their young, but the long gestation may also reflect metabolic differences between soricoids and tenrecs which do not necessarily bear on phylogenetic arguments.

Litter size varies widely from over 30 in *Tenrec ecaudatus* to 1 or 2 in *Microgale talazaci*. Litter size reflects predation levels and survivorship as well as potential longevity and does not offer any light concerning "primitive" characteristics. It is noteworthy that of rainforest-adapted tenrecids, *Microgale talazaci* holds the record for longevity in captivity at 5 years 6 months, while *Hemicentetes semispinosus* has survived for only 2 years 7 months. The long-lived *Microgale* not only has the smallest litter size but also the

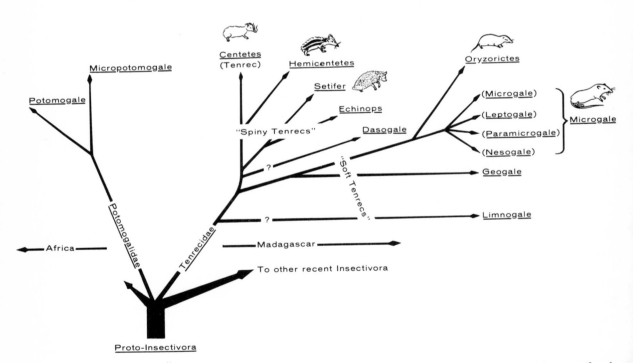

FIG. 1. Dendrogram illustrating probable relationships of the tenrecoid insectivores. The Tenrecinae are referred to as "spiny tenrecs" and the Oryzorictinae as "soft tenrecs." The status of *Dasogale* is uncertain. The genus *Microgale* is broken into 4 subgenera which probably represent a recent radiation. The Madagascar radiation shows the greatest range of adaptations. Although the early radiation on Africa may have been equally diverse, the fossil record is not complete enough to fill in the picture.

longest developmental time, and the highest encephalization quotient (EQ) for any terrestrial tenrecid (Jerison, 1973). This correlation of brain development, litter size, parental care, and longevity is a complex which shows itself again and again in other taxa having evolved convergently in response to several interrelated selective pressures.

Finally, a word about brain and behavior. As pointed out in a recent review (Eisenberg, 1973), the gross behavioral repertoires of most mammals are roughly equivalent in complexity. Maintenance behaviors as well as postures, sounds, and movements involved in social interactions are of a similar level of magnitude when primates are compared with insectivores. It remains to be proved, however, whether primates utilize more of the potential information contained in social exchanges than do insectivores. What is of interest is the fact that small insectivores, such as *Hemicentetes semispinosus*, can accomplish much integrative behavior with an extremely small brain. Some tenrecids share with didelphid marsupials and soricoid insectivores the lowest brain–body-weight ratios of living mammals (Bauchot and Stephan, 1966; Jerison, 1973). Among the tenrecs, the genera *Hemicentetes*, *Tenrec*, *Setifer*, and *Echinops* are considered to possess a basic brainstem structure irrespective of EQ (Bauchot and Stephan, 1966). Thus, it seems reasonable to assume that these particular genera represent an eutherian base line in terms of brain structure. Yet the behavior repertoire and signaling system of *Hemicentetes* is among the most specialized of all the Tenrecidae. The littermate group may not leave the mother and, since her daughters can conceive at 35 days of age, the possibility for a multigeneration extended family utilizing the same burrow is ever present in this species. The use of stridulating quills as a means of coordinating the movements of the mother–juvenile foraging group is certainly a specialized communication system (Eisenberg and Gould, 1970).

Thus a corollary to our review of the Tenrecidae would include the principle that possession of a low EQ does not necessarily set limits on the complexity of coordination mechanisms or signal systems, nor does it set limits on the size of social groupings. What it may mean, however, is that an animal with a low EQ is more limited in the plasticity of its behavior and most nearly approaches the "Cartesian ideal of classical ethology" (Jerison, 1973).

III. Lessons from the Edentata

The living edentates represent the last survivors of an ancient lineage well documented by Patterson and Pasqual (1972). The radiation of the Edentata was confined to South America in relative isolation from the extensive eutherian radiations of the northern hemisphere and Africa. Starting from a generalized insectivorous ancestral form, the edentates radiated to fill herbivore niches. The successful terrestrial herbivores included the ground sloths (Megalonychoidea and Mylodontoidea) of the infraorder Pilosa and the Glyptodontidae of the infraorder Cingulata. All these taxa were extinct or nearly so by the end of the Pleistocene (Patterson and Pasqual, 1972). The three surviving families, Dasypodidae, Bradypodidae, and Myrmecophagidae, represent three rather separate adaptations. The Dasypodidae (armadillos) are the most diverse with respect to feeding habits, antipredator mechanisms, and litter size (Kühlhorn, 1936; Moeller, 1968b). In general they are insectivores or omnivores adapted for a terrestrial life with reduced climbing ability. Some forms are almost entirely fossorial (e.g., *Burmeisteria*), and have occupied a niche similar to that of the Holarctic Talpidae.

The Myrmecophagidae (anteaters) are specialized for feeding on ants and termites. The three living genera are graded in size and arboreal ability and represent niche specializations comparable to the African pangolins (*Manis*) described by Pagès (1970, 1972a,b).

When we consider the Bradypodidae (sloths), we encounter a set of specializations which are duplicated in all the other major continental radiations of mammals. The sloths have evolved dental and gut adaptations permitting the use of plant cellulose as a nutrient source (Goffart, 1971). To be a browsing herbivore in the canopy of a rainforest permits a species to reach extremely high densities. The biomass of sloths in neotropical rainforests may be the major component of the mammalian fauna (Eisenberg and Thorington, 1973). This trend is paralleled in the evolution of colobine primates in Africa and Asia (Eisenberg et al., 1972).

Another concept may be introduced from a study of living edentates; this involves a consideration of r and K selection (MacArthur, 1972). If we consider the equation which describes the natural growth of a population as:

$$\frac{dN}{dt} = rN - \frac{(K - N)}{K},$$

where N is the population number at time t and dN/dt is the unit change per unit time, then r represents the intrinsic rate of population increase if unchecked and K represents the maximum density compatible with sustained resource utilization (i.e., the carrying capacity). When two alleles in a population or two populations are under selection and the effect of selection is to increase r, then the population is said to have undergone r selection. On the other hand, selection favoring a population which does not increase r or has a lower r is said to be K selected. The concept has great heuristic value, but also certain limitations (for a full discussion, see Pianka, 1972). For example, a species, such as *Dasypus novemcinctus*, reproduces at an early age and typically has 4–8 young. Many members of the tribe Dasypodini have similarly high reproductive rates. This suggests that these species are adapted to make use of transient resources which have rapidly fluctuating densities. The specialized armadillos of the genera *Tolypeutes* and *Priodontes*, as well as the highly specialized Myrmecophagidae and Bradypodidae, typically have a single young. The latter two families show a prolonged gestation, and a long period of intensive parental care. In both their capacity to reproduce and their potential longevity, sloths and anteaters suggest a K strategy in which a species is adapted to a stable niche and there is no special advantage to having a high reproductive rate. Instead, emphasis is placed upon the ability of an individual to retain a home range for its long-term use. In brief, then, reproduction is geared to maintain the population at a stable carrying capacity.

As recently shown (Montgomery and Sunquist, 1975), each individual 3-toed sloth (*Bradypus infuscatus*) has a set of trees in its home range which it visits with varying frequency. Over 40 species of trees and vines may be involved in the diet of the sloth and in some manner the sloth retains a memory of the location of its "modal" feeding trees and visits them on a schedule which is yet poorly understood in a causal sense. Although sloth home ranges overlap, each sloth tends to move and feed alone unless paired for mating or in the special case of a mother accompanied by her dependent young (Montgomery and Sunquist, 1975).

The mode of tree use suggests that the sloth is able to remember characteristics of its home range permitting it to move to preferred feeding trees in an orderly fashion. It is not surprising that sloths have a respectable brain size which no doubt reflects the necessity for integrating considerable information concerning spatial location of trees that provides for the effective utilization of its home range for a period of many years (Britton and Kline, 1939). It is entirely possible that in the wild, once established as adults, sloths live for a period exceeding 15 years, perhaps even longer.

The arboreal anteaters, *Tamandua* and *Cyclopes*, similarly must navigate in a 3-dimensional environment feeding on ant and, to a lesser extent, termite nests in a rotational fashion. This presupposes an ability once again to retain masses of individually acquired information concerning the structure of food sources within their home range. When the Edentata are compared with respect to brain size, sloths and anteaters have larger brains relative to their body weight than do armadillos (Eliot-Smith, 1898; Röhrs, 1966). This may very possibly reflect the particular needs for such long-lived animals as anteaters and sloths to acquire and store information concerning the location of food within the discrete 3-dimensional structure of their home range. Note again the correlation between small litter size, long life, and relatively high EQ (Fig. 2).

Undeniably, the Edentata are conservative in certain aspects of their morphology and physiology. They have a lower basal metabolic rate than more "advanced" eutherians, the sloths much more so than the anteaters. Certain features of the reproductive system in sloths may be considered morphologically conservative (Goffart, 1971). In none of the Edentata do we find specializations for great visual acuity, although their auditory systems appear to function about as well as many other eutherian mammals (Peterson and Heaton, 1968). The olfactory system is apparently well developed and, in the case of *Tamandua*, hunting by scent trailing is the major feeding strategy (Montgomery and Lubin, personal communication). If this order be considered somewhat morphologically conservative, surely all the members represent certain degrees of specialization. It is interesting to note that, in their adaptive radiation, they have replicated trends observable in other mammalian radiations in the Old World tropics.

IV. Lessons from the Marsupialia

In isolation from the major eutherian radiations, Australian marsupials and to some extent the South American marsupials adapted to fill a multiplicity of niches. In Australia the radiation reached its maximum flowering and, although many larger herbivorous marsupials are now extinct, the record would indicate that in Australia marsupials showed a range of diversity parallel in complexity to early eutherian radiations in the

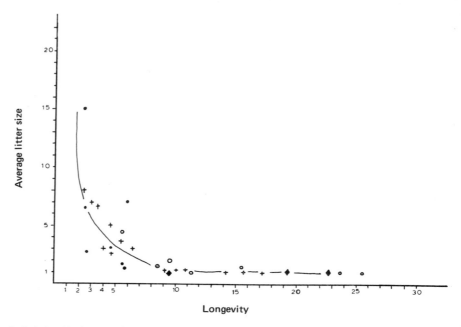

FIG. 2. Relationship between litter size and longevity for selected mammals. ◆ = Monotremata, + = Marsupialia, ○ = Edentata, ● = Tenrecidae, ⊙ = *Solenodon*. Data concerning reproduction and longevity in the Monotremata and Marsupialia taken from Collins (1973); data for the Tenrecidae and Solenodontidae taken from the publications of Eisenberg and Gould; data for the Edentata come from unpublished records of the National Zoological Park. Longevity records are in most cases based on averages of 3 or more individuals. Two-toed sloths (*Choloepus*) have reproduced when at least 22 years of age in captivity. Maximum longevities for members of the family Tachyglossidae have exceeded 49 years in captivity.

more continuous land masses. As Tyndale-Biscoe (1973) has documented in his recent book, marsupials should not be thought of as an inferior type of mammal. Rather, they have gone about the business of adaptive radiation with a slightly different basic body plan. The Australian marsupials apparently show a lower metabolic rate than their eutherian counterparts (MacMillen and Nelson, 1969). Certainly, the reproductive system of marsupials is their most unifying morphological and physiological feature, and their process of producing young is somewhat different from that developed by eutherians (Sharman *et al.*, 1966; Wooley, 1966). Yet when reproductive patterns of living marsupials are compared, the following trends in the evolution of marsupial viviparity can be noted. When we compare morphologically conservative forms in so-called conservative niches with more advanced forms, then specialization often shows that: (1) the number of ova shed from the ovary is reduced, (2) the number of young is reduced, (3) the weight of the neonatus is increased, and (4) the gestation period is lengthened (Sharman, 1965). These trends were convergently followed in many eutherian orders, if one accepts the proposition by Portman (1965) that the "primitive" reproductive methods of eutherians involved the production of several rather altricial young initially reared in a nest.

Without doubt most modern marsupials trace their lineage to a common ancestor strongly adapted for arboreality. This appears to be true of both the New World and Old World radiations (Dollo, 1899; Gregory, 1951). Although New World marsupials adapted to fill a semiherbivorous niche in the extinct Polydolopidae (Patterson and Pasqual, 1972), the use of plant cellulose as a food base has strongly evolved at least twice within the Australian marsupials (e.g., the Phalangeridae and Macropodidae). The evolution of adaptations for feeding on grasses in macropodid marsupials was accompanied by a trend toward the formation of cohesive social groupings as well as larger, temporary groups forming from more cohesive subgroups based upon a female and her descendant offspring. Such trends in the structuring of group size also accompanied the evolution of grazing in the Artiodactyla (Eisenberg, 1966; Kaufmann, 1974).

Some aspects of marsupial behavior which appear to be unique center around the profound differences between the marsupial and eutherian mode of reproduction. Marsupials are characterized by having a rather short gestation period relative to their size, followed by a prolonged period where the young are helpless and completely dependent upon female transport and nutrition. As is well known, the altricial marsupial young transport themselves to a pouch or marsupium (or teat area in the case of some Dasyuridae and Didelphidae). Attachment to the teat and further development outside the female reproductive tract may be prolonged, and in general the developmental rate of young marsupials is somewhat slower than that of comparably sized eutherians (see Fig. 3). The evolution of litter size within the order Marsupialia shows trends similar to those displayed by continental eutherians. Large litters are characteristic of small insectivorous and carnivorous forms; smaller litters are typical of larger herbivorous forms, with the smallest litters produced by the arboreal folivore, *Phascolarctos*, and the large grazing macropodids, *Megaleia* and *Macropus*.

During the neonatal and early postnatal developmental stages of the young marsupial, there is an intimate association with the mother. The young are attached to teats in the pouch or teat area. When the sense organs (i.e., eyes and ears) of the young become functional, the mother's body is the primary environment for the young. Although the mother may have a nest into which she retires, it is not the nest, but her body, which has a strong valence for the young. During the period subsequent to development of the sense organs, often termed the "socialization period" in eutherian mammals, young marsupials become capable of some independent locomotion. The young now may still be transported at times on the mother's body or in the pouch, although in the case of dasyurid marsupials the young often remain in the nest when the mother forages. Even so, the female dasyure will still continue to nurse the young in the nest and the young may ride on her for short periods (Collins, 1973; Ewer, 1968).

The mother marsupial is, in a sense, the "nest" during the initial phase of development for the young. It is only when the young reach the stage of socialization that they begin to approximate the eutherian condition and, in the case of nest-building forms, the nest attains an importance as a source of conditioning stimuli. Littermates serve as social partners during development while the mother is an important source of socializing stimuli in the larger macropodids which bear a single young (Russell, 1973).

In terms of the development of brain and sense organs, marsupials exhibit trends similar to those of eutherians. While it is true that living didelphids do not show brain development beyond the level of that attained by tenrecid insectivores, the carnivorous Dasyuridae of Australia do show brain development at a much higher level. Nevertheless, the Australian carnivorous marsupials show levels of brain development below those of eutherian carnivores in comparable niches in the main continental areas (Moeller, 1968a; Wirz, 1950). Similarly, the grazing macropodids of Australia show the lowest brain–body weight ratios of any terrestrial herbivore. These latter comments on the brains of marsupials give rise to two major points:

1. When a morphologically conservative species specializes for a niche with certain key demands for information storage and retrieval and muscular coordination, the brain undergoes appropriate selection and increases to the size necessary for receiving, intercorrelating, and retrieving data from the relevant sense organs. Thus, carnivores that pursue fast, mobile prey generally tend to have larger brains than insectivores which feed upon abundant, slow or stationary prey which are located by chemical or tactile means. (An exception, of course, would be bats which must locate winged insects using echolocation data.)

2. Morphologically conservative forms in relative isolation often evolve brain size at slower rates; thus, the Marsupialia and Tenrecidae exhibit brain–body ratios and brain structures which are more conservative than those of mammals in comparable niches on the continental land masses (Andrew, 1962; Jerison, 1973).

V. Lessons from the Tupaiidae and Macroscelididae*

The tree shrews and elephant shrews represent specialized taxa which share many conservative morphological characters found in the classical lipotyphlan insectivores (Weber, 1928). Both taxa show certain parallel trends in brain and sense-organ structure departing from the patterns of the lipotyphlan insectivores, yet neither family is closely related to the other (Patterson, 1965). If we assume that early eutherians were nocturnally adapted, then it follows that olfactory, tactile, and auditory senses were most important for obtaining information about their external environment. The eye would be of reduced utility, although our hypothetical ancestral mammal may, in fact, have had moderately sized eyes with a rod retina. Nevertheless, as a distance receptor, the ears, eighth cranial nerve, and all associated projection areas for the eighth nerve, began to assume primary importance.

* I do not consider the Tupaiidae as a family of the order Primates; see McKenna (1963) and Van Valen (1965).

FIG. 3. Maternal care patterns for selected mammals. M = mating; L = laying; H = hatching; P = partus; ✳ = beginning of lactation; W = end of weaning period; E = eye opening; d = days; ///// = period of teat attachment; ■■■ = period of female and young in same nest; and ⌁⌁⌁ = period of absentee parental care. Data for Monotremata and Marsupialia taken from Collins (1973); data for *Hemicentetes* from Eisenberg and Gould (1970); data for *Macroscelides* from Sauer and Sauer (1972); data for *Tupaia* from Martin (1968); and data for caviomorph rodents from Kleiman (1974). The sequences for certain events in the Monotremata are imperfectly known, but the "absentee system" for *Tachyglossus* is established. Question marks indicate uncertainties in the timing of events (see Collins, 1973).

20 d.

Tachyglossus aculeatus

27 d.   ?   10 d.   *   50 d   90 d.   ?

M

L   H   Young in Pouch   Young in Nest   W

Ornithorhynchus anatinus

15 d.   ?   10 d.   120 d.   ?

M

L   H
Egg in Nest   Young in Nest   W

Young "Riding"
Young Following

Marmosa robinsoni

14 d.   *   36 d.   E   4 d   26 d.   W

Dispersal

M

P

Young Attached
to Teat

Young May Be
Left in Nest

Young Following

Hemicentetes semispinosus

57 d.   *   8 d   E   10 d   W

M

P
Young in Nest

E

Dispersal

Macroscelides proboscideus

56 d.   *   8 d.

M

P   W

Tupaia belangeri

45 d.   *   20 d.   E   10 d.   W

M

P

Young Following

Proechimys semispinosus

64 d.   E   11 d.   25 d.   ?   W

M

P
Young in Nest

Young Following

Dasyprocta punctata

115 d.   E   22 d.   23 d.   W

M

P

The tupaiids and macroscelidids took two routes in their evolution. On the one hand, the elephant shrews retained a terrestrial niche, whereas the tupaiids began to develop increasing specializations for arboreality. In both lines, however, the animals show trends toward reliance on vision, if nocturnal, and ultimately specializations of the eye during the invasion of diurnal niches. Becoming diurnal must have placed demands upon the eye as a receptor of more and more fine-grained detail. Central projection areas for visual information began to become more important, a sequence that can still be seen when *Ptilocercus* is compared with *Tupaia*. With this dependence on visual information, a corresponding increase in brain size relative to body weight can be demonstrated together with the development of appropriate projection areas in the neopallium (Clark, 1924, 1926).

The diurnal tupaiids began to utilize much more information concerning depth perception and the structure of the 3-dimensional, arboreal world that they were invading; this may again have accounted for some of the increase in brain size which they show. Certainly an increased brain size in correlation with diurnality and arboreality is reflected in the evolution of the brains of sciurid rodents (Pilleri, 1959; Wirz, 1950). It is difficult, however, to tease these two variables apart unless by inference we assume that the demands of navigation in a 3-dimensional environment always presuppose some increase in brain size relative to body weight as is reflected in the rather large brains of the aquatic insectivores *Potomogale*, *Limnogale*, *Desmana*, and *Galemys*, all four of which do not show any extraordinary development in eye size, but do show increased development in the cerebellum and cerebral cortex (Bauchot and Stephan, 1966). A similar argument could be applied to the Pinnipedia and Cetacea. The latter also evolved an echolocation ability which parallels the bats. Dependence on such auditory integration may account for the initial increase in brain size of the Cetacea (Jerison, 1973) (see Fig. 4).

It would seem that, in an attempt to understand the steps from nocturnal life to diurnal life and all that this entailed in the evolution of mammals, one should look more closely at the behavior, reproductive biology, thermoregulation, and trophic specializations of *Ptilocercus*, for it truly is a connecting link between *Tupaia* and whatever ancestral form gave rise to them. Speculations concerning the ancestors of tupaiids and primates are perhaps best deferred until the biology of *Ptilocercus* is better understood.

Certainly no speculations concerning the evolution of the tupaiid way of life should be made without reference to parallel developments in the Macroscelididae. The unique method of parental care demonstrated by Martin (1968) in *Tupaia* has now been confirmed for *Macroscelides* by Sauer and Sauer (1971, 1972). In *Tupaia*, nests are constructed by both the male and female, although the female specifically constructs a natal nest into which the young are *altricially* born. While nesting separately from the male, the female only returns to the natal nest to nurse the young at intervals of 24–48 hr. In *Macroscelides*, the female gives birth to 2 *precocial* young in a secluded spot, returning to nurse them at 24-hr intervals. This demonstrated behavioral similarity need *not* be homologous. Similar forms of maternal care are demonstrated in the Monotremata (Tachyglossidae), Lagomorpha (Leporidae), Artiodactyla (Cervidae, Antilocapridae, and some Bovidae), Perissodactyla (Tapiridae), and Rodentia (*Dasyprocta*, *Pediolagus*, and *Myoprocta*)

(Kleiman, 1972, 1974; Lent, 1974). It would appear that the trend toward the production of precocial or semiprococial young often involves the evolution of the "absentee system" of parental care. The unique attribute of the *Tupaia* system is that it involves relatively altricial young (Martin, 1968) (see Fig. 3).

It has been suggested (Martin, 1968) that the lack of a retrieval pattern shown by the female *Tupaia* plus the "absentee" maternal care system could be part of a conservative mammalian pattern of reproductive behavior. If *Macroscelides* and *Tupaia* are indeed only distantly related, then the possibility exists that the "absentee" parental care system is a plesiomorph character. Yet the presence of maternal retrieving behavior plus a conventional nesting attendance phase by such ecologically and morphologically

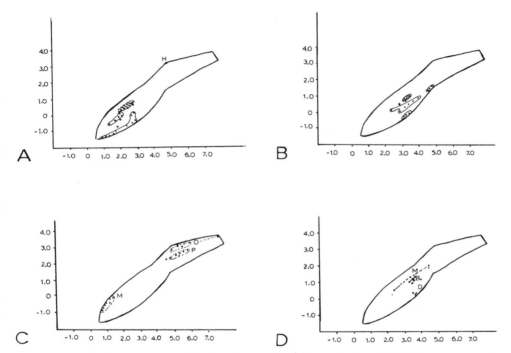

FIG. 4. Brain–body weight relationships for selected mammals. Ordinate = brain weight (E_c) in grams expressed as $\log_{10}$. Abscissa = body weight in grams expressed as $\log_{10}$. Boundary of diagram outlines the total set of brain–body weight points for the class Mammalia. Method of plotting adapted from Jerison (1973). All absolute brain weights converted to E_c where E_c = absolute brain weight (E) minus E_v and $E_v = (0.03) \times$ (body weight)$^{0.66}$ (see Jerison, pp. 78–81). The transformation in no way effects the relationships as shown here. (A) Brain–body weight relationships for insectivores, tree shrews, and elephant shrews (the latter two groups are cross-hatched). Points adjacent to the cross-hatching represent the set including the aquatic insectivores, *Limnogale*, Potamogalidae, and Desmaninae. H = the average point for the genus *Homo*. (B) Brain–body weight relationships for monotremes and marsupials. The cross-hatched set is the Monotremata. (C) Large brains for rapid navigation in three dimensions. M is the set for several Microchiroptera; O encloses selected species of the Odontoceti; P bounds several species of Pinnipedia. (D) Brain–body weight relationships for the Edentata. M denotes the Myrmecophagidae (connected by a dotted line). B shows three values for the Bradypodidae (connected by a solid line). D includes several points for the Dasypodidae. Note that the long-lived Euphractini (open circles) approximate the Bradypodidae.

conservative mammals as the edentate genera *Euphractus* and *Chaetophractus* and all genera of the Tenrecidae render the plesiomorph hypothesis dubious (Encke, 1965; Gucwinska, 1971; Eisenberg and Gould, 1970).

VI. The Evolution of Mammalian Patterns: Attempt at a Reconstruction

The brilliant synthesis by Pearson (1964) clearly indicates a marked climatic deterioration at the close of the Cretaceous followed by return to widespread tropical conditions in the early Tertiary. Subsequent to that time, there is evidence of continued change in climate accompanied by alterations in vegetation. Several cyclic patterns of climatic change have been proposed. With the decline of the reptiles at the end of the Cretaceous, the ancestral Eutheria and Marsupialia were offered the opportunity of filling unoccupied ecological niches (Olson, 1961). With such a changing pattern of selection pressures, rapid radiation occurred with the ordinal groups of mammals becoming clearly established in the Paleocene. Superimposed on the initial mammalian radiation was the concomitant breaking up of the earth's land masses which contributed to the geographical isolation of certain key groups of morphologically conservative mammals (Fooden, 1972). Australia, South America, and Madagascar were able to develop mammalian faunas which were distinctive until more recent invasions of eutherians from the more contiguous land masses.

Subsequent to this initial radiation, some living mammals existing in present-day tropical forests are relatively unmodified morphologically when compared with their ancestral stocks. This conforms to the hypothesis advanced by Simpson (1944) that bradytely or persistence of slowly evolving taxa results from an original adaptation of a widespread population to a relatively homogeneous environment, which may then persist through time as isolated populations in those areas which undergo little environmental modification. Thus, it seems legitimate to infer that certain morphologically conservative species inhabiting niches in the tropics represent forms that are, in a sense, adapted to a "conservative niche." Therefore a phylogenetic reconstruction of mammalian behavioral evolution must use as its starting point generalizations developed from a comparison of morphologically conservative forms occupying what might be considered to be the niches of the Paleocene–Eocene boundary (Jerison, 1973).

If we assume a polyphyletic origin of mammals (Simpson, 1959), it is even more difficult to reconstruct the probable evolutionary steps leading back from the pantotheres to the divergence from the stock giving rise to the monotremes. No doubt the trend away from oviparity and the evolution of mammary glands in the monotremes indicates either a parallel analogous with developments which took place in the stock giving rise to marsupials and eutherians or a primitive mammalian condition representing the ancestral pattern for both Metatheria and Eutheria. Concomitant with the evolution of mammary glands came a reorganization in methods of parental care. A comparison of the eutherian and marsupial lines indicates that parallel developments may have taken place in the evolution of reproductive patterns. The stock for both lines must have diverged from a precursor which possessed mammary glands to nourish the neonate by means of milk rather than by a large store of maternally derived yolk. The therian ancestor had departed from the pattern of laying large-yolked eggs and was either laying smaller eggs with little yolk or was ovoviviparous with a choriovitelline placenta (see Luckett, this volume).

If we agree that the early niches exploited by eutherians and marsupials were in forested habitat with a moderately stable climate, that the activity patterns of these mammals were confined to darkness, and further, that the animals were specialized as insectivores, then it follows that the evolution of thermoregulation and the reliance on audition for gaining distance information went hand in hand with the exploitation of a nocturnal niche. Invasion of an arboreal niche, and yet remaining nocturnal, perhaps placed further demands on the eye as an additional source of distance information. This would be especially true of nocturnal forms that moved rapidly in a 3-dimensional environment and employed leaps to get from one position in space to another. Either the echolocating ability would have to be refined so that a spatial map could be constructed based on echo information as in the case of the Chiroptera, or the eye would have to assume additional importance as a distance receptor, as is the case with the Phalangeridae, Lorisidae, and Lemuridae, to mention only a few. The most dramatic departure, then, once mammals had evolved to this point of agile use of trees and enhanced auditory and visual perception, would be for the organism to undergo selection to occupy diurnal niches with the eye continuing to be refined as a distance receptor. No doubt the Chiroptera passed through a stage of reliance on the eye while springing in trees at night. Sophisticated use of echolocation for maneuvering around obstacles or catching prey could only come *after* the development of some controlled flight. Presumably echolocation in the buoyant aquatic realm was accomplished much easier in the evolution of Cetacea.

The evolution of physiological and morphological adaptations for obtaining plant material and processing it in order to extract energy from cellulose with the aid of bacterial symbionts was another great step forward taken convergently by the Marsupialia and Eutheria. Convergent and/or parallel trends are widespread in modern taxa. Terrestrial use of plants dominates the evolution of the Artiodactyla and Perissodactyla in the Eutheria and the Macropodidae in the Marsupialia. Arboreal browsing as an alternative route for the exploitation of plants reaches its height in the colobine primates, bradypodid edentates, and phalangerid marsupials. Such evolutionary steps in the different mammalian lines often involved alternative solutions to the same problem. At the same time, when such similar niches are occupied by representatives of different radiations, we see repeated convergences in methods of antipredator behavior, rearing of young, and form of social structure.

The validity of comparative studies utilizing behaviors as phenotypic expressions of genotype (for the reconstruction of phylogenies extending back over 60 million years) is severely limited when the investigations of modern-day forms are used as points of reference. Comparative studies of behavior are most useful at the present stage of our understanding when they are confined to closely related species in taxa well defined on morphological grounds.

In order to assess what restrictions the phylogenetic heritage of a given radiation places upon the forms of the adaptations that it evolves, one must continue to analyze the various aspects of ecology and behavior of those species which have evolved in relative isolation from the competition of many species which has been imposed on the evolution of the Eutheria within the contiguous continental land masses. Thus, the analysis of behavior in the Malagasy tenrecids and Australian marsupials becomes more pressing and relevant.

To summarize some of the foregoing conclusions, I offer the following set of hypotheses:

A. Brains and Perceptual Worlds

Throughout the text I have made reference to various indices of brain enlargement. It is true that mammals and birds have larger brains in proportion to their body weight than do reptiles and amphibians (Jerison, 1973). The causes for this difference are imperfectly understood, but the differences are readily available for empirical verification. Furthermore, Jerison (1973) argues rather successfully that there has been a progressive trend in the enlargement of the brain throughout the history of the Mammalia. This increase in absolute and relative brain size has not occurred at equal rates in all taxa, but seems to accompany increasing specialization and refinement in the occupancy of new niches. Thus certain groups of mammals, such as tenrecid and soricoid insectivores, may retain brain structure essentially little modified from the brain structure of Cretaceous mammals.

Increase in brain size need not be confined to only one area of the brain. An increased brain size in one taxon may not reflect increase in the size of the neocortex, but only increase in the size of some subcortical structures. Those areas of the brain that do increase in size are often related to specializations in certain sense organs and thus correlate with the collection, storage, and retrieval of specific kinds of data relevant to the niches which the animals exploit. This principle is extremely well illustrated in the publication by Mann (1963) concerning the brain of the Chiroptera. He points out that "the cortical circuits are not empowered to replace advantageously, in every case, the subcortical mechanisms. Specialized phylogenetic trends may involve, therefore, increase of such subcortical centers without a corresponding increase in the neocortex" (Fig. 4).

B. Arboreality

Specialization for arboreal niches may have come early in the case of the Marsupialia and somewhat later in Eutheria. To be arboreal and nocturnal need not require any great specializations in the sense organs beyond the primitive mammalian level if the organism locomotes slowly and carefully in the branches and is specialized for capturing slow or relatively immobile prey. Great specialization in vision and hearing seem to accompany departures in locomotion which involve springing from one point to another in a truly 3-dimensional environment. The range of adaptations and morphological correlates has been reviewed by Cartmill (1972).

C. Diurnality

The evolution of diurnal habits seemed to involve greater reliance on vision as a means of obtaining information concerning distant objects. As they evolved in the Mammalia, the neural mechanisms for the integration of visual information in the central nervous system seemed to involve structural parallels with the mechanisms by which auditory data from the eighth cranial nerve were organized at the subcortical and cortical levels. This parallel development of brain mechanisms involving the second cranial nerve resulted in a greater encephalization by all those mammals which became diurnal (Jerison, 1973).

D. Herbivory

Utilization of plant cellulose as a food substrate occurred in every major mammalian radiation. Yet these tasks of trophic adaptation were in some cases parallel, e.g., the Rodentia (Vorontsov, 1960), or convergent, e.g., Marsupialia and Eutheria. The consequences of becoming a terrestrial herbivore and, in particular, utilization of grasses resulted in convergent evolution of social tolerance mechanisms which permitted formation of larger cohesive groupings. Such trends have been reviewed previously (Eisenberg, 1966).

E. Longevity and Litter Size

Increase in length of life and, in particular, effective reproductive life always seems to accompany reduction in litter size. Yet reduction in litter size may be a response to relatively short-term selective pressures. In the invasion of certain niches, such as the grazing niche, we have seen a convergent trend in reduction of litter size and the production of relatively mobile young; for example, when the leporid lagomorphs or caviomorph rodents are compared with bovid ungulates (Kleiman, 1972). Adaptations for these types of niches, then, do involve some predictable trends in litter size, but these are convergences at best and more often than not the longevity–litter size correlation is the result of particular factors related to the relative efficiency of replacement which vary greatly from one taxon to another. The importance of so-called K and r selection is related to a host of factors which does not permit the establishment of anything but trends and simple correlations among closely related species.

F. Parental Care

Viviparity and the production of altricial young demanded some form of parental care for maintenance of a high temperature for the developing young and provision of extrauterine nourishment. In the Marsupialia both objectives were accomplished with the evolution of the marsupium; in the Eutheria, apparently the first objective was obtained by bearing the young in a protected nest site where the parent could "brood" them. The latter problem was solved in the same manner as in the Marsupialia, i.e., by the evolution of the mammary gland–teat complex.

Without any necessary reduction in the complexity of parental care, more precocial young could be produced, thus shortening the time necessary for the parent to maintain a high temperature within a nest for the developing young. If the young were born precocial enough to maintain their own body temperature or born in a sufficiently insulated nest, the parent need only tend the young during periods of lactation. The so-called "absentee" system of parental care (Martin, 1968) could then have evolved in a convergent or parallel manner many times. If the young were born precocial enough to follow the mother, the absentee system could be eliminated almost entirely, provided the young were agile enough to participate in appropriate antipredator responses (see Fig. 3). This condition has been attained in some Artiodactyla.

On the other hand, bearing young in an arboreal habitat presents special problems. The young must have sufficient muscular coordination so that they will not fall before they leave the nest area. If a trend toward precocial young should be selected for, then the transition phase must be bridged from, at one extreme, a young that can immediately follow the mother to, at the other extreme, a young that is helpless in a nest. In general, this has been done either by retaining a nest phase in the rearing cycle of the young or eliminating it by producing a small number (e.g., 1–2) of very precocial young which can cling to the mother. The mother thus assumes the function of a "moveable nest." A convergent solution to the problem is seen in arboreal marsupials, but this shows only functional analogs since in this case altricial young transfer themselves to a pouch area and in this manner the mother becomes a "moveable nest."

G. The Formation of Social Groupings

Incipient pair-bond formation can be seen in many morphologically conservative mammals, such as the insectivores *Cryptotis parva* and *Microgale talazaci*. Such pair bonds are probably formed because of greater efficiency in mating when the presence of the opposite sexed partner is known and available to the other. This system is permissible in those conditions where a given pair can occupy and exploit resources efficiently in the same home range. Where joint exploitation is not possible or is less efficient, permanent pair bonds do not seem to exist and separate foraging areas for the male and female are the rule (Smith, 1968).

Where the male and female occupy the same home range and show some tolerance for their progeny, the possibility exists for the formation of small temporary inbreeding groups. Such units of population may provide the beginnings of group selection (Wilson, 1973). That such systems evolve in response to the peculiarities of each exploitation system is no doubt the case. Behavioral convergences in the form of social structure occur again and again in the Mammalia and only a generalized phylogenetic pattern can be detected (Eisenberg, 1966; Eisenberg *et al.*, 1972). Trends can be discerned, however, although the correlations apply only to cases where the evolutionary history of the group is reasonably well known (Kleiman and Eisenberg, 1973). Differences between mammalian social systems surely reflect overall adaptations to the physical habitat and the species' mode of exploitation.

The home range or area which an animal utilizes is very much a function of the distribution pattern of its foodstuffs (Altman and Altman, 1970), and home-range size is roughly correlated to the size of the species in question. But, even within body-size ranges that are similar, vast differences exist between home-range size for, let us say, carnivores on the one hand and herbivores on the other (McNab, 1963).

The form of social organization is often profoundly influenced by the nature of the shelter that the animal constructs. If a shelter is constructed and a considerable amount of energy is expended in the construction of such shelters, then defense of the shelter against a conspecific competitor is most essential. In addition to shelter construction, one must also consider the caching of foodstuffs. The formation of a food supply in space for later utilization by the assembler or its progeny very often involves an intraspecific defense system so that use of such an assemblage of food is restricted.

The mobility of a species has a profound influence on the form of its social organization (Eisenberg and Lockhart, 1972; McKay and Eisenberg, 1974). For example, a top carnivore may still form social groups, if the group itself possesses a sufficient ability to allow movement over a wide enough home range to effectively utilize its resources without overcropping (Kleiman and Eisenberg, 1973). Finally, antipredator behavior of prey animals profoundly influences whether or not groups can be formed. Small forms whose most adequate predator defense is inconspicuousness cannot afford to form groups which render their concealment impossible. Such species may become communal and utilize the same resources as long as they can do so while at the same time remain relatively invulnerable to predation (Eisenberg and Lockhart, 1972; Eisenberg and McKay, 1974).

To sum up then, in general those morphologically conservative mammals which are alive today exhibit the following trends in ecology and social behavior: They require a high energy content in their diet and are either specialized as small carnivores, insectivores, or generalized omnivores. As a result, they tend to range over a wide area relative to their body size and some spacing mechanisms are necessary to promote efficient utilization of habitat and further to reduce overutilization of restricted food sources. A given male and female may show overlap in home range and yet show very limited contact except at the time of mating. Parental care may fall entirely to the female; nevertheless, the same male and female may mate during consecutive seasons as the result of the proximity of their home ranges and as a result of their own agonistic tendencies toward conspecifics of the same sex. The adult female repels other females and the male repels contending males. Through such a behavioral mechanism the dispersal of their own progeny is assured. However, if family clustering is to be promoted and some selective advantage is retained by such a group (e.g., *Hemicentetes semispinosus*), it is generally tolerance of a female for her daughter rather than a male for his son that leads to the formation of nuclear family groupings. Such family groups consist of several females related by descent with their progeny and a dominant, single male utilizing roughly the same home range. Thus a so-called solitary species can exhibit a "family" structure which demonstrates a minimum of direct social contact; this can be taken as the simplest type of inbreeding population unit in the sense of Anderson (1970).

VII. References

ALTMAN, S., and J. ALTMAN. 1970. *Baboon Ecology.* University of Chicago Press, Chicago.

ANDERSON, P. K. 1970. Ecological structure and gene flow in small mammals, pp. 299–325. *In* R. Berry and H. N. Southern, eds., *Variation in Mammalian Populations.* Academic Press, New York.

ANDREW, R. J. 1962. Evolution of intelligence and vocal mimicking. *Science* **137**:585–589.

ANDREW, R. J. 1963. The origins and evolution of the calls and facial expressions of the primates. *Behavior* **20**:1–109.

ANDREW, R. J. 1964. The displays of primates, pp. 227–309. *In* J. Buettner-Janusch, ed., *Evolutionary and Genetic Biology of Primates.* Academic Press, New York.

ASDELL, S. A. 1965. Reproduction and development, pp. 2–43. *In* R. Mayer and R. VanGelder, eds., *Physiological Mammalogy,* Vol. II. Academic Press, New York.

BAUCHOT, R., and H. STEPHAN. 1966. Donees nouvelles sur l'encephalisation des Insectivores et des Prosimiens. *Mammalia* **30**:160–196.

BRITTON, S. W. and R. F. KLINE. 1939. Augmentation of activity in the sloth by adrenal extract, emotion, and other conditions. *Am. J. Physiol.* **127**:127–130.

CARTMILL, M. 1972. Arboreal adaptations and the origin of the Order Primates, pp. 97–122. *In* R. Tuttle, ed., *The Function and Evolutionary Biology of Primates.* Aldine-Atherton, Chicago/New York.

CLARK, W. E. LeGros. 1924. On the brain of *Tupaia minor. Proc. Zool. Soc. London* **1924**:1053–1074.

Clark, W. E. LeGros. 1926. On the anatomy of the pen-tailed tree shrew (*Ptilocercus lowii*). *Proc. Zool. Soc. London* **1926**:1179–1309.

COLLINS, L. 1973. *Monotremes and Marsupials: A Reference for Zoological Institutions.* Smithsonian Press, Washington, D.C.

CROOK, J. H. 1970. Social organization and the environment: Aspects of contemporary social ecology. *Anim. Behav.* **18**:197–209.

DOLLO, L. 1899. Les ancetres des marsupiaux etaient-ils arboricoles? *Travaux Station Zoologie de Wimeraux,* **7**:588 (as cited in Gregory, 1951).

EISENBERG, J. F. 1966. The social organizations of mammals. *Handbuch der Zoologie, VIII(10/7), Lieferung 39,* W. De Gruyter, Berlin, 92 pp.

EISENBERG, J. F. 1973. Mammalian social systems: Are primate social systems unique? pp. 232–249. *In* E. Menzel, ed., *Symposium of the Fourth International Congress of Primatology,* Vol. 1. S. Karger, Basel.

EISENBERG, J. F., and E. GOULD. 1970. The tenrecs: A study in mammalian behavior and evolution. *Smithson. Contrib. Zool.* **27**:1–137.

EISENBERG, J. F., and M. LOCKHART. 1972. An ecological reconnaissance of Wilpattu National Park, Ceylon. *Smithson. Contrib. Zool.* **101**:1–118.

EISENBERG, J. F., and G. McKAY. 1974. Comparison of ungulate adaptation in the New World and Old World tropical forests with special reference to Ceylon and the rainforests of Central America, pp. 585–602. *In* V. Geist and F. Walther, eds., *The Behavior of Ungulates and Its Relation to Management,* Vol. 2. I.U.C.N. Publication 24, Morges.

EISENBERG, J. F., and R. W. THORINGTON, Jr. 1973. A preliminary analysis of a neotropical mammal fauna. *Biotropica* **5**:150–161.

EISENBERG, J. F., A. MUCKENHIRN, and R. RUDRAN. 1972. The relationship between ecology and social structure in primates. *Science* **176**:863–874.

ELIOT-SMITH, G. 1898. The brain in the Edentata. *Trans. Linn. Soc. London, Ser. II, (Zoology)* **7**:277–394.

ENCKE, W. 1965. Aufzucht von Borestengürteltieren, *Chaetophractus villosus. Der Zoologische Garten* (NF) **31**(1/2):88–90.

EWER, R. F. 1968. A preliminary survey of the behavior in captivity of the dasyurid marsupial, *Sminthopsis crassicaudata. Z. Tierpsychol.* **25**:319–365.

FOODEN, J. 1972. Breakup of Pangaea and isolation of relict mammals in Australia, South America, and Madagascar. *Science* **175**:894–898.

GOFFART, M. 1971. *Function and Form in the Sloth.* Pergamon Press, New York.

GOULD, E. 1965. Evidence for echolocation in the Tenrecidae of Madagascar. *Proc. Am. Philos. Soc.* **109**(6):352–360.

GOULD, E., and J. F. EISENBERG. 1966. Notes on the biology of the Tenrecidae. *J. Mammal.* **47**(4):660–686.

GREGORY, W. K. 1951. *Evolution Emerging,* Vols. 1 and 2. Macmillan Co., New York.

GUCWINSKA, H. 1971. Development of six-banded armadillos, *Euphractus sexcinctus,* at Wroclaw Zoo. *Int. Zoo Yearb.* **11**:88–89.

JERISON, H. J. 1973. *Evolution of the Brain and Intelligence.* Academic Press, New York.

KAUFMANN, J. 1974. Social ethology of the whiptail wallaby, *Macropus parryi,* in northeastern New South Wales. *Anim. Behav.* **22**:281–307.

KLEIMAN, D. G. 1972. Maternal behavior of the green acouchi (*Myoprocta pratti* Pocock), a South American caviomorph rodent. *Behaviour* **43**:48–84.

KLEIMAN, D. G. 1974. Patterns of behaviour in hystricomorph rodents, pp. 171–209. *In* I. W. Rowlands and B. Wier, eds., *The Biology of Hystricomorph Rodents. Symp. Zool. Soc. London, No. 34.* Academic Press, New York.

KLEIMAN, D. G., and J. F. EISENBERG. 1973. Comparisons of canid and felid social systems from an evolutionary perspective. *Anim. Behav.* **21**:637–659.

KÜHLHORN, F. 1936. Die anpassungstypen der Gürteltiere. *Z. Saeugetierk.* **12**:245–303.

LENT, P. 1974. Mother–infant relationships in ungulates, pp. 14–55. *In* V. Geist and F. Walther, eds., *The Behavior of Ungulates and Its Relation to Management,* Vol. 1. I.U.C.N. Publication 24, Morges.

MACARTHUR, R. H. 1972. *Geographical Ecology: Patterns in the Distribution of Species.* Harper and Row, New York.

MACMILLEN, R. E., and J. E. NELSON. 1969. Bioenergetics and body size in dasyurid marsupials. *Am. J. Physiol.* **217**(4):1246–1251.

MANN, G. 1963. Phylogeny and cortical evolution in Chiroptera. *Evolution* **17**:589–591.

MARLER, P. 1961. The logical analysis of animal communication. *J. Theoret. Biol.* **1**:295–317.

MARTIN, R. D. 1968. Reproduction and ontogeny in tree shrews (*Tupaia belangeri*), with reference to their general behaviour and taxonomic relationships. *Z. Tierpsychol.* **25**(4):409–495.

MCKAY, G., and J. F. EISENBERG. 1974. Movement patterns and habitat utilization of ungulates in Ceylon, pp. 708–721. *In* V. Geist and F. Walther, eds., *The Behavior of Ungulates and Its Relation to Management,* Vol. 2. I.U.C.N. Publication 24, Morges.

MCKENNA, M. C. 1963. New evidence against tupaioid affinities of the mammalian family Anagalidae. *Am. Mus. Novit.* **2158**:1–16.

MCNAB, B. K. 1963. Bioenergetics and the determination of home range size. *Am. Nat.* **97**:133–140.

MOELLER, H. 1968a. Zur Frage der Parallelerscheinungen bei Metatheria und Eutheria. *Z. Wissenschaft. Zool.* **177**:282–392.

MOELLER, H. 1968b. Allometrische Analyse der Gürteltierschädel. Ein Beitrag zur Phylogenie der Dasypodidae. *Zool. Jahrb.* **85**(3):411–528.

MONTGOMERY, G. G., and M. E. SUNQUIST. 1975. Impact of sloths on neotropical forest energy flow and nutrient cycling, pp. 69–111. *In* F. Golley and E. Medina, eds., *Tropical Ecological Systems.* Springer-Verlag, New York.

OLSON, E. C. 1961. The food chain and the origin of mammals, pp. 97–116. *In* International Colloquium on the Evolution of Lower and Nonspecialized Mammals, Vol. 1. Kon. VI Acad. Wetemsch. Lett. Sch. Kunsten Belgïe, Brussels.

PAGÈS, E. 1970. Sur l'ecologie et les adaptations de l'orycterope et des pangolins sympatriques du Gabon. *Biol. Gabonica* **6**:27–92.

PAGÈS, E. 1972a. Comportement agressif et sexuel chez les pangolins arboricoles (*Manis tricuspis* et *M. longicaudata*). *Biol. Gabonica* **8**(1):3–62.

PAGÈS, E. 1972b. Comportement maternel et developpement du jeune chez un pangolin arboricole (*M. tricuspis*). *Biol. Gabonica* **8**(1):63–120.

PATTERSON, B. 1965. The fossil elephant shrews (Family Macroscelididae). *Bull. Mus. Comp. Zool. (Harvard)* **133**(6):297–335.

PATTERSON, B., and R. PASQUAL. 1972. The fossil mammal fauna of South America, pp. 247–310. *In* A. Keast, F. Erk, and B. Glass, eds., *Evolution, Mammals, and Southern Continents.* State University of New York, Stony Brook.

PEARSON, R. 1964. *Animals and Plants of the Cenozoic Era.* Butterworths, London.

PEPPER, S. C. 1961. *World Hypotheses.* University of California Press, Berkeley and Los Angeles.

PETERSON, E. A., and W. C. HEATON. 1968. Peripheral auditory responses in representative Edentates. *J. Audit. Res.* **8**:171–184.

PETTER, F. 1972. The rodents of Madagascar, pp. 661–666. *In* G. Battistini and G. Richard-Vindard, eds., *Biogeography and Ecology in Madagascar.* Junk, The Hague.

PIANKA, E. R. 1972. r and K selection or b and d selection. *Am. Nat.* **106**:581–588.

PILLERI, G. 1959. Beiträge zur vergleichenden Morphologie des Nagetiergehirnes. *Acta Anat.* **39**(Suppl. 38):1–124.

PODUSCHKA, W. 1974. Das Paarungsverhalten des Grossen Igel-Tenrek (*Setifer setosus*, Froriep 1806) und die Frage des phylogenetischen Alters einiger Parrungseinzelheiten. *Z. Tierpsychol.* **34**:345–358.

PORTMAN, A. 1965. Über die Evolution der Tragzeit bei Säugetieren. *Rev. Suisse Zool.* **72**:658–666.

RÖHRS L. M. 1966. Vergleichende Untersuchungen zur Evolution der Gehirne von Edentaten. I. Hirngewicht-Körpergewicht. *Z. Zool. Syst. Evolutionforsch.* **4**:196–207.

RUSSELL, E. M. 1973. Mother–young relations and early behavioural development in the marsupials, *Macropus eugenii* and *Megaleia rufa. Z. Tierpsychol.* **33**:163–203.

SAUER, E. G. F., and E. M. SAUER. 1971. Die kurzohrigen Elefantenspitzmaus in der Namib. *Namib Meer* **2**:5–43.

SAUER, E. G. F., and E. M. SAUER. 1972. Zur Biologie der Kurzohrigen Elefantenspitzmaus. *Z. Kölner Zoo* **15**(4):119–139.

SHARMAN, G. B. 1965. Marsupials and the evolution of viviparity, pp. 1–28. *In* J. D. Carthy and C. L. Diddington, eds., *Viewpoints in Biology*, Vol. 4. Butterworths, London.

SHARMAN, G. B., J. H. CALABY, and W. E. POOLE. 1966. Patterns of reproduction in female diprotodont marsupials, pp. 205–232. *In* I. W. Rowlands, ed., *Comparative Biology of Reproduction in Mammals*. Academic Press, New York.

SIMPSON, G. G. 1940. Mammals and land bridges. *J. Wash. Acad. Sci.* **30**:137–163.

SIMPSON, G. G. 1944. *Tempo and Mode in Evolution*. Columbia University Press, New York.

SIMPSON, G. G. 1959. The nature and origin of supraspecific taxa. *Cold Spring Harbor Symp. Quant. Biol.* **24**:255–272.

SMITH, C. 1968. The adaptive nature of social organization in the genus of tree squirrels, *Tamiasciurus. Ecol. Monogr.* **38**:31–63.

TYNDALE-BISCOE, H. 1973. *Life of Marsupials*. American Elsevier, New York.

VAN VALEN, L. 1965. Tree shrews, primates and fossils. *Evolution* **19**(2):137–151.

VORONTSOV, N. N. 1960. The ways of food specialization and evolution of the alimentary system in Muroidea, pp. 360–371. *In* J. Kratochvil, ed., *Symposium Theriologicum*. Ceskoslovenska Akademie Ved., Brno.

WEBER, M. 1928. *Die Säugetiere*, Vol. 2. Systematics, with O. Abel. Gustav Fischer, Jena.

WICKLER, W. 1961. Ökologie und Stammesgeschichte von Verhaltensweisen. *Fortschr. Zool.* **13**:303–365.

WILSON, E. O. 1973. Group selection and its significance for ecology. *Bioscience* **23**:631–638.

WIRZ, K. 1950. Studien über die Cerebralisation: Zur Quantitativen Bestimmung der Rangordnung bei Säugetieren. *Acta Anat.* **9**:134–196.

WOOLEY, P. 1966. Reproduction in *Antechinus sp.* and other dasyurid marsupials, pp. 281–294. *In* I. W. Rowlands, ed., *Comparative Biology of Reproduction in Mammals*. Academic Press, New York.

4

Nocturnality and Diurnality

An Ecological Interpretation of These Two Modes of Life by an Analysis of the Higher Vertebrate Fauna in Tropical Forest Ecosystems

PIERRE CHARLES-DOMINIQUE

I. Introduction

Nocturnal and diurnal life styles have profound implications for the behavior and neuro-physiology of vertebrates. Moreover, in numerous animal groups these two modes of life coincide with taxonomic subdivisions, thereby indicating the roles that they may have played during the course of vertebrate evolution.

Among the Primates, the Anthropoidea (Hominoidea, Cercopithecoidea, and Ceboidea) are all diurnal (with the exception of the cebid *Aotus trivirgatus*), whereas more variation occurs among the Prosimii. Tarsiidae, Lorisidae, Galagidae, Daubentoniidae, Cheirogaleinae, and Lepilemurinae are all nocturnal, while only Lemurinae and Indriidae (except *Avahi laniger*) are diurnal. Similar divisions also exist in other orders of Mammalia and Aves, and we will return to this point later.

Such different physical conditions place nocturnal and diurnal animals in sensory environments that are clearly different; this is responsible in great part for the orientation

PIERRE CHARLES-DOMINIQUE · Laboratoire de Primatologie et d'Écologie des Forêts Equatoriales, C.N.R.S., and Muséum National d'Histoire Naturelle, Laboratoire d'Écologie Générale, 91800 Brunoy, France.

of their evolution. Within the order Primates, the diurnal species are the ones that exhibit the most evolved cerebral development and social structures, whereas the nocturnal species are solitary. In relation to this, it is important to clarify certain terms that are often used incorrectly in the literature; *solitary* is not the opposite of *social*, but of *gregarious*. The nocturnal primates that I have studied (Charles-Dominique, 1972, 1974, 1975) are solitary and social; they travel independently but communicate with each other by vocal signals (cries), and especially by odoriferous signals (in the urine) which permit indirect communication, staggered in time. These odoriferous signals, imperceptible to our sense organs, may be quite complex, but we have an anthropomorphic tendency to consider a species of diurnal primate as "more social" when its means of communication are more similar to our own.

Thus, nocturnal and diurnal primates communicate by different methods, and only diurnal life seems to have permitted the development of gregariousness, responsible in great part for the complexity of interrelations that exist among individuals of the higher primates.

TABLE I

ESSENTIAL CHARACTERISTICS OF DIURNAL AND NOCTURNAL PRIMATES

Diurnal life	Nocturnal life
1. Gregarious animals	1. Solitary animals
2. Social communications principally by vocal and visual signals (plus olfactory signals in diurnal lemuriforms and platyrrhines)	2. Social communications principally by olfactory and vocal signals
3. Color vision present; absence of a tapetum lucidum (except for the special case of diurnal lemuriforms)	3. Color vision poor or absent; presence of a tapetum lucidum (except in *Tarsius* and *Aotus*)
4. Numerous anatomical and physiological specializations; greater behavioral complexity than in nocturnal forms	4. Conservation of numerous primitive anatomical and physiological characters

Table 1 summarizes the essential differences between the diurnal and nocturnal primates; those particular cases which do not enter completely into this scheme will be discussed in more detail.

In an ecosystem, the major problem facing all living organisms is the acquisition of nourishment for which it must become specialized in order to survive in competition with other sympatric species. Thus, each organism utilizes those of the available food categories for which it is best equipped in order to obtain maximum profit. This is the concept of an *ecological niche* as defined by Odum (1953). For any organism, the efficiency of the search for nourishment varies during the course of each day/night cycle. Generally, for the same dietary category, two different sympatric species are active at different times, one by day and the other by night, without competing with each other since plant growth and the maturation of fruits are continuous processes. Thus, in addition to dietary specializations, each ecological niche involves a distinct activity rhythm (generally diurnal or nocturnal) which ensures ecological separation from certain other species whose dietary needs are identical.

Ecologically speaking, the "diurnal world" and "nocturnal world" are completely different; they involve completely different means of communication and techniques of hunting or of locating fruits (a predominance of olfaction occurs in nocturnal species, in contrast to an emphasis on vision in diurnal forms). Therefore, these conditions would seem to be of particular importance for investigation, as much for the study of evolution of social structures as for the numerous aspects of primate biology that are closely linked with it.

This report is an attempt to interpret ecologically the modes of diurnal and nocturnal life. We will proceed by an analysis of tropical forest ecosystems, representative of the milieu in which the primates have evolved, as have numerous other orders of higher vertebrates.

II. The Forest Ecosystem of Gabon (Equatorial Africa)

First, we shall consider the higher vertebrates (birds and mammals) living in the same ecosystem—the primary forest of equatorial Africa in the region of Makokou, Gabon. The presence of the Laboratoire de Primatologie et d'Écologie des Forêts Equatoriales, du C.N.R.S., established in this area more than 10 years ago, permitted a reasonably complete inventory of the fauna to be made (see the different studies that have appeared in the journal *Biologia Gabonica*). Considering only the aspects of the primary forest (excluding migratory birds and savannah species which live in the clearings), we have counted 121 species of mammals and at least 216 species of birds, all sympatric, which occupy between them the different ecological niches of the equatorial forest. For both the birds and mammals, we have distinguished diurnal species, nocturnal species, and species which are both nocturnal and diurnal. For each of these categories we have made a classification by weight: (1) less than 100 g, (2) between 100 g and 1 kg, (3) between 1 kg and 5 kg, (4) between 5 kg and 50 kg, and (5) more than 50 kg (Figs. 1 and 2).

We can make the following comments concerning this study area:

1). 96% of the bird species are diurnal and 70% of the mammalian species are nocturnal.
2). The diurnal mammalian species are generally heavier than the nocturnal mammals.

III. The Forest Ecosystems of Panama (Tropical America)

The presence of the Smithsonian Institution Laboratory on Barro Colorado Island permits a rather complete knowledge of the fauna of this region. Utilizing the publication of Eisenberg and Thorington (1973) we constructed a graph for the mammalian fauna comparable to that of equatorial Africa (Fig. 3). However, there are several notable differences between the two tropical ecosystems. There is less diversity in the American mammalian fauna (86 species in Panama vs. 121 in Gabon); only a small number of large

PIERRE CHARLES-
DOMINIQUE

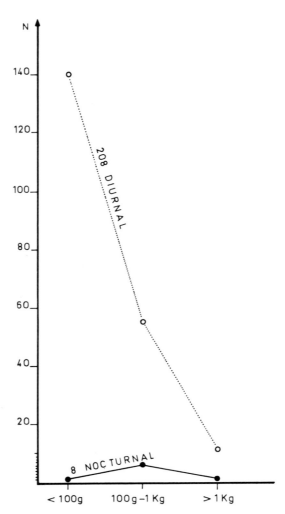

FIG. 1. Weight categories of sympatric diurnal (208 species) and nocturnal (8 species) birds from the equatorial primary forest of Gabon. Ordinate = number of species in each category.

American mammals (1 species exceeding 50 kg in Panama vs. 9 in Gabon); a large number of bat species in Panama, with 47 species comprising 55% of the mammalian fauna, as compared with 30 species of bats in Gabon, comprising 25% of the mammalian fauna.

IV. Interpretation of the Data Obtained from These Two Tropical Ecosystems

The graphs illustrated in Figs. 2 and 3 incorporate data from arboreal, terrestrial, and aquatic mammals. By considering only arboreal mammals (that is, those which obtain their nourishment in the trees), one obtains the same type of curve in both Gabon and in Panama (Figs. 4 and 5). The distribution curve for nocturnal mammals descends and that for diurnal mammals ascends for the higher weight categories. The majority of the forest avifauna lives in the canopy, with the exception of a few species which feed in the litter of the forest floor or in the water. As an overall impression, one can conclude that birds occupy primarily the diurnal ecological niches and that mammals occupy the nocturnal ecological niches.

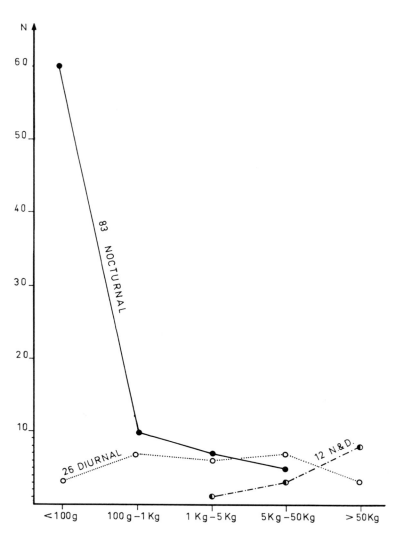

FIG. 2. Weight categories of sympatric nocturnal (83 species), diurnal (26 species), and nocturnal–diurnal (12 species) mammals of the equatorial primary forest of Gabon. Ordinate = number of species in each category. Note the symmetry between diurnal birds and nocturnal mammals on the one hand, and nocturnal birds and diurnal mammals on the other hand (cf. Figs. 1 and 2).

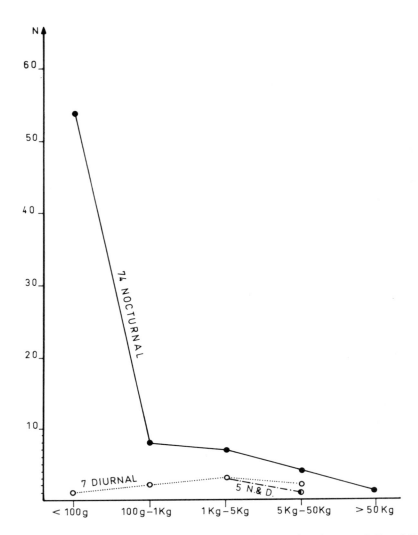

Fig. 3. Weight categories of the nocturnal (74 species), diurnal (7 species), and nocturnal–diurnal (5 species) mammals of Barro Colorado Island, Panama (after Eisenberg and Thorington, 1973).

One can postulate that this distribution pattern, at least in the equatorial forest environment, has existed since the appearance and radiation of these two vertebrate classes at the beginning of the Tertiary. The first forest mammals would have been derived from nocturnal forms which, as a result of their homeothermy, would have been able to supplant the nocturnal reptiles (if they existed there) that were probably handicapped by the nocturnal coolness. Moreover, nocturnal movements through the branches are more readily accomplished by climbing mammals with well-developed olfaction than by birds whose flight prohibits the slow movements necessary in an encumbered environment with poor light conditions. In this same forest milieu, birds would have conquered the diurnal ecological niches by utilizing flight, a condition which demands a very fine and precise perception of the environment and is possible with vision under good lighting conditions.

The fact that the sense of *vision* is predominant in birds, whereas that of *olfaction* may predominate in mammals (at least in the more "primitive" forms that are all macrosmatic) suggests in general that the adaptive conditions of diurnal birds and nocturnal mammals may have remained identical up to the present day, although with secondary adaptations which will be discussed further.

A. Nocturnal Birds

Among birds there are two crepuscular or nocturnal groups. The crepuscular goatsuckers and nightjars hunt in relatively good light conditions, thanks to their well-developed eyes specialized for nocturnal vision (presence of a reflecting *tapetum lucidum*). Owls hunt or fish under fully nocturnal conditions, but in relatively open environments. In addition to their specialized vision for nocturnal life, certain species have developed a

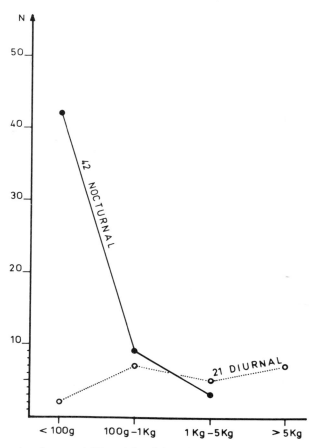

FIG. 4. Weight categories of nocturnal (42 species) and diurnal (21 species) arboreal mammals in the equatorial primary forest of Gabon (region of Makokou).

highly refined system of sound localization (acoustical parabolas surrounding the orbits which permit them to locate their prey very precisely).

B. Flying Mammals

Flying mammals (bats, which are all nocturnal) have developed a very efficient sonar mechanism that serves to locate obstacles and prey. The species which do not echolocate (most Megachiroptera) are mediocre navigators which travel generally above the canopy, or at least in relatively unencumbered spaces.

C. Gliding Mammals

These are all nocturnal (flying squirrels and dermopterans) and utilize gliding flight only to move from one tree to another. Otherwise, they move around in the branches by climbing.

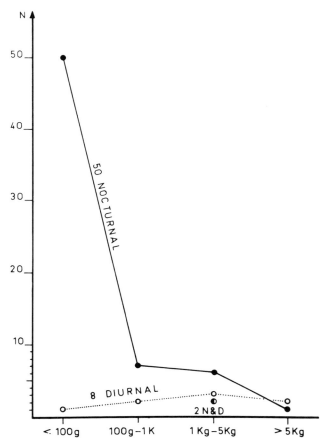

FIG. 5. Weight categories of nocturnal (50 species), diurnal (8 species), and nocturnal-diurnal (2 species) arboreal mammals from Barro Colorado Island, Panama.

D. Diurnal Mammals

There are only a few completely diurnal forest mammals. In Gabon, they constitute only 20% of the mammalian species, represented by 9 squirrel species, 11 monkeys and apes, 4 antelopes (out of 15 species of ungulates), 1 mouse (as opposed to 13 nocturnal species), and 1 pangolin (out of 3 species present). Only the monkeys and squirrels constitute groups that are truly adapted to an arboreal diurnal life; the others represent only special cases, consisting of species which feed in the litter of the forest floor (with the exception of the pangolin).

In Panama, the situation is quite comparable. The completely diurnal forest mammals constitute only 9.5% of the mammalian fauna (4 monkey species, 2 squirrels, 1 anteater, 1 carnivore, and 1 agouti). If one examines these diurnal mammals, it is evident that all have developed specializations which permit them to exploit food sources inaccessible to birds (Fig. 6) by utilizing several different "solutions."

1. Increase in Body Size. The most widespread "solution" has consisted of an increase in body size which permits the occupation of new ecological niches that are inaccessible to birds. Indeed, for physical reasons, flight is incompatible with excessive body weight, and 5 kg appears to be the upper limit. The large mammals (artiodactyls, pongids, most monkeys, some carnivores) possess teeth and digestive tracts proportional to their size, and this allows them to break down foodstuffs that are either tough, corrosive, difficult to digest, nutritionally poor, or bulky to the animal.

2. Continuously Growing Incisors. Another ecological "solution" for diurnal life has been provided by the specialization of continuously growing teeth. The incisors of rodents can serve as a "tool"; they enable squirrels to open tough fruits and nuts, to scrape bark, etc., without entering into competition with birds.

3. Claws. Different specializations of the claws enable several mammalian groups (squirrels, marmosets, pangolins, anteaters, certain carnivores) to dig out hidden prey.

4. Intelligence. The potential for intelligent behavior of mammals, considerably superior to that of birds, allows certain species (notably the primates) to exploit new ecological niches.

In general, one finds a combination of several of the above factors, but the majority of diurnal mammals in both Gabon and Panama are of medium or large body size. Thus, the first "solution" for diurnal life (increased body size) is the most common. One might be tempted to believe that there is a relationship here with predation. However, I do not think that this is the case (except of course for the very large species such as elephants), since predators occur in all sizes, adapted for most prey species. In Gabon, at least in the region of Makokou, there have been recorded 9 species of diurnal forest birds of prey (goshawks and eagles) whose body weights range from 80 g to 5 kg (Brosset, 1973). The largest of these, the monkey eagle, attacks large squirrels, monkeys, and the smallest species of diurnal antelope. The 10 species of carnivorous mammals present in the same region exhibit a range in body weights from 500 g to more than 50 kg. The leopard, which is the largest predator, attacks the large nocturnal and diurnal terrestrial game animals.

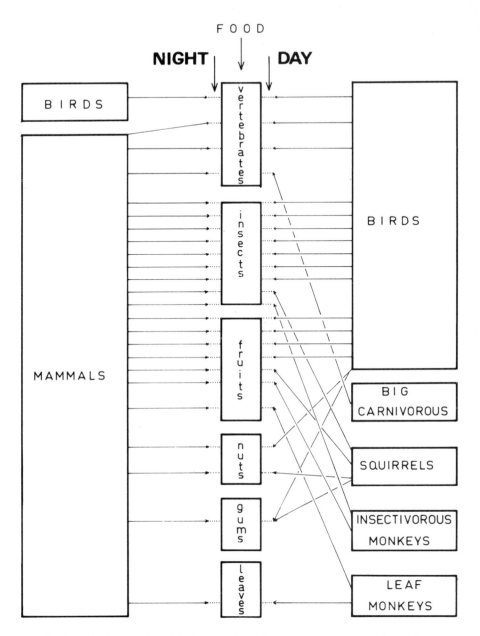

FIG. 6. Utilization of principal food sources by birds and mammals during night and day.

E. Small Body Size of Arboreal Nocturnal Mammals

Most arboreal nocturnal mammals can detect insects over short distances, generally by smell, although sometimes by hearing; this does not enable an individual to collect them in great quantity during the night, since the insects are generally dispersed. Only the Pholidota in the Old World and the Myrmecophagidae in the New World, specialized for the collection of colonial insects (ants and termites), whose nests they open by means of their specialized claws, can attain large body sizes. The other arboreal nocturnal mammals, of small body size, generally complement their insectivorous diet with fruits or gums (Charles-Dominique, 1971) which they detect by smell, although this technique permits the discovery of trees in fruit only over short distances. In a previous publication (Charles-Dominique, 1971), it was demonstrated that trees with a meager production of fruit were abundant in the forest of Gabon, whereas those which produced extensive amounts of fruit were rare and widely scattered (Fig. 7). Among arboreal species, only diurnal forms (aided in part by color vision and sometimes by observing the movements of frugivorous birds) can detect at a great distance such trees that are capable of nourishing numerous large and often gregarious frugivores.

The detection of fruits by the sense of smell enables only small- or medium-sized species with modest food requirements to obtain adequate nourishment. Solitary animals can discover trees with only a small or moderate amount of fruit production by this method, but such trees are of very high density and thus easy to locate. The limitation in body size of numerous arboreal nocturnal mammals therefore appears to be correlated with the modest amount of nourishment that they are able to obtain at night.

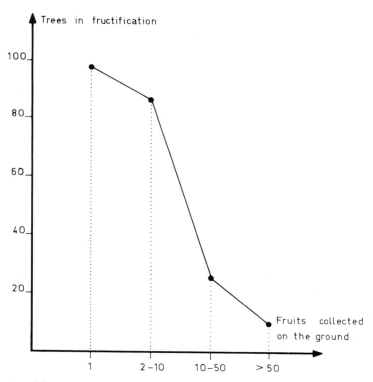

FIG. 7. Distribution of fruit trees according to the abundance of fruit production (primary forest of Gabon). The counts were carried out on fruits that have fallen on the ground; it is the trees which produce fruit in small quantities that are most numerous. (Modified from Charles-Dominique, 1971.)

PIERRE CHARLES-
DOMINIQUE

V. The Different Mammalian Orders in Their Ecosystems

Following this broad analysis of the avian and mammalian faunas, we can now examine the situation in the principal orders of eutherian mammals by considering their ecological situation in relation to nocturnal or diurnal activity. More extensive consideration will be given to the analysis of the order Primates.

A. Artiodactyla

This group is specialized for the consumption of coarse vegetation (necessitating specialized molars) which is digested in a complex and voluminous digestive tract that increases the body weight. Thus, they do not enter into ecological competition with birds. Among 14 sympatric species in Gabon, 6 are active during both day and night, 4 are exclusively diurnal, and 4 are exclusively nocturnal. In numerous other ecosystems (tropical and nontropical), artiodactyls occupy analogous ecological situations both nocturnally and diurnally.

B. Pholidota

The pangolins eat exclusively termites and ants which they extricate and collect by utilizing their enormous claws and specialized tongue. They do not enter into competition with birds, since the latter cannot occupy such an ecological niche. Among the 3 sympatric species of pangolin in Gabon, 2 are nocturnal and 1 is diurnal (Pagès, 1970). The situation is analogous for the South American Myrmecophagidae (Eisenberg and Thorington, 1973). Thus, these different species occupy successively, by day and by night, ecological niches provided by the exploitation of ant and termite colonies.

C. Insectivora

Practically all the Insectivora are nocturnal and, with the exception of a few fossorial or aquatic species, all hunt on the ground in the forest floor litter or under the grassy cover of the prairies and savannahs. Their vision is considerably reduced, and it is predominantly by olfaction, and perhaps also by echolocation, that they detect their prey. Insectivores occupy the ecological niches of small terrestrial predators by night; these niches are occupied during the day primarily by birds that hunt by vision, and also by reptiles and some diurnal mammalian species with well-developed vision (certain monkeys and squirrels), capable of dislodging hidden prey by scraping and stirring up the litter.

D. Chiroptera

All bats are nocturnal, and it is the members of this group which, because of their mode of locomotion, present the most comparable ecological parallel (night/day) with birds. The order is divided into two suborders: Megachiroptera and Microchiroptera.

The Megachiroptera, with the exception of *Rousettus*, lack the ability to echolocate. These bats are distributed in the intertropical zone of the Old World; in general they have evolved for unspectacular flight above the canopy or in relatively open environments where they navigate by vision (thanks to well-developed eyes with a tapetum lucidum). They eat fruits, pollen, or nectar which they detect by olfaction.

The Microchiroptera possess complex systems of ultrasonic emitters and receptors which enable them to navigate and to locate their prey with great precision in total darkness. This group, comprising the majority of bat species, has undergone adaptive radiation to occupy numerous ecological niches (insectivorous, carnivorous, piscivorous, sanguinivorous, frugivorous, nectarivorous, and pollenivorous) (Brosset, 1966). Such a great diversity of dietary preferences is analogous to the condition in birds; indeed, bats occupy a great number of the ecological niches at night that are filled during the day by birds.

E. Carnivora

Carnivores are likely to enter into dietary competition with predatory birds. Among the 8 sympatric mammalian species of forest carnivores in Gabon, 6 are purely nocturnal and 2 (a mongoose and the leopard) are both nocturnal and diurnal. In the same ecosystem there are 9 species of sympatric hawks and eagles, all diurnal. Among the 6 species of nocturnal birds of prey (owls), 2 are piscivorous, 3 are insectivorous, and only 1 is truly a predator of small forest vertebrates. An analogous situation occurs in numerous ecosystems: in general, predatory animals can be divided into *diurnal birds* and *nocturnal mammals*.

If one considers mammalian carnivores in general, body size appears to be a decisive factor. The largest predators (lion, tiger, leopard, cheetah, bears, large canids) occupy both nocturnal and diurnal ecological niches; they attack large prey animals that are inaccessible to predatory birds. However, some small carnivores also occupy diurnal ecological niches; this probably entails adaptations for hunting prey species which are inaccessible to birds (for example, small mustelids and herpestines hunt rodents in their burrows).

Thus, mammalian carnivores should be considered as a primitively nocturnal group, with only a few representatives adapted secondarily to a diurnal life. They all possess eyes with a tapetum lucidum adapted for nocturnal life.

F. Rodentia

The continually growing incisors of rodents, a "tool" serving to open or gnaw grains or hard fruits, plant stems, roots, bark, wood, etc., have enabled this group to occupy a great number of unique ecological niches. The order, distributed over all the continents except Antarctica, is represented by a very great number of species (3000, more than half of all extant mammalian species) that are adapted to very diversified life styles: terrestrial, burrowing, arboreal, gliding, or amphibious. Although generally of small body size, different taxonomic groups of rodents may be adapted to a nocturnal (the majority of rodents) or diurnal way of life (all squirrels except the flying squirrels).

G. Tupaiidae

With the exception of one species that is nocturnal, the tree shrews are diurnal and arboreal, semiarboreal, or terrestrial. In the absence of sufficient information concerning their ecology, one must simply suppose that these small-bodied mammals possess certain peculiarities which permit them to occupy diurnal ecological niches.

H. Primates

1. Anthropoidea. The Anthropoidea, all diurnal with the exception of *Aotus*, constitute 70% of the 175 species of living primates. We have chosen several tropical ecosystems (Africa, America, Madagascar, and Asia) in order to compare the ecological niches occupied by primates (Figs. 8–12).

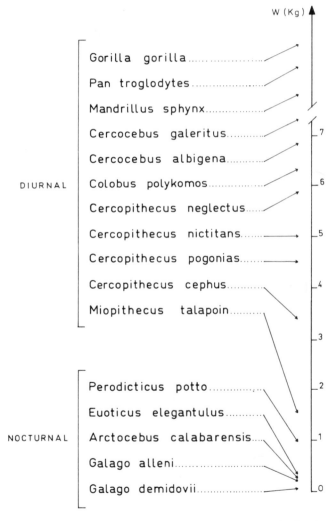

FIG. 8. Weight categories of sympatric primates in the equatorial primary forest of Gabon (region of Makokou).

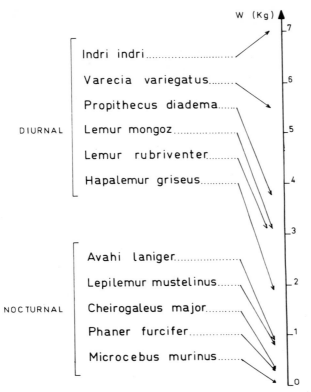

FIG. 9. Weight categories of sympatric primates living in Madagascar, in the Eastern rain forest (Perinet). To these can be added *Daubentonia madagascariensis*, 1–2 kg, living in the coastal forest.

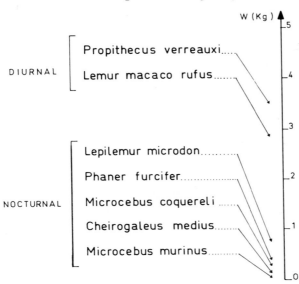

FIG. 10. Weight categories of sympatric primates living in the dry forests of western Madagascar (region of Morondava).

PIERRE CHARLES-
DOMINIQUE

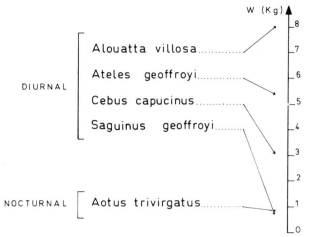

FIG. 11. Weight categories of sympatric primates living in the forest of Panama (Barro Colorado Island).

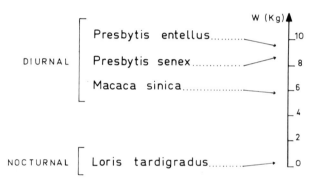

FIG. 12. Weight categories of sympatric primates living in Ceylon.

It is important to note that for diurnal species, the rule of body size is applicable in most cases; 1 kg appears to be the upper limit for nocturnal primates and the lower limit for diurnal primates. The only exceptions to this rule are the Callitrichidae, all diurnal forms, with one species in each of the different forest ecosystems of South America. They are distinguished from other Anthropoidea by their small size (200–500 g) and by the presence of claws, in contrast to the occurrence of nails in all other Anthropoidea. Marmosets eat insects, lizards, and birds (Napier and Napier, 1967), but there is very little information available on their hunting methods. Eisenberg (personal communication) has observed some callitrichids digging and extricating insects from bark with the aid of their teeth and claws. In contrast, Hladik and Hladik (1969) have observed *Saguinus geoffroyi* hunting visually along branches and bushes. Thus, the ecological adaptations of this family still remain unclear.

In general, the Anthropoidea can rummage in the vegetation, under bark, or in the forest litter in search of insects or small hidden vertebrates, thanks to their hands specialized for precise and complex movements. Their different hunting techniques, which utilize both great dexterity and "intelligent" behaviors for the search or discovery of prey (*Miopithecus talapoin*, see Gautier-Hion, 1971; *Pan troglodytes*, see Goodall, 1963; *Cebus capucinus*, see Hladik and Hladik, 1969; *Saimiri sciureus*, see Thorington, 1967), provide the means for obtaining the necessary dietary protein. However, all of them complement this diet with fruits (drupes or stone-fruits), rich in sugars but deficient in proteins, which provide them with a caloric source (Hladik, 1975). Their relatively large body size, correlated with a robust dentition (in particular the well-developed incisors),* enable them to open certain large fruits with tough rinds which birds cannot open.* Other small fruits, for which they enter into competition with birds, are collected with great efficiency by the Anthropoidea; prehension and modes of locomotion on the terminal branches of trees enable them in effect to gather fruits situated in a large radius, from one or two bases of support (Grand, 1972).

Other species of Anthropoidea obtain both their dietary protein and their energy needs by consuming green vegetation (stems, leaves, flowers, green fruits) which, following mastication by a powerful dentition, requires the presence of a specialized digestive tract. This condition characterizes colobus monkeys, langurs, gorillas, siamangs, howler monkeys, and spider monkeys (certain intermediate species obtain their proteins from both prey and green vegetation). These nonpredatory folivores require great quantities of food in order to compensate for the poor nutritional value of their diet. Moreover, the digestion of cellulose and hemicellulose, requiring the intermediary activity of an intestinal bacterial flora, is a time-consuming process and hence necessitates the accumulation of foodstuffs which make the animal heavy. Such ecological niches, inaccessible to birds for multiple reasons, are correlated with a relatively large body size.

2. Strepsirhini. In the ecosystems of Madagascar, the diurnal Lemuriformes exceed 1 kg in body weight, whereas the nocturnal species, with the exception of the highly specialized aye-aye, all weigh less than 1 kg. It is interesting to note that no diurnal lemuriforms are predatory (Petter, 1962),† with the exception of *Lemur catta*, which is occasionally seen in the act of eating insects during the rainy season (Jolly, 1966). In contrast, the nocturnal lemuriformes are nearly all predators, although they never capture their prey by rummaging through the vegetation as the Anthropoidea do.‡ Moving insects that are spotted are seized with a sudden grasp, often in flight, by a stereotyped movement, during which the two hands (or only one hand) close over the prey which is subsequently transported to the mouth (Charles-Dominique, 1971; Martin, 1972*a*).

Bishop (1964) has shown that the Lemuriformes and Lorisiformes have poor

---

* Parrots can feed upon nuts and hard fruits, but for many of them, their horny beak, although strong, is less efficient than a dentition consisting of incisors, premolars, and molars, such as occurs in anthropoids or squirrels.

† In captivity in Brunoy, *Lemur fulvus*, *Lemur mongoz*, and *Hapalemur griseus* can be trained to eat insects (to which they quickly become partial), whereas in the wild they never eat insects. In such a case, the capture is accomplished by the same stereotyped movements observed in the nocturnal insectivorous strepsirhines (*Galago, Microcebus*).

‡ Experimentally, we have trained a *Galago demidovii* to seize crickets located in a small box, with access via a circular opening 2 cm in diameter. The galago always extended its arm in an identical, rapid, and stereotyped movement, then immediately drew back its closed fist. It continued these movements until it captured the prey, simply by changing the direction of its rapid "coups de mains." It never explored the interior of the box by touching.

PIERRE CHARLES-
DOMINIQUE

neuromuscular control of their hands, and that they cannot separately control the movements of their fingers. This lack of manual "dexterity" does not allow diurnal lemuriforms to rummage through vegetation in search of insects as numerous anthropoids do. Competition with birds for the visual capture of prey would have forced diurnal lemuriforms toward ecological niches for eating leaves, pods, and green fruits, the only foods capable of giving them the protein balance they need.

Another peculiarity of diurnal lemuriforms is that nearly all of them possess a nocturnal type of eye with a reflecting tapetum lucidum, more or less marked by a pigmented layer, according to the species (Pariente, 1970). The nature of this tapetum is identical in the strepsirhine species tested by Alfieri *et al.* (1974) (*Microcebus murinus, Hapalemur griseus, Perodicticus potto*). The tapeta of strepsirhines contain riboflavin (or elements of this family); this pigment has not been found in the tapeta of other mammals that have been tested up to the present time.

In addition to this reflecting tapetum lucidum, the eyes of diurnal lemuriforms have a poorly defined or nonexistent fovea centralis, and even if the perception of some colors seems possible in certain species, these conditions are not at all comparable with those that exist in the Anthropoidea.

VI. Discussion

If one attempts to "reconstruct" the common ancestor of Primates, one should admit that this animal may have had an eye slightly differentiated for both nocturnal and diurnal vision, capable of evolving toward the type seen in the Anthropoidea on the one hand, and in the Strepsirhini on the other hand. Just as the eye of Anthropoidea represents a stage specialized for diurnal vision, the eye of living strepsirhines should not be considered as a primitive type, but rather as a type specialized for nocturnal vision. In the same manner, the morphology and neuromuscular control of the hands have evolved in two different directions: (1) in the Anthropoidea, fleshy fingers sensitive over all of their palmar surface (dermatoglyphics with complex patterns), associated with a neuromuscular coordination permitting great dexterity; and (2) in the Strepsirhini, long, slender fingers, sensitive primarily on the palmar surface of the flattened terminal phalange (dermatoglyphics arranged in parallel longitudinal ridges); a "reversed" articulation between the second and third phalanges, permitting contact of only the terminal phalanges in opposition with the palm.

The first of these evolved conditions of the hand and of the neural centers which are associated with it (anthropoid type) is adapted to a diurnal mode of life, necessitating considerable finesse and precision for searching through vegetation for concealed prey. The second evolutionary pathway (strepsirhine type) is adapted to a nocturnal mode of life; this necessitates rapid stereotyped movements of the hands and fingers for the capture of unconcealed prey, grabbed while fleeing or in flight.

These different characteristics, associated with the fact that many diurnal lemuriforms have a certain disposition toward a crepuscular, or even occasionally nocturnal, way of life (Petter, 1962), suggest that diurnal lemuriforms may have been derived from a nocturnal ancestor. Martin (1972b) has reached a similar conclusion, based on the fact that extant diurnal Malagasy lemuriforms retain the reflective tapetum lucidum characteristic of nocturnal forms. By itself, the ecological "vacancy" resulting from the isolation of Madagascar would have permitted the evolutionary radiation that has occurred from a stock already quite specialized for nocturnal life.

One might imagine that the dichotomy between nocturnal and diurnal Primates began to take place very early (probably before the colonization of Madagascar), and that the living species represent forms that are highly specialized to these two modes of life. Nevertheless, it seems likely that a nocturnal mode of life may be more "conservative," if one judges it by the greater number of primitive characters retained in the living lemuriforms.

In contrast, a diurnal mode of life would offer many more adaptive possibilities to the Primates: (1) a complexity of social structures, nearly all based on gregariousness, and (2) cerebral development, which, associated with increase in body size, contributed to the occupation by Anthropoidea of ecological niches inaccessible to birds.

ACKNOWLEDGMENT

The author wishes to thank Dr. W. Patrick Luckett for kindly providing this English translation from the original French manuscript.

VII. References

ALFIERI, R., G. PARIENTE, and P. SOLE. 1974. Dynamic electroretinography in monochromatic lights and fluorescence electroretinography in lemurs. *12th Int. Soc. Clin. E.R.G. Symp.*, May 1974, Clermont-Ferrand.

BISHOP, A. 1964. Use of the hand in lower primates, pp. 133–225. *In* J. Buettner-Janusch, ed., *Evolutionary and Genetic Biology of Primates*, Vol. 2. Academic Press, New York.

BROSSET, A. 1966. *La Biologie des Chiroptères*. Mason et Cie, Paris.

BROSSET, A. 1973. Evolution des *Accipiter* forestiers de l'Est du Gabon. *Alauda* **41**:185–202.

CHARLES-DOMINIQUE, P. 1971. Eco-ethologie des prosimiens du Gabon. *Biol. Gabonica* **7**:121–228.

CHARLES-DOMINIQUE, P. 1972. Ecologie et vie sociale de *Galago demidovii* (Fischer 1808; Prosimii). *Z. Tierpsychol.* Suppl. 9:7–41.

CHARLES-DOMINIQUE, P. 1974. Vie sociale de *Perodicticus potto* (Primates, Lorisidés). *Étude de terrain en Forêt Equatoriale de l'Ouest Africain au Gabon. Mammalia* 38:355–379.

CHARLES-DOMINIQUE, P. 1976. *Field Studies of Nocturnal Primates.* Duckworth, London, in press.

EISENBERG, J. F., and J. R. THORINGTON. 1973. A preliminary analysis of a Neotropical mammal fauna. *Biotropica* 5:150–161.

GAUTIER-HION, A. 1971. Répertoire comportemental du Talapoin (*Miopithecus talapoin*). *Biol. Gabonica* **7**:295–391.

GOODALL, J. 1963. Feeding behaviour of wild chimpanzees. *Symp. Zool. Soc. London* **10**:39–47.

GRAND, T. I. 1972. A mechanical interpretation of terminal branch feeding. *J. Mammal.* **53**:198–201.

HLADIK, A., and C. M. HLADIK. 1969. Rapports trophiques entre végétation et primates dans la forêt de Barro Colorado (Panama). *Terre et Vie* **1**:25–117.

HLADIK, C. M. 1975. Ecology, diet and social patterning in Old and New World Primates, in press. *In* R. Tuttle, ed., *World Anthropology: Primates Socioecology and Psychology*. Mouton, The Hague.

JOLLY, A. 1966. *Lemur Behavior: A Madagascar Field Study.* University of Chicago Press, Chicago.

MARTIN, R. D. 1972a. A preliminary field-study of the lesser mouse lemur (*Microcebus murinus* J. F. Miller 1777). *Z. Tierpsychol.* Suppl. 9:43–89.

MARTIN, R. D. 1972b. Adaptive radiation and behaviour of the Malagasy lemurs. *Phil. Trans. Roy. Soc. London* **264**:295–352.

88

PIERRE CHARLES-
DOMINIQUE

NAPIER, J. R., and P. H. NAPIER. 1967. *A Handbook of Living Primates*. Academic Press, New York.

ODUM, E. P. 1953. *Fundamentals of Ecology*. Saunders, Philadelphia.

PAGÈS, E. 1970. Sur l'écologie et les adaptations de l'Oryctérope et des Pangolins sympatriques du Gabon. *Biol. Gabonica* **6**:27–92.

PARIENTE, G. 1970. Rétinographie comparée des Lémuriens malgaches. *C.R. Acad. Sci.* **270**:1404–1407.

PETTER, J. J. 1962. Recherches sur l'écologie et l'éthologie des Lémuriens malgaches. *Mem. Mus. Hist. Nat., A.* **27**:1–146.

THORINGTON, R. W. 1967. Feeding and activity of *Cebus* and *Saimiri* in a Columbian forest, pp. 180–187. *In* D. Starck, R. Schneider, and H.-J. Kuhn, eds., *Neue Ergebnisse der Primatologie*. International Primatological Society, Frankfurt.

PHYLOGENY OF
PRIMATE HIGHER TAXA

5

Phylogeny of Primate Higher Taxa

The Basicranial Evidence

FREDERICK S. SZALAY

I. Introduction

There is a consensus among mammalian systematists insofar as most hold the composition of the bulla and, to a lesser degree, the circulatory patterns associated with the middle ear to be both conservative in some features and sufficiently diversified in others to be extremely useful. It is this combination that renders basicranial evidence valuable in formulating phylogenetic hypotheses for taxa above the generic and family rank.

Exact reasons for evolutionary alterations of the basicranium are not understood in any one group of mammals known to me. Such factors as increased demand for auditory acuity, circulatory demands, changes in relative brain size and jaw mechanics surely play roles of varying importance in the transformation of this area of the skull. Of equal significance is the fact that lineage-specific mutations, without specific adaptational significance, clearly helped determine the particular morphology of various groups, particularly the homologies of the bulla. As this area of the skull is considerably complex, detection of convergences can be relatively easy.

FREDERICK S. SZALAY · Department of Anthropology, Hunter College of the City University of New York, and Department of Vertebrate Paleontology, the American Museum of Natural History, New York, New York.

Just as bullar hypertrophy may be correlated with increased auditory sensitivity, the selective demands for increasing or decreasing the size of arteries may be explained by the augmented need for nutrients and oxygen. As noted, in a tube of large diameter the velocity of flowing blood is greater than the flow in smaller vessels (Szalay, 1972), and therefore increase in blood supply is best accomplished by increasing one vessel at the expense of others. In primates, as in other mammals, there seem to have been several particularly important areas of the head that required varying supplies of fresh blood. I believe that increase in relative brain size was the most important along with ophthalmic, masticatory, and other factors which become responsible for many of the circulatory changes found in mammals. Added to these selective forces were undoubtedly the resulting necessary biochemical changes brought about in the cranium as determined by mastication or sensory emphasis of one sort or another.

This chapter will not give a comprehensive morphological account of the basicrania of the groups in question (but see Tandler, 1899, 1902; Saban, 1963; Bugge, 1974; and references in these). Rather an analysis will be made of the phylogenetic significance of divergent basicranial characters of the immediate ancestry of primates and the subsequent major modifications within the order.

II. Methodology

At the risk of being trite I would like to state briefly the methodological biases that influenced my evaluation of differences and resemblances, results of which led to the conclusions advocated here and in my paper on the fossil tarsiiforms (Chapter 15, this volume). My analytical approach is based largely on what has been known as the cladistic methodology outlined by Hennig (1950, 1965, 1966). Considering some of the qualifications noted below, the methodology used by me may be labeled as "stratocladistic." In spite of the recognized importance of the temporal position of a taxon, I maintain that character analysis is the supreme arbiter in testing phylogenetic hypotheses. Hennig's central contribution, in which he distinguishes between the relative merits for genealogy of the similarities between taxa owing to sharing of (1) primitive (ancestral, plesiomorph) features, (2) advanced (derived, apomorph) features, and (3) convergences, has strongly influenced my analysis of homologous character states and subsequently my decisions in inferring phylogeny. As most systematists know, single characters often cannot be isolated; rather, the total gestalt of given character complexes is judged on the basis of one of the three categories of resemblance. As advocated by Hennig most recent relationships between two or more taxa are recognized on the basis of shared derived characters (synapomorphies) from a common ancestry, whereas shared primitive features (symplesiomorphies) between several taxa merely indicate that they all shared a common ancestor. What is significant, however, is that one group's symplesiomorphies may be the advanced characters of its morphotype when the latter is compared to its sister taxa.

To establish phylogeny within a group studied, either with individual or with a gestalt of biological features chosen, the taxonomic characters are evaluated in terms of these three categories of similarities. The obligatory evaluation of homologies, the basis of most phylogenetic reconstructions, is profoundly influenced by the clear understanding of the differences between the three categories of similarities. There are a variety of ways to determine whether given features are primitive or advanced, i.e., the morphocline polarity of the character states. Usually, if a character is widespread in a higher category, such as an order, in many lower categorical levels it is often taken as merely representing the ancestral state. On the other hand the relative rarity of certain features (preferably unique acquisitions) confined to a particular group, is usually considered derived, barring explicable special conditions related to past diversity of the group or other factors. The inferred derived nature of characters is clearly strengthened if it correlates with others, based also on the criterion of exclusive presence, equally derived in the very same taxa.

If a biological feature has at least three successive character states, one may often decide with confidence on the morphocline polarity which, at least for that character, clearly suggests an historical sequence. As systematists recognize more and more separable biological characters with each having at least three successive states, the corroboration of a hypothesis of cladistic relationships becomes correspondingly more significant. It is of particular importance to have a character that in its derived state represents an acquisition of a new or the disproportionate enlargement of an existing feature (e.g., petromastoid inflation or a newly emphasized circulatory pattern, respectively) correlated with another character that is equally derived but represents a structural or functional loss (e.g., "vestigial" remnants). If the successive stages of the two or more characters in the respective morphoclines correspond positively in the same taxa, one increasing and the other becoming reduced or eliminated, then a clear direction of change may emerge. Such considerations lead to the establishment of cladistic relationships or, when temporal information accompanying the fossil evidence is available, to the more complete evolutionary assessment, the phylogenetic hypotheses.

One cannot ignore the fact that primitive characters must temporally precede their advanced versions and that derived features cannot be antecedent to ancestral forms within a recognized taxon. If the model of successive adaptive radiations is valid, then the existence of ancestral characters of higher categories in earlier radiations could well be regarded as more of a probability than the reverse. In this context I would like to state my own bias, and perhaps those of many other evolutionary morphologists (see Simpson, this volume), in contrast to some aspects of the recent stance of Schaeffer et al. (1972) and the pure cladistic methodologies of some other students of phylogeny. Perhaps purity is not always synonymous with virtue, and therefore both biostratigraphic (temporal) and paleogeographic information usually associated with fossil taxa can be especially useful when employed in perspective with judicious character analysis.

Evolutionary studies of organisms, even those undertaken to determine phylogenetic relationships, are more than the result of a cladistic character analysis. Temporal and geographical information tied to fossil taxa, when decisive morphological criteria are lacking, may be an important aspect of their evolutionary significance. The time dimension supplied by those fossils that may be the actual ancestors or closely resemble the morphotype obtained in a character analysis allows the only realistic approximations of the phylogeny (*sensu stricto*) of a group. Not only do the special attributes of the fossil record provide the phylogenetic analyses with the factual reality of temporal and geographical dimensions, but their significance extends to the study of evolutionary meaning beyond genealogical events or biostratigraphical factors as well.

In explaining my methodological bias I should state that I do not concur with the prevalent philosophy of taxonomists who maintain there is little or nothing to be gained from functional studies for the construction of phylogenies. This view is not uncommon, in spite of the generally recognized fact that reliability of these hypotheses is dependent on the diversity and abundance of weightable characters. It is my personal experience that even limited inquiry into mechanical function and associated biological roles (*fide* Bock and von Wahlert, 1965) usually yields a great number of "new" characters and a more thorough understanding of "known" ones. Thus, not only is the repertoire of usable characters increased, but also much sounder criteria for weighting and for establishing homologies vs. convergences are determined, and therefore decisions on the polarity of morphoclines are facilitated. It follows that functional studies should be especially profitable in cases where a sufficient number of recognized characters are lacking for phylogenetic analysis.

III. Basicranial Morphology

A. The Eutherian Basicranial Morphotype

The usefulness of a model for a primitive eutherian basicranium may be explained as follows. Only a few characters in this region of the skull may be called truly primate characters; these are some of the diagnostic innovations of the primate ancestor in contrast to its own ancestry. In order to identify precisely the group of eutherians which gave rise to the primates, without the usual phraseology of referring to primitive eutherians, we must seek the taxon that shares derived eutherian characters with the primitive morphotype. In order to identify which features are primitive or derived for the early Eutheria we need to infer, however, a eutherian morphotype for the basicranium. The best possible amalgam of the various primitive basicranial character states of all known higher categories is necessary to construct this eutherian morphotype. Because of space limitations I am restricting my comments to selected eutherians although some problems clearly involve metatherian morphology as well.

Primitive representatives of most eutherian orders are not very well known. The complex morphological patterns offered by a smattering of skulls and petrosal bones from the early Tertiary, however, have long held the attention of students of mammalian systematics (see, for example, Kampen, 1905; Gregory, 1910, 1920; Klaauw, 1931). Although lack of space prevents a discussion of the eutherian morphotype, a reconstruction of the general conformation of the bony morphology and some of the probable paths of arterial and venous circulation is possible (Fig. 1). Note that neither the bulla nor the shape and position of the ectotympanic is shown on this pictorial essay. Although the usual assumption is that a cartilaginous precursor of the ossified bullae of later taxa was the condition in the eutherian morphotype, this is by no means adequately demonstrated. The bullae of some primitive metatherians and various Insectivora have contributions from the basisphenoid and alisphenoid, the latter also from the petrosal, and it is very likely that a number of very ancient Cretaceous therians also had bony bullae.

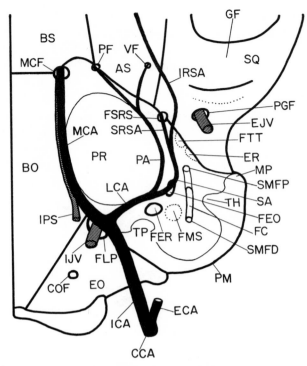

FIG. 1. Schematized reconstruction of various aspects of the basicranium and some of the associated circulatory patterns in the eutherian morphotype. For abbreviations see Table 1.

Morphologists studying early eutherian or extant basicranial remains (MacDowell, 1958; Van Valen, 1966; McKenna, 1966; MacIntyre, 1972; Szalay, 1972; Bugge, 1974; etc.) generally agree with Matthew's (1909) thesis that the primitive eutherian internal carotid was originally divided into a medial internal carotid artery★ coursing its way between the petrosal and the basioccipital and basisphenoid bones, and a lateral branch, the lateral internal carotid artery,★ which is divided into the large stapedial and the small promontory branch (MacIntyre, 1972). The evidence for the presence of this probably primitive eutherian and perhaps also therian character is from living as well as fossil Eutheria.

In terms of basicranial morphology and circulation, the morphotype of the Eutheria perhaps may be characterized by either the "unguiculate" or "ferungulate" petrosals described by MacIntyre (1972) and the circulatory patterns shown by Butler (1956), McKenna (1963), Van Valen (1966), and Szalay (1969). Most of the taxa derived from this group retain the carotid circulation of their morphotypes, a pattern here considered primitive for the Eutheria. The ancestral miacoid carnivorans have an essentially similar, primitive eutherian basicranial circulatory pattern (G. MacIntyre, and R. Tedford, personal communication), although most of the "neocarnivorans" lose either the promontory or stapedial arteries while retaining the medial internal carotid.

The medial internal carotid artery has been retained in the lagomorphs (Bugge,

★ Nomenclature used is that suggested by H. Wagner (personal communication).

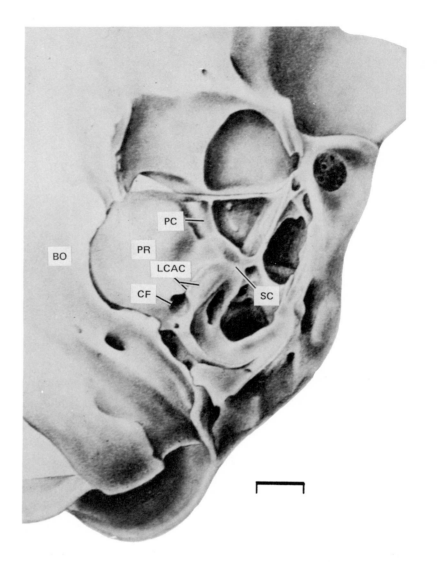

Fig. 2. Right basicranium of the tupaiid *Ptilocercus lowii* (USNM-DM-311313). The highly derived condition of the middle-ear cavity makes assessment difficult, although the canal for a large promontory artery (see text) is clearly visible. For abbreviations see Table 1. Scale represents 1 mm.

1967, 1974), the majority of hystricomorph rodents (Bugge, 1971, 1974), and can be assumed to have been present in primitive Rodentia (Wood, 1962; Guthrie, 1963; Wahlert, 1973).

With the exception of the basisphenoid, alisphenoid, and petrosal, which are utilized in the bulla construction of various insectivorans, the ectotympanic and entotympanic are the most common components for bulla formation among eutherians. The ectotympanic is a membrane bone that ossifies from a single element and is probably a derivative of the reptilian angular bone. In contrast the entotympanic is a cartilage bone, probably a neomorph among mammals. Depending on the location of the ossification the entotympanic can be either rostral or caudal (Klaauw, 1929). Although it is often assumed that the primitive bulla cover in the eutherian morphotype was cartilaginous, I know of no conclusive evidence or detailed discussion that supports this conclusion.

With this too brief discussion and somewhat schematic reconstruction of what may be assumed to be the ancestral eutherian pattern for the basicranium and intrabullar carotid circulation, I can now proceed to those known eutherians that may share derived characters and explain my concept of the primitive primate basicranial morphology. As expected, difficulties arise on all categorical levels because of uncertainty as to whether some of these character states are homologous or convergent, and if homologous, what their morphocline polarity is.

B. Some Derived Basicranial Characters of Nonprimate Eutherians Shared with Primates

As previously noted we are to identify the nonprimate taxon that shared the most recent common ancestor with the primates. It is all too often assumed that whatever is primitive for the Eutheria (i.e., a character of the eutherian morphotype) is also primitive for the Primates. This inaccurate assumption, sometimes employed by all systematists (including this one) to some degree, is often extended to both morphological and behavioral–ecological assumptions about the systematically precise, that is, contextually defined, meaning of the concept primitive. Are we to assume, in the case of the primates, that what we infer to be the first diagnostic characters (i.e., derived ones) in the morphology of the ancestral primate were directly transformed from a primitive eutherian base? The reason we most often shortcut our analysis, jumping from a poorly understood eutherian morphotype to the inferred ancestral characters of an ordinal or other category, is rooted in lack of information on evolutionary transformations that have clearly taken place in between.

The following, admittedly brief, list of character states may be considered primitive for the primates. However, some of these may also be part of the primitive features of other eutherian groups related to the primates to an unknown degree: (1) petrosal bulla; (2) medial internal carotid artery lost in the ontogenetic stage beginning with cranial ossification (prior ontogenetic conditions are clearly not inferrable in fossils); (3) a rounded promontorium; (4) bony canal for the entire intrabullar carotid circulation; and (5) fenestra rotunda ventrally "shielded" by the lateral internal carotid canal.

There are not many known groups that can be meaningfully compared with this inferred ancestral primate morphology. Many students have scrutinized the tupaiid ear region for phylogenetic information. Most recently, Van Valen (1965) and McKenna (1966) have extensively commented on it and have concluded that no shared derived characters are likely to be uniquely possessed by tupaiids and primates alone. In examining the basicranial morphology, particularly of *Ptilocercus*, I consider only the homology of the bulla, the position of the ectotympanic, and the promontory canals as noteworthy. The significance of the totally bulla-enclosed ectotympanic shared by tupaiids and lemuriforms is enigmatic. I suspect it to be a primitive rather than a shared derived character. The differences in the construction of the bulla in tupaiids and primates may not be

Frederick S.
Szalay

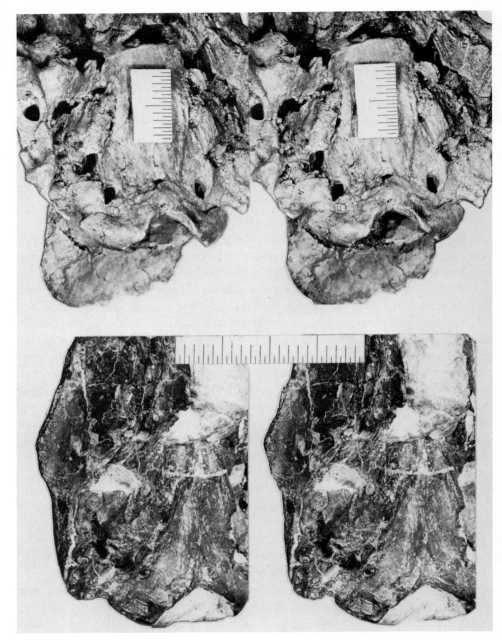

Fig. 3. Basicrania of two early Tertiary eutherians, representatives of near primitive conditions of their respective higher taxa. The late Paleocene European arctocyonid condylarth *Arctocyonides arenae* (MNHN CR 733), above, and the late early Eocene North American paramyid rodent *Paramys copei* (AMNH 4755), below. Neither of these specimens nor other representatives of their orders show any known derived characters shared by any known primate. Significantly, however, both of these genera, and presumably the ancestors of each of the orders, show a basicranial morphology somewhat just as primitive, or perhaps even more so, as the known basicranial remains of microsyopids.

significant. MacDowell (1958) has suggested that the primate bulla may in fact be the entotympanic of tupaiids ontogenetically fused to the petrosals. Embryological evidence for this, as far as I know, has not been published. It has been brought to my attention, however, by both D. Starck (this volume) and R. Martin (personal communication), that young specimens of *Tarsius* in their possession show a separate caudal ossification on the bulla that they interpret to be homologous with one of the entotympanic ossifications.

As noted by numerous students the bony canal housing the promontory artery is present in diverse mammalian groups. There is a clear, recognizable distinction between the relative size of the basicranial arterial pathways in the Tupaiidae and in known Paromomyiformes and Lemuriformes. In *Ptilocercus* and various species of *Tupaia* the tube for the promontory artery is relatively very large and, as Bugge (1972) reports, the brain is largely supplied by the internal carotid, assisted by the vertebral artery. Judged by the complex qualitative information that may be gleaned from the whole basicranium, however, it appears likely that the enlargement of the tupaiid and tarsiiform promontory is not a homologous condition but an independently acquired one.

In conclusion, the basicranial characters listed by Van Valen (1965) and McKenna (1966) and their arguments against a close tupaiid–primate relationship based on these remain valid.*

Primitive primates share more significant similarities with the erinaceotan (*sensu* Van Valen, 1967) morphotype than with any other group. A large petrosal component is characteristic of most known erinaceotan bullae, whether fossil or extant; in some cases the petrosal contributes up to half the entire structure (e.g., Rich and Rich, 1971). Like primates, both living and fossil erinaceids tend to be characterized by a roundly shaped promontorium. By itself, this last character is not a convincing similarity as it has converged in this fashion in many groups of mammals.

I believe that the bone-enclosed arteries of the middle ear were primitive in erinaceids and probably in the Erinaceota, *sensu lato*, although, admittedly, the known fossils (Gawne, 1968; Rich and Rich, 1971) that may be used to support this conclusion are not older than Oligocene. No ear regions are known in the broadly defined group of the Adapisoricidae (Van Valen, 1967), a species of which was the likely ancestor of the Erinaceidae.

The medial internal carotid artery is not known in erinaceotans. The erinaceids that have their intrabullar carotid circulation enclosed in bony tubes are the Oligocene *Proterix* and the Miocene *Brachyerix* (reported by Gawne, 1968; and Rich and Rich, 1971, respectively) and the living genus *Paraechinus* (redescribed by Rich and Rich, 1971). In these forms the fenestra rotunda is shielded by the carotid canal, in a fashion virtually

---

* I would like to caution against the total abandonment of the tupaiid–primate hypothesis; it has been useful and extremely stimulating for those fields of comparative biology dealing with primates and other eutherians. Furthermore, our knowledge of early primates is not even remotely satisfactory for anything other than allowing some choice between a number of alternative hypotheses. I do, however, take exception to renewed statements in the literature that tupaiids are the most primitive living eutherians, serving as satisfactory models for a host of behavioral, morphological, or locomotor studies that attempt to understand a eutherian ancestor. Tupaiid studies should not be employed as a replacement for the exacting and sometimes frustrating method of morphotype construction in either behavioral or morphological studies.

identical to that of *Phenacolemur* and adapids. Origins of the primates, then, from a eutherian I would dub, *faute de mieux*, an adapisoricid insectivoran, with a basicranium not very different from that of the erinaceotan morphotype, appears a likely possibility. The dental evidence, particularly that of *Purgatorius*, is also corroborative, suggesting a derivation of the order from a genus phenetically not far removed from *Leptacodon* (for details of the morphology see McKenna, 1968).*

C. Paromomyiformes†

The basicranial evidence from taxa of the Paromomyiformes is quite incomplete; only two genera, *Phenacolemur* and *Plesiadapis*, are known. *Plesiadapis* has received detailed treatment by D. E. Russell (1964) and Saban (1963), but the relatively large number of known petrosals lead to contradictory interpretations in some areas of the middle ear. A new basicranium of *Plesiadapis tricuspidens* (see Cartmill, this volume) clarifies previously unknown aspects of the genus. *Phenacolemur* has been recently described (Fig. 4) (Szalay, 1972).

A review of the evidence of *Phenacolemur* and *Plesiadapis* shows that the entry of the carotid into the bulla was posterior and the promontory artery was relatively very insignificant as, apparently, in all primitive Eutheria. A bony canal was probably present, at least for the portion of the internal carotid inside the bulla prior to the branching off of the stapedial. In *Phenacolemur* the canal appears to be continuous, but in *Plesiadapis tricuspidens* it only runs a short distance, as shown on Fig. 5. The fenestra rotunda appears to be obstructed in *Phenacolemur* when viewed ventrally. The ventrally shielded fenestra rotunda is a conspicuous feature of not only *Phenocalemur* but also of lemuriforms, and I suggest that this was also a diagnostic character of the primate morphotype. There is no

---

* I am currently studying a series of synapomorphous features of paromomyiforms, lemuriforms and the Dermoptera which indicate the origins of the colugos from early primates. This will be reported elsewhere.

† Bown and Gingerich (1973) recently challenged the exclusion of the Microsyopidae from the Primates, and, as Van Valen (1969), they regarded this family as primitive primates. Their arguments to demonstrate that the Microsyopidae are primates are unconvincing for two reasons. First, they have chosen to disregard the basicranial evidence (Fig. 4) that is available for the microsyopids and paromomyiforms, making no attempt to incorporate it into their final analysis. Second, they have violated some basic rules of phylogenetic assessment by making their comparison of *bona fide* primates, such as *Plesiolestes*, not with the two oldest and (independently of age) most primitive of microsyopids, *Cynodontomys wilsoni* and *C. alfi*, but with the younger and more evolved *C. latidens*. Both McKenna (1960) and Szalay (1969) show in specimens of *C. angustidens* and *C. wilsoni*, respectively, that the fourth premolars, associated with molars, are premolariform and not semi-molariform as are those of the more derived *C. latidens*. Thus, Bown and Gingerich (1973) demonstrated that the similarities which exist between *Plesiolestes* and *Cynodontomys latidens* are convergent. They have, inadvertently, made the strong point that microsyopid similarities to the earliest primates are not due to common inheritance. Using their approach, and if we ignored the basicranial and pedal evidence, we would be obliged to consider several genera of undoubted artiodactyls as primates on the basis of their dental resemblances to some genera of Eocene lemuriforms.

FIG. 4. Comparison of the left ear region of the alleged primate, the microsyopid *Cynodontomys* (top) with that of the primate *Phenacolemur* (bottom). Scrutiny of the figures and perusal of the pertinent literature where detailed descriptions are provided (Szalay, 1969, 1972) reveal that no shared derived characters may be recognized which are present in both. The generally "more primitive," very leptictid and rodent-like basicranium of *Cynodontomys* does not show even incipient stages of such primitive primate features as canals for the arteries, petrosal contributions to the bulla, or a shielded fenestra rotunda. The microsyopid does show an apparently derived condition, enlargement, but not inflation, of the petromastoid. For abbreviations see Table 1.

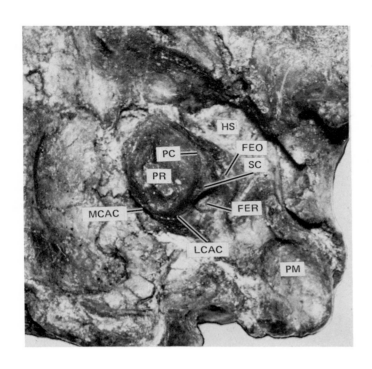

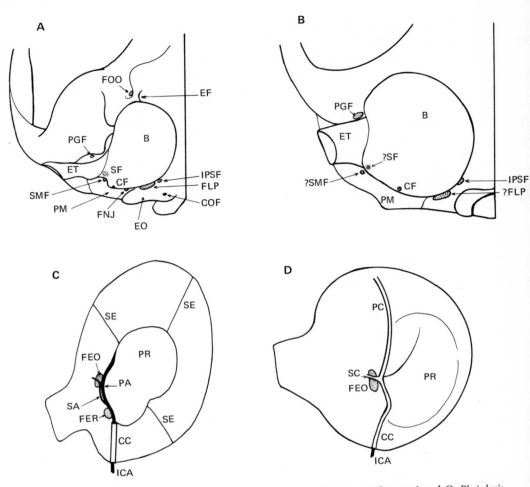

FIG. 5. Simplified schema of the basicranial morphology in known Paromomyiformes. A and C, *Plesiadapis tricuspidens* (late Paleocene), and B and D, *Phenacolemur jepseni* (early Eocene). Above: basicranium intact; below: the bulla and ectotympanic removed to show carotid circulation.

appreciable petromastoid enlargement in either *Plesiadapis* or *Phenacolemur*, and in both genera the ectotympanic is extrabullar, although, significantly, to different degrees. This bone, although it is not ringlike in either genus, extends into the bulla in *Plesiadapis* and in both genera forms the base of the external auditory canal (see Cartmill, this volume).

How can the difference in the conformation of the ectotympanics be explained in these two paromomyiforms, and which of these conditions is most likely to be primitive? As noted, the middle ear morphology of the two genera differ inasmuch as *Plesiadapis* has a large number of septae and in general a more inflated bulla. *Phenacolemur*, on the other hand, has only a longitudinal septum which houses the lateral carotid artery and its craniad continuation, the promontory artery. In these respects, therefore, *Phenacolemur* is probably primitive. The great inflation of the bulla in *Plesiadapis* may supply the clue to the developmental events which led to the condition of the ectotympanic displayed by that genus. We may postulate an ancestral condition shown in Fig. 16 and view the development in *Plesiadapis* as the result of the lateral displacement of the petrosal during the general hypertrophy of the bulla itself. As in many other mammalian lineages, selection for increase in auditory sensitivity would have been accomplished by an increase in middle-ear cavity volume. An extension of the middle-ear cavity laterally, below the ear drum, would surround the medial end of the auditory meatus by the tympanic cavity, and thus make a large portion of the ectotympanic (true ectotympanic plus the ossified anulus membrane) intrabullar. Such a mechanism has, in fact, long been known in developing lemuriforms and it is shown schematically by Cartmill (this volume, Chapter 14, Fig. 8).

Phenacolemur, unlike *Plesiadapis*, apparently has only a small portion of its ectotympanic within the tympanic cavity, the crista tympani, supporting the tympanic membrane. This condition, as that of *Plesiadapis*, may be derived as postulated in Fig. 16G, contra my previous suggestion (Szalay, 1972). This view hypothesizes that the extrabullar condition of *Phenacolemur* and *Plesiadapis* has been derived along pathways *analogous* (!) to those of *Tarsius* from an ancestor with an ectotympanic enclosed by the middle-ear cavity.

D. Strepsirhini

In addition to the wealth of extant and subfossil species, fossil strepsirhines are reasonably well known cranially. Skulls or skull fragments of *Pelycodus*, *Notharctus*, *Smilodectes*, *Adapis*, *Leptadapis*, and *Pronycticebus* give a fair idea of cranial diversity. The two Miocene crania of lorisids, those of *Komba* and *Mioeuoticus*, suggest both a degree of anagenetic advance and at least an upper limit for the time of origin of lorisid skull morphologies.

It is possible that, in general, much of the evidence we consider primitive lemuriform in both adapids and later lemuroids is largely primitive retention from paromomyiforms.

The only measure of the relative importance for blood transport in arteries of fossils is the relative size of the bony canals that housed them. Gregory's (1920) assertion about the differences in the relative sizes of the promontory and stapedial arteries in primitive strepsirhines and haplorhines was confirmed on new, and in the case of adapids, more primitive fossil material. I have drawn the diameters of the lateral internal carotid and the promontory and stapedial canals from AMNH 88802, a petrosal from the early Eocene East Alheit Pocket that I identify to be of (?)*Pelycodus* sp. (Figs. 6 and 7), and from the Montauban 9 skull of *Necrolemur antiquus* from the (?)late Eocene phosphorites of Quercy. The relative diameters are shown on Fig. 8. Although the size of the promontory artery is by no means insignificant in *Pelycodus*, in the primitive notharctine the stapedial artery appears to have been the larger vessel immediately after the separation of the lateral internal carotid artery into the two former vessels. This is the condition overwhelmingly established for *Adapis*, *Leptadapis*, *Notharctus* (but see Gingerich, 1973), and extant Lemuridae (*sensu stricto*), Indriidae, and Daubentoniidae. As stated by Gregory, *Necrolemur* and *Tetonius* clearly show the greater emphasis of the promontory as opposed to the stapedial canals, and therefore presumably of the arteries also. Thus the recognition

Frederick S.
Szalay

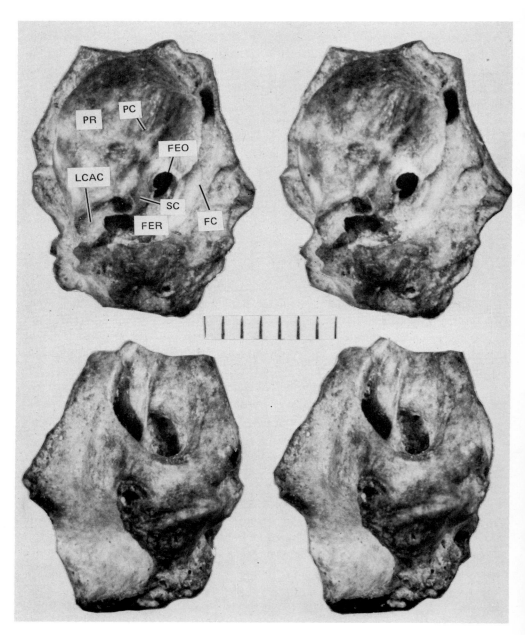

Fig. 6. ?*Pelycodus* sp., AMNH 88802, early Eocene, East Alheit Pocket. Stereopairs of left petrosal fragment: tympanic (above) and cerebellar (below) sides. For abbreviations see Table 1. Subdivisions on scale represent 0.5 mm.

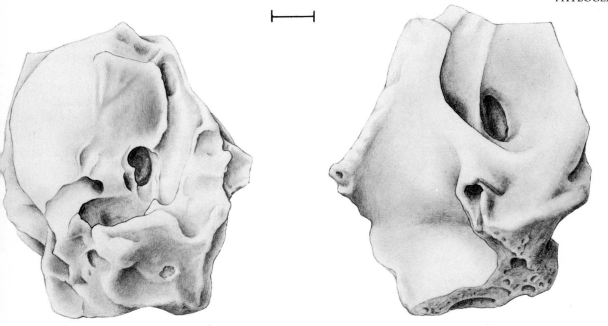

FIG. 7. ?*Pelycodus* sp., AMNH 88802, early Eocene, East Alheit Pocket. Left petrosal fragment. Tympanic view on left and cerebellar view on right. Scale represents 1 mm.

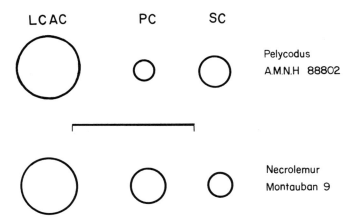

FIG. 8. Comparison of diameters of the lateral carotid arterial canal, promontory canal, and stapedial canal in ?*Pelycodus* sp. and *Necrolemur antiquus*. Scale represents 1 mm.

of the early dichotomy between strepsirhine and haplorhine intrabullar circulation is perhaps justified (see Szalay, Chapter 15, this volume).

The diversity of basicranial transformations among strepsirhines offers an excellent opportunity for character analysis. This fact, I believe, is of considerable significance for evaluating a variety of phylogenetic hypotheses, some recent, of both strepsirhine phylogeny and various aspects of strepsirhine evolution.

The hypothesis that the cheirogaleids and lorisids are on the whole more primitive than the lemuriforms (Charles-Dominique and Martin, 1970; Martin, 1972) was recently embraced and further elaborated by Groves (1972), who states (p. 13):

> The Lorisiformes are more like the higher Primates than the Lemuriformes. It seems likely that their similarities are primitive, not advanced, and that the Lemuriformes are more specialized . . .
>
> In other ways . . . the Lorisiformes show special features, less specialized than the Lemuriformes but unlike the Anthropoidea. In the anatomy of the vascular system of the ear, the main artery, the entocarotid, divides into the promontory and stapedial arteries before reaching the skull; moreover the stapedial is very small, and the promontory very big: the latter does not pass through the tympanic cavity but enters the skull directly through the foramen lacerum. In the Lemuriformes the division takes place inside the tympanic cavity and the stapedial is the larger branch; in other Primates also the stapedial is the larger, but the promontory is less reduced, and the undivided artery enters the bulla near its posterior margin—not centrally as in the Lemuriformes.

Cartmill (1972, p. 120), like the previously cited authors, expressed belief that the stem stock of Malagasy primates was a form very similar to extant cheirogaleids, and that this animal resembled *Microcebus* in particular.

The diversified basicranial morphology of the strepsirhines offers evidence for a basically different hypothesis from that advocated by the above authors. This in turn is not significantly different from a scheme advocated by Gregory (1920), and appears compatible with other aspects of strepsirhine biology.

For purposes of this discussion★ there are three major basicranial categories among the strepsirhines. The adapids, lemurids, indriids, and daubentoniids share basic similarities, which can be described as those of a primitive strepsirhine pattern. The Cheirogaleidae, the second group, are uniform in some of their distinctive characters (see Cartmill, Chapter 14, this volume). Similarly, the lorisids, the third group, are closely knit in sharing a number of clearly derived character states.

A review of strepsirhine basicrania indicates to me that the characters shown by the known Adapidae, and to some degree by the Lemuridae and Indriidae, represent relatively unmodified versions of the primitive pattern for strepsirhines, although not necessarily for the order. Whenever we choose to determine polarity for any of the several characters found in two or more conditions, it is invariably the adapid–lemurid pattern that appears to be primitive, and the lorisid one derived. The cheirogaleids show an intermediate condition in some features although they possess some primitive and some derived character states like those found in the lorisids. The most conspicuous

---

★ I will not consider here the basicranial patterns of the Archaeolemuridae, Palaeopropithecidae, and Megaladapidae as there are no hypothesis advanced for their special ties with Lorisiformes (for details of morphology see Lamberton, 1941 and Saban, 1956, 1963).

advanced feature is the presence of an enlarged ascending pharyngeal artery (see Bugge, 1972) and an anterior carotid foramen in all forms here called lorisoids.

The following basicranial characters are particularly pertinent to this discussion of significant differences among the strepsirhines: (1) place of entry of the carotid and ascending pharyngeal arteries into the bulla; (2) absence or presence of an anterior carotid foramen and enlarged ascending pharyngeal artery; (3) relative size, or presence or absence, of the stapedial canal and artery; (4) relative size of the promontory canal and artery; (5) relative size of the petromastoid; and (6) relative size and position of the ectotympanic in relation to the petrosal.

Neither the details of the morphology nor some of the problems of function and homology surrounding them are discussed here. Lamberton's (1941) and Saban's (1956, 1963) contributions give exquisite detail of strepsirhine basicranial morphology, and

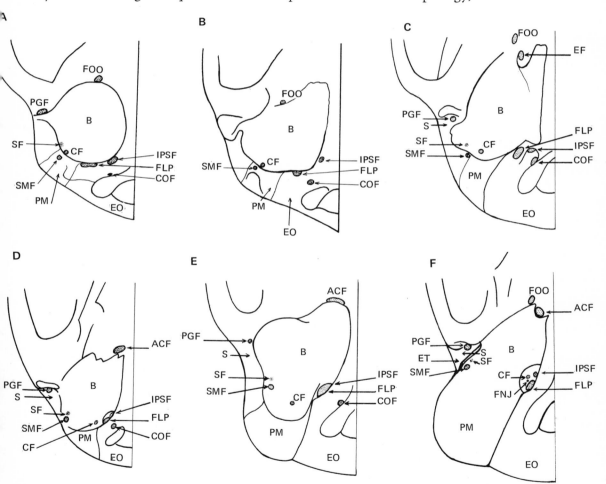

FIG. 9. Simplified schema of intact basicranial morphology of some representative Strepsirhini. (A) *Pronycticebus* (?late Eocene); (B) *Adapis parisiensis* (medial or late Eocene); (C) *Lemur* sp. (Recent); (D) *Microcebus* sp. (Recent); (E) *Komba* sp. (early Miocene); and (F) *Galago crassicaudatus* (Recent). A–C represent relatively primitive lemuroids whereas D–E represent progressively more advanced lorisiforms. For abbreviations see Table 1.

Szalay and Katz (1973) and Cartmill (this volume) recently discussed some issues of homology and morphocline polarity as they relate to strepsirhine phylogeny. The pertinent evidence and my interpretation of the homologies of vessels and openings in the basicrania are shown on Figs. 9 and 10, and interpretations of the morphocline polarities are briefly stated below. As mere factual information, it is evident (see the figures) that (1) the entry of the internal carotid artery into the bulla is either central as in the Miocene cranium of ?*Komba*, KNMR 1005-50 (Fig. 9E) or it is on the posteromedial side of the bulla as in extant genera; (2) the ascending pharyngeal artery of cheirogaleids and lorisids is not the promontory artery, as the latter is present in at least some stages of the ontogeny, coursing intrabullarly across the promontorium (Fig. 10); (3) the petromastoid is greatly inflated in the extant lorisids; and (4) the stapedial circulation is insignificant in the lorisids in contrast with the lemuroids.

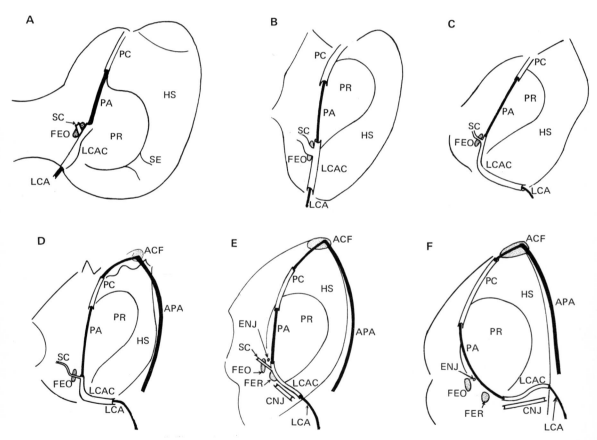

FIG. 10. Simplified schema of middle-ear morphology and associated carotid circulation in (A) *Adapis parisiensis* (medial or late Eocene); (B) *Lemur* sp. (Recent); (C) *Lepilemur* sp. (Recent); (D) *Microcebus* sp. (Recent); (E) hypothetical condition based on the external morphology of *Komba* sp. (early Miocene); and (F) *Galago crassicaudatus* (Recent). The bulla and ectotympanic are removed to show the carotid circulation. As in Fig. 6, the various conditions depicted in A–F approximate the polarity of a morphocline for the character complex shown. For abbreviations see Table 1.

According to the criteria listed under phylogeny, therefore, the basicranial evidence is compelling in that it points to well-understood morphocline polarities. This is one of the better known and most diagnostic morphological complexes of strepsirhines, inasmuch as it clearly differentiates the major groups, and I have given it much weight in deciding their cladistic as well as anagenetic relationships. We may conclude, therefore, that the most ancient and ancestral strepsirhines were animals that can be characterized in terms of the known morphology of the adapids. Apart from the relatively larger brain and a tooth comb, the most recent common ancestor of the Lemuridae and Indriidae was either the most recent common ancestor of all the known tooth-combed strepsirhines or was, at least, more like a lemurid than a cheirogaleid or lorisid. The Cheirogaleidae were derived from a lemuroid, a form not unlike *Lepilemur*. The derived characters shared by cheirogaleids and lorisids, and the clear evidence of the entire skeleton favoring the primitiveness of the cheirogaleids compared with the lorisids indicate that a strepsirhine species of cheirogaleid affinity and of a similar level of organization was the ancestor of lorisids. I hold that the Lorisiformes, including both the Cheirogaleidae and Lorisidae, are derived descendants of much more primitive *bona fide* lemuroids, and most of the characters shared by all lorisiforms are advanced features not only among primates but also within the tooth-combed strepsirhines.

E. Haplorhini

1. Tarsiiformes. It is my working hypothesis that a tarsiiform was more recently related to the morphotype of the anthropoid primates than the latter to any other known groups of primates. I take the view, therefore, that it is heuristically justified to refer to the common ancestor of platyrrhines and catarrhines as having been derived from a tarsiiform.

It was Gregory (1915, pp. 430–431) who recognized the significance of the basicranium of *Tetonius* and stated that

> The basicranial region, as a whole, is remarkably similar to that of *Tarsius* save that the trochlea, or auditory prominence, is much smaller. The bulla was greatly inflated, as in *Tarsius* and *Necrolemur*, and its anterointernal extension likewise completely covered over the region where the foramen lacerum medium is located in the Nycticebidae [i.e., Lorisidae]. The internal carotid must surely have traversed the tympanic chamber, but its exact course is doubtful. In *Tarsius* it pierces the middle of the bulla on the lower surface, then passes directly upward (craniad) through the margin of the septum of the cavum bullae, passing into the cranial cavity at the apex of the enlarged cochlea (Kampen, 1905, p. 676). In the only known skull of '*Anaptomorphus*' *homunculus* the whole lower wall of the bulla is broken away, so the place of entry of the carotid into the cavum bullae is not indicated. To the small cochlea is attached a remnant of a long septum, which may have carried the carotid canal.

Further study of the specimen confirms Gregory's suggestion that the septum housed the promontory artery. This condition, then, is identical to that of *Necrolemur*. It is equally clear that virtually all other preserved details of the middle-ear cavity in *Tetonius* match those of *Necrolemur*.

In all known paromomyiforms, most strepsirhines, and known haplorhines, the foramen lacerum posterior appears to be distinct from the usually smaller and more anterior inferior petrous sinus foramen. Clearly, this represents a shared primitive condition.

Tarsiiform basicrania containing varying amounts of information are known for *Rooneyia* (Fig. 11), *Tetonius*, *Necrolemur* (Fig. 12), and *Tarsius* (Fig. 13). The sample is far from adequate, and it appears sufficient only to allow assessment for a few characters that may diagnose the tarsiiform morphotype. Work is now in progress to further describe in detail the middle-ear morphology of *Rooneyia* and the up-to-now-unrecognized aspects of *Necrolemur*. Cleaning of specimens of these taxa has revealed hitherto unknown characters of the middle ear and the ectotympanic.

When we view the middle-ear cavity of the known tarsiiforms and compare them to specimens of adapids, such as *Notharctus tenebrosus*, *Adapis parisiensis*, *Leptadapis magnus*, or that of a new early Oligocene adapid (Wilson and Szalay, in press), it becomes evident that a series of morphological stages, at least in a number of characters, are represented when we arrange these along a morphocline.

The most unique aspect of the middle-ear morphology of *Rooneyia* is the presence of an ectotympanic deep within the tympanic cavity. This bone is continuous with a shelf, clearly an ossified annulus membrane, supported from below by two transversely

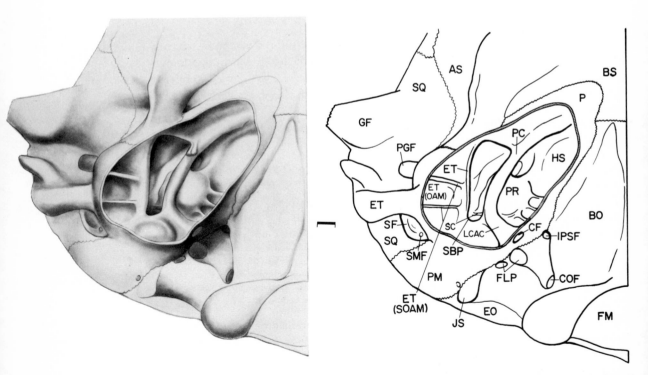

FIG. 11. Basicranial morphology of *Rooneyia viejaensis* with only the ventral floor of the bulla removed. Ventral and slightly medial view. Identification of structures is on the right side of the figure. Drawing is based on University of Texas No. 40688-7. For abbreviations see Table 1. Scale represents 1 mm.

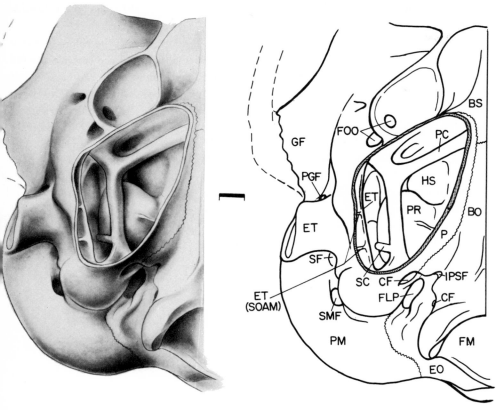

FIG. 12. Basicranial morphology of *Necrolemur antiquus* with only ventral floor of the bulla removed. Ventro-medial view. Identification of structures is on the right side of the figure. Drawing is based on Montauban No. 9. For abbreviations see Table 1. Scale represents 1 mm.

long, stout struts. In *Necrolemur* a basically identical condition exists except that its ectotympanic is closely abutted against the bulla proper.

In *Rooneyia*, therefore, the middle-ear cavity extends pervasively below the ossified annulus membrane and well lateral to the ectotympanic proper, whereas in *Necrolemur* this is only true to a small degree. In *Tarsius*, on the other hand, the ectotympanic is in the most lateral extreme known within tarsiiform bullae.

Rather than only viewing the three tarsiiforms as representing three phylogenetic stages and thus as an aid in gaining a picture of the phylogeny of this structure, the characteristic morphologies of these middle-ear cavities supply a partial adaptational explanation for the ectotympanic conformation. The middle-ear cavities of all three tarsiiforms are exceptionally enlarged, yet in a divergent manner, suggesting independent evolution. As selection independently favored an increased relative size of the middle-ear cavity for increased auditory sensitivity, divergent pathways were emphasized, perhaps as a result of different base (heritage) morphologies available for these taxa. *Rooneyia* extended the tympanic cavity below and lateral to the ectotympanic, *Necrolemur* enormously inflated the petromastoid, whereas *Tarsius* hypertrophied the hypotympanic sinus.

The point of entry of the carotid into the bulla is coupled with the enlargement of the promontory artery in known tarsiiforms. The medial entry of the internal carotid is a primitive condition for the tarsiiforms and other haplorhines, but it is derived when compared with the primitive strepsirhine and primate conditions. The path of the lateral internal carotid artery in *Tarsius* is clearly a more advanced version of what is seen in *Rooneyia*, *Necrolemur*, and *Tetonius*. In all these forms the promontory artery is relatively very large and relatively much more important than in any known strepsirhines. The comparative dimensions of the promontory and stapedial arteries in some fossils have been already noted above (see Fig. 8).

The known omomyid basicrania, when integrated with postcranial evidence, suggest that the major characters which may be considered derived, having diverged after their origin from a primitive ancestral condition that can best be called lemuriform,

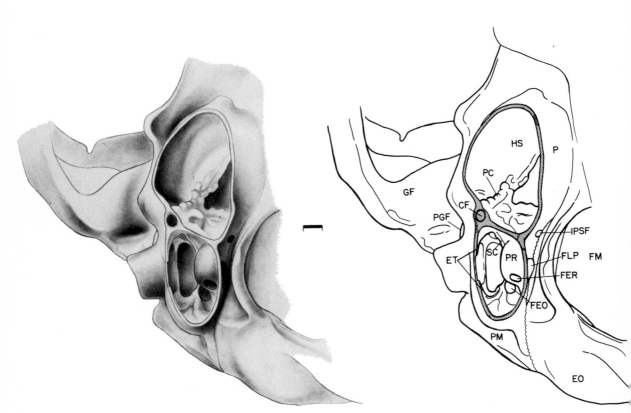

FIG. 13. Basicranial morphology of *Tarsius syrichta*, with only ventral floor of the bulla removed. Ventromedial view. Identification of structures is on the right side of the figure. Drawing is based on American Museum of Natural History, Department of Mammalogy No. 150448. For abbreviations see Table 1. Scale represents 1 mm.

were (1) an increase in the relative size of the promontory artery compared with the stapedial artery; (2) a medial shift in the entry of the carotid artery into the bulla; and (3) an increasingly extrabullar position of the ectotympanic, particularly as epitomized in *Tarsius*, derived from an intrabullar ectotympanic ring through the ossification of the annulus membrane.

   2. Anthropoidea. A gap exists between the tarsiiform morphotype and what can be judged the more primitive basicrania of known anthropoids. Considering the tarsiiform evidence, the most primitive living anthropoid basicranial morphology appears to be possessed by the cebids, forms relatively conservative among themselves in this respect. Nevertheless, a detailed survey of the anthropoids would greatly aid determination of possibly significant morphocline polarities among platyrrhines and catarrhines.

   Until now, fossil anthropoid basicrania have been largely lacking. Recently, however, important evidence with bearing on ancestral anthropoid morphology has been described by Gingerich (1973). The right petrosal and squamosal (YPM 23968) reported by him are allocated to *Apidium phiomense* from the Upper Fossil Wood zone of the Jebel el Qautrani Fm. In addition to describing the specimen (Gingerich, 1973), Gingerich (1974) has formalized his views by erecting the suborder Plesitarsioidea to show the special ties of archaic, nonlemuriform primates with the tarsiiforms and the Anthrolemuroidea to suggest the origin of the Anthropoidea from a lemuriform rather than a tarsiiform stock. His arguments for the validity of the lemuriform–anthropoid tie (Gingerich, 1973) are as follows:

1. Characters of soft anatomy are not known from fossils and therefore it is nearly impossible to tell whether suggested haplorhine similarities are synapomorphies, merely primate symplesiomorphies, or independently acquired characters; hence soft anatomical features bearing on haplorhine monophyly can be dismissed.
2. Only osteological characters should be used to test the validity of the Haplorhini.
3. There is no evidence for any derived osteological features shared by *Tarsius* and anthropoids which can be shown to be the result of common inheritance.
4. Postorbital closure in *Tarsius* is different from anthropoids; the promontory arteries of primitive lemuroids are not as relatively small as has been reported, hence the diagnostic differences between the carotid circulation of living lemuroids and *Tarsius* were not as yet established by the medial Eocene.
5. Presence of a free ectotympanic in tupaiids proves that the primitive primate condition of the bone was as seen in lemuroids.
6. Because *Necrolemur* and *Rooneyia* are widely separated geographically and morphologically, a tubular ectotympanic was an early acquisition of tarsiiforms.
7. Early catarrhines and platyrrhines lack a tubular ectotympanic; hence anthropoids were derived from primates more primitive than any known tarsioids.
8. "*Apidium* provides positive evidence that anthropoid primates evolved directly from a lemuroid ancestor" (Gingerich, 1973, p. 335).
9. Because *Pelycodus* and close relatives can be found during the early Eocene in both North America and Europe, the distribution pattern of anthropoids does not require rafting hypotheses.

It is emphasized in the following assessment of the views and arguments expressed by Gingerich that I cannot corroborate his proposed primate phylogeny. His hypothesis, incidentally, is not unlike the one proposed by Saban (1963, p. 333. Fig. 84).

1 and 2. Numerous if not all, osteological characters are mechanically linked to soft anatomy, and therefore soft anatomical characters cannot be rejected for phyletics. The great diversity of many soft anatomical features allows construction of more extensive morphocline polarities than usually permitted by the available fossils, a fact favourably balancing their absence in the fossil record (see Luckett, this volume).

3. Lack of osteological (but not soft anatomical!) synapomorphies between *Tarsius* and anthropoids is no argument against inferred synapomorphies between the most recent relatives of *Tarsius* (e.g., omomyids, or their morphotype) and the common ancestor of platyrrhines and catarrhines.

4. An assessment of the relative sizes of the stapedial artery in (?) *Pelycodus* sp. and *Necrolemur antiquus* (see above, and Fig. 8) indicates, contra Gingerich, that at least in these taxa, after the subdivision of the lateral carotid artery, the stapedial canal is emphasized in the former and the promontory one in the latter. In these two lineages the differences were likely to have been established prior to the early Eocene.

5. A free intrabullar ectotympanic in tupaiids does not prove or disprove that this condition was present in the primate morphotype.

6. A tubular ectotympanic on the outside of the petrosal is clearly present in known tarsiiforms, but this cannot be causally linked with the three distinct stages displayed by the intrabullar portions of the ectotympanic and ossified annulus membrane in *Rooneyia*, *Necrolemur*, and *Tarsius*. In *Tarsius*, the tarsiiform ectotympanic is in its most derived condition, closely juxtaposed against the lateral wall of the bulla. A condition similar to this without a wide extrabullar component is essentially the one we find in platyrrhines. Both *Apidium* and *Aegyptopithecus*, unlike later cercopithecoids and hominids, have a platyrrhine-like ring rather than a tube.

7. A ringlike ectotympanic may have been primitive for the anthropoids, and as argued under (6), this configuration was probably present in *bona fide* advanced tarsiiforms. Lorisids show the fully differentiated features of their family, inherited from a cheirogaleid diverged from a lemuroid ancestry, yet they display a variety of conformations of the ectotympanic (contrast *Galago* vs. *Nycticebus*, for example). This character, as noted, is apparently quite independent from major alterations in the circulatory pattern or the degree of petromastoid inflation.

8. I cannot ascertain the positive evidence for anthropoid origins directly from the lemuroids. There are, I believe, a few incorrect observations in Gingerich's description; consequently, his only evidence for relating the *Apidium* basicranium to lemuroids appears to be of doubtful significance. The broken posterior crus of the ectotympanic is probably correctly identified, but a small depression he asserts to be a facet for the anterior crus may or may not be for the ectotympanic. The anterior crus probably fitted immediately medial to the postglenoid process into a discernible groove that extended posteriorly medial to the postglenoid foramen. This is exactly what one finds in many specimens of platyrrhines, such as *Saimiri*. The lack of fusion of the anterior crus would not

be unexpected in a relatively young individual. Hershkovitz (1974) has recently shown that an unfused anterior crus can be found among immature *Tarsius* and even among some mature cebids.

9. The statement that the presence of *Pelycodus* in Europe and North America does not suggest dispersal of early anthropoids from either Africa or South America implies that (a) *Pelycodus* is the common ancestor of the Anthropoidea, and (b) the Platyrrhini and Catarrhini independently evolved a number of shared characters clearly not present in *Pelycodus*. I consider the evidence nonexistent and therefore this phylogenetic conclusion fallacious.

The evidence from the petrosal of *Apidium* is important for both cladistic and anagenetic assessment of anthropoid origins. An anthropoid specialization, contrasted

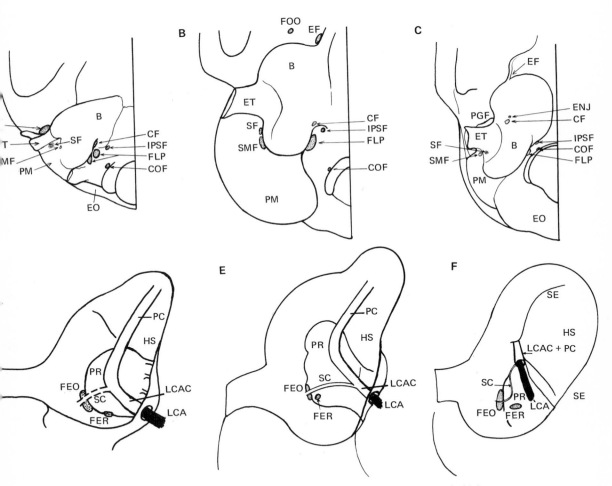

FIG. 14. Simplified schema of some aspects of basicranial morphology in a few known tarsiiform Haplorhini. A and D) *Rooneyia viejaensis* (late Eocene); (B and E) *Necrolemur antiquus* (medial or late Eocene); and (C and F) *Tarsius* sp. (Recent). Above: basicrania intact, below: the bulla and ectotympanic are removed to show carotid circulation. Broken lines represent reconstruction. For abbreviations see Table 1.

to those of other primates, is the extreme enlargement of the promontory artery. This vessel plunges into the petrosal quite medially and appears, in its probably primitive form, both in *Apidium* and some platyrrhines and catarrhines as a raised surface on the medial side of the promontorium (Fig. 15). Another character that appears to be a synapomorphy between *Apidium*, platyrrhines, and catarrhines is the pneumatized nature of both the petrosal and squamosal. This feature could link an anthropoid morphotype with a group of early tarsiiforms that had an enlarged petromastoid and an ectotympanic ring at the edge of the bulla proper. Furthermore, the sinus-filled bullar portion of anthropoids (Fig. 15C–E) corresponds to the area in tarsiiforms that usually has a large hypotympanic sinus.[*] As noted elsewhere (Szalay, in press) the cancellous condition of the anthropoid basicranium should be distinguished from the highly inflated and partitioned middle-ear cavity and the inflated petromastoid of lorisids. The latter contain numerous larger cavities partly continuous with the middle-ear chamber.

F. Ectotympanic Evolution

In a final note I will summarily emphasize that the changes in the primate ectotympanic in various groups of the order present a complex and still unresolved series of problems, as the foregoing discussions should indicate. I have published (Szalay, 1972) a rather preliminary interpretive scheme to explain diversity in this structure within the order, but new facts and assessments require modifications of some of the proposed hypotheses. One must bear in mind that each of the three subordinal categories of the order have various types of ectotympanics, both narrow and wide, and both the Strepsirhini and Haplorhini have ringlike as well as tubelike ectotympanics. Knowledge of the known extant and extinct conditions must be carefully balanced to arrive at morphotype conditions and subsequent hypotheses of transformations for the various taxa considered, even if an adequate developmental mechanism is available.

The hypothetical primitive eutherian condition (Fig. 16A) is one in which the bulla may have been either cartilaginous, perhaps with a center of ossification at some stage of the ontogeny, an entotympanic (rostral or caudal), or with one or several contributions to the bulla from the surrounding bones and with the ectotympanic partly lateral to the cartilage or bone. The configuration shown in Fig. 16A, which is not necessarily the immediate ancestral condition to that in Fig. 16B, is a remote structural stage. In B the petrosal is continuous with an ossified bulla, with the ectotympanic largely lateral to and in articulation with a petrosal bulla. This condition may or may not represent the primate morphotype.

---

[*] This conspicuous aspect of anthropoid basicranial morphology might reflect (1) the ancestral condition, and (2) the manner of subsequent transformation in the anthropoid morphotype. Pneumatization of the petrosal in anthropoids might have been the mechanism to alter the hearing-related morphology and function of an inflated hypotympanic sinus and petromastoid.

FIG. 15. Simplified schema of some aspects of basicranial morphology in a few representative genera of anthropoid Haplorhini: (A and C) *Saimiri* sp. (Recent); (B and E) *Cercopithecus* sp. (Recent); (D) *Apidium* sp. (Oligocene). Above: basicrania intact, below: the bulla and ectotympanic are removed and the internal architecture of the pneumatized regions is schematically shown. Heavy line around middle ear cavity shows where bone is cut to expose the morphology inside. Broken lines represent reconstruction. For abbreviations see Table 1.

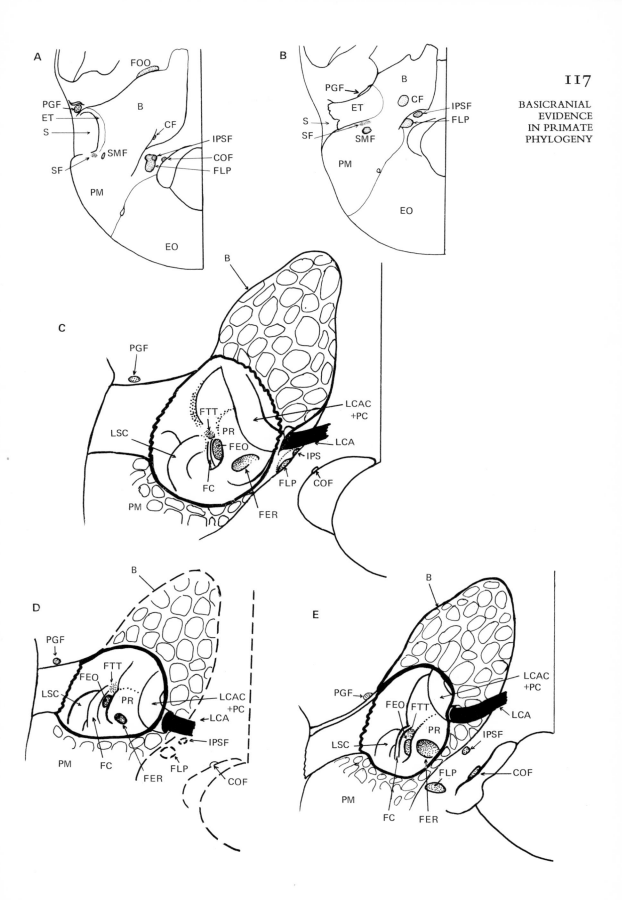

BASICRANIAL
EVIDENCE
IN PRIMATE
PHYLOGENY

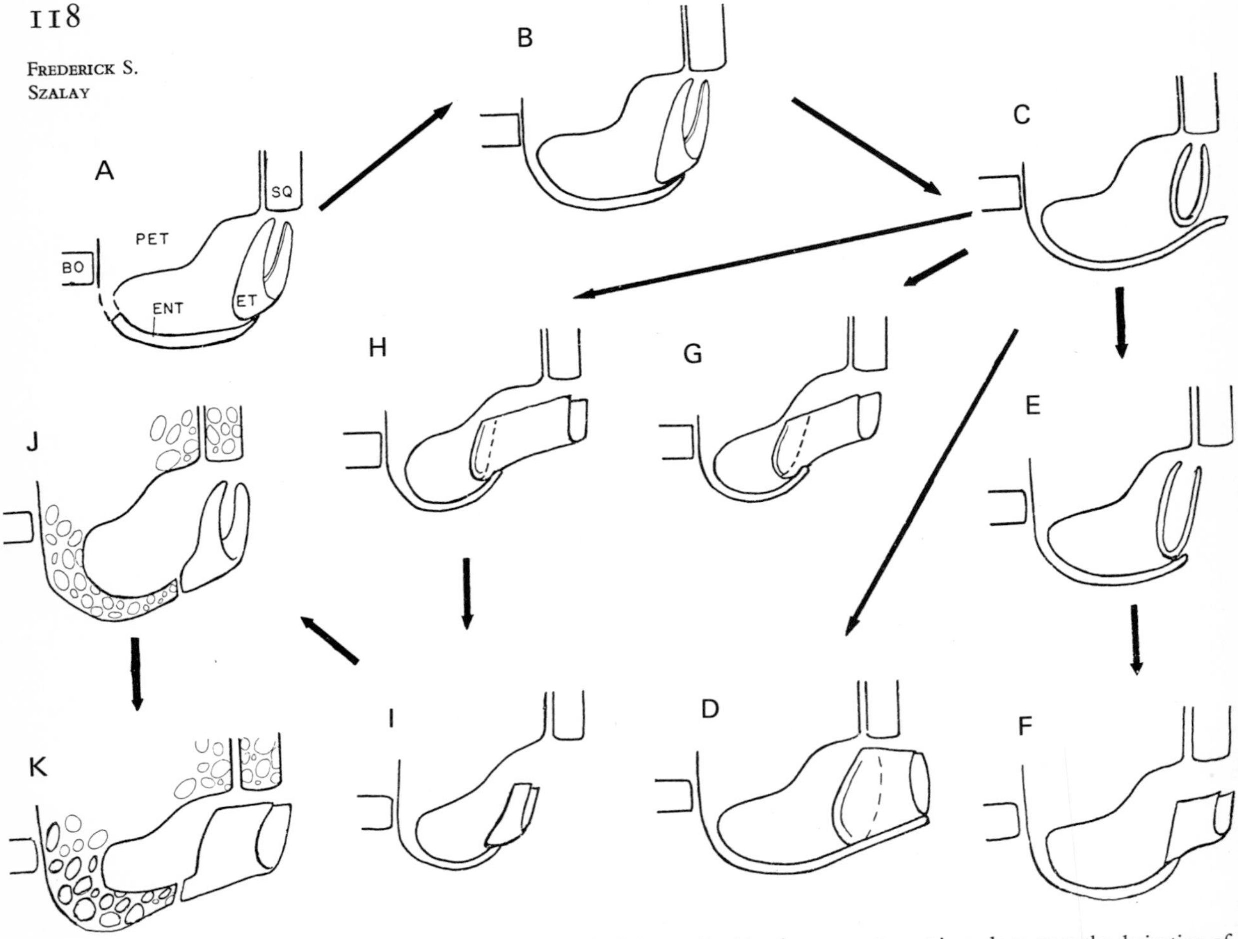

FREDERICK S.
SZALAY

FIG. 16. Ectotympanic evolution in the Primates. In this scheme, an attempt is made to trace the derivation of various ectotympanic conditions (other aspects of the basicranium are ignored for the sake of clarity) from inferred antecedent states. Note that in this overview primates have evolved a bony "tube" at least three times independently from a condition that can be termed primitive lemuroid. As the mechanism for the construction of this tube is extremely simple, involving only the ossification of the annulus membrane, parallel evolutions are postulated for this derived condition in some paromomyiforms, tarsiiforms, and lorisiforms. In catarrhines the external bony tube does not involve the ossification of the annulus membrane in the same manner as in the three previous groups.

A. Hypothetical ancestor. B. Intermediate between eutherian ancestor and primate morphotype. C. Primate morphotype, also probably the primitive condition in the Strepsirhini. D. A modified lemuriform condition (e.g., *Megaladapis*) where the petrosal enlongated laterally and the annulus membrane of the ectotympanic also ossifies inside the laterally expanded petrosal. E. Lorisid morphotype (e.g., *Galago*), slightly more advanced than that of cheirogaleids. F. A derived lorisiform condition (e.g., *Nycticebus*). G. Condition shown by *Plesiadapis*. Although morphology of the ectotympanic is extremely similar to that of *Rooneyia*, other aspects of the basicranium strongly suggest that the ectotympanic conformations are convergent. H. Condition shown by *Rooneyia*, possibly representative of the tarsiiform morphotype. I. Advanced (?)tarsiiform condition shown by *Necrolemur*. J. Platyrrhine morphotype. K. Condition known among living cercopithecoids and hominoids, possibly representing the catarrhine morphotype, although parallel evolution of this form of ectotympanic tube may have evolved independently from a catarrhine ancestor with a condition similar to that shown on J.

For abbreviations see Table 1.

As noted, derivation of the tarsiiform condition(s) is from a *bona fide* lemuriform. The sequence from a primitive tarsiiform (Fig. 16H) and from the latter to I and then to J is most probable. The morphotypes of the living cercopithecids and hominoids may have independently reached stage K from J.

The strepsirhine and primate morphotype pattern is judged to be that shown in Fig. 16C, as exemplified by most lemuroids, and is most likely to have given rise to E, one of the lorisid conditions.

Figure 16D represents a modified lemuriform condition, that of *Megaladapis*. Lamberton's (1941) studies on the bony-ear region of the Madagascan primates revealed that the island species fall into two morphological groups as far as the position of the ectotympanic is concerned. He characterizes the first group (made up of *Archaeolemur*, *Hadropithecus*, *Mesopropithecus*, *Neopropithecus*, *Propithecus*, and *Lemur*), what is considered here the primitive lemuroid condition, as having either a very wide and short or non-existent auditory canal (formed by the petrosal), an annulus tympanicus inside the bulla, free on part of its circumference, and a stapedial artery enclosed in a bony tube. Genera of his second group (*Megaladapis*, *Palaeopropithecus*, and *Archaeoindris*, i.e., the largest forms), have a bony and narrow external auditory canal, an annulus tympanicus attached entirely to the bulla, and a reduced (or possibly missing) stapedial branch not enclosed in a bony tube. Lamberton did not mention, however, that the external auditory tube is petrosal derived,★ although internally an extension of the annulus membrane is also ossified.

IV. Summary and Conclusions

The basicranial construction of the eutherian morphotype was briefly essayed. It was concluded that the most significant morphological resemblance of the primitive primate basicranial morphology is to those of early erinaceoids among the known groups of Eutheria. The petrosal component of the bulla and the tube-enclosed carotid circulation of the middle ear are particularly significant aspects of this similarity. It is still not clear whether the similarity of the tupaiid–lemurid ectotympanic is homologous or not, and if homologous, how widespread this condition might be among ancient Eutheria.

In attempting to establish the morphotype of the primate basicranium, difficulties arise concerning the condition of the ectotympanic in relation to the petrosal. Of the two paromomyiforms known by adequate ear regions (*Phenacolemur* and *Plesiadapis*), *Phenacolemur* has its ectotympanic entirely outside the bulla proper and the fenestra rotunda is "shielded." In *Plesiadapis* the ectotympanic is well inside the bulla, but it extends far out from an external auditory tube. The morphotype for the basicranium of the strepsirhines, typified by the adapids, has the internal ectotympanic ring and a "shielded" fenestra rotunda. The medial entry of the internal carotid into the bulla, the enlargement of the promontory artery, and the auditory tube comprised of an intrabullar

★ This is unlike an extrabullar ectotympanic tube, the condition Simons (1972, p. 95) and Tattersall (1973) attributed to *Megaladapis*.

ectotympanic and an ossified annulus membrane of the tarsiiform morphotype are judged to be derived from adapids. The anthropoid basicranial morphotype, typified in this study by most living anthropoids and also the fossil *Apidium*, can be most easily derived from a hypothetical tarsiiform with an enlarged promontory artery, enlarged hypotympanic sinus, and a greatly inflated petromastoid. Recent suggestions that the Anthropoidea is derived from some unknown stocks of lemuriforms cannot be corroborated by the phylogenetic analysis of the basicranial morphology or other morphological systems (both soft and hard anatomy). This survey of basicranial morphology, aided by other evidence known to me, suggests then that of the possible relationships on the specified *subordinal level*, part A of both Figs. 17 and 18 is the most plausible. In Fig. 19, a phylogenetic hypothesis, unlike the cladograms, it is suggested that the branching

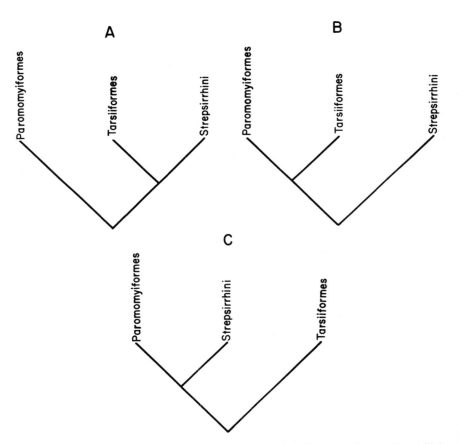

FIG. 17. Three possible different hypotheses of cladistic relationships of the Paromomyiformes, Strepsirhini, and Tarsiiformes. Most of the evidence corroborates hypothesis A, whereas hypothesis C is probably the least likely.

points, although poorly understood, have morphological, i.e., phenetic and anagenetic, significance. Even though the evidence is still poor and only minimum time of divergences can be ascertained, an attempt is made to reflect some estimated temporal values for splitting on the level of the categories shown.

The monophyly of the order Primates (the Paromomyiformes included), of the Strepsirhini, Haplorhini, and Anthropoidea is supported by basicranial characters. The evidence for a common source of origin for the Strepsirhini–Haplorhini may still seem equivocal if only the basicranial morphology is examined. Numerous postcranial features of the omomyids, however, and to a lesser degree dental evidence, most strongly indicate haplorhine origins from *bona fide* lemuriforms which, if known, might be referred to the Adapidae.

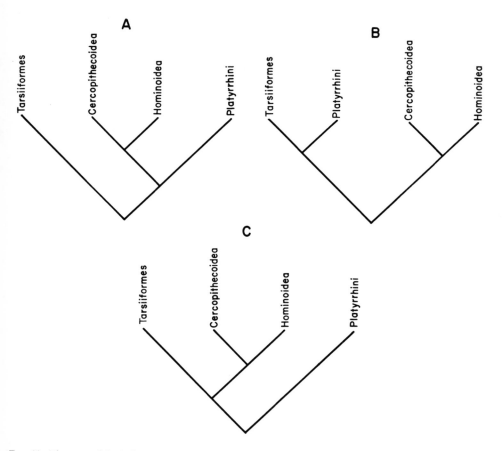

FIG. 18. Three possible different hypotheses of cladistic relationships of the Tarsiiformes, Platyrrhini, and Catarrhini (Cercopithecoidea and Hominoidea). Most of the evidence corroborates hypothesis A, whereas hypothesis C is probably the least likely.

Table i

ABBREVIATIONS USED IN TEXT AND FIGURES FOR ANATOMICAL TERMS EMPLOYED FOR THE
BASICRANIUM

ACF	anterior carotid foramen	IPSF	inferior petrous sinus foramen
AM	annulus membrane	IRSA	inferior ramus of stapedial artery
APA	ascending pharyngeal artery	JS	jugular spine
AS	alisphenoid		
B	bulla	LCA	lateral carotid artery
BO	basioccipital	LCAC	lateral carotid arterial canal
BS	basisphenoid	MCA	medial carotid artery
		MCAC	medial carotid artery canal
CCA	common carotid artery	MCF	medial carotid foramen
CF	carotid foramen	MP	mastoid process
CNJ	canal for nerve of Jacobson		
COF	condyloid foramen	OA	ophthalmic artery
		OAM	ossified annulus membrane
ECA	external carotid artery	P	petrosal
EF	eustachian foramen	PA	promontory artery
EJV	external jugular vein	PC	promontory canal
ENJ	exit for nerve of Jacobson	PF	promontory foramen (=foramen lacerum medium)
ENT	entotympanic		
EO	exoccipital	PGF	postglenoid foramen
ER	epitympanic recess	PGP	postglenoid process
ET	ectotympanic	PM	petramastoid
FC	facial canal	PR	promontorium of petrosal
FEO	fenestra ovale	RI	ramus inferior
FER	fenestra rotunda	RS	ramus superior
FLA	foramen lacerum anterior		
FLP	foramen lacerum posterior	SA	stapedial artery
FM	foramen magnum	SBP	superior border of the petrosal
FMS	fossa muscularis stapedius	SC	stapedial canal
FNJ	foramen for nerve of Jacobson	SE	septum
FOO	foramen ovale	SF	styloid fossa
FSRS	foramen for superior ramus of stapedial artery	SMF	stylomastoid foramen
		SMFD	stylomastoid foramen definitivum
FTT	fossa tensor tympani	SMFP	stylomastoid foramen primitivum
GF	glenoid fossa	SOAM	support for ossified annulus membrane
		SQ	squamosal
HS	hypotympanic sinus	SRSA	superior ramus of stepedial artery
ICA	internal carotid artery	TH	attachment for hyoid arch
IJV	internal jugular vein	TP	tympanic process
IPS	inferior petrous sinus	VF	vidian foramen

The review underlines the fact that some uncertainties exist as to the homologies of some basicranial characters, as well as the morphocline polarities of others. Because important features (inasmuch as they are often recognizable in fossils) cannot as yet be unambiguously interpreted, some doubts about the recency of relationships do exist. It appears to me, however, that the broad outlines of primate phylogeny advocated here can be corroborated by other lines of evidence (see Luckett, this volume; Starck, this volume).

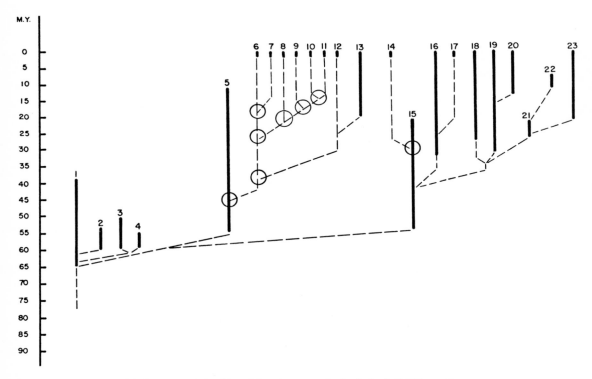

FIG. 19. An hypothesis of phylogenetic relationships of the primates on the family level. Solid heavy lines represent known ranges and broken lines depict recency of relationships: (1) Paromomyidae; (2) Picrodontidae; (3) Plesiadapidae; (4) Carpolestidae; (5) Adapidae; (6) Lemuridae; (7) Megaladapidae; (8) Archaeolemuridae; (9) Palaeopropithecidae; (10) Daubentoniidae; (11) Indriidae; (12) Cheirogaleidae; (13) Lorisidae; (14) Tarsiidae; (15) Omomyidae; (16) Cebidae; (17) Callithricidae; (18) Hylobatidae; (19) Pongidae; (20) Hominidae; (21) Parapithecidae; (22) Oreopithecidae; (23) Cercopithecidae. Circles represent uncertainty as to even an approximate time of divergence.

V. References

BOCK, W. J., and J. H. VON WAHLERT. 1965. Adaptation and the form-function complex. *Evolution* **19**:269–299.

BOWN, T. M., and P. D. GINGERICH. 1973. The Paleocene primate *Plesiolestes* and the origin of Microsyopidae. *Folia Primat.* **19**:1–8.

BUGGE, J. 1967. The arterial supply of the middle ear of the rabbit with special reference to the contribution of the stapedial artery to the development of the superior tympanic artery and the petrosal branch. *Acta Anat.* **67**:280–220.

BUGGE, J. 1971. The cephalic arterial system in New and Old World hystricomorphs, and in bathyergoids, with special reference to the systematic classification of rodents. *Acta Anat.* **80**:516–536.

BUGGE, J. 1972. The cephalic arterial system in the insectivores and the primates with special reference to the Macroscelidoidea and Tupaioidea and the insectivore–primate boundary. *Z. Anat. Entwg.* **135**:279–300.

BUGGE, J. 1974. The cephalic arterial system in insectivores, primates, rodents and lagomorphs, with special reference to the systematic classification. *Acta Anat.* **87** (Supply. 62):1–160.

BUTLER, P. M. 1956. Skull of *Ictops* and the classification of the Insectivora. *Proc. Zool. Soc. London* **126**:453–481.

CARTMILL, M. 1972. Arboreal adaptations and the origin of the Order Primates, pp. 99–122. *In* R. Tuttle, ed., *The Functional and Evolutionary Biology of Primates.* Aldine-Atherton, Chicago/New York.

CHARLES-DOMINIQUE, P., and R. D. MARTIN. 1970. Evolution of the lorises and lemurs. *Nature* **227**:257–260.

GAWNE, C. E. 1968. The genus *Proterix* (Insectivora, Erinaceidae) of the upper Oligocene of North America. *Am. Mus. Novit.* **2315**:1–26.

GINGERICH, P. D. 1973. Anatomy of the temporal bone in the Oligocene anthropoid *Apidium* and the origin of Anthropoidea. *Folia Primat.* **19**:329–337.

GINGERICH, P. D. 1974. Cranial anatomy and evolution of early tertiary Plesiadapidae (Mammalia, Primates). *Dissertation Abstracts International* Vol. 35, no. 5.

GREGORY, W. K. 1910. The orders of mammals. *Bull. Am. Mus. Nat. Hist.* **27**:3–524.

GREGORY, W. K. 1915. On the classification and phylogeny of the Lemuroidea. *Bull. Geol. Soc. Am.* **26**:426–446.

GREGORY, W. K. 1920. On the structure and relations of *Notharctus*, an American Eocene primate. *Mem. Am. Mus. Nat. Hist.* **3**:51–243.

GROVES, C. P. 1972. Phylogeny and classification of primates, pp. 11–57. *Pathology of Simian Primates*, Part I. Karger AG., Basel.

GUTHRIE, D. A. 1963. The carotid circulation in the Rodentia. *Bull. Mus. Comp. Zool.* **128**:455–581.

HENNIG, W. 1950. *Grundzuge einer Theorie der phylogenetischen Systematik.* Deutscher Zentralverlag, Berlin.

HENNIG, W. 1965. Phylogenetic systematics. *Ann. Rev. Entomol.* **10**:97–115.

HENNIG, W. 1966. *Phylogenetic Systematics.* University of Illinois Press, Chicago.

HERSHKOVITZ, P. 1974. The ectotympanic bone and origin of higher primates. *Folia Primat.* **22**:237–242.

KAMPEN, P. N. VAN. 1905. Die Tympanalgegend des Säugetierschädels. *Morphol. Jahrb.* **34**:322–722.

KLAAUW, C. J. VAN DER. 1929. On the development of the tympanic region of the skull in the Macroscelididae. *Proc. Zool. Soc. London* **37**:491–560.

KLAAUW, C. J. VAN DER. 1931. The auditory bulla in some fossil mammals. *Bull. Am. Mus. Nat. Hist.* **62**:1–352.

LAMBERTON, C. 1941. Contribution à la connaissance de la faune subfossile de Madagascar. Note IX. Oreille osseuse des Lémuriens. *Mem. Acad. Malgache* **35**:1–132.

MacDOWELL, S. B., Jr. 1958. The Greater Antillean insectivores. *Bull. Am. Mus. Nat. Hist.* **115**:113–214.

MacINTYRE, G. T. 1972. The trisulcate petrosal pattern of mammals. *In* T. Dobzhansky, M. K. Hecht, and W. C. Steere, eds., *Evolutionary Biology*, Vol. 6. Appleton-Century-Crofts, New York.

MARTIN, R. D. 1972. Adaptive radiation and behaviour of the Malagasy lemurs. *Philos. Trans. Royal Soc. London,* Ser. B **264**:295–352.

MATTHEW, W. D. 1909. The Carnivora and Insectivora of the Bridger Basin, middle Eocene. *Mem. Am. Mus. Nat. Hist.* **9**:289–567.

McKENNA, M. C. 1960. Fossil Mammalia from the early Waastchian Four Mile Fauna, Eocene of northwest Colorado. *Univ. Calif. Publ., in Geol. Sci.* **37**:1–130.

McKENNA, M. C. 1963. The early Tertiary primates and their ancestors, pp. 69–74. *Proc. XVI Int. Cong. Zool.*

McKENNA, M. C. 1966. Paleontology and the origin of the primates. *Folia Primat.* **4**:1–25.

McKENNA, M. C. 1968. *Leptacodon*, an American Paleocene nyctithere (Mammalia, Insectivora). *Am. Mus. Novit.* **2317**:1–12.

RICH, T. H. V., and P. V. RICH. 1971. *Brachyerix*, a Miocene hedgehog from western North America, with a description of the tympanic regions of *Paraechinus* and *Podogymnura*. *Am. Mus. Novit.* **2477**:1–58.

RUSSELL, D. E. 1964. Les mammifères Paléocènes d'Europe. *Mem. Mus. Nat. Hist. Nat.* **13**:1–324.

SABAN, R. 1956. L'os temporal et ses rapports chez les Lumeriens sufossiles de Madagascar. *Mem. Inst. Sci. Mad., Ser. A* **10**:251–297.

SABAN, R. 1963. Contribution à l'étude de l'os temporal des Primates. *Mem. Mus. Nat. Hist. Nat., Ser. A* **29**:1–378.

SCHAEFFER, B., M. K. HECHT, and N. ELDREDGE. 1972. Phylogeny and paleontology. *In* T. Dobzhansky, M. K. Hecht, and W. C. Steere, eds., *Evolutionary Biology*, Vol. 6. Appleton-Century-Crofts, New York.

SIMONS, E. L. 1972. *Primate Evolution: An Introduction to Man's Place in Nature.* Macmillan, New York.

SZALAY, F. S. 1969. Mixodectidae, Microsyopidae, and the insectivore–primate transition. *Bull. Am. Mus. Nat. Hist.* **140**:193–330.

SZALAY, F. S. 1972. Cranial morphology of the early Tertiary *Phenacolemur* and its bearing on primate phylogeny. *Am. J. Phys. Anthropol.* **36**:59–76.

SZALAY, F. S. (in press). Systematics of the Omomyidae (Tarsiiformes, Primates): Taxonomy, phylogeny, and adaptations. *Bull. Am. Mus. Nat. Hist.*

SZALAY, F. S., and C. C. KATZ. 1973. Phylogeny of lemurs, galagos and lorises. *Folia Primat.* **19**:88–103.

TANDLER, J. 1899. Zur vergleichenden Anatomie der Kopfarterien bei den Mammalia. *Denkschr. Acad. Wiss. Wien* **67**:677–784.

TANDLER, J. 1902. Zur Entwicklungsgeschichte der Kopfarterien bei den Mammalia. *Morph. Jb.* **30**:275–373.

TATTERSALL, I. 1973. Cranial anatomy of the Archaeolemurinae (Lemuroidea, Primates). *Anthrop. Papers Amer. Mus. Nat. Hist.* **52**:1–110.

VAN VALEN, L. 1965. Treeshrews, primates and vossils. *Evolution* **19**:137–151.

VAN VALEN, L. 1966. Deltatheridia, a new order of mammals. *Bull. Am. Mus. Nat. Hist.* **132**:1–126.

VAN VALEN, L. 1967. New Paleocene insectivores and insectivore classification. *Bull. Am. Mus. Nat. Hist.* **135**:217–284.

VAN VALEN, L. 1969. A classification of the Primates. *Amer. J. Phys. Anthro.* **30**:295–296.

WAHLERT, J. H. 1973. *Protoptychus*, a hystricomorphous rodent from the late Eocene of North America. *Breviora* **419**:1–14.

WILSON, J. A., and F. S. SZALAY. In press. New adapid primate of European affinities from Texas. *Folia Primatologica.*

WOOD, A. E. 1962. The early Tertiary rodents of the family Paramyidae. *Trans. Am. Philos. Soc.* **52**:1–261.

6

The Development of the Chondrocranium in Primates

Dietrich Starck

I. Introduction

Because of its richness of characters the adult vertebrate skull has been a principal subject of taxonomic and phylogenetic investigations for a long time. Thus, it is logical that there should also have been many attempts to include the morphology of embryonic stages of the cranium for the elucidation of phylogenetic problems.

Contrary to widespread opinion, however, there is no fixed stage of chondrocranium development, during which it is stable for some period of time and can be said to represent an initial model for the developing osteocranium. The growth of the chondrocranium starts at an earlier embryonic age than that of the osteocranium, but even in late stages it is still differentiating in some places, while it has already become reduced or ossified in others (Augier, 1931; Frick, 1954; Kuhn, 1971; Starck, 1961, 1962, 1967). Hence, there does not exist a so-called "stadium optimum" of the chondrocranium; to get a reliable understanding of its development, one has to investigate whole ontogenetic series. There are many well-known cases of incorrect interpretation of facts which were derived from single embryonic stages. Henckel (1928a), for example, concluded from one very young stage of *Tupaia* that its chondrocranium ought to be "platybasic," and that these animals

---

Dietrich Starck · Anatomisches Institut der Universität, 7 Theodor-Stern Kai, 6-Frankfurt am Main, West Germany.

therefore should be excluded from the Primates. Recent studies of more advanced stages revealed, however, that the cranium becomes "tropibasic" later on (Spatz, 1964; Starck, 1967). As for *Galeopithecus*, it has been stated that it has no optic foramen, because the radix postoptica is missing; however, our embryonic stage of 45-mm total length still shows a postoptic bar. Therefore, ontogenetic heterochronies may lead to erroneous conclusions, if no complete series are available.

The study of embryonic series of the skull is complicated by the very time-consuming technique of serial sectioning and wax-plate reconstruction. Therefore, we are still far from a satisfactory knowledge of the developmental morphology of the chondrocranium of mammalian orders. It is especially regrettable that we do not know enough about the craniogenesis of Primates; it is my purpose to overview the more recent and detailed investigations of this group. Quite a few of my own unpublished results have been incorporated into this text.

To incorporate cranial structures into phylogenetic considerations one must not only give detailed descriptions but also a careful weighting of their morphological value. The cranium forms a complex which is composed of very heterogeneous elements; this is especially true for the "syncranium" of the Mammalia, which is the product of highly complicated transformations. The evaluation of its different components makes it necessary to study their phylogenetic derivations. Comparability of structures should not be based primarily on sheer similarity or topographic coordination, but on homology alone; hence, membrane bones can not be compared directly with cartilage bones.

For an understanding of morphological relationships one must keep in mind that all extant taxa are the result of long-lasting historical processes and that all are specialized in one direction or another. It is important to comprehend and separate these specializations from more conservative characters. Real phyletic lines can never be reconstructed by merely aligning extant forms; therefore we will never be able to reconstruct the phylogeny of the chondrocranium by direct methods. A comprehensive knowledge of the present chondrocrania, in light of the fossil osteocrania, will allow some conclusions, however. Primarily, the skull is composed of endocranial and exocranial elements. The endocranium differentiates in deep connective tissue layers and forms the capsules of the brain and the main sense organs (neurocranium) as well as the skeleton of the gill apparatus (splanchnocranium or viscerocranium); in ontogenesis it primarily consists of cartilage and becomes ossified during later stages (enchondral ossification) (Fig. 1).

The exocranium is formed by the dermal or membrane bones, which lie superficially and which develop directly out of mesenchymal connective tissue. The exoskeleton covered large parts of the body of lower vertebrates, but became reduced during the course of vertebrate evolution. In mammals it is only represented by the clavicle and in the membrane bones of the skull. The number of these bones has been reduced during phylogeny, but generally their size increased. The individuality of most of the dermal bones remained very constant over long time spans, and so their homologies can easily be assessed. Secondary division or primordial fusion of dermal bones (premaxillary–maxillary in Hominoidea) has occurred only in rare cases. In late stages of ontogenesis some dermal bones may exhibit cartilaginous tissue (dentary, palatine, maxillary, squamosal in mammals); they ossify according to the endochondral mode. Cartilages of

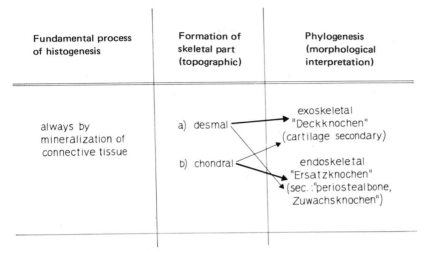

Fundamental process of histogenesis	Formation of skeletal part (topographic)	Phylogenesis (morphological interpretation)
always by mineralization of connective tissue	a) desmal b) chondral	exoskeletal "Deckknochen" (cartilage secondary) endoskeletal "Ersatzknochen" (sec. :"periosteal bone, Zuwachsknochen")

FIG. 1. Relations between histogenesis, formation of skeletal part, and morphological interpretation.

the chondrocranium may be covered by dermal bones, and ossify in a similar manner (vomer–paraseptal cartilage). So-called "secondary cartilage" normally appears late and primarily is separated from the chondrocranium; in most cases it can easily be interpreted as a functional adaptation to rapid growth or mechanical stress (cartilago pterygoidea: M. tensor veli palatini). As Romer (1942) stated, cartilaginous tissue generally should be understood as an adaptation to the physiological needs of the fast-growing fetus; this is especially true for the "secondary cartilage."

Morphological evaluation of a skeletal element therefore does not depend on its histology and histogenesis, but only on its phylogenetical connections. This statement holds true for endochondral bones as well, in which the cartilage phase may be suppressed. The neural arches in Scardinius (Teleostei) differentiate according to the intramembranous ossification mode (Matveev, 1929), but nobody will doubt their homology with the arches of related species, in which they are preformed by cartilage. In late stages of ontogenesis of the mammalian skull, completion to cartilage bones can be observed which do not run through a cartilaginous phase of development. This phenomenon is called "periosteal bone" (Romer, 1942) or "Zuwachsknochen" (German terminology), and it is probably an adaptation to some special problems of growth. In man, this kind of bone occurs mainly in the middle-ear region (canal of the facial nerve, solum, and tegmen tympani). In younger stages of mammalian craniogenesis, endoskeletal and exoskeletal parts show some independence, and therefore their examination proves valuable.

II. The Chondrocranium of the Mammalia

The chondrocranium of mammals is distinguished from that of the nonmammals by a number of characters. All mammals share many common features, so it is difficult to define those which are characteristic for special orders or families. The shape of the main organs of the head (brain, eyes, nose, masticatory apparatus, labyrinth) very often indicates adaptive specializations. Because of their close correlation with the formation and function of the skull, these organs largely influence the shape of the skull as well (brain–eye ratio, reduction of the nasal capsule, construction of the base of the skull, etc.). Highly specialized skulls exhibit most of their adaptive features even in very early embryonic ages (cf. *Tarsius*, *Daubentonia*); generalized forms, on the other hand, present generalized chondrocranial features (*Tupaia*, *Lemur*). Because of the fundamental similarity of nearly all mammalian chondrocrania, special attention must be paid to details (nasal cartilages, structure of the sidewalls, commissural bars, etc.). First we will discuss the basic and typical characters of the mammalian cranium and show how they differ from lower vertebrates. It should be mentioned in this context that birds independently reached a high degree of skull differentiation; this group also shows a considerable increase of brain size, which resulted in a number of identical structures of the ear and occipital region (Müller, 1961). Because of the wide taxonomic gap between mammals and birds, these features must be interpreted as convergent or parallel developments.

The evolutionary changes from the synapsid reptile cranium to that of the mammals are fairly well documented by fossil records (Starck, 1967; Frick and Starck, 1963; Kermack and Kermack, 1971).

The key characters of the mammalian cranium are as follows:

1. The increase of the brain volume makes necessary an enlargement of the neurocranium. This is achieved by means of reductions of the primary roof and sidewalls of the endocranium. Parts of the visceral skeleton (epipterygoid) become incorporated into the sidewall; thereby a new room, the epipteric cave, is added to the braincase. Its new sidewall at this place is formed by the alisphenoid, which is derived from the visceral epipterygoid. A similar process takes place near the primary olfactory foramen, where the cavum supracribrosum is included in the braincase. Still other consequences of increased brain size were a shifting of the optic capsule to the base of the neurocranium, a change of the position of the occipital region and the foramen magnum, as well as an expansion of the dermal roof bones (frontal, squamosal, parietal, interparietal).

2. The evolution of a heterodont dentition (chewing mechanism) and the development of lips and cheeks (suckling) and secondary palate and tongue equipped with musculature, are all correlated with a fundamental reconstruction of the jaws and the auditory ossicles. The aboral part of the primitive lower jaw, the articular, was transformed into the malleus, and the quadrate into the incus. The secondary jaw articulation between dental and squamosal is a new acquisition. The remaining dermal bones of the lower jaw became reduced or changed their functions: the prearticular was incorporated in the malleus, the angular formed the tympanic. All of these transformations are functionally correlated with the development of the mammary apparatus and homeothermic mechanisms.

3. The dominance of the olfactory system is significant for all primitive mammals. It is manifested by an enormous aboral extension of the nasal capsule (regio ethmotur-

binalis). The interorbital septum thereby is included in the ethmoidal region as part of the nasal septum; consequently, the skull is characterized as platybasic. A secondary reduction of the nose or an increase in size of the eyes may result in the reappearance of an interorbital septum.

4. The total loss of kinetics makes possible a strong fixation of the facial skull to the neurocranium.

5. The increasing importance of the auditory system causes an enlargement of the capsula cochleae and its separation from the base of the skull.

6. The ringlike occipital region exhibits rudimentary roof elements whose exact position depends on brain size. The craniovertebral boundary is characterized by the formation of two condyles, the morphology of which is strongly influenced by functional agencies.

Not all characters of the cranium appear at the same stage or remain stable throughout development. Therefore, it is impossible to draw general conclusions solely from single stages; craniogenesis has to be studied as a complete process.

Embryonic crania of most Primates are known only from a few single stages; in consequence, comparisons are difficult to make because most of them are not equivalent. For many years we have endeavored to close these gaps by studying complete series or by completing the published data. At the moment we possess several stages of *Tupaia*, *Aotus*, *Alouatta*, and *Papio*. Unfortunately, the Cercopithecidae and the Pongidae are not well known presently.

A survey of all available material allows us to make some generalizations. It will be shown that Primates in general stand close to the "Bauplan" of the Eutheria; this is especially true for *Tupaia*, which, as a basal form, proves valuable as a starting point for morphological considerations. Among Primates, the Prosimii are characterized by a slight reduction of the ethmoidal region and of the sidewall of the braincase. True monkeys exhibit considerable diminution of the ethmoid and in some cases a total loss of the cartilages from the nasal capsule floor. Cebidae and Cercopithecidae represent nearly equal evolutionary grades and show manifold parallels (or homologies). As far as we know, Pongidae seem to have undergone further evolutionary change, and they differ from man only quantitatively. Our investigations have shown that it is essential to know more about late fetal stages of craniogenesis; so far only a few species are adequately known in this respect (*Propithecus*, *Tupaia*, *Alouatta*).

In Monotremata the lateral wall of the cranium between the nasal and the otic capsule is very completely developed. In the Theria this wall is much reduced and fenestrated for the passage of cranial nerves II–VII; between the foramina, bars of cartilage are preserved. In its dorsal parts the primary sidewall of the braincase is very complete in mammals, whereas it may be confined to a narrow taenia marginalis in reptiles. Parallel with the enlargement of the brain, in synapsids and mammals the dorsal part of the sidewall grew out as a broad cartilaginous plate which continues caudally into the lamina supracapsularis of the otic region. In therapsid reptiles, this lateral wall showed much coherency, but in Eutheria some reduction is evident. This results in a segregation into a platelike ala orbitalis and the narrow, striplike commissures (rostrally: comm. orbitonasalis; caudally: comm. orbitoparietalis) (Fig. 2). Near its origin from the trabeculae, the ala orbitalis is penetrated by the optic foramen, leaving a radix praeoptica and metoptica. Extreme reductions of the dorsolateral walls occur in Primates; this feature is foreshadowed in *Tupaia*. In young stages of *Tarsius* considerable remains of the commissures are preserved, and in Lemuriformes and Anthropoidea (= Simiae) scarce vestiges have been described.

From the anterior margin of the preoptic pillar, a narrow bar of cartilage may originate, which fuses with the septum or the cupula nasi and separates a supraseptal foramen from the common orbitonasal foramen. Reinbach (1952a, b) homologized this structure with the planum supraseptale of the reptiles, but Kuhn (1971) demonstrated that it is no ancestral feature. It is the orbitotemporal region where many variations and specializations occur, and it seems reasonable to call this cartilage simply the radix anterior of the pila praeoptica.

The mode of closure of the gap between ala orbitalis and capsula otica is principally different in Prototheria and Theria (for details see Kuhn, 1971). In lower vertebrates, two lateral processes arise from the basal plate and form the processus basipterygoidei (proc. alares); they may have a true articulation with the basal process of the palato-quadrate (Crossopterygii, Pelycosauria). The loss of kinetics of the skull leaves this articulation without function; the basipterygoid process and the palatoquadrate fuse and become incorporated secondarily in the wall of the neurocranium. This is the manner in which extracranial space, the cavum epiptericum with its semilunar ganglion, is added to the primary cavum cerebri. The ala temporalis of the mammals can be derived partly from these structures (Fig. 2).

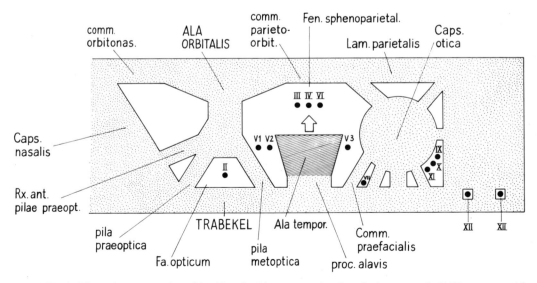

FIG. 2. Schematic representation of the sidewall of the neurocranium in eutherian mammals. II, III, etc. = cranial nerves.

A. Tupaiiformes

Van Valen (1965), Campbell (1966), McKenna (1966), and Szalay (1968) have recently argued against a close relationship between Tupaiiformes and Primates. Although they offer many good reasons for their leptictid hypothesis, they must admit that owing to the lack of fossils not all systematic problems are settled. We have discussed these

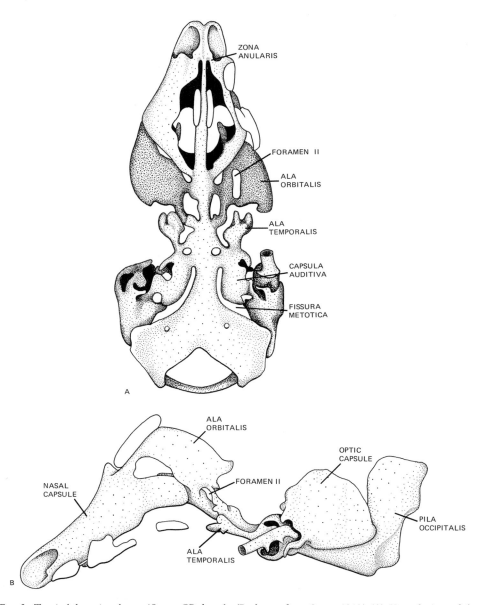

FIG. 3. *Tupaia belangeri* embryo, 15 mm CR length. (Redrawn from Spatz, 1964.) (A) Ventral view of the chondrocranium. (B) Side view of the chondrocranium; dermal bones removed.

questions at some length (Starck, 1974) and are inclined to accept some parallel evolution; we still think a comparative morphological approach to be worthwhile.

The skull of *Tupaia* is well known from several developmental stages. Chondrocrania have been reconstructed of *Tupaia belangeri* (15 mm, 59 mm; Spatz, 1964) (Figs. 3A and B; 4A and B) and *T. javanica* (20 mm; Henckel, 1928a). In many respects, the embryonic cranium of *Tupaia* resembles that of other primitive mammals. Henckel (1928a) has pointed out the strong resemblance with corresponding stages of Insectivora, but denied any close relationship between *Tupaia* and the Primates. A more detailed analysis reveals, however, that most characters in which *Tupaia* differs from the Primates are primitive features which are to be expected to occur in a true forbear of this order (Spatz, 1964; Starck, 1974). Since Henckel was not able to study more advanced stages of prosimians, reliable comparisons were not possible. Our observations on *Propithecus* (Fig. 7A and B) suggest that it can easily be derived from a morphological type like *Tupaia*.

The extant Tupaiidae are not direct forerunners of the Primates, of course, since they have had their independent phylogeny for more than 60 million years. The tupaiids have developed quite a few specializations of their own, but they also exhibit an adaptive pattern which seems to be symplesiomorphic with basic primates.

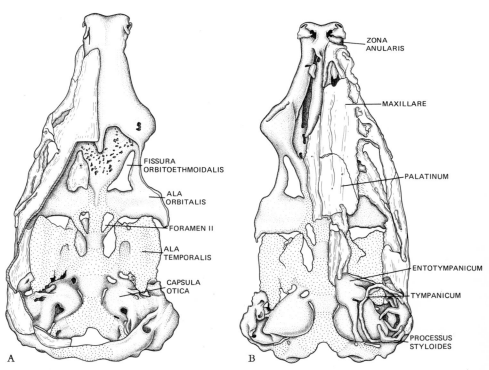

FIG. 4. *Tupaia belangeri* neonate, 59-mm CR length (redrawn from the original reconstruction by Spatz, 1964). (A) Dorsal view of the skull; dermal bones on the right side removed; endochondral bones stippled. (B) Same as above, ventral view.

The cranium of *Tupaia* is elongated and shows a division in two parts. The length-ratio of the nasal region to the entire cranium is 1:2.3 in the young stage, 1:2 in the neonatus, and 1:1.6 in the adult (Figs. 3 and 4). The maximum breadth is across the pars canalicularis of the otic capsule. The chondrocranium shows a marked declination of the facial skeleton against the base of the braincase, but in adults the skull is straightened. The orientation of the foramen magnum is vertical in early stages but is turned downward later on. In the 15-mm stage the sidewalls are still very incomplete and consist only of the ala orbitalis and the commissura orbitonasalis; a narrow commissura orbitoparietalis only exists for a short time (20 mm; Henckel, 1928*a*). This tendency to reduce the wall elements is distinctive of Primates and atypical for Insectivora. The primordial ala temporalis is very small at the beginning, but is considerably enlarged by periosteal bone deposition in order to close the lateral wall of the braincase.

The regio nasalis exhibits all characters typical for the chondrocranium of the macrosmatic Eutheria; there are only few signs of reduction (Fig. 5). The cupula nasi anterior is completely developed; the floor of the nasal capsule has all of its canonical elements: a broad lamina transversalis anterior with a narrow zona anularis, cartilagines ductus nasopalatini, paired cartilagines palatinae, cart. paraseptales, and in neonates even a cartilago papillae palatinae. The interior of the nasal capsule resembles the typical

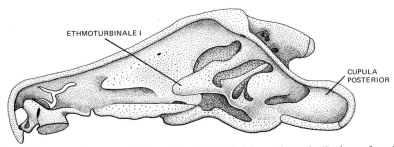

FIG. 5. *Tupaia belangeri* newborn, medial view of the right half of the nasal capsule. (Redrawn from Spatz, 1964.)

eutherian structure; each of the 3 ethmoturbinals has developed an epiturbinal, and 2 frontoturbinals are present as well. The cupula posterior extends far backward, and a lamina transversalis posterior is well developed (Fig. 5).

Caudally the nasal septum forms an unmistakable interorbital septum (Fig. 4); it is already present in the 15-mm stage (Spatz, 1964) and persists through to adult stages. Henckel (1928*a*) did not find an interorbital septum in his 20-mm embryo, and concluded from that a systematic remoteness of *Tupaia* from the Primates. However, De Beer (1937) found that in Henckel's reconstruction a septum interorbitale was discernible, and Henckel's statement seems to be a misinterpretation of the facts. There can be no doubt about the existence of an interorbital septum in the Tupaiidae (Spatz, 1964; Starck, 1974), although there might be some differences between various species.

We were able to demonstrate that the interorbital septum is a plastic structure which depends on the size and orientation of the eyes, the development of the nose, and the declination of the skull base. The relations of the different head organs vary throughout ontogenesis, and therefore significant morphological changes may occur as well. *Alouatta*, for example, does not show a clearly defined interorbital septum in younger stages (20, 36, and 43 mm); it is well developed at 130 mm, however, and lost again later on. *Propithecus* has medium-sized, laterally directed eyes and an enormous nasal capsule; hence, an interorbital septum is missing—a rare case among Primates. In species with large eyes and reduced olfactory system, the septum occurs regularly (*Tarsius*, *Cercopithecus talapoin*, etc.) (Fig. 6). From this it is clear that such a variable structure cannot serve as a valuable diagnostic character.

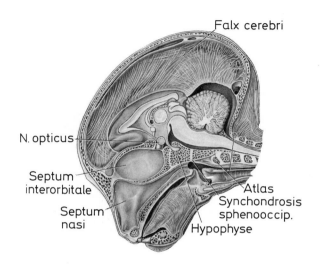

FIG. 6. *Cercopithecus (Miopithecus) talapoin.* Sagittal section of the head of a subadult specimen to show the interorbital septum. (From Starck, 1974.)

The optic capsule is linked with its surrounding by 6 commissures; among these a commissura alicochlearis and a comm. basicapsularis anterior are developed very early. The tegmen tympani is comparatively large, and it is pierced by the stapedial artery. The existence of a true entotympanic is proven in *Tupaia*. Saban (1956/1957) was not able to find an independent entotympanic in this genus, and he suspected that the bulla might be formed by the petrosal alone. It was Spatz (1964) who demonstrated a connection between the cartilage of the auditory tube and the entotympanic (Fig. 4); in 8-day-old embryos it is linked with the tegmen tympani. It does not show any relationship with the hyoid arch or other parts of the visceral skeleton. The entotympanic encloses the tympanic ventromedially, and is connected with the processus mastoideus of the oticum posteriorly.

To summarize, it may be stated that in general the chondrocranium of *Tupaia* does not exhibit pronounced specializations but is characterized more by its plesiomorphies. Compared with the Insectivora it is close to the primates in the following characters:

1. The nasal capsule still shows all of the basic structures but seems to be slightly reduced. There is a well-developed spina mesethmoidalis; the crista galli is low and short as in the prosimians.

2. By the formation of an interorbital septum and by the morphology of the whole orbital region.
3. The connections of the otic capsule are very comparable with those of the Malagasy prosimians.
4. The tegmen tympani is more advanced and resembles that of the Lemuridae.
5. The reduction of the primary sidewall of the braincase begins at an early stage and results in the disappearance of the orbitoparietal commissure.

B. Strepsirhini

The cranium of the Strepsirhini is unsatisfactorily known; some stages are not adequately discussed in the literature (*Microcebus*, *Avahi*), and some are scarcely mentioned. Therefore our considerations have to be based mainly on facts derived from an older fetus of *Propithecus* (Starck, 1962) (Fig. 7A and B).

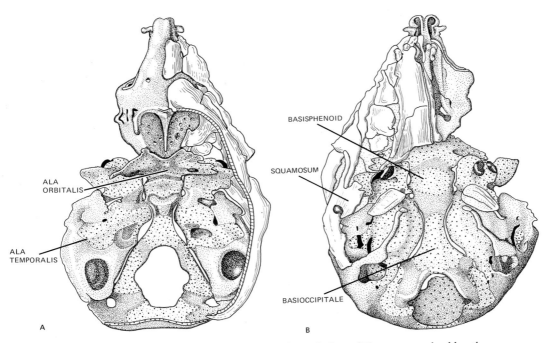

Fig. 7. *Propithecus sp.* Dorsal (A) and ventral (B) views of the skull base of a fetus of 26-mm greatest head length. (Redrawn from Starck, 1962.)

The skull base of this late fetus is only slightly declined, and differs from adult stages by a few degrees only; the early embryonic declination seems to be compensated for at a very young age, whereas in monkeys this process takes a longer time. The narrow nasal capsule is contrasted by the rounded braincase, which also overlays the posterior third of the regio nasalis. The regio ethmoturbinalis of the nose is very broad, especially in the Indriidae. The forehead is vaulted and forms an angle with the dorsum of the nose;

this angle may be straightened by the development of frontal sinuses in later stages. The base of the cranium is narrowed between the two large otic capsules which form part of the lateral walls. In *Propithecus* we find a long fissura basicochlearis, whereas *Nycticebus* shows a solid connection between otic capsule and basal plate. The hypophyseal fossa is well developed, and it is bound posteriorly by a dorsum sellae with processus clinoidei posteriores. The commissura alicochlearis is missing in the young stage of *Nycticebus*, but is always encountered in *Microcebus*, *Avahi*, and *Propithecus*.

In *Nycticebus* and *Propithecus* the alae temporales are more extended than the alae orbitales (Fig. 7); hence, they are true alae majores. The tegmen tympani is larger than in *Tupaia*. The interorbital septum is present in *Nycticebus*, but is not as defined as in monkeys; this is also true for *Microcebus*. In *Avahi* it is very short, and it is entirely missing in our stage of *Propithecus*; this is probably a consequence of the enormous extension of the rear parts of the nasal capsule. According to Frets (1914), the interorbital septum is discernible in a fetus of 55 mm.

Characteristic of the Primates is a nearly complete reduction of the primary sidewalls of the braincase. *Nycticebus* shows a much-extended lamina parietalis and a small triangular ala orbitalis, which is connected with the nasal capsule by a narrow commissura spheno-ethmoidalis. An orbitoparietal commissure is missing entirely, but a short processus orbitoparietalis anterior occurs as a rudiment. In *Microcebus* the ala orbitalis is comparatively large, and a commissura sphenoethmoidalis is also present; the lamina parietalis contributes much to the formation of the lateral sidewalls. In *Avahi* and *Propithecus* we encounter similar conditions: the primary lateral border of the braincase is built by comparatively small alae orbitales and laminae parietales (which are quite large in *Avahi*). Both commissures are missing in these species. The nasal capsule of *Propithecus* shows a cartilaginous process, which should be considered as an atavism.

Lemur catta is known by a brief description of a young stage of 25 mm (Henckel, 1928b). It shows a commissura sphenoethmoidalis, but the orbitoparietal connection is lacking. It still exhibits a clearly distinct interorbital septum, which is much reduced in the newborn.

The nasal region deserves consideration with regard to the structure of the floor cartilages and its general construction. In prosimians the former elements are much more developed than those of the Cercopithecoidea and the Hominoidea; their ethmoturbinal apparatus is intermediate between the Tupaiiformes and the Anthropoidea (= Simiae).

The floor of the primary nasal capsule of mammals is always incomplete; it is formed by the lamina transversalis posterior and anterior, as well as by the paraseptal cartilage. The latter element differentiates independently and then fuses with the lamina transversalis anterior (*Tupaia*, *Propithecus*, *Lemur*, *Galago*). The paraseptal cartilage occurs in prosimians and simians, but it is much more elaborated in the former group. In lacertilian reptiles the paraseptal cartilage forms a continuous bar which connects the lamina transversalis anterior with the planum antorbitale. It is difficult to explain the morphology of this cartilage simply by the differentiation of the vomeronasal organ of Jacobson; this olfactory organ is primarily connected with the lamina transversalis anterior and with the anterior parts of the paraseptal cartilage (Gaupp, 1906; Seydel, 1891). In mammals, a continuous bar is only encountered in the kangaroo *Macropus*. In marsupials there is a strong functional and morphological interdependence between the organ and the cartilage of Jacobson; but this seems to be a derived condition. Here the cartilage is closely connected with the vomer and may have supportive tasks as well. The complicated morphology of this skeletal element in mammals therefore has to be regarded as a progressive feature. After the loss of the vomeronasal organ in Sirenia and Catarrhini, the paraseptal cartilage may persist as an insignificant bar.

The aboral part of the solum nasi is formed by the lamina transversalis posterior; it may give origin to a processus paraseptalis posterior (*Tupaia, Galago, Loris*), but this structure is missing in *Propithecus* and *Nycticebus*. The lamina transversalis anterior usually connects the paries nasi with the lower margin of the septum nasi and thus separates the fenestra narina from the fenestra basalis. This ringlike structure (zona anularis) is present in *Tupaia, Loris, Avahi,* and *Propithecus,* but it is lacking in *Galago* and *Microcebus*. The lamina transversalis anterior, which is very constant, is connected with more cartilaginous elements of the nasal floor. The cartilago ductus nasopalatini encloses the nasopalatinal duct dorsally, laterally, and anteriorly; it is normally linked with the paraseptal cartilage and the lamina transversalis anterior, and it is shifted below the processus palatinus of the premaxilla. Since this cartilage is missing in monotremes and marsupials, but usually occurs in the Eutheria, it might well be a secondary acquisition of this group. Obviously it is a special structure connected with the nasopalatinal duct, which is derived from the paraseptal cartilage. It is usually well developed in the prosimians (*Tupaia, Galago, Propithecus,* and *Avahi?*), but often missing in advanced Anthropoidea.

An ancient and independent structure, the cartilago palatina, is situated behind the nasopalatinal duct, and may be fused with the lamina transversalis anterior or the cartilago ductus nasopalatini. It is derived from the cartilago ectochoanalis of the reptiles, and it regularly occurs in prosimian taxa. The cartilago papillae palatinae is an unpaired element which, histogenetically, is formed by secondary cartilage; it occurs in many primitive mammals and was observed in a newborn *Tupaia*.

The nasal cavity is organized in three parts: pars anterior, pars intermedia, and pars posterior. The latter is situated below the cerebrum, and it is a new acquisition of the mammals. It contains the recessus ethmoturbinalis and frontoturbinalis. The roof of the pars posterior is formed by the lamina cribrosa. The newly acquired ethmo- and frontoturbinalia are very constant as to their number and arrangement. The ethmoturbinal I is formed at the borderline of the partes intermedia and posterior. In *Tupaia* we still find 3 ethmoturbinals, of which the first is large and extended anteriorly; ethmoturbinale I and II have additional epiturbinals. Two frontoturbinals are to be found in *Tupaia*. The free lower margin of the anterior paries nasi is turned inward, and it is called maxilloturbinal; in front of the lamina transversalis anterior it is continued as atrioturbinal. The nasal morphology of the prosimians is easily derived from these conditions in *Tupaia*; the differences are mainly of a quantitative nature. However, only the nose of *Propithecus* has been described in some detail; here, besides the large maxilloturbinal, only two ethmoturbinals and one frontoturbinal are present. The nasoturbinal, which is well developed in *Tupaia*, only consists of a ridge.

C. Tarsiiformes

Tarsius is an extremely specialized primate and many of its special adaptations are manifested in its cranial morphology. Henckel (1927*a*, *b*) and Fischer (1905) have described the cranium of a young embryo of 24 mm. We are able to add some new information from a stage of 55 mm (reconstructed by D. Wünsch, unpublished) (Fig. 8A and B). These new facts demonstrate that it is very difficult to draw conclusions solely from a single stage, and that craniogenesis must be conceived as a process.

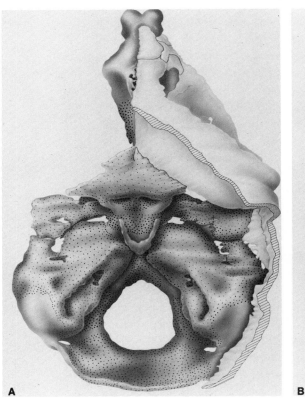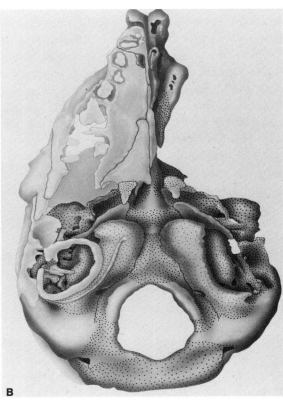

A   B

FIG. 8. *Tarsius bancanus borneanus*, 55-mm CR length. Reconstruction of the skull, dorsal (A) and ventral (B) views. Dermal bones removed on left side. (Courtesy D. Wünsch, unpublished reconstruction.)

Tarsius differs from all other mammals by the very unusual mass relations of its head organs. The volume of one bulbus oculi surpasses that of the whole brain (Starck, 1953). The eyes are oriented in a frontal plane. The brain is characterized by its primitiveness and at the same time, by the dominance and elaboration of all optic systems. It is to be expected, of course, that these peculiarities should influence the whole construction of the skull.

In the younger stage of Henckel, the basal plate is strongly narrowed in a rostral direction and declined ventrally. The pars vestibularis of the otic capsule forms part of the lateral wall of the braincase, and the occipital arch is turned backward. The brain is elevated rostrally, and hence, belongs to the clinocranial type (Hofer, 1952). The nasal capsule is small and scarcely developed in its aboral parts. Therefore, we find a very large septum interorbitale. A singular feature among primates is the presence of a very complete primary sidewall of the braincase, which shows only short interruptions of the taenia orbitoparietalis. The orbitoethmoidal commissure is missing. The lamina transversalis anterior forms a narrow zona anularis; cartilago paraseptalis and cartilago ductis naso-palatini are well developed, whereas the lamina cribrosa and the conchae are not yet differentiated.

In the older stage, the sidewalls are much reduced (Fig. 8) and are only formed by the small ala orbitalis. The ala temporalis is large and nearly ossified; its base is pierced by the foramen rotundum. The radices praeopticae are fused to form a planum supraseptale of some extension. The interorbital septum is of considerable size. Henckel (1927a) stated that in *Tarsius* a lamina cribrosa is missing and that there only exists a foramen olfactorium evehens. So far, even in adult skulls a cribral plate has been described only by Frets (1913), but his findings are overlooked in recent publications. Our new investigations have proven the existence of this structure, although it is very small and delicate; there is a considerable distance between brain and lamina cribrosa, however (Figs. 9–11). Since the bulbus olfactorius is small and scarcely rises above the lobus frontalis, this distance has to be bridged by a long nervus olfactorius; it runs through an osseous canal mainly formed by the frontal bones and is called tubus olfactorius (Spatz, 1968) (Figs. 9–11).

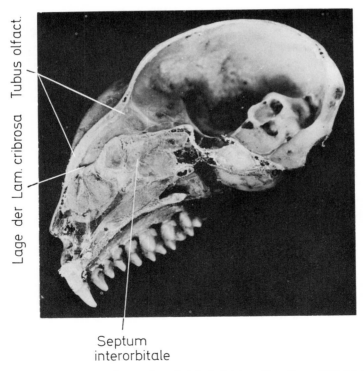

FIG. 9. *Tarsius bancanus borneanus.* Median section of adult skull (photo from Spatz, 1968, labeling changed).

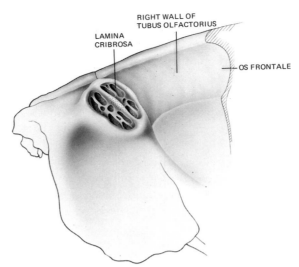

FIG. 10. *Tarsius bancanus borneanus*, 55-mm CR length. Same reconstruction as Fig. 8 (courtesy of D. Wünsch). Lamina cribrosa and right side of olfactory tube. The left frontal bone is removed.

The enormous size of the eyeballs causes much shifting and extension of the skull regions between the nasal capsule and the braincase. The tubus olfactorius belongs to the cavum cranii and corresponds to the spatium supracribrosum, which is bound by a secondary floor, the lamina cribrosa. This lamina is newly evolved in mammals. Spatz (1968) was wrong when assuming for *Tarsius* a topography of this region entirely different from other Eutheria; it is easily explained by the general transformations of the orbito-nasal region. In spite of the small size of the nasal capsule, the older cranium shows well-developed floor cartilages. There persists a narrow zona anularis, and the cartilago paraseptalis is nearly as long as the whole fenestra basalis; it is connected with the lamina transversalis anterior and the large-sized cartilago ductus nasopalatini. Lateral to the anterior end of the paraseptal cartilage, a small, isolated cartilage is present, which corresponds to the "outer bar" of Broom (1915) (= fibula reuniens Reinbach); it is regularly encountered in basal mammals, but in primates is only known in *Tupaia* and *Galago senegalensis*.

Jacobson's organ has been studied with regard to both fetal and adult stages (the findings will soon be published by D. Wünsch). The organ possesses a well-functioning sensory epithelium sporadically forming island-like zones. It is well equipped with glands and nerves (Fig. 12). The paraseptal cartilage levels off rostrally and only forms a base above the organ, ending directly above the incisive duct (Fig. 12C). Also with respect to the adult, the incisive duct is wide open and empties into the roof of the oral cavity beside an incisive papilla (Fig. 12D).

The posterior end of the nasal capsule is compressed by the large eyes, and it is open; the cupula posterior is incomplete and a lamina transversalis posterior is missing. The interior of the capsule is narrowed and the conchae are much reduced; maxilloturbinale and nasoturbinale are present as well as the ethmoturbinal I, but there are no recessus except the r. maxillaris.

The existence of an entotympanic is not yet proven. In the younger stage the tympanic region does not show any peculiarities and an entotympanic is lacking. In the older specimen there exists a floor of the middle-ear cavity, formed by enchondral bone; it is closely connected with the otic capsule, but it shows a rostral extension near the auditory tube. The ossified bone still contains some cartilaginous tissue, and this would indicate the presence of a true entotympanic, since the petrosal components of the floor are usually formed by periosteal bone. These results are in accordance with the findings of Van Kampen (1905).

To summarize, it may be stated that the cranium of *Tarsius* is mainly characterized by the structural influences of the optic and otic sense organs. Despite the general reduction and compression of the nose, its floor elements are as completely developed as those of other prosimians. All peculiarities of the cranium of *Tarsius* are more constructive modifications of a basic type, which is very similar to that of the Lemuriformes.

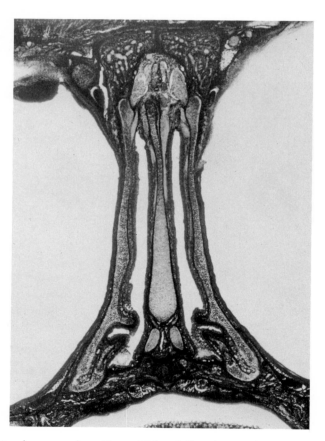

FIG. 11. *Tarsius bancanus borneanus* embryo, 55-mm CR length. Posterior part of nasal cavity and lamina cribrosa
(× 50).

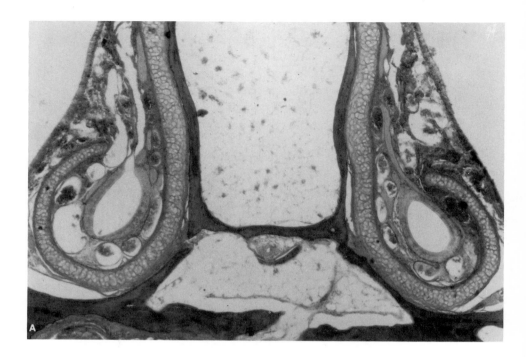

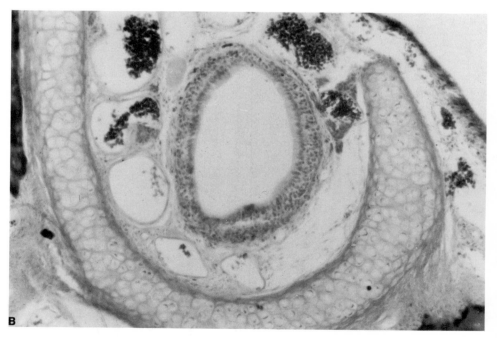

Fig. 12. *Tarsius bancanus borneanus* adult. (A) Jacobson's organ and cartilago paraseptalis; (B) Jacobson's organ and cartilago paraseptalis; (C) Opening of Jacobson's organ into the anterior part of the nasal cavity; (D) Ductus incisivus. Open communication between the nasal cavity just below the opening of Jacobson's organ and the oral cavity.

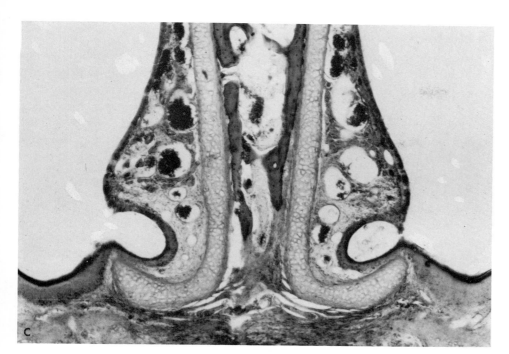

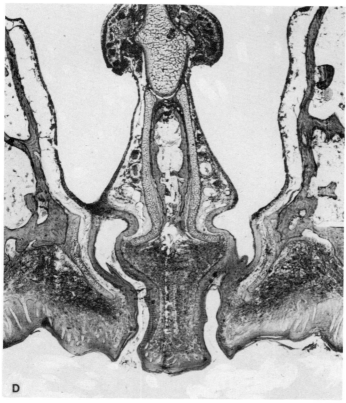

D. Platyrrhini

At the present moment there is no information available as to the cranial develop-ment of Callitrichidae; one reconstruction of *Callithrix* is to be finished soon, however. Here, we have to confine our discussion to the Cebidae. In all embryos we notice a different proportioning of the cranium in comparison with those of the prosimians. The nose is absolutely and relatively shorter in the South American monkeys, and it is situated subcerebrally. In young stages the facial skull is much declined, but the base of the skull is straightened with increasing age. The braincase is broadened and its primary sidewall is never complete.

Some of the more important features may be enumerated: The tectum posterius and the foramen magnum show a vertical position in young stages of *Aotus* (Fig. 13A), but are tilted backward later on. The base of the skull in between the otic capsules is narrow in *Aotus*, but wide in the other taxa. The commissura alicochlearis is missing in *Aotus*, *Saimiri*, and *Pithecia*; in *Alouatta* there are short processes which may be remin-iscent of this commissure. The fissura basicochlearis usually is very short.

The septum interorbitale is always met with in Callitrichidae and in *Saimiri*, *Pithecia*, *Aotus*, and *Ateles* as well; it is not developed in *Cebus*, according to De Beer (1937), but I have seen it in subadult specimens. The rapidly changing conditions in *Alouatta* have been discussed above.

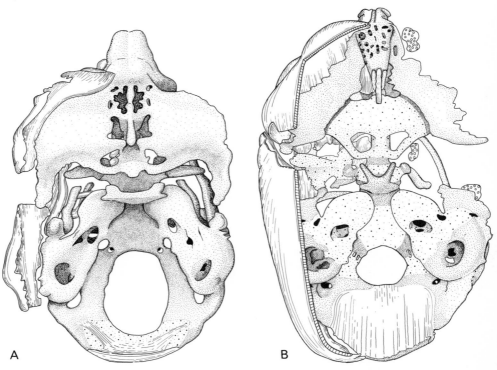

FIG. 13. *Aotus trivirgatus*. (A) Embryo, 25.5 mm; dorsal view of skull base. (B) Embryo 76 mm; dorsal view of skull base.

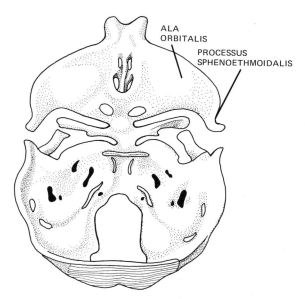

ALA
ORBITALIS

PROCESSUS
SPHENOETHMOIDALIS

FIG. 14. *Saimiri sciureus* embryo, 24-mm CR length. Dorsal view of the base of the chondrocranium. (Redrawn from Henckel, 1928*b*.)

The primary sidewalls are most complete in young stages of *Saimiri* (24 mm, Henckel) (Fig. 13) and *Aotus* (25 mm) (Fig. 14); here it is formed by a lamina parietalis, a large ala orbitalis, and a very broad commissura orbitoethmoidalis. *Aotus* shows an anterior orbitoparietal process; it may also be present in *Saimiri*. In *Pithecia* and *Alouatta* these structures appear to be more reduced. The tegmen tympani is well developed in older stages.

The structure of the nasal region is very noteworthy in the platyrrhines. Generally speaking, all floor elements are as completely developed as in the prosimians; moreover, the cupula anterior is very complete and may be even secondarily enlarged; hence, the nasal openings are pushed to the sides. The great distance between the external nasal apertures (which gave the name Platyrrhini to the group) therefore does not depend on a broadening of the nasal septum, but on the progressive enlargement of the capsula nasi.

The lamina transversalis anterior is strongly developed, and it forms a zona anularis in *Aotus*, *Callicebus*, *Saimiri*, and *Alouatta*; this is in contradiction to a statement of Henckel (1928*b*). Cartilago paraseptalis, cart. ductus nasopalatini, and cart. palatinae are always present and show a highly complicated structure in some species (W. Maier, personal communication). Contrary to a widespread misunderstanding, Jacobson's organ is generally well developed with regard to platyrrhines and functions satisfactorily in the adult phase. This is possibly correlated with the formation of various skin glands in platyrrhines (perigenital glands in Callitrichidae, *Pithecia*, and *Alouatta*; sternal glands in *Ateles*). Wide-open incisive ducts into the oral cavity are found (W. Maier, unpublished results) in fetal, postnatal, and adult Platyrrhini (demonstrated with respect to *Callithrix*, *Oedipomidas*, *Pithecia*, *Saimiri*, *Alouatta*, and *Ateles*). This is clearly contrary to the formation of the corresponding structures in Catarrhini. In Platyrrhini the sensory epithelium of Jacobson's organ is lower and spread more island-like in comparison with

Strepsirhini. The interior architecture of the nasal cavity exhibits the basic plan of the Eutheria, but is much reduced, as is the whole olfactory system. But there are still well-developed maxillo-, atrio-, and nasoturbinalia, as well as two ethmoturbinalia, the second of which is very small.

E. Catarrhini

1. Cercopithecidae. The chondrocranium of the cercopithecids is not well known. Reconstructions of total crania are only available for *Macaca fascicularis*, *Presbytis auratus* (Fig. 15), and *Papio hamadryas*; all of them are very young stages. *Macaca* and *Presbytis* in many respects are similar, whereas *Papio* differs in many points from these taxa. This suggests that one should be very cautious about systematic conclusions from single specimens.

All crania are compressed rostrooccipitally, which causes rounded contours as in human chondrocrania. The facial skull is much declined in young stages. The basal plate is broader at the otic capsules than in the platyrrhines; that is especially true for *Papio*. An interorbital septum is present in all known cercopithecoids; it is a small area in the young

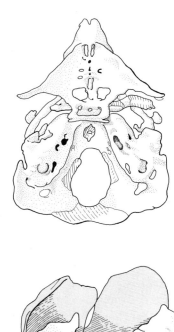

FIG. 15. *Presbytis* (= *Semnopithecus maurus*) embryo, 53 mm. Dorsal and left side view of the skull. (Redrawn from Fischer, 1903.)

baboon, but much more extended in an older embryo. The size of the interorbital septum seems to be correlated with the decrease in declination of the facial skull.

The fissura basicochlearis has become very insignificant, as compared with the prosimians and platyrrhines. A commissura alicochlearis has only been demonstrated in *Papio* so far. A tegmen tympani is always present, at least in older stages. The sidewalls are very much reduced in catarrhines. The ala temporalis is smaller than in the platyrrhines; an orbitonasal commissure is present. In *Presbytis* and *Papio* a foramen praechiasmaticum and a radix anterior pilae praeopticae are encountered; these structures are missing in other primates. The lamina parietalis is not well developed, and an orbitoparietal commissure is always missing.

Henckel (1928*b*) noticed that in the catarrhines the cartilages of the nasal floor are mostly reduced; in fact, this is indeed true, but older stages show the presence of more elements than might be gathered from the literature. The final differentiation of the anterior nasal cartilages may well occur in very late stages of pregnancy. In all material so far studied, the nasal capsule is completely open from below; however, a lamina transversalis anterior exists in both stages of *Papio*, which also show a cartilago ductus nasopalatini and a short cartilago paraseptalis (*Macaca*; *Papio*, 33 mm). The latter was missing in the older stage of *Papio anubis*, as was the entire organ of Jacobson.

The interior of the nasal cavity has been studied in some detail only in *Papio*. A semicircular crest separates the anterior from the intermediate part. The maxilloturbinal and 2 ethmoturbinals are well developed; the ethmoturbinale I separates the interior into a recessus ethmoturbinalis and a rec. frontoturbinalis. The latter is connected with the recessus lateralis, which in older stages is equipped with a large glandula nasalis lateralis.

Frets (1914) has emphasized that the nose of the catarrhines generally exhibits a greater amount of reduction than that of the platyrrhines; this holds true especially for the floor cartilages and the cupula anterior.

2. Pongidae. The investigation of human crania has shown that there may exist a high variability in the development of some structures. Therefore it is highly regrettable that only one single cranium of pongids has been studied so far (*Pan troglodytes*; Starck, 1960, 1973). It shows many similarities with corresponding stages of human embryonic skulls. The differences are mainly quantitative. Compared with the crania of cercopithecids, the skull is much more rounded, and the nose is situated subcerebrally to a high degree.

The primary sidewalls of the chondrocranium are reduced: the alae orbitales are small in comparison with the alae temporales, which are enlarged by periosteal bone formation. The orbitonasal commissure is broader than in *Homo*; the orbitoparietal commissure is missing. The fissura basicochlearis is longer than in most of the other primates and in man. Anterior and posterior basicochlear commissures are present, but a commissura alicochlearis is lacking. The otic capsule is relatively larger than in *Homo*, and it is oriented more vertically. Near the parotic crest a foramen is formed for the chorda tympani; the tegmen tympani is very large. The canalis craniopharyngeus, which persists in 40% of adult chimpanzees and in 0.5% of humans, is still opened. In *Pan* the os frontale contributes relatively more to the formation of the cranial vault than in *Homo*.

The nasal capsule is more reduced than in the cercopithecids. The cupula anterior usually is more differentiated than in *Homo*, but a 75-mm stage of the latter exhibited a very complete cupula as well (Peter, 1913); a high variability is to be expected for this region. The cartilages of the nasal prominence (cart. alaris) are progressively developed in *Homo*. The solum nasi is very incomplete and both laminae transversales are missing. The organ of Jacobson and the paraseptal cartilages are present, but both structures are vastly independent from each other; the paraseptal cartilage shows a rounded cross section and does not form a groove anymore. The rudimentary organ of Jacobson is situated more dorsally at the side of the nasal septum; the paraseptal is continued rostrally by a cartilago ductus nasopalatini, although there is no ductus nasopalatini. The maxilloturbinal and 2 ethmoturbinals are differentiated; the recessus lateralis is separated into a large rec. maxillaris (anteriorly) and a small rec. frontalis (posteriorly).

III. Discussion

The presently known facts of chondrocranial morphology do not allow for many conclusions about phylogenetic relationships. The chondrocranium of the primates does not show basic differences from the general eutherian pattern. Peculiarities are derived from quantitative shifts of proportions, caused by specializations of sense organs, brain, and masticatory apparatus. Nevertheless, different evolutionary *grades* within this order can be characterized without difficulty.

Tupaia closely resembles other basic mammalian forms; its skull morphology is dominated by many plesiomorphic eutherian characters, and it could well serve as a starting point for several adaptive lines. The whole genus still shows many macrosmatic specializations, but there are also signs of reduction of this sense; the optical system shows a progressive tendency. Both features are characteristic of the primates. Contrary to statements in the older literature, *Tupaia* has an interorbital septum. The cranial morphology would suggest a connection between Lemuridae and the Tupaiidae. In this context we should recall the reduction of the primary walls of the braincase and the structure of the ectotympanic; here, the inclusion of this bone and of the tympanic membrane inside the bulla is unusual, but both groups have this character in common. It would be useful, however, to know more about the physiology of the middle-ear region, but there can be no doubt that both tupaiids and lemurids assume an exceptional position. We now have proof for the participation of the entotympanic in forming the tympanic bulla of tupaiids; however, there is no certainty, so far, about its occurrence in prosimians. Therefore, the bulla of tupaiids does not appear to be homologous with that of Lemuriformes. The lemur skull stands close to that of the tupaiids, but the reduction of the nasal skeleton is more advanced, and the progressive expansion of the brain results in changes of skull proportions.

Our comprehension of the chondrocranium of *Tarsius* is considerably extended by the study of more advanced stages. *Tarsius* is characterized by the enormous enlargement of its eyes and by one-sided specializations of the brain. The brain shows a low grade of cerebralization, on which hypertrophized optic systems are superimposed. The nose exhibits reductions, mainly of its posterior parts; nevertheless, the olfactory sense must play an important role in the orientation of this primate (skin glands!). Therefore, the anterior parts of the nose and the organ of Jacobson are well developed. Most of the peculiarities of the skull structure of *Tarsius*, namely the formation of the tubus olfactorius, the orbit, and the sidewalls of the braincase, are to be explained by the mass relations of the head organs.

The crania of the platyrrhines and catarrhines show many similarities, which may indicate analogous adaptations. The most dominant features are connected with the progressive expansion of the brain and with the reduction of the olfactory system. The reduction of the primary sidewalls, enlargement of the ala temporalis, and the subcerebral position of the nose may well be parallel developments. A subtle analysis reveals numerous divergent evolutionary trends; we refer to the different morphology of the rostral nasal regions, which certainly depends partly on the presence of the vomeronasal organ in platyrrhines, and its absence in catarrhines (an apomorphic condition). Systematists of the old days have rightly chosen these features as diagnostic for both groups. In spite of many parallel and convergent characters, both taxa must have evolved independently for a very long time.

Our knowledge of the pongid cranium is still too incomplete to allow conclusions to be drawn on the phylogenetic relationships of these primates.

Finally, we have to correct Henckel's (1928b) view of the chondrocranium as a functionless organ which ought to be well suited to the study of phylogenetic questions. Instead, the chondrocranium strongly depends on the formation of the other head organs, and it must be conceived as a transitory stage of skull development. Hence, it is impossible to make general judgements which are based on single stages and which do not consider the adaptations of the primordial head skeleton itself. We have tried to contrast the primate chondrocranium with the basal eutherian pattern (insectivores), and to elaborate plesiomorphic and apomorphic features. Many parallelisms and convergent developments further complicate the whole setup, which may be illustrated by the comparison between platyrrhines and catarrhines. Well-founded phylogenetic conclusions will have to take into consideration the whole complex of morphological characters, as well as their constructions and functional analysis.

IV. Conclusions

The above discussion illustrates that phylogenetic relations of the different primate groups should not be derived from a single complex of features. Phylogenetic conclusions from the morphology and development of the chondrocranium are only admissible if they can be inserted into the larger scope of a comprehensive analysis. Special adaptations may be manifested very impressively in the cranium and have to be taken into consideration. The findings of the cranium completely correspond to the concept derived from other systems on the phylogenesis of the primates. In detail we refer to the following conclusions:

1. The cranium of the Tupaiidae is on the whole closely connected with that of the

basal Eutheria. Some findings (auricular region, lateral wall, beginning reduction of the nose, tegmen tympani) point at specializations also to be found in the primates. It is advisable to keep the family Tupaiidae as a separate taxon beside the lipotyphlous Insectivora and the Macroscelididae and not to include it in the classification of primates, although the total organization of the family may serve as a clear model of ancestral primates. The exceptional position of the Tupaiidae is based on new findings of the early ontogenesis (Kuhn and Schwaier, 1973; Luckett, 1974) and the biology of behavior (Holst, 1969; Martin, 1972).

2. The classification at the subordinal level of Strepsirhini and Haplorhini may also be supported from the findings of the cranium. A superficial consideration suggests that the Strepsirhini assume an intermediate position between Tupaiidae and Anthropoidea with respect to the cranium. The basal position of this group on the whole and the abundance of plesiomorphic features make this impression understandable, although there is no doubt (considering all the available biological evidence) that the Anthropoidea are not descendants of actual Strepsirhini in the present meaning (contrary to Gingerich, 1973).

3. The hypothesis of Martin (1973), that only the Tupaiidae and the Strepsirhini possess an open ductus nasopalatinus and a functional Jacobson's organ, is not confirmed. An open incisive duct and a functional Jacobson's organ also occur in *Tarsius* and many Platyrrhini, and are consequently no key features of the Strepsirhini. Physiological and ethological investigations (sexual behavior) will have to clarify the significance of this organ.

4. The recent genus *Tarsius* is specialized to such an extent that the phylogenetic evaluation of its skull is considerably complicated. The enormous enlargement of the eyes and the formation of extended central and optical regions with regard to relatively inferior cerebral level on the whole lead to a distortion of the cranial contours, requiring at the same time new constructive solutions. The regression of the nose in the olfactory region, correlated with the formation of immense eyes, leaves the rostral, basal part of the nose surprisingly intact. Open incisive ducts with respect to a functional Jacobson's organ will still be found in adults. Evidence for the occurrence of an entotympanic requires the discovery of new findings in suitable material. A lamina cribrosa is present. The distortion of the skull with regard to reduced olfactory nerves involves the formation of the special tubus olfactorius. The findings of the skull justify the exceptional position of *Tarsius* opposite Strepsirhini and its association with Haplorhini.

5. The cranium of Platyrrhini and Catarrhini is on principle constructed in a very similar manner, because they represent similar adaptive types and comparable evolutionary levels. More important are decisive differences with regard to representatives of both groups. Such differences concern the more complete formation of the lateral wall with regard to young stages, and especially the complete formation of the rostral cartilage of the nasal base with respect to a functional Jacobson's organ with open incisive duct, characteristic of all platyrrhines investigated until now. In contrast, the regression of the nose in Cercopithecidae and Pongidae is more pronounced, and the cartilages of the nasal base are considerably simplified. A completion of that occurs, however, in late developmental stages (heterochrony). Paraseptal cartilages and Jacobson's organ regress

early during ontogeny. Peculiarities of the Pongidae (only one stage of *Pan* is known now) result from increasing neencephalization and from a further regression of the nose as an olfactory organ.

TABLE I

DEVELOPMENTAL STAGES OF THE CHONDROCRANIUM UTILIZED IN THE STUDY

	Crown–rump length (mm)	Head length (mm)	Reference[a]	
Tupaiiformes				
Tupaia javanica	20		Henckel, 1928a	R
Tupaia belangeri	15	8	Spatz, 1964	R
	59 (neonat)	25	Spatz, 1964	R
Strepsirhini				
Lemuridae				
Microcebus murinus	13, 25		Bähler, 1938	R
Lemur catta	25		Henckel, 1928b	S
Indriidae				
Avahi laniger	39	17	Frei, 1938	R
Propithecus spec.		26	Starck, 1962	R
Lorisidae				
Nycticebus coucang	30		Henckel, 1927	R
Loris tardigradus	6, 5–30 (10 stages)		Ramaswami, 1957 clarified	
Galago senegalensis	30 only nasal region		Eloff, 1951	S
Tarsiiformes				
Tarsius spectrum (species ?)	24		Fischer, 1905	
			Henckel, 1927	R
Tarsius bancanus borneanus	55		Wünsch, unpublished	R
Platyrrhini				
Callitrichidae				
Callithrix jacchus	34		Henckel, 1928b	S
Cebidae				
Saimiri sciureus	24		Henckel, 1928b	R
Aotus trivirgatus	25, 5 76		Starck, unpublished	R
Pithecia monacha		29	Tobias, unpublished	R
Alouatta spec.	29		Henckel, 1928b	S
Alouatta caraya	36		Kummer in Starck, 1967	R
Ateles geoffroyi	51	19	Maier, unpublished	R
Saimiri				
Alouatta				
Ateles	only nasal region		Frets, 1913	
Catarrhini				
Cercopithecinae				
Macaca fascicularis	25		Fischer, 1903	R
M. fascicularis	only nasal region		Frets, 1914	
Papio hamadryas	33		Reinhard, 1958	R
Colobinae				
Presbytis auratus ("Semnopithecus maurus")	53		Fischer, 1903	R
Nasalis larvatus	only nasal region		Frets, 1914	
Pongidae				
Pan troglodytes	71	31	Starck, 1960, 1973	R

[a] R = reconstruction; S = slides.

V. References

DIETRICH
STARCK

AUGIER, M. 1931. Squelette céphalique, pp. 89–654. *In* P. Poirier and A. Charpy, eds., *Traité d'Anatomie Humaine*, Vol. 1, 4th ed. Masson, Paris.

BÄHLER, H. 1938. Das Primordialcranium des Halbaffen *Microcebus murinus*. *Diss. Auszug Med. Fak. Bern.*

BROOM, R. 1915. On the organ of Jacobson and its relations in the Insectivora. *Proc. Zool. Soc. London* **1915**:157–347.

CAMPBELL, C. B. G. 1966. Taxonomic status of tree shrews. *Science* **153**:436.

DE BEER, G. R. 1937. *The Development of the Vertebrate Skull.* Clarendon Press, Oxford.

ELOFF, F. C. 1951. On the organ of Jacobson and the nasal-floor cartilages, in the chondrocranium of *Galago senegalensis*. *Proc. Zool. Soc. London* **121**:651–655.

FISCHER, E. 1903. Zur Entwicklungsgeschichte des Affenschädels. *Z. Morphol. Anthropol.* **5**:383–414.

FISCHER, E. 1905. On the primordial cranium of *Tarsius spectrum*. *Proc. K. Nedr. Akad. Wet.* **8**:397–400.

FREI, H. 1938. Das Primordialcranium eines Foetus von *Avahis laniger*. *Inaug. Diss. Bern.*

FRETS, G. P. 1913. Beiträge zur vergleichenden Anatomie und Ontogenie der Nase der Primaten. II. *Morphol. Jahrb.* **45**:557–726.

FRETS, G. P. 1914. Beiträge zur vergleichenden Anatomie und Ontogenie der Nase der Primaten. III. *Morphol. Jahrb.* **48**:239–279.

FRICK, H. 1954. *Die Entwicklung und Morphologie des Chondrocraniums von Myotis Kaup.* G. Thieme Verlag, Stuttgart.

FRICK, H., and D. STARCK. 1963. Vom Reptil- zum Säugerschädel. *Z. Saeugetierkd* **28**:321–341.

GAUPP, E. 1906. Die Entwicklung des Kopfskelettes, pp. 573–874. *In* O. Hertwig, ed., *Handbuch der Entwicklungslehre der Wirbeltiere*, Vol. 3, Fischer, Jena.

GINGERICH, P. D. 1973. Anatomy of the temporal bone in the Oligocene anthropoid *Apidium* and the origin of Anthropoidea. *Folia Primat.* **19**:329–337.

HENCKEL, K. O. 1927a. Zur Entwicklungsgeschichte des Halbaffenschädels. *Z. Morphol. Anthropol.* **26**:365–383.

HENCKEL, K. O. 1927b. Das Primordialcranium der Halbaffen und die Abstammung der höheren Primaten. *Verh. Anat. Ges.* **36**:108–116.

HENCKEL, K. O. 1928a. Das Primordialcranium von *Tupaia* und der Ursprung der Primaten. *Z. Anat. Entwg.* **86**:204–227.

HENCKEL, K. O. 1928b. Studien uber das Primordialcranium und die Stammesgeschichte der Primaten. *Morphol. Jahrb.* **59**:105–178.

HOFER, H. 1952. Der Gestaltwandel des Schädels der Säuger und Vögel, nebst Bemerkungen über die Schädelbasis. *Verh. Anat. Ges.* **50**:102–113.

HOLST, D. VON. 1969. Sozialer Stress bei Tupajas (*Tupaia belangeri*). Die Aktivierung des sympathischen Nervensystems und ihre Beziehung zu hormonal ausgelösten ethologischen und physiologischen Veränderungen. *Z. Vergl. Physiol.* **63**:1–58.

KERMACK, D. M., and K. A. KERMACK, eds., 1971. *Early Mammals.* Academic Press, London.

KUHN, H. -J. 1971. Die Entwicklung und Morphologie des Schädels von *Tachyglossus aculeatus*. *Abh. Senckenb. Naturforsch. Ges.* **528**:1–224.

KUHN, H.-J., and A. SCHWAIER. 1973. Implantation, early placentation, and the chronology of embryogenesis in *Tupaia belangeri*. *Z. Anat. Entwg.* **142**:315–340.

LUCKETT, W. P. 1974. Comparative development and evolution of the placenta in Primates, pp. 142–234. *In* W. P. Luckett, ed., *Reproductive Biology of the Primates, Contributions to Primatology*, Vol. 3. S. Karger, Basel.

MARTIN, R. D. 1972. Adaptive radiation and behaviour of the Malagasy lemurs. *Philos. Trans. Roy. Soc. London*, Ser. B **264**:295–352.

MARTIN, R. D. 1973. Comparative anatomy and primate systematics. *Symp. Zool. Soc. London* **33**:301–337.

MATVEEV, B. 1929. Die Entwicklung der vorderen Wirbel und des Weberschen Apparates bei Cyprinidae. *Zool. Jahrb. Anat.* **51**:463–530.

MCKENNA, M. C. 1966. Paleontology and the origin of Primates. *Folia Primat.* **4**:1–25.

MÜLLER, H. J. 1961. Über strukturelle Ähnlichkeiten der Ohr- und Occipitalregion bei Vogel und Säuger. *Zool. Anz.* **166**:391–402.

PETER, K. 1913. *Atlas der Entwicklung der Nase und des Gaumens beim Menschen.* G. Fischer Verlag, Jena.

RAMASWAMI, L. S. 1957. The development of the skull in the slender loris, *Loris tardigradus lydekkerianus* Cabr. *Acta Zool.* **38**:27–68.

REINBACH, W. 1952a. Zur Entwicklung des Primordialcraniums von *Dasypus novemcinctus* Linné (*Tatusia novemcincta* Lesson). I. *Z. Morphol. Anthropol.* **44**:375–444.

REINBACH, W. 1952b. Zur Entwicklung des Primordialcraniums von *Dasypus novemcinctus* Linné (*Tatusia novemcincta* Lesson). II. *Z. Morphol. Anthropol.* **45**:1–72.

REINHARD, W. 1958. Das Cranium eines 33 mm langen Embryos des Mantelpavians, *Papio hamadryas* L. *Z. Anat. Entwg.* **120**:427–455.

ROMER, A. S. 1942. Cartilage: An embryonic adaptation. *Am. Nat.* **76**:394–404.

SABAN, R. 1956/1957. Les affinitiés du genre *Tupaia* Raffles 1821, d'après les caractères morphologiques de la tête osseuse. *Ann. Paléontol.* **42**:170–224; **43**:1–44.

SEYDEL, O. 1891. Uber die Nasenhöhle der höheren Säugetiere und des Menschen. *Morphol. Jahrb.* **12**:1–60.

SPATZ, W. 1964. Beitrag zur Kenntnis der Ontogenese des Cranium von *Tupaia glis* (Diard 1820). *Morphol. Jahrb.* **106**:321–416.

SPATZ, W. 1968. Die Bedeutung der Augen fur die sagittale Gestaltung des Schädels von *Tarsius* (Prosimiae, Tarsiiformes). *Folia Primat.* **9**:22–40.

STARCK, D. 1953. Morphologische Untersuchungen am Kopf der Säugetiere, besonders der Prosimier, ein Beitrag zum Problem des Formwandels des Säugetierschädels. *Z. Wiss Zool.* **157**:169–219.

STARCK, D. 1960. Das Cranium eines Schimpansenfetus (*Pan troglodytes* Blumenbach, 1799) von 71 mm SchStlg., nebst Bemerkungen über die Körperform von Schimpansenfeten. *Morphol. Jahrb.* **100**:559–647.

STARCK, D. 1961. Ontogenetic development of the skull in Primates. *International Colloquium on the Evolution of Mammals.* Kon. Vlaamse Acad. Wetensch. Lett. Sch. Kunsten Belgie, Brussels. **1**:205–214.

STARCK, D. 1962. Das Cranium von *Propithecus spec.* (Prosimiae, Lemuriformes, Indriidae). *Bibl. Primatol.* **1**:163–196.

STARCK, D. 1967. Le crâne des mammifères, pp. 405–549, 1095–1102. *In* P. Grassé, ed., *Traité de Zoologie*, Vol. 16, Part 1. Masson, Paris.

STARCK, D. 1973. The skull of the fetal chimpanzee, pp. 1–33. *In* G. H. Bourne, ed., *The Chimpanzee*, Vol. 6. S. Karger, Basel.

STARCK, D. 1974. Die Stellung der Hominiden im Rahmen der Säugetiere, pp. 1–131. *In* G. Heberer, ed., *Die Evolution der Organismen*, Vol. 3. G. Fischer Verlag, Stuttgart.

SZALAY, F. S. 1968. The beginnings of primates. *Evolution* **22**:19–36.

VAN KAMPEN, P. N. 1905. Die Tympanalgegend des Säugetierschädels. *Morphol. Jahrb.* **34**:321–722.

VAN VALEN, L. 1965. Treeshrews, primates, and fossils. *Evolution.* **19**:137–151.

7

Ontogeny of the Fetal Membranes and Placenta

Their Bearing on Primate Phylogeny

W. Patrick Luckett

I. Introduction

The mammalian placenta is usually defined as "an apposition or fusion of the fetal membranes to the uterine mucosa for physiological exchange" (Mossman, 1937), and this definition is equally true for the placenta established independently within several genera of reptiles. Mammalian placentation is initiated by the attachment of the blastocyst to the uterine endometrium and is terminated by the delivery of the newborn at the time of parturition. In spite of its relatively brief life-span during the ontogeny of the individual, the placenta is the most important and most physiologically complex organ during intra-uterine development. During its life history it performs functions analogous to those of the lung, intestine, kidney, liver, and, in some species, it is involved in endocrine functions comparable to those of the pituitary and gonads.

The extraembryonic or fetal membranes involved in mammalian placentation are identical to those that occur in reptiles and birds: (1) yolk sac, (2) amnion, (3) chorion,

W. Patrick Luckett · Department of Anatomy, Creighton University, School of Medicine, Omaha, Nebraska.

157

and (4) allantois. The last three membranes develop only in reptiles, birds, and mammals, and these vertebrate classes are frequently grouped as a superclass Amniota, in contrast to the lower vertebrates (Anamniota).

All vertebrate eggs develop in an aqueous environment, regardless of whether the eggs are laid in water, deposited on land, or retained within the mother's body. The self-contained *cleidoic* or shelled egg of reptiles evolved as a prerequisite for the occurrence of development on land rather than in water; Romer (1967) considered this to be "the most marvelous 'invention' in vertebrate history."

The origin and evolution of the fetal membranes were concomitant with the development of a yolk-laden cleidoic egg in primitive reptiles, and their developmental pattern and function have remained relatively constant within reptiles and their descendants—birds and prototherian mammals.

The *yolk sac* is the most primitive of the vertebrate fetal membranes; it occurs in all taxa (including nonamniotes) that possess yolk-rich or megalecithal eggs. All three embryonic germ layers (ectoderm, mesoderm, and endoderm) spread over the yolk mass and envelop it to form a trilaminar yolk sac. The endodermal lining of the yolk sac is in continuity with the developing midgut, and the vascularized yolk sac serves as a fetal nutritive organ in all vertebrates with megalecithal eggs. The yolk sac mesoderm also serves as the initial site of hematopoiesis in all vertebrates. In amniotes, the peripheral expansion of the developing exocoelom (Figs. 1 and 2) separates the trilaminar yolk sac into an inner splanchnopleuric layer (endoderm plus vascular splanchnic mesoderm) and an outer somatopleuric layer (ectoderm plus avascular somatic mesoderm). It is the yolk sac splanchnopleure that becomes the definitive yolk sac of mammals.

The *amnion* provides an aqueous environment in which the embryo can develop

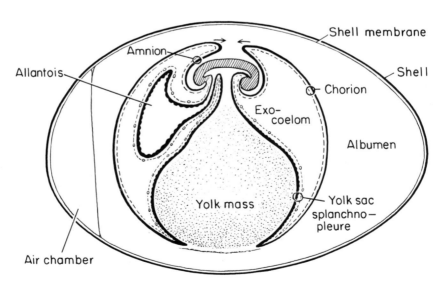

FIG. 1. Idealized sauropsid egg showing amniogenesis by folding (arrows) and the nature of the other fetal membranes.

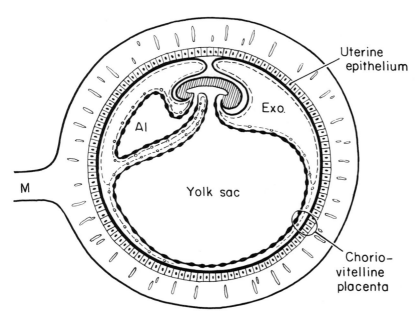

FIG. 2. Idealized strepsirhine embryo and fetal membranes *in utero*. The uterine mesometrium (M) is oriented to the left to facilitate comparison with the sauropsid condition. The chorion and yolk sac splanchnopleure are intimately apposed to the uterine epithelium to form a choriovitelline placenta. A region of mesoderm-free bilaminar omphalopleure persists temporarily at the abembryonic pole. Al = allantois; Exo = exoceolom.

symmetrically, free from desiccation or distortion by pressures from surrounding structures (Mossman, 1937). It develops by a similar process of folding of the extra-embryonic somatopleure (ectoderm plus mesoderm) during somite stages in reptiles, birds, monotremes, marsupials, and most eutherian mammals (Figs. 1 and 2). However, amniogenesis occurs by a process of cavitation within the embryonic epiblast in some eutherian taxa, apparently correlated with the mechanism of blastocyst implantation (Figs. 9 and 10). This specialized condition will be discussed in more detail later.

The somatopleuric *chorion* develops in continuity with the amnion when amniogenesis occurs by folding, and it becomes separated from the amnion by the closure of the amniotic folds and the expansion of the exocoelom (Figs. 1 and 2). Thus, the entire embryo and the other extraembryonic membranes—yolk sac, amnion, and allantois—come to be completely surrounded by the chorion. It is the chorion that comes into direct contact with the surrounding environment; in the cleidoic egg this consists of the albumen, shell membrane, and shell, whereas in metatherians and eutherians the chorion contacts the uterine endometrium (Figs. 1 and 2).

The splanchnopleuric *allantois* is essentially an outgrowth of the embryonic hindgut, and it becomes continuous secondarily with the urogenital sinus or urinary bladder. Initially, it may serve as a receptacle for nitrogenous waste products (uric acid or urea) derived from the fetal mesonephric kidney. During later somite stages, the vascular allantois expands into the exocoelom (Figs. 1 and 2) and eventually comes to line much of the inner surface of the chorion, fusing with it to form a vascularized chorioallantoic membrane (Fig. 3). This membrane plays an important role in the exchange of respiratory gases between the fetus and the external environment, across the shell of the cleidoic egg. Moreover, it is the vascularized chorioallantoic membrane that becomes modified to form the fetal component of the definitive or chorioallantoic placenta characteristic of all eutherian mammals.

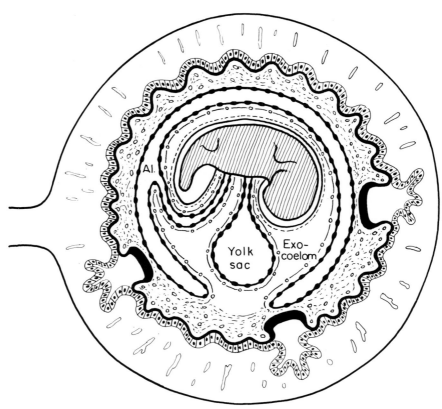

Fig. 3. Early stage of chorioallantoic placentation in Strepsirhini. The allantoic vesicle (Al) has greatly expanded, and its mesoderm now vascularizes the chorion to form a chorioallantoic placenta. Chorionic vesicles occur opposite the openings of uterine glands. The yolk sac has been removed from its contact with the chorion by the expanded allantois; consequently, it is greatly reduced.

II. Mammalian Fetal Membranes and Placentation

In contrast to the fundamental similarity of fetal membrane morphogenesis in reptiles, birds, and prototherians, considerable variation occurs in their developmental and functional relationships within eutherian mammals. Although viviparity has evolved independently among various representatives of elasmobranchs, teleosts, and reptiles (see Amoroso, 1952), eutherians are the only higher taxonomic category of vertebrates to be characterized by viviparity and the development of a placental relationship between mother and fetus. The fetal component of the vascular placentae of eutherian mammals consists of the modified chorion vascularized by either the yolk sac splanchnopleure or the allantois, and a consideration of the early development of the fetal membranes in monotremes and marsupials provides insight into the possible primitive conditions present in the ancestral eutherian mammals.

A. Prototheria

Studies on the earliest developmental stages of monotremes have demonstrated certain advances on the sauropsid (reptile and bird) condition that foreshadow the development of the therian blastocyst. The ovulated monotreme egg is megalecithal, although considerably smaller (± 4 mm) than that of a comparably sized reptile, and it does not contain enough stored nutrients to carry the developing embryo through the intrauterine and incubation periods of development (Hill, 1910; Flynn and Hill, 1947). A concomitant of this reduction in size is the relatively precocious growth and differentiation of the extraembryonic ectoderm. It spreads peripherally to completely enclose the yolk mass by the late preprimitive streak stage of development in both *Tachyglossus* and *Ornithorhynchus* (Flynn and Hill, 1947). There is essentially no growth in diameter of the developing embryo from cleavage until the enclosure of the yolk mass. Following this enclosure, however, there is rapid expansion of the thin-walled "blastocyst" as a result of the absorption of uterine gland secretions by the extraembryonic ectoderm and concomitant with the liquefaction of the enclosed yolk. Thus the shelled egg of *Tachyglossus* measures 11.5 mm at the 3–4 somite stage of intrauterine development, and about 15–17 mm (19 somite stage) when it is laid (Flynn and Hill, 1939). Such intrauterine growth of the egg is unknown in oviparous reptiles and birds. This marked precocious differentiation of the extraembryonic ectoderm as a nutritive organ in monotremes foreshadows its increased functional importance and even more precocious differentiation as the trophoblast of therian mammals (Hill, 1910; Flynn and Hill, 1947). Thus, monotremes exhibit "the beginnings of that process of substitution of uterine for ovarian nutriment which reaches its culmination in the Eutheria" (Hill, 1910).

Despite this marked period of intrauterine growth, the development of the monotreme fetal membranes follows the typical sauropsid pattern in later stages: amniogenesis by folding, a large vascular yolk sac involved with fetal nutrition and respiration, and an allantoic vesicle that functions initially as a urinary receptacle, then expands and fuses with the chorion after laying to serve as a respiratory membrane (Griffiths, 1968; Luckett, unpublished).

B. Metatheria

The ovulated egg of marsupials is intermediate in size between that of prototherians and eutherians, and in at least some species it contains a moderate amount of stored yolk. Although a shell is not deposited, there are an "albumen" layer and a thick *shell membrane* homologous to those of monotremes (Hill, 1910). A striking feature of phylogenetic significance occurs during the first cleavage stage: yolk granules are extruded from the dividing cells and come to lie between the resultant blastomeres (Hill, 1910). The yolk elimination permits a more uniform or holoblastic cleavage of all of the blastomeres, so that a hollow unilaminar blastocyst forms rapidly. All these features of the ovulated egg and early development provide ontogenetic evidence that the ancestors of metatherians (or therians) were oviparous and presumably prototherian (Hill, 1910). A concomitant of the extrusion of yolk during cleavage is the precocious differentiation of the extra-embryonic ectoderm (trophoblast) of the blastocyst as a nutritive organ, foreshadowing the condition in eutherian mammals.

All marsupials exhibit three fundamental differences in the further development of their fetal membranes, when compared to oviparous monotremes and sauropsids. (1) Mesoderm extends over only about half the surface of the endodermal yolk sac to form a trilaminar omphalopleure, leaving a persisting abembryonic region of mesoderm-free bilaminar omphalopleure (trophoblast plus endoderm). (2) The vesicular allantois projects into the exocoelom, but ceases to expand and fails to contact the chorion in most marsupials. (3) The disruption of the shell membrane during late intrauterine life brings the vascular trilaminar omphalopleure into intimate apposition with the uterine endometrium to establish a functional choriovitelline placenta, characteristic of all marsupials. This latter condition is only a slight advance on that of late intrauterine stages of monotremes, since the vascular choriovitelline membrane is involved in nutrient absorption and respiration in both taxa.

Strictly speaking, the marsupial embryo is ovoviviparous, since the "shelled" egg undergoes considerable intrauterine development before it "hatches" *in utero* to undergo a brief period of attachment and choriovitelline placentation before birth. The developmental state of the vascular yolk sac and allantois in marsupials corresponds to the condition at the end of intrauterine life in monotremes (Tyndale-Biscoe, 1973), and it is possible that parturition prevents the further differentiation of these two fetal membranes. Furthermore, in many marsupials the embryo invaginates into the yolk sac splanchnopleure, so that folds of the yolk sac may secondarily envelop the embryo and prevent the allantois from contacting the overlying chorion (Sharman, 1961).

The retention of a bilaminar omphalopleure and functional choriovitelline placenta, correlated with a modest and incomplete expansion of the allantoic vesicle, probably represent paedomorphic conditions in marsupials. In contrast, these are only transitory ontogenetic stages in monotremes and most eutherians (cf. Figs. 2 and 3) (for a further discussion of paedomorphosis, see Beer, 1958).

A vascular chorioallantoic placenta develops in the marsupial family Peramelidae, and there has been considerable debate as to whether this represents the primitive marsupial or therian condition, or whether it is a derived feature unrelated to the ancestral condition (see Hill, 1949; Pearson, 1949; Sharman, 1961). Based on both ontogenetic and comparative considerations, I agree with the concept (Pearson, 1949; Sharman, 1961) that the establishment of a choriovitelline placenta represents the primitive therian condition, and this has been retained as the definitive vascular placenta in all marsupials. Chorioallantoic placentation has evolved independently within several families of reptiles (Bauchot, 1965) and in peramelids, whereas it is a diagnostic feature of all eutherian mammals and must have been established in the ancestral eutherian.

The independent establishment of a vascular chorioallantoic placenta during late stages of peramelid development does not prolong the period of intrauterine life; instead, this family exhibits one of the shortest gestation periods known for marsupials (12 days). Tyndale-Biscoe (1973) has emphasized the importance of prevention of an allograft rejection by the mother against the fetus as a crucial adaptive factor in the evolution of prolonged intrauterine gestation in Eutheria, and he suggested that such a protective mechanism did not evolve in marsupials. Support for this hypothesis is provided by the occurrence of increased numbers of neutrophils and mononuclear cells immediately beneath the invasive placental zone of peramelids (Luckett, unpublished), and by the apparent destruction of the invasive trophoblast in near-term stages (Padykula and Taylor, 1974). These observations suggest the possibility that the precocious parturition characteristic of marsupials may be the result of an immunological homograft rejection.

In summary, the fetal membranes of marsupials exhibit some functional advances on the monotreme condition (see Tyndale-Biscoe, 1973):

1. Nutrition is accomplished primarily by the absorption of uterine gland secretions (histotrophe) through the bilaminar and trilaminar omphalopleures, following the elimination of most of the stored yolk during cleavage and concomitant with the precocious differentiation of a bilaminar blastocyst.

2. The vascular splanchnopleure of the yolk sac, and subsequently the chorio-vitelline placenta of late stages, are probably the principal sites of respiratory exchange, since the allantois does not contact the chorion in most marsupials. This is homologous to the intrauterine condition of monotremes, where the allantois does not differentiate and expand to serve as a respiratory membrane until after laying.

3. The allantois serves primarily as a urinary reservoir in marsupials, as it does in early stages of monotremes and sauropsids. The chorioallantoic placenta of peramelids exists during a very brief period of gestation, but it must be relatively efficient in the exchange of nutrients and gases between maternal and fetal blood (hemotrophe), since there is considerable growth of the embryo during this brief period.

C. Eutheria

All Eutheria are viviparous and develop a definitive chorioallantoic placenta, advances over the presumed primitive condition of ovoviviparity and choriovitelline placentation of the ancestral therian stock. The eutherian blastocyst differs from that of marsupials in that two distinct cell types are evident at the time of its initial differentiation: (1) A peripheral layer of cuboidal or squamous cells, the trophoblast, forms the complete outer wall of the blastocyst, and (2) a small cluster of cells, the inner cell mass, is attached to the inner surface of one pole of the trophoblast (Fig. 4). The inner cell mass will give rise to all of the tissues of the embryo proper, and it is completely segregated from the intrauterine environment by the surrounding trophoblast. This specialized eutherian blastocyst is a direct result of the formation of a solid cluster of blastomeres, the morula, in preceding stages, a condition unknown in noneutherian mammals. The early differentiation of a complete layer of trophoblast is doubtlessly related to its primary role as a nutritive organ.

In contrast to the fundamental similarity of early blastocyst formation in all eutherians, there is considerable variation in the further developmental relationships of the fetal membranes and placenta among the higher taxonomic categories (orders and suborders). Variation most frequently occurs in the (1) fate of the polar (embryonic) trophoblast, (2) pole of the trophoblast involved in the attachment (implantation) on the uterine endometrium; (3) orientation of the embryonic disc at the time of implantation, (4) region of the endometrium on which the blastocyst initially attaches, (5) depth of blastocyst implantation, (6) mechanism of amnion formation, (7) fate of the bilaminar omphalopleure and abembryonic wall of the yolk sac splanchnopleure, (8) participation of the yolk sac splanchnopleure in formation of a temporary choriovitelline placenta, (9) growth and expansion of the allantoic vesicle, and (10) location, shape, and finer morphology of the definitive chorioallantoic placenta.

Much of the available data on the comparative development of the fetal membranes and placenta in eutherians has been summarized previously (Mossman, 1937; Amoroso, 1952; Starck, 1959). Mossman's (1937) comprehensive survey provided a basis for evaluating the relative conservatism of different fetal membrane characters, as judged by their constancy within superfamilies, suborders, and orders. The most conservative characters are: (1) orientation of the embryonic disc to the endometrium at the time of

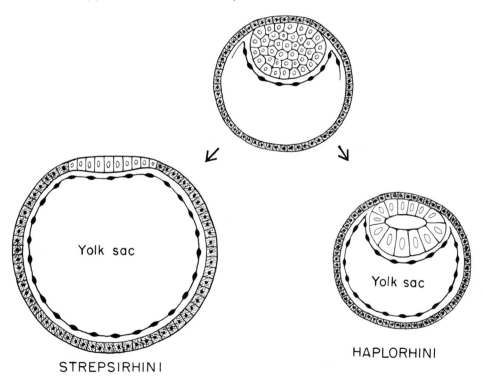

STREPSIRHINI

HAPLORHINI

FIG. 4. Generalized eutherian blastocyst (top), with trophoblast, inner cell mass, and early endoderm differentiation. The polar trophoblast is subsequently disrupted in strepsirhines, and the embryonic mass is exposed at the surface as a flattened disc. In contrast, all haplorhines exhibit initial cavitation of the inner cell mass to form a primordial amniotic cavity.

implantation, (2) the nature of the yolk sac and its vascular splanchnopleure, (3) the nature of the allantoic vesicle, and (4) the detailed structure of the definitive chorio-allantoic placenta. Mossman (1937, 1953) further suggested that the conservative nature of the mammalian fetal membranes may be caused by their relative isolation from the selective effects of the external environment. While it may be true that external selective forces, such as diet, habitat preference, and climate, have had little, if any, apparent selective effect on the evolutionary development of the fetal membranes, it seems likely that there have been intense centripetal selective forces acting upon them to maintain a relatively constant pattern (Luckett, 1974). Selection is considered to be centripetal when there is a tendency to retain a relatively static optimal condition of a character and to eliminate variant forms, and such conservative characters are probably of great adaptive importance to the organism, perhaps even essential to its survival (Farris, 1966). Thus, the vascularization of the chorion by either the yolk sac or allantois is essential for the functional development of a vascular placenta in therian mammals, and there are probably fundamental interrelationships in their development. For instance, the lack of a chorio-vitelline placenta in haplorhine primates is consistently associated with the precocious development of the chorioallantoic placenta and the vestigial nature of the allantoic diverticulum.

III. Use of Fetal Membrane Characters in Assessing Phylogenetic Relationships

Developmental information is available on the fetal membranes and placenta of all mammalian orders and suborders, most superfamilies, and many families. Furthermore, the conservative nature of the fetal membranes permits generalizations about their developmental relationships within families and superfamilies to be based on a knowledge of relatively few species, since significant differences are rare within families (Mossman, 1937). Such conservative characters are extremely valuable in assessing phylogenetic relationships, because they are easier to trace across the gaps in the evolutionary record (Farris, 1966).

The mammalian fetal membranes and placenta develop from all three embryonic germ layers and comprise an interrelated complex of genetic information; this complexity minimizes the possibility of convergent evolution in their ontogenetic relationships. Mossman (1967) emphasized that it is the entire developmental pattern of the fetal membranes and placenta that is available for assessing phylogenetic relationships, not just the morphology of the definitive chorioallantoic placenta. Such developmental studies permit recognition of parallel or convergent evolution of *individual* characters within different taxa, such as the convergent evolution of a hemochorial placenta in some primates, insectivores, and bats. Thus, detailed ontogenetic and comparative studies of a complex organ system such as the placenta and fetal membranes, coupled with their conservative evolutionary nature, serve to offset the absence of these characters in the fossil record and to increase their usefulness in assessing phylogenetic relationships among higher taxonomic categories.

W. Patrick
Luckett

While neontological studies cannot provide direct evidence of the ancestor–descendant relationships in a phylogeny, the wide range of morphological, biochemical, and behavioral characters available permits an extensive analysis of cladistic relationships among extant taxa. Cladistic relationships are defined in terms of relative recency of common ancestry; this can be expressed as follows: two or more taxa are more closely related to each other than to any other taxa (and together form a "sister group" or monophyletic group of higher rank) when it can be demonstrated that they possess shared derived homologous characters (synapomorphies) inherited from a common ancestor not shared with other taxa (Hennig, 1966). The methodology used to reconstruct the branching sequences of a cladogram involves the identification of all alternative states of homologous characters, and the subsequent arrangement of a series of character states in a sequence (morphocline or transformation series) from the most primitive to the most derived or specialized (Hennig, 1966; Schaeffer *et al.*, 1972).

Extensive comparative analyses of the morphology and ontogeny of characters among a wide spectrum of taxa provide the basic data concerning the range of character states that have evolved in living (and whenever possible, fossil) taxa. Following this, the two essential tasks of cladistic analysis are to establish the homology of the characters under study and to assess the relative primitiveness or derivedness of each character state (the polarity of the morphocline). Most students of phylogeny or cladistics (Hennig, 1966; Schaeffer *et al.*, 1972) would agree that such character analyses are best accomplished by: (1) detailed ontogenetic studies of each character, whenever feasible, in order to establish homology and "ontogenetic character precedence" (Hennig, 1966), and (2) assessing the frequency of particular character states within a particular taxon and its closely related sister groups.

Homology (the inheritance of a character from a common ancestor) and convergence (the development of similar characters independently in lineages without a common ancestry) can be distinguished by ontogenetic studies and by the demonstration of detailed similarities (both structurally and functionally) by means of wide-ranging comparative analyses (Simpson, 1961). After character homologies are established, it is essential for cladistic analysis that homology resulting from retention of ancestral or primitive (plesiomorph) characters be distinguished from homology resulting from the acquisition of unique shared derived (apomorph) characters (Hennig, 1966). Primitive character states can be determined by (1) the common occurrence of a character state within a particular taxon and its sister groups (as well as in distantly related taxa), and (2) developmental studies, particularly of early embryonic stages, which provide evidence of ontogenetic character precedence or "logical antecedents" (see Martin, this volume).

As an example, this method of cladistic analysis can be utilized to underscore the strong probability that a vascular choriovitelline placenta evolved in ancestral therian mammals and represents the primitive condition for the subclass. Broad comparative studies reveal that this is the only type of vascular placentation in all marsupials (with the exception of the Peramelidae), and it develops in a homologous pattern in all of them. Within the Eutheria (the sister group of the Metatheria), a homologous (although transitory) choriovitelline placenta develops in less "specialized" members of all living orders (with the possible exception of the Edentata). Furthermore, a choriovitelline placenta develops earlier during ontogeny in all Eutheria (as well as in Peramelidae) than does the chorioallantoic placenta (a derived feature in Theria) which replaces it both temporally and functionally. Thus, both ontogenetic character precedence and extensive comparative considerations support the primitive character state of choriovitelline placentation in Theria. It is important to note here that character states are relatively primitive or relatively derived depending on the taxonomic categories being compared. Although the development of a choriovitelline placenta is a primitive feature in Theria, its occurrence is a derived condition within the class Mammalia, when Theria are compared with their sister group, the Prototheria. Furthermore, within several eutherian orders (Primates, Rodentia), the absence of a transitory choriovitelline placenta appears to be a derived condition (see below). Clearly, one must define the taxonomic categories being compared when discussing primitive and derived character states; the lack of a choriovitelline placenta appears to be a primitive condition in the class Mammalia, but a derived condition in haplorhine primates and muroid and hystricognath rodents.

Utilizing the above methodology, the relative primitive and derived conditions for each of the mammalian fetal membrane and placental characters can be assessed, and the polarity of a morphocline can be established for each character. The fetal membranes of reptiles, birds, prototherians, and metatherians are also considered in evaluating the primitive and derived states of each character and in reconstructing the ancestral morphotype of the eutherian and primate fetal membranes. This comparative analysis permits the construction of a cladogram based on the possession of shared derived fetal membrane characters in monophyletic sister groups, following Hennig's (1966) scheme of argumentation of phylogenetic systematics. The retention of shared primitive character states can occur in distantly related taxa and is of little significance in establishing phylogenetic relationships (Hennig, 1966). Unfortunately, this important concept has been frequently ignored by comparative placentologists in discussing the phylogenetic significance of fetal membrane characters.

IV. Basic Developmental Pattern of Primate Fetal Membranes and the Suggested Primitive Eutherian Condition

Detailed studies of the morphogenesis of the fetal membranes and placenta of Primates have been published previously (Hill, 1932; Starck, 1956; Luckett, 1974) and will only be summarized here, with particular emphasis on those aspects that are significant in establishing the morphocline polarity of each character. Table 1 summarizes these data, and they are arranged so that the primitive condition is listed first and the most derived condition last for each of the characters.

A. Implantation and Amniogenesis

Early preimplantation blastocysts are basically similar in all eutherians, and many of the developmental differences in placentation are the result of the mechanism of implantation. These differences relate in part to the pole of the blastocyst involved in the

TABLE I

SUMMARY OF THE MAJOR FEATURES OF FETAL MEMBRANE MORPHOGENESIS IN PRIMATE SUPERFAMILIES

	Lemuroidea and Lorisoidea	Tarsioidea	Ceboidea	Cercopithecoidea	Hominoidea
Implantation					
Lateral orientation of disc	+	+	+	+	+
Attachment trophoblast					
Paraembryonic pole	+	+			
Embryonic pole			+	+	+
Location					
Central	+				
Eccentric		+	+	+	
Interstitial					+
Amniogenesis					
Primordial amniotic cavity					
Absent	+				
Present		+	+	+	+
Definitive amniogenesis					
Folding	+	+			
Cavitation			+	+	+
Yolk sac					
Large, free, reduced later	+				
Small, free, reduced later		+	+	+	+
Choriovitelline placenta					
Present	+				
Absent		+	+	+·	+
Allantoic vesicle					
Large, permanent	+				
Small, rudimentary		+	+	+	+
Chorioallantoic placenta					
Diffuse, epitheliochorial	+				
Discoid, hemochorial		+	+	+	+·
Labyrinthine		+			
Trabecular			+		
Villous				+	+

initial attachment and to the degree of invasive activity of the attached trophoblast. Figure 5 illustrates the regions of the blastocyst and the endometrium that may be involved in implantation, and all possible combinations are known within Eutheria. The orientation of the embryonic disc to the endometrium at the time of implantation is the most conservative fetal membrane character, being constant in most orders except the Insectivora and Chiroptera. An antimesometrial orientation of the embryonic disc (Fig. 6A) occurs in the majority of orders, including Carnivora, Artiodactyla, Perissodactyla, Pholidota, and most Insectivora. The significance of this constant pattern of orientation is unclear, although Mossman (1971) has emphasized that the orientation of the embryonic disc is correlated with the subsequent location and further development of the yolk sac, amnion, chorion, and allantois, and ultimately with the developmental site of the chorioallantoic placenta.

In all Primates, the embryonic disc is oriented orthomesometrially (laterally) (Fig. 6B), but attachment may be effected by the paraembryonic trophoblast, as in strepsirhines and *Tarsius* (Figs. 7 and 8), or by the polar (embryonic) trophoblast, as in Platyrrhini and Catarrhini (Figs. 9 and 10). An orthomesometrial orientation of the disc is rare in other eutherians, although it occurs in many families of Chiroptera and may represent the primitive condition in that order (Luckett, unpublished). Superficial, noninvasive, and central attachment of a relatively expanded bilaminar blastocyst by its paraembryonic trophoblast probably represents the primitive eutherian condition, based on both ontogenetic and comparative data (Luckett, 1974). This condition is retained in strepsirhine Primates, Artiodactyla, Perissodactyla, Pholidota, and presumably Cetacea (Fig. 7). Limited invasive activity of the trophoblast has developed independently in at least two species of galagids (Butler, 1967), but it is unknown in other strepsirhines.

A concomitant of the mechanism of implantation is the fate of the polar trophoblast,

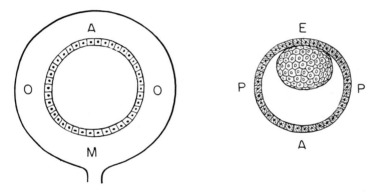

Fig. 5. Diagram of the poles of the uterine endometrium (left) and blastocyst (right) that may participate in blastocyst attachment. A = antimesometrial, M = mesometrial, O = orthomesometrial poles of endometrium. E = embryonic, A = abembryonic, P = paraembryonic poles of blastocyst.

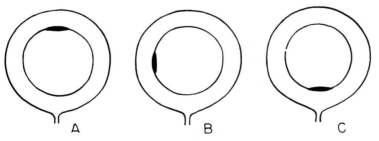

Fig. 6. Patterns of orientation of embryonic disc to endometrium during implantation. (A) Antimesometrial orientation, characteristic of majority of Eutheria orders, including most Insectivora; (B) orthomesometrial orientation, characteristic of Primates and many Chiroptera; (C) Mesometrial orientation, characteristic of Lagomorpha and Rodentia.

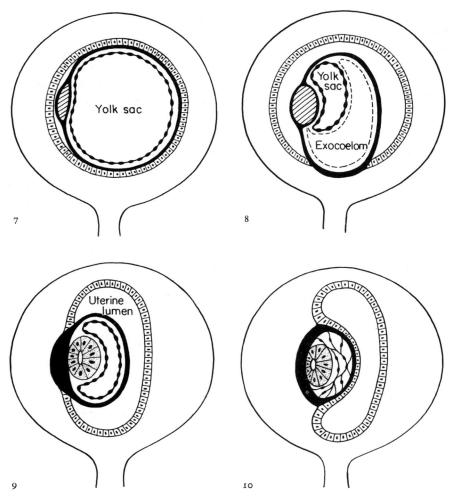

Fig. 7. Noninvasive, central implantation of expanded bilaminar blastocyst, characteristic of most strepsirhines.

Fig. 8. Invasive, eccentric implantation of *Tarsius* blastocyst by paraembryonic trophoblast to mesometrial pole. Early differentiation of exocoelom separates yolk sac from contact with trophoblast.

Fig. 9. Invasive, eccentric implantation characteristic of ceboids and cercopithecids. Attachment occurs orthomesometrially by the embryonic trophoblast, with the subsequent persistence of the primordial amniotic cavity.

Fig. 10. Interstitially implanted hominoid blastocyst, with persisting primordial amniotic cavity. The meshlike modification of the yolk sac endoderm is associated with interstitial implantation.

and its subsequent influence on the process of amnion formation. Paraembryonic attachment of an expanded blastocyst is invariably associated with the loss of the polar trophoblast overlying the embryonic disc, so that the disc is secondarily exposed at the surface of the blastocyst (Fig. 7). Amniogenesis occurs by folding in all eutherians in which the polar trophoblast is lost (Figs. 2 and 11), and this primitive condition is characteristic of all noneutherian amniotes (Fig. 1). On the other hand, amniogenesis occurs by a process of cavitation within the embryonic mass (Fig. 4) of all eutherians in which the polar trophoblast persists, usually as the result of precocious blastocyst implantation at the embryonic pole (Figs. 9 and 10). Platyrrhine and catarrhine Primates, tenrecid and erinaceid Insectivora, Macroscelidea, Dermoptera, and some Chiroptera are characterized by this derived condition. An intermediate type of amniogenesis occurs within several eutherian families, including tupaiid and soricoid Insectivora and *Tarsius*. In these, a *primordial amniotic cavity* develops by cavitation, but the roof of the amniotic cavity and the overlying polar trophoblast rupture at about the time of implantation. As a result, the embryonic mass is again exposed at the surface of the blastocyst, and the definitive amnion subsequently forms by folding (Figs. 8 and 11). A useful generalization is that definitive amniogenesis occurs by folding in all eutherians in which the polar trophoblast is lost, whereas it occurs by cavitation when the polar trophoblast persists. The development of a primordial amniotic cavity provides a free epithelial surface for the embryonic epiblast, and this may be a prerequisite for the morphogenetic movements that occur during primitive streak formation (Luckett, 1975).

In most mammalian taxa with invasive implantation, there is a proliferation and limited invasion of trophoblast into the underlying maternal tissue at the initial attachment site. This is particularly true for those taxa that subsequently develop a hemochorial

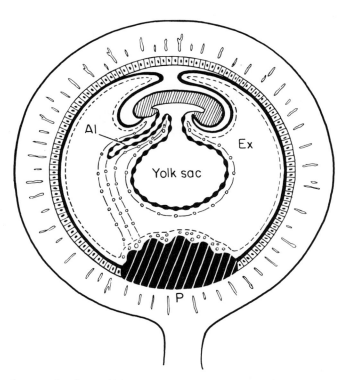

FIG. 11. Fetal membranes and embryo of *Tarsius spectrum in utero*, with short tubular allantois (Al) projecting into body stalk, reduced vascular yolk sac, mesometrial placental disc (P), and amniogenesis by folding. Ex = exocoelom.

placenta; the trophoblastic proliferation may establish a preplacenta in which maternal blood begins to circulate (Figs. 8–10). The preplacenta undergoes further differentiation and becomes the definitive chorioallantoic placenta when it receives its fetal circulation via the invasion of vascular allantoic mesodermal villi (Figs. 11 and 12). In contrast to the constant orientation of the embryonic disc (orthomesometrial) in all Primates, the trophoblastic pole involved in initial attachment and preplacental development differs in *Tarsius* (paraembryonic) and the Anthropoidea (embryonic); therefore, the chorioallantoic placenta is situated mesometrially in *Tarsius*, but orthomesometrially in Anthropoidea (cf. Figs. 8 and 11 and Figs. 9, 10, and 12). The anterior and posterior walls of the

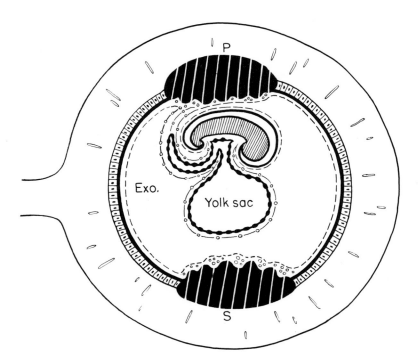

FIG. 12. Fetal membranes and embryo of *Macaca mulatta in utero*, characteristic of the ceboid and cercopithecid condition. The secondary placental disc (S) of Anthropoidea is homologous with the single placental disc of *Tarsius;* these differences are concomitant with the different mechanisms of implantation. The hominoid condition is fundamentally similar to that of cercopithecids, except that there is no secondary placental disc and implantation is interstitial.

simplex uterus of Anthropoidea are considered to be homologous with the orthomesometrial walls of the bicornuate uterus of other eutherians, including strepsirhines and *Tarsius* (Luckett, 1974), and the relationships illustrated in Figs. 9, 10, and 12 reflect this homology in order to facilitate comparisons among primates.

The localized invasive trophoblast of *Galago senegalensis* serves as a temporary holdfast during implantation (Butler, 1967) but does not exhibit further invasive activity; instead it resembles the transitory "attachment cones" of lagomorphs and rodents (Mossman, 1937). Although undoubtedly a derived condition within galagids, this temporary holdfast suggests a possible ontogenetic mechanism for the initial step in the evolution of hemochorial placentation in haplorhines. The further invasion and proliferation of trophoblast at such a localized attachment site would bring maternal blood into intimate contact with the trophoblast and establish a preplacenta, a characteristic feature of all haplorhines and a prerequisite for the development of their discoidal hemochorial placenta. The ontogeny of preplacental differentiation in all superfamilies of haplorhine primates suggests the probability that the haplorhine hemochorial placenta evolved directly from an epitheliochorial condition, without an intermediate endotheliochorial ancestor (Luckett, 1974).

Whereas central attachment of the expanded bilaminar blastocyst of Strepsirhini, Artiodactyla, and Carnivora doubtlessly represents the primitive eutherian condition (Fig. 7), there is a tendency for attachment to be eccentrically located in those taxa (*Tarsius*, Ceboidea, Cercopithecidae) in which there is a more precocious, invasive attachment of a smaller, less-differentiated blastocyst (Figs. 8 and 9). The most derived condition, interstitial implantation, occurs in Hominoidea and hystricognath Rodentia; following attachment, the blastocyst rapidly invades the endometrium and comes to lie completely encapsulated within it (Fig. 10).

B. Choriovitelline Placentation

The vascular splanchnopleure of the yolk sac fuses with the chorion to establish a choriovitelline placenta in Strepsirhini, Artiodactyla, Carnivora, Insectivora, Tupaiidae, Chiroptera, and less-specialized members of most other eutherian orders (Fig. 2). This primitive eutherian condition is shared with all Metatheria. The choriovitelline placenta is the definitive placenta of all Metatheria except Peramelidae, whereas it is only a temporary stage of eutherian development, being replaced during ontogeny by the chorioallantoic placenta. A choriovitelline placenta is lacking in Haplorhini (a derived condition), because the precocious differentiation of the exocoelom separates the chorion from the yolk sac prior to its vascularization (Figs. 8, 11, and 12). The absence of a transitory choriovitelline placenta in haplorhines (as well as in muroid and hystricognath rodents) appears to be correlated with the precocious differentiation of allantoic (body stalk) mesoderm and the precocious establishment of the chorioallantoic placenta (Luckett, 1974).

The early development of the exocoelom is concomitant with precocious differentiation of extraembryonic mesoderm and neoformation of a body stalk in all haplorhines, and further differences of yolk sac development within Haplorhini appear to be the result of (1) formation of a simplex uterus in Anthropoidea, and (2) interstitial implantation in Hominoidea (Luckett, 1974). A secondary yolk sac develops in all Anthropoidea by the pinching off and degeneration of the abembryonic portion of the primary yolk sac; this may be a result of the limited expandability of the simplex uterus. Edentates are also characterized by a simplex uterus, and the abembryonic wall of the yolk sac is apparently lost in all of them. A further complication occurs in the development of the primary yolk sac in Hominoidea; a meshlike array of extraembryonic endoderm fills much of the blastocyst cavity in preprimitive streak blastocysts (Fig. 10). It is suggested that growth differences in the endoderm and blastocyst owing to interstitial implantation are a causative factor in the temporary development of this "reticulated" endoderm. A similar derived pattern develops convergently in early blastocysts of *Galagoides demidovii* which exhibit an analogous and transitory type of "interstitial" implantation, but not in other strepsirhines or haplorhines with superficial or eccentric implantation.

C. Allantoic Vesicle

The development of a large allantoic vesicle characterizes Strepsirhini (Figs. 2 and 3) and all eutherian orders that possess a diffuse or zonary chorioallantoic placenta; this primitive condition is shared with sauropsids and prototherians (Fig. 1). The vascular mesoderm of the expanding allantoic vesicle fuses with the chorion to initiate the development of the chorioallantoic placenta. The further expansion of the exocoelom and allantois gradually displaces the yolk sac splanchnopleure from its fusion with the chorion (cf. Figs. 2 and 3), so that there is a temporal and functional replacement of the chorio-vitelline placenta by the chorioallantoic placenta during ontogeny. In striking contrast to the strepsirhine condition, the allantoic diverticulum is tubular and reduced in haplorhines; this is concomitant with the precocious differentiation of the extraembryonic mesoderm and body stalk (Figs. 11 and 12). The body stalk mesoderm of haplorhines is homologous to the allantoic mesoderm of strepsirhines and other mammals (Hill, 1932; Luckett, 1974), and the reduction in allantoic endoderm is correlated with the precocious vascularization of the chorion by the body stalk. This emphasizes the important concept that it is the vascular allantoic mesoderm, rather than the endodermal allantoic vesicle, that is the essential component of the allantois for the functional development of the chorioallantoic placenta.

D. Chorioallantoic Placenta

Eutherian chorioallantoic placentae are categorized according to their gross shape, location, invasive activity, and finer morphology (see Mossman, 1937). The finer morphology is classified according to the number of layers that separate maternal and fetal blood in the chorioallantoic placenta (Grosser, 1909). In general, the fetal component (chorion) remains intact, whereas differences are the result of reduction in the number of the maternal layers. In the epitheliochorial placenta there is no loss of maternal tissue, and the uterine epithelium is closely apposed to the chorion. Loss of maternal epithelium and the underlying connective tissue results in the intimate apposition of maternal capillary endothelium and the chorion to form an endotheliochorial placenta. A hemochorial relationship is established when maternal blood directly bathes the chorionic trophoblast, without intervening maternal tissue.

There is a striking dichotomy in the nature of the chorioallantoic placenta in Primates; Strepsirhini possess a diffuse, epitheliochorial placenta (Fig. 3), whereas an invasive, discoidal, hemochorial placenta characterizes all Haplorhini (Figs. 11 and 12). These differences are a reflection of the mechanism of implantation and trophoblastic invasion within the two suborders. A modified zone of endotheliochorial placenta exists within the otherwise diffuse epitheliochorial placenta of *Galagoides demidovii*, and it has been suggested that this may represent an intermediate stage in the evolution of the haplorhine hemochorial placenta from the epitheliochorial strepsirhine condition (J. P. Hill, 1932; W. C. O. Hill, 1953). However, developmental evidence suggests that this is a secondarily derived condition in *Galagoides*, resulting from precocious attachment of the blastocyst in an "implantation" chamber, and as such it has no bearing on the evolution of a hemochorial placenta in Haplorhini (Luckett, 1974).

Structural variations in the placental disc among haplorhine superfamilies result in part from differences in invasive activity of the cytotrophoblast. This activity is minimal in *Tarsius* and ceboids, whereas the placentae of all catarrhines are characterized by extensive invasion and peripheral spread of cytotrophoblast to form a trophoblastic shell. Wislocki (1929) demonstrated a transformation series in the morphology of the hemochorial placenta from a primitive labyrinthine condition in *Tarsius*, an intermediate trabecular arrangement in Ceboidea, to the derived villous condition in Catarrhini. A fundamentally labyrinthine or trabecular arrangement occurs during early developmental stages of catarrhines, followed by the secondary formation of free villi. Thus, ontogenetic evidence supports the morphocline polarity suggested by Wislocki.

The presumed primitive and derived conditions of chorioallantoic placentation have generated more discussion and controversy than any other aspect of mammalian placentation, but unfortunately sound principles of phylogenetic inference have been commonly ignored. The diffuse epitheliochorial placenta, characteristic of Strepsirhini, Artiodactyla, Perissodactyla, Cetacea, and Pholidota, is the simplest condition developmentally and has been considered the primitive eutherian condition (Turner, 1877; Grosser, 1909; Hill, 1932). However, Hubrecht (1908) and Wislocki (1929) believed that an invasive hemochorial placenta represents the primitive condition, since this type occurs in the most "archaic" mammals: insectivores, bats, rodents, edentates, hyracoids, and anthropoid primates. They emphasized that the "advanced" orders Artiodactyla, Perissodactyla, and Cetacea would not have a primitive placentation and that their diffuse placenta must be a secondary simplification from a more primitive invasive type. This hypothesis ignores the concept of mosaic evolution, the occurrence of both primitive and derived characters within every taxa, and it also fails to consider the nature of the other fetal membrane characters associated with these two placental types.

Mossman (1937) has suggested that a zonary endotheliochorial placenta (similar to that which occurs in most Carnivora) arose early during the evolution of viviparous (= therian) mammals and that this may have been the ancestral condition for both the diffuse epitheliochorial type of placentation on the one hand, and the discoidal hemochorial placenta on the other hand. This hypothesis has been revived recently by Martin (1969, and this volume). Thus, we have an unusual situation in which the three principal character states of a morphocline have each been considered to represent the primitive condition. However, utilization of the principles of cladistic analysis outlined above strongly supports the primitive nature of diffuse epithel; ochorial placentation in eutherian mammals and the relatively derived states of endotheliochorial and hemochorial placentation, particularly when the morphogenesis of each fetal membrane character (including ontogenetic character precedence) and detailed homologies are analyzed on a broad comparative basis.

1. All unrelated eutherian taxa with a diffuse (or cotyledonary) epitheliochorial placenta (including Strepsirhini, Artiodactyla, Perissodactyla, Pholidota, and Cetacea) exhibit a fundamental and detailed homology in the morphogenesis of all their fetal membrane characters. Such detailed homologies would be expected in cases of primitive retention (symplesiomorphy) of the ancestral condition in distantly related descendant taxa, but would be highly unlikely if similarities owing to convergence had occurred. Conversely, detailed homologies of fetal membrane character states are not found in different higher taxa that develop either endotheliochorial or hemochorial placentation.

2. The endotheliochorial placentas that develop in Carnivora, Proboscidea, Tupaiidae, Soricidae, Bradypodidae, and several families of Chiroptera are the results of differing patterns of morphogenesis, and there may be considerable differences in the developmental pattern of their other fetal membranes (including the yolk sac, allantois, amniogenesis, and mechanisms of implantation). As an example, a large allantoic vesicle develops in carnivores, tupaiids, and elephants, whereas it is small in soricids and vestigial in bradypodids. Conversely, the allantoic vesicle is very large in all eutherian taxa with an epitheliochorial placenta, as it is in all monotremes, reptiles, and birds. These varying patterns of fetal membrane morphogenesis associated with endotheliochorial placentation support the hypothesis that this placental type is a derived character state which has evolved convergently in several unrelated mammalian taxa, contrary to the assertions of Mossman (1937) and Martin (1969, and this volume). Similar arguments are valid for the convergent evolution of hemochorial placentation (and the associated variation in fetal membrane morphogenesis) in unrelated taxa such as Macroscelidea, Erinaceidae, Dasypodidae, Haplorhini, Dermoptera, and most Rodentia.

3. There is no unequivocal ontogenetic evidence for the derivation of a diffuse epitheliochorial placenta from a zonary endotheliochorial condition, but there is clear developmental evidence for the differentiation of a zonary endotheliochorial placentation from a diffuse condition during the early ontogeny of Carnivora (as noted by Mossman in 1937). Nevertheless, Martin (this volume) violates his own "principle of the logical antecedent" by suggesting the evolutionary derivation of a diffuse epitheliochorial placenta from an ancestral zonary endotheliochorial condition, despite ontogenetic evidence to the contrary.

4. The Mossman–Martin hypothesis of an ancestral endotheliochorial placenta giving rise to all hemochorial placental types is not supported by ontogenetic evidence. Although there is compelling developmental and ultrastructural evidence for the phyletic derivation of the hemochorial placenta in vespertilionid and desmodontid bats from an ancestral endotheliochorial condition (Enders and Wimsatt, 1968; Björkman and Wimsatt, 1968), this is not true for many other hemochorial placentas. Early developmental stages of Haplorhini, Macroscelidea, Dermoptera, and some Insectivora suggest that their hemochorial placentation was derived directly from an ancestral epitheliochorial condition, without an intermediate endotheliochorial stage (Lange, 1933; Luckett, 1974; contra Luckett, 1969). Therefore, ontogenetic evidence supports the phyletic origin of endotheliochorial placentation from an epitheliochorial ancestor (as in Carni-

vora) and the origin of hemochorial placentation from either an epitheliochorial or endotheliochorial ancestor.

In summary, cladistic analysis of the entire developmental pattern of the fetal membranes and placenta in sauropsids, prototherians, metatherians, and all eutherian orders suggests that the primitive eutherian condition included: (1) paraembryonic and noninvasive attachment of an expanded bilaminar blastocyst, (2) a large vascular yolk sac in early stages, (3) development of a temporary choriovitelline placenta, (4) amniogensis by folding, (5) a large allantoic vesicle, and (6) a noninvasive, diffuse epitheliochorial placenta. All of these character states occur in Strepsirhini and are believed to represent the primitive primate condition (Luckett, 1974).

V. Phylogenetic Relationships among Primates Suggested by Fetal Membrane Development

The assessment of primitive and derived states of each fetal membrane character permits the construction of a cladogram based on the possession of shared derived characters of sister groups, according to Hennig's (1966) scheme of argumentation of phylogenetic systematics. Because of the conservative nature of fetal membrane characters, differences are minimal below the superfamily level, and therefore the sister groups evaluated in this study are the commonly recognized superfamilies of Primates. Thus, fetal membranes provide valuable evidence for the assessment of phylogenetic relationships among higher categories of Primates, particularly the relationship of the Tarsioidea and the possible relationships of the Tupaiidae.

The suggested primitive and derived states of primate fetal membrane characters are listed in Table 2 and illustrated in the derived phylogeny (Fig. 13). Because of the fundamental differences in fetal membrane morphogenesis between Strepsirhini and Haplorhini, there are only two shared character states common to all Primates: the orthomesometrial orientation of the embryonic disc, a derived condition, and the reduced, free vascular yolk sac of later stages, a retention of the primitive eutherian condition. These shared characters are not included in the phylogeny. An assessment of these data reveals the following: (1) All listed fetal membrane characters of Lemuroidea and Lorisoidea are primitive retentions of the suggested ancestral primate (and ancestral eutherian) condition. As such, they provide no criteria for distinguishing Lemuroidea from Lorisoidea. (2) Fetal membrane character states shared by *Tarsius* and Strepsirhini are all primitive retentions and therefore provide no evidence of a special phylogenetic relationship. (3) Character states shared by all haplorhine superfamilies are derived when compared to the primitive condition, and these shared derived character states support the monophyletic classification of the suborder Haplorhini. (4) All shared derived characters of Anthropoidea (Ceboidea, Cercopithecoidea, Hominoidea) that do not occur in *Tarsius* are thought to be the result of development of a simplex uterus (Luckett, 1974). (5) Derived characters of Hominoidea not shared with other Anthropoidea are all the result of interstitial implantation.

Some of the derived character states of different transformation series that occur in

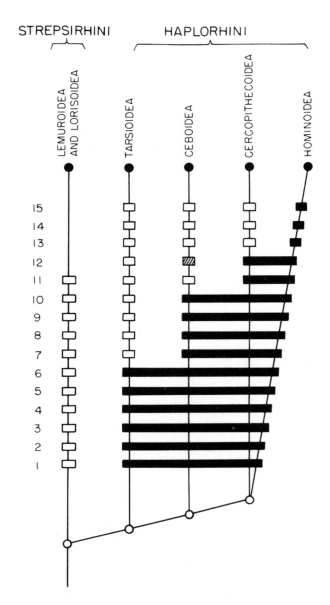

Fig. 13. Cladogram of primate superfamilies derived from cladistic analysis of fetal membrane and placental characters listed in Table 2, utilizing Hennig's (1966) scheme of argumentation of phylogenetic systematics. Note: Characters 12–15 in the cladogram and in Table 2 are applicable only for the hemochorial placentae of Haplorhini.

Haplorhini appear to be correlated, and it might be argued that these characters should not be considered separately in a cladistic analysis of sister-group relationships. For instance, rudimentary development of an allantoic diverticulum is correlated with the precocious differentiation of a mesodermal body stalk, and these developmental processes are correlated in turn with the absence of a choriovitelline placenta. These three derived character states are correlated in haplorhine primates (characters 4–6 in Table 2 and Fig. 13), dasypodid edentates, and muroid and hystricognath rodents. However, other character states which at first glance appear to be correlated are not necessarily so, particularly when nonprimate taxa are also considered. Thus, amniogenesis by cavitation necessitates the development of a primordial amniotic cavity, but a primordial amniotic cavity can also develop in concert with definitive amniogenesis by folding, as in *Tarsius*, tupaiids, and many ungulates. Also, a discoidal hemochorial placenta requires invasive attachment or implantation, but invasive attachment does not always lead to hemochorial placentation; it is also associated with endotheliochorial placentas.

Preliminary assessment of structural and functional character correlations in the fetal membrane development of primates and other eutherian mammals suggests that developmental acceleration of certain characters (precocious extraembryonic mesoderm formation) is generally associated or correlated with developmental retardation or reduction of other characters (reduced allantoic diverticulum and absence of a choriovitelline placenta). Both of these modes of differential rates of ontogeny appear to be important causative factors of phylogeny (Beer, 1958).

Evaluation of an additional character complex provides further evidence for the cladistic relationship of *Tarsius* and the Anthropoidea. A double discoidal placenta characterizes most Ceboidea and Cercopithecidae (Wislocki, 1929), and this doubtlessly represents a primitive condition in Anthropoidea. A double discoidal placenta is rare among eutherian mammals; it is found only in Anthropoidea, Tupaiidae, and some species of the carnivore family Mustelidae (see Mossman, 1937). The development of a double discoidal placenta in both Tupaiidae and Anthropoidea has been cited as evidence for the inclusion of tupaiids in the order Primates (Clark, 1971), despite the fact that there is distinct ontogenetic evidence for the nonhomologous and convergent evolution of double discoidal placentation in these two taxa (Luckett, 1969, 1974). In tupaiids (as in

TABLE 2

PRIMITIVE (PLESIOMORPH) AND DERIVED (APOMORPH) CHARACTER STATES OF PRIMATE FETAL MEMBRANE CHARACTERISTICS UTILIZED IN FIG. 13

Primitive □	Derived ■
1. Diffuse, epitheliochorial placenta	1. Discoidal, hemochorial placenta
2. Noninvasive attachment	2. Invasive attachment
3. No primordial amniotic cavity	3. Primordial amniotic cavity
4. Choriovitelline placenta	4. No choriovitelline placenta
5. No body stalk	5. Body stalk
6. Large, vesicular allantois	6. Rudimentary allantois
7. Paraembryonic pole attachment	7. Embryonic pole attachment
8. Amniogenesis by folding	8. Amniogenesis by cavitation
9. Bicornuate uterus	9. Simplex uterus
10. Primary yolk sac	10. Secondary yolk sac
11. No cytotrophoblastic shell	11. Cytotrophoblastic shell
12. Labyrinthine placental disc	12. Villous placental disc
13. Uterine symplasma	13. No uterine symplasma
14. Superficial implantation	14. Interstitial implantation
15. No decidua capsularis	15. Decidua capsularis

mustelids) both placental discs are primary, that is, they develop simultaneously at the initial sites of implantation, and both receive their fetal vascularization by a primary division of the umbilical vessels. In contrast, one placental disc in Old and New World monkeys is always primary, that is, it develops at the initial implantation site (the embryonic pole of the blastocyst), whereas the other disc is always secondary (and frequently smaller) in that it develops slightly later during ontogeny at the abembryonic pole of the blastocyst. Furthermore, the secondary disc receives its fetal blood supply only from vessels that cross from the primary disc, and never directly from the umbilical cord (Wislocki, 1929). The secondary placental disc (but never the primary one) has been lost independently in several genera of Anthropoidea (*Callithrix*, *Alouatta*, *Papio*). Moreover, there may be an evolutionary trend toward its loss in other genera; Chez *et al.* (1972) observed that only a single discoidal placenta occurred in 22% of their examined pregnancies of *Macaca mulatta*.

There has been little, if any, discussion of the possible selective factors that have led to the evolution of the unique pattern of double discoidal placentation in Anthropoidea. However, a comparison of the temporal sequence of development of both placental discs in *Macaca mulatta* with that of the single placental disc in *Tarsius* (cf. Figs. 11 and 12) suggests the possibility that the secondary disc of Old and New World monkeys is homologous with the single placental disc of *Tarsius* and that the primary placental disc of Anthropoidea is a neomorph that develops as a direct result of precocious implantation by the embryonic pole of the blastocyst (Luckett, 1974). If this proposed homology is true, it furnishes a plausible explanation for the origin of double discoidal placentation in Anthropoidea, and it provides additional evidence for the phyletic derivation of Anthropoidea from a tarsiiform ancestor.

VI. Phylogenetic Relationships of the Tupaiidae Suggested by Fetal Membrane Development

The development and phylogenetic significance of placentation in Tupaiidae have been detailed elsewhere (Luckett, 1968, 1969, 1974) and are only summarized here. Their placental characters are a combination of primitive eutherian retentions (amniogenesis by folding, large yolk sac and allantois, temporary choriovitelline placenta) and tupaiid specializations that do not occur in Primates (specialized bilateral endometrial pads, antimesometrial implantation chamber, decidual knot formation, and double discoidal, endotheliochorial placenta). Furthermore, neither of the two character states shared by all Primates (orthomesometrial orientation of the embryonic disc, and reduced yolk sac in late stages) occurs in tupaiids. Thus, the fetal membrane evidence excludes the Tupaiidae from any special relationship with Primates.

VII. Conclusions

Individual fetal membrane characters have clearly undergone convergent evolution in unrelated eutherian taxa; however, there is no evidence that an entire interrelated pattern of shared derived characters (such as occurs in Haplorhini) has evolved convergently. The relatively conservative fetal membrane characters provide valuable evidence of the phylogenetic relationships among higher taxa, and they strongly support the subordinal divisions of Strepsirhini and Haplorhini in a phylogenetic classification, in agreement with basicranial and immunodiffusion analyses (see Szalay and Goodman, both this volume). The grouping of strepsirhines and tarsiiforms as Prosimii appears to represent a grade or paraphyletic classification (Hennig, 1966) based on the retention of many primitive characters; this is clearly evident in an analysis of their fetal membrane characters. The fetal membrane evidence also supports the concept of Tarsiiformes and Anthropoidea as sister groups within the suborder Haplorhini and of the monophyletic origin of the Anthropoidea (Luckett, 1974). Within Anthropoidea, the Platyrrhini are relatively more primitive, and the Catarrhini, relatively more derived, in reference to their fetal membrane characters, particularly in the nature of invasive implantation and differentiation of the placental disc. Within Catarrhini, the superfamily Hominoidea is the more derived sister group of the Cercopithecoidea, with regard to the changes associated with interstitial implantation.

VIII. References

AMOROSO, E. C. 1952. Placentation, pp. 127–311. *In* A. S. Parkes, ed., *Marshall's Physiology of Reproduction*, 3rd ed., Vol. II. Longmans, Green and Co., London.

BAUCHOT, R. 1965. La placentation chez les reptiles. *Ann. Biol.* **4**:547–575.

BEER, G. DE 1958. *Embryos and Ancestors*. Oxford University Press, London.

BJÖRKMAN, N. H., and W. A. WIMSATT. 1968. The allantoic placenta of the vampire bat (*Desmodus rotundus murinus*): A reinterpretation of its structure based on electron microscopic observations. *Anat. Rec.* **162**:83–98.

BUTLER, H. 1967. The giant cell trophoblast of the Senegal galago (*Galago senegalensis senegalensis*) and its bearing on the evolution of the primate placenta. *J. Zool.* **152**:195–207.

CHEZ, R. A., J. J. SCHLESSELMAN, H. SALAZAR, and R. FOX. 1972. Single placentas in the rhesus monkey. *J. Med. Primat.* **1**:230–240.

CLARK, W. E. LE GROS. 1971. *The Antecedents of Man*, 3rd ed. Edinburgh University Press, Edinburgh.

ENDERS, A. C., and W. A. WIMSATT. 1968. Formation and structure of the hemodichorial chorio-allantoic placenta of the bat (*Myotis lucifugus lucifugus*). *Am. J. Anat.* **122**:453–490.

FARRIS, J. S. 1966. Estimation of conservatism of characters by constancy within biological populations. *Evolution* **20**:587–591.

FLYNN, T. T., and J. P. HILL. 1939. The development of the Monotremata. Part IV. Growth of the ovarian ovum, maturation, fertilisation, and early cleavage. *Trans. Zool. Soc. London* **24**:445–623.

FLYNN, T. T., and J. P. HILL. 1947. The development of the Monotremata. Part VI. The later stages of cleavage and the formation of the primary germ-layers. *Trans. Zool. Soc. London* **26**:1–151.

GRIFFITHS, M. 1968. *Echidnas*. Pergamon Press, Oxford.

GROSSER, O. 1909. *Vergleichende Anatomie und Entwicklungsgeschichte der Eihäute und der Placenta*. Wilhelm Braumüller, Vienna.

HENNIG, W. 1966. *Phylogenetic Systematics*. University of Illinois Press, Urbana.

HILL, J. P. 1910. The early development of Marsupialia, with special reference to the native cat (*Dasyurus viverrinus*). *Quart. J. Microsc. Sci.* **56**:1–134.

HILL, J. P. 1932. The developmental history of the primates. *Philos. Trans. Roy. Soc.*, Ser. B. **221**:45–178.

HILL, J. P. 1949. The allantoic placenta of *Perameles*. *Proc. Linn. Soc. London* **161**:3–7.

HILL W. C. O. 1953. *Primates*, Vol. 1, Strepsirhini. Edinburgh University Press, Edinburgh.

HUBRECHT, A. A. W. 1908. Early ontogenetic phenomena in mammals and their bearing on our interpretation of the phylogeny of the vertebrates. *Quart. J. Microsc. Sci.* **53**:1–181.

Lange, D. de. 1933. Plazentarbildung. *Handb. Vergleich. Anat. Wirbelt.* **6**:155–234.

Luckett, W. P. 1968. Morphogenesis of the placenta and fetal membranes of the tree shrews (family Tupaiidae). *Am. J. Anat.* **123**:385–428.

Luckett, W. P. 1969. Evidence for the phylogenetic relationships of tree shrews (family Tupaiidae) based on the placenta and foetal membranes. *J. Reprod. Fertil.* Suppl. 6:419–433.

Luckett, W. P. 1974. Comparative development and evolution of the placenta in primates, pp. 142–234. *In* W. P. Luckett, ed., *Reproductive Biology of the Primates, Contributions to Primatology*, Vol. 3. S. Karger, Basel.

Luckett, W. P. 1975. Causal relations in mammalian amniogenesis. *Anat. Rec.* **181**:415.

Martin, R. D. 1969. Evolution of reproductive mechanisms in primates. *J. Reprod. Fertil.* Suppl. 6:49–66.

Mossman, H. W. 1937. Comparative morphogenesis of the fetal membranes and accessory uterine structures. *Contrib. Embyrol. Carneg. Inst.* **26**:129–246.

Mossman, H. W. 1953. The genital system and the fetal membranes as criteria for mammalian phylogeny and taxonomy. *J. Mammal.* **34**:289–298.

Mossman, H. W. 1967. Comparative biology of the placenta and fetal membranes, pp. 13–97. *In* R. M. Wynn, ed., *Fetal Homeostasis*, Vol. 2. New York Academy of Science, New York.

Mossman, H. W. 1971. Orientation and site of attachment of the blastocyst, pp. 49–57. *In* R. J. Blandau, ed., *The Biology of the Blastocyst*. University of Chicago Press, Chicago.

Padykula, H. A., and J. M. Taylor. 1974. Cytological observations on marsupial placentation: The Australian bandicoots (*Perameles* and *Isoodon*). *Anat. Rec.* **178**:434.

Pearson, J. 1949. Placentation of the Marsupialia. *Proc. Linn. Soc. London* **161**:1–3.

Romer, A. S. 1967. Major steps in vertebrate evolution. *Science* **158**:1629–1637.

Schaeffer, B., M. K. Hecht, and N. Eldridge. 1972. Phylogeny and paleontology, pp. 31–46. *In* T. Dobzhansky, M. K. Hecht, and W. C. Steere, eds., *Evolutionary Biology*, Vol. 6. Appleton-Century-Crofts, New York.

Sharman, G. B. 1961. The embryonic membranes and placentation in five genera of diprotodont marsupials. *Proc. Zool. Soc. London* **137**:197–220.

Simpson, G. G. 1961. *Principles of Animal Taxonomy*. Columbia University Press, New York.

Starck, D. 1956. Primitiventwicklung und plazentation der Primaten, pp. 723–886. *In* H. Hofer, A. H. Schultz, and D. Starck, eds., *Primatologia*, Vol. 1. S. Karger, Basel.

Starck, D. 1959. Ontogenie und Entwicklungsphysiologie der Säugetiere. *Handb. Zool.* **8**:1–276.

Turner, W. 1877. Some general observations on the placenta, with especial reference to the theory of evolution. *J. Anat. Physiol.*, **11**:33–53.

Tyndale-Biscoe, H. 1973. *Life of Marsupials*. Arnold, London.

Wislocki, G. B. 1929. On the placentation of primates, with a consideration of the phylogeny of the placenta. *Contrib. Embryol. Carneg. Inst.* **20**:51–80.

8

The Central Nervous System

Its Uses and Limitations in Assessing Phylogenetic Relationships

C. B. G. CAMPBELL

I. The Organization of the Nervous System: Methods of Study

The central nervous system has played a significant role in the definition and characterization of the order Primates. In order to discuss the uses and limitations inherent in using the nervous system for assessing phylogenetic relationships, it will be necessary to briefly review some aspects of the anatomy of the nervous system and the technical methods formerly and presently available for studying it. Our understanding of this organ system is markedly limited by the methods used for its study.

The vertebrate central nervous system is composed of populations of nerve cells, called neurons, which communicate with other neurons, sensory end-organs, gland cells, or muscles by way of extensions of their protoplasm called axons and dendrites. In general, dendritic processes receive information from other neurons or sensory end-organs, while axons carry information away from the cell soma as a nerve impulse to contacts with other neurons, glands, or muscles. Neurons with like functions tend to be grouped together into clusters called "nuclei." Groups of axons in bundles are called tracts, fascicles, or commissures. Within the central nervous system, nuclear groups

C. B. G. CAMPBELL · Departments of Anatomy and Radiological Sciences, University of California, Irvine, California College of Medicine, Irvine, California.

comprise the so-called gray matter, and myelinated axon bundles comprise the white matter seen on gross inspection of a cut surface of nervous tissue.

It has proven to be very difficult to chart in detail the organization of the central nervous system pathways formed by the neuron–neuron chains. There are several reasons why this has been true. First, most routine staining methods used in histology deposit stain in either the cell soma, where the cell nucleus lies, *or* in the cell processes, but not in both simultaneously. One of the commonly used methods for demonstrating fiber (axon) bundles is the Weigert method and its modifications (Jones, 1950). This stain demonstrates only myelinated axons (axons enveloped in several layers of the modified cytoplasm of certain supporting cells called neuroglia). Since many of the pathways within the brain and spinal cord are composed of nonmyelinated or poorly myelinated axons, an incomplete picture results from the use of these staining methods alone. The precise site of termination of an axon bundle cannot be known with certainty and, since all the myelinated pathways stain simultaneously in areas in which various pathways intermingle, the pathway of special interest becomes lost in a maze of cell processes. The bulk of the older literature on fiber pathways (especially in animals other than the cat) is based on this inadequate method.

An experimental method of tracing pathways devised by Marchi (Marchi and Algeri, 1897) was based on the selective impregnation of degenerating myelin after damage to the neuron with which it was associated. This method has three main drawbacks. Firstly, only myelinated pathways could be stained in this way. Secondly, the same problem of determining the precise termination of the pathway still existed. Thirdly, the method was plagued with artifact which often led to erroneous results. The method, nevertheless, had its uses and could have yielded more information than it did. Unfortunately, many investigators preferred not to use experimental methods for various reasons, and the method was not widely exploited by comparative neuroanatomists.

Until approximately 15 years ago our understanding of central nervous system anatomy was based primarily on gross morphology and histological studies utilizing the Weigert methods for fiber bundles and the Nissl methods (Jones, 1950) for nuclear groups. At about this time a much improved method for tracing degenerating axons and axon terminals was introduced by Nauta and his co-workers (Nauta and Ryan, 1952; Nauta and Gygax, 1954) and was then further refined by Fink and Heimer (1967). Some radioactive tracer methods have been introduced even more recently which show great promise (Cowan *et al.*, 1972).

II. Reassessing the Dogma

These newer methods have been applied to animals other than the rat or cat only since the mid-1960s. It has become increasingly apparent that the new information coming from a number of laboratories is forcing a reevaluation of our fundamental concepts concerning the evolution of the vertebrate nervous system. Most of these concepts, which have achieved the status of dogma, are derived from work done around the turn of the century. Discussions of the nature and evolutionary history of the primate brain found in modern textbooks and reviews sound very much like the writings of Grafton Elliot Smith (e.g., 1924) in the late 1920s.

One of the major elements of the dogma has been that the brains of anamniotes have forebrains dominated by the olfactory sense. It was believed that the other great sensory systems, visual, auditory, and somatic sensory, were scarcely or not at all represented in the forebrain of fishes, amphibians, and reptiles. Vision and audition were thought to be controlled in the midbrain. Birds were considered to represent the first departure from olfactory domination of the forebrain because of their reduced dependence on this sense as evidenced by their small olfactory bulbs. Their obviously increased use of vision was thought to foreshadow in some way the rise of the mammals. The fact that mammals did not evolve from birds has sometimes been forgotten. There has been a tendency to search for uniqueness in order to explain the "superiority" of one animal group over others. Another major element of the dogma has been that mammals are clearly separable from anamniotes, reptiles, and birds by virtue of their unique forebrain with its highly organized thalamus and 6-layered neocortex. The forebrain of mammals was thought to capture control of vision, audition, somatic sensation, and motor movement from the centers lower in the neuraxis which apparently controlled these modalities in non-mammalian vertebrates.

The so-called dorsal cortex of reptiles has been considered to be the only possible precursor of the mammalian neocortex (Ariëns Kappers *et al.*, 1936). The neocortex is supposedly the most important element in the "superiority" of the mammalian brain and has played a major role in establishing mammalian dominance. The primates, of course, have been thought to have carried the development of the neocortex even further. The neocortex has supposedly been better differentiated in the primates, and it occupies so much more area on the surface of the cerebral hemispheres that it had to be thrown into folds or gyri in order to fit within the cranium. Elliot Smith (1924) said:

> If all the factors in his emergence are not yet known, there is one unquestionable, tangible factor that we can seize hold of and examine—the steady and uniform development of the brain along a well-defined course throughout the Primates right up to Man—which must give us the fundamental reason for "Man's emergence and ascent," whatever other factors may contribute toward that consummation.

Elliot Smith believed that the adaptation to an arboreal habitat was responsible for the diminished importance of olfaction in the primates, as well as for the increased importance of vision in this group. Again, this view is an extension of the general notion that in mammals as a group, more space is left in the forebrain for other modalities to enter as olfaction diminishes. It is now beginning to appear, however, that olfaction may never have dominated the forebrain to the extent formerly believed. Recent studies on anamniotes using modern neuroanatomical methods have shown that only a small portion of the forebrain is directly concerned with this sense. Ebbesson and Heimer (1970) have shown this to be so in the nurse shark. In addition, experimental studies have revealed the course of the visual pathways in this species to include a projection from the principal visual nucleus of the thalamus to a distinct area in the forebrain (Ebbesson, 1972; Ebbesson and Schroeder, 1971).

The first studies to indicate that the old concepts concerning the nature of nonmammalian forebrains might not be entirely correct were conducted by Karten and his co-workers on birds (Karten, 1967, 1968; Karten and Hodos, 1970; Karten and Revsin, 1966; Karten et al., 1973). The bulk of the forebrain in these animals is composed of masses of neurons organized into nuclear groups. Similar masses of neurons are to be found buried deep in the forebrain of mammals beneath the neocortical mantle. These "basal ganglia" are involved in certain patterns of motor activity. Because of this superficial similarity, the nuclear groups comprising nearly all of the forebrain of birds (and anamniotes as well) have been considered to be homologs of the basal ganglia of mammals.

Experimental studies of the central auditory and visual pathways of birds (Karten, 1967, 1968; Karten and Hodos, 1970; Karten et al., 1973) have revealed that distinct areas in the forebrain receive auditory and visual projections from the thalamus, just as the neocortex does in mammals. Further, the efferent projections of these forebrain areas are like the projections of comparable areas of the neocortex in mammals. Similar work (Hall and Ebner, 1970a, b) has since been performed in reptiles and revealed precisely the same general pattern of organization. In addition, there is now evidence that somatic sensory pathways project to the forebrain in reptiles (R. G. Northcutt, personal communication).

The results of these studies suggest that the same fundamental forebrain organization may be found in all major vertebrate groups. The mammalian forebrain then did not spring forth from its reptilian ancestry utterly without precedent. It has been suggested (Ebbesson et al., 1972) that not only was olfaction not as predominant as formerly thought, but vision may have always played the leading role. There is some evidence that motor pathways resembling the supposedly unique pyramidal system of mammals, which originates in the forebrain and projects to sensory and motor nuclei of the brain stem and spinal cord, have a counterpart in birds and sharks (Ebbesson and Schroeder, 1971; Karten, 1971). Does this new information imply that the mammalian neocortex is not a significant departure from the pattern of forebrain organization found in other vertebrates? Not at all. Clearly this unique 6-layered structure with its input–output pattern and intrinsic connection patterns is a significant morphological novelty and implies some functional novelty as well. Some neuroanatomists have felt these new concepts imply that all animals are really the same. On the contrary, they merely suggest that the differences between the nervous systems of major animal groups, including primates, are perhaps more subtle than the mere presence vs. absence of major components.

III. Homology and Homoplasy in Comparative Neurology

The discussions concerning the validity of these new views of forebrain organization (e.g., Ebbesson et al., 1972), as well as discussions concerning the postulated relationships of the Tupaiidae to the order Primates (Campbell, 1966a), have brought the concepts of homology and homoplasy into an unaccustomed prominence. Nieuwenhuys and Bodenheimer (1966) have pointed out a general lack of concern on the part of comparative neuroanatomists in regard to the criteria for the establishment of homologies in the nervous system. In addition, there has been little concern with the problems created by homoplastic similarities in assessing phylogenetic relationships using neuroanatomical data. Indeed, the terms "homoplasy," "parallelism," and "convergence" have been largely lacking in the comparative neuroanatomical literature.

Homology and homoplasy have been central issues in that part of evolutionary biology concerned with phylogenetic inference. Comparative neurologists have not been unwilling to make phylogenetic inferences from their data, but they have sometimes been unaware of the importance of these concepts in regard to the validity of their hypotheses. Perhaps the fact that neuroanatomical studies are largely pursued in medical schools, where close contact with the methods and principles of zoology is lacking, serves to explain this situation.

In addition to the above, there has been a significant problem in assessing the role that homoplastic similarity plays in the nervous system. Neuroanatomical methods, especially the modern experimental ones, are tedious and often require surgical procedures and postoperative care. This has reduced the number of species examined and favored some animals which can be readily obtained and easily kept in the laboratory. A result of this is that we have a prototypical marsupial (*Didelphis*), reptile (*Tupinambis*), bird (*Columba*), anthropoid primate (*Macaca*), etc. The hedgehog (*Erinaceus*) now seems to be the prototypical insectivore. There is always a significant danger in choosing a single animal as representative of a large taxon. It must be admitted, however, that there is much to be gained by the thorough study of a single species, as the work of Karten and Hodos (e.g., Karten, 1967, 1968; Karten and Hodos, 1970) on *Columba* and Ebbesson (e.g., Ebbesson, 1967, 1972) on *Tupinambis* and *Gingylmostoma* demonstrates.

Nevertheless, these kinds of studies do not address themselves to the problems of intragroup variation and adaptation within the nervous system to various modes of life. Very few neuroanatomists have pursued these kinds of problems. In the older literature anatomical differences between animals are often ascribed to "species difference," with no attempt to analyze the differences in regard to their meaning.

Studies on the corticospinal tracts of anthropoid primates using modern experimental methods (Kuypers, 1958, 1960a, b, 1964) showed that some neurons, whose cell somata lie in the motor area of the neocortex, send their axons down the neuraxis to synapse on the cell somata of spinal motor neurons responsible for fine motor movements of the extremities. A comparable study on at least one prosimian primate, *Nycticebus coucang* (Campbell *et al.*, 1966) suggested that this may be true of prosimians as well, although to a lesser degree. No other mammalian studies prior to 1966 demonstrated such direct corticomotoneuronal connections. Consequently, it was believed that this was a primate characteristic not found in other mammalian orders, presumably related to the fine control found in the limbs of primates, and under voluntary neocortical command.

Experimental studies of the corticospinal tracts of the raccoon, *Procyon lotor* (Buxton and Goodman, 1967; Petras and Lehman, 1966), a carnivore which manipulates its food with its forepaws, have shown that this carnivore also possesses direct corticomotoneuronal connections on those neurons which control the digits of the forepaws. The domestic cat and dog (Buxton and Goodman, 1967; Chambers and Liu, 1957), also carnivores, do not possess such connections. This appears to be a clear case of convergence.

The dorsal lateral geniculate nucleus (the principal thalamic visual nucleus) of anthropoid and prosimian primates is seen to be organized into layers or laminae in preparations stained to show neuron cell bodies. Carnivores such as the cat and dog also have such laminated nuclei, although organized somewhat differently (Guillery, 1970; Rioch, 1929). Since primates and carnivores have been considered to be "advanced" mammals, lamination has been considered to be an "advanced" characteristic. Recent experimental studies have shown that there are species with laminated dorsal lateral geniculate nuclei in 8 orders, and less than half the extant mammalian orders have been examined (Campbell, 1972). In addition, lamination is sometimes "concealed," i.e., only the experimental degeneration methods demonstrate the lamination which results from differential projection of retinal axons from the two eyes to the same nucleus. Interestingly, in the Marsupialia, the opossum *Didelphis marsupialis virginiana* lacks lamination in this nucleus, while it is well developed in the phalanger *Trichosurus* (Hayhow, 1967) and is found in the sugar glider *Petaurus* and the kangaroo and its allies (Johnson and Marsh, 1969). Among the Chiroptera it is lacking in the microchiropterans examined, but present in the megachiropteran *Pteropus* (Crowle, 1974). If one examines the modes of life of the animals possessing this character, the conclusion is inescapable that lamination has arisen independently many times in animals which must rapidly evaluate spatial relations. These are animals that fly, are arboreal, glide, or move rapidly on the ground (Campbell, 1972).

Primates are well known for their excellent vision. Their visual cortex is highly differentiated into layers and sublayers and is characterized by a layer of myelinated axons which can be seen with the unaided eye in the unfixed, unstained brain. This layer is termed the stria of Gennari. The brain of *Tupaia glis* has a visual cortex which resembles the primate visual cortex. A recent study has shown that the visual cortex of the arboreal grey squirrel is almost identical to that of *Tupaia*. Such an elaborate visual cortex is not found in terrestrial rodents. This suggests that an elaborate visual cortex can appear convergently in divergent lineages and need not imply affinity to the primates as advocated by Clark (1971). I have suggested elsewhere that the primate resemblances found in the brain of *Tupaia* are homoplastic and are essentially all reflections of an elaborate visual system (Campbell, 1966a,b).

These are just three examples of homoplastic resemblances to be found among the nervous systems of mammals, including primates. If the central nervous system is to be used as a source of characters for drawing phylogenetic inferences, then we must concern ourselves with the criteria available for the recognition of homology and homoplasy in this organ. This subject has been discussed in some detail elsewhere (Campbell and Hodos, 1970). It was pointed out at that time that many neuroanatomists were not aware that there were two major approaches to the concept of homology: the phylogenetic and the structuralist. In general, European neuroanatomists have given more thought and concern to these topics than have their North American colleagues. Europeans have most often adopted a structuralist approach. Campbell and Hodos (1970) prefer a phylogenetic approach and accept definitions of homology, homogeny, homoplasy, convergence, parallelism, chance similarity, and analogy based upon Ghiselin (1966a,b), Bock (1969), Lankester (1870), and Simpson (1961). Campbell and Hodos suggest that comparative neurologists should at least make clear in what sense they are using these terms. They further suggest that of the several criteria for the recognition of homology put forth by Simpson (1961), minuteness of similarities, multiplicity of similarities, and ontogenetic similarities are perhaps most useful in neuroanatomical studies. The fossil record is less useful but has a distinct role to play.

In discussions of which features of the central nervous system might be utilized for inferring homologies, it has been commonplace to advocate a single feature as the most useful. Other features are considered either less useful or totally meaningless. Kuhlenbeck (1929) thought that in anamniotes the sulci are the important determinants of brain organization. Bergquist (1932) and Källen (1951) suggested that the topographic relations of nuclear groups as observed in the embryo were most useful. Nieuwenhuys and Bodenheimer (1966) advocated the topologic relations of nuclear groups as the most reliable feature, while Herrick (1933) believed that sulci, nuclear structure, and fiber connections would all be in agreement.

Nieuwenhuys and Bodenheimer (1966) studied the organization of the diencephalon in some fishes and amphibians. They concluded that the ventricular sulci were too variable in presence, position, and distinctness to be useful in determining the homologies of diencephalic regions. They deplored the use of fiber (axon) connections in establishing homologies because they considered such use to be based on circular reasoning. They argue that since fiber tracts are identified (and, therefore, homologized) on the basis of the nuclear groups that they interconnect, it is unreasonable to then use the relations of nuclear groups to fiber pathways in order to homologize nuclear groups. It was their belief that the topologic relationship of nuclear groups is more reliable than any other feature and should be *the* criterion of homology. It should be noted that Nieuwenhuys and Bodenheimer (1966) have adopted a structuralist definition of homology.

Campbell and Hodos (1970) disagreed with the view that fiber connections are useless in determining homologies. Nuclear groups and axon (fiber) bundles are not really separate entities. They are component parts of the same entities: neurons or nerve cells. If the brain, spinal cord, and peripheral nerves are considered to be a "system," then certain populations of neurons with their processes are related to other neurons, sensory end-organs, etc., by contacts termed "synapses." If they all are more or less involved in the performance of the same broad function, e.g., vision or audition, then they form a subsystem. One can only determine to which subsystem a given population of neurons may belong by determining its relation to other neuron populations, sensory end-organs, etc.

It is not intended to imply that a knowledge of the fiber connections of populations of neurons in two divergent animal species will always enable one to determine if they are homologous, but it is at least an important first step. For example, in a recent study of the retinal efferent pathways of the duckbill platypus, *Ornithorhynchus anatinus* (Campbell and Hayhow, 1972), it was found that a nucleus designated as the nucleus of the stria terminalis by Hines (1929), on the basis of topographic relations, is apparently a major diencephalic visual nucleus (one of the divisions of the lateral geniculate nucleus). Whether it is the homolog of the dorsal or the ventral division of this nucleus is not yet known. More evidence is needed before this can be determined.

The platypus is a species in which the diencephalon is not well differentiated into clear-cut nuclear groups on the basis of the Nissl stain. To further complicate matters, those portions of the brain devoted to somatic sensation from the soft, leathery bill are greatly expanded so that the brainstem is distorted, and the usual relations of one area with another are disturbed. Since the lateral geniculate nucleus is usually on the dorso-lateral surface of the diencephalon, Hines designated a nucleus in that position as possibly representing that structure in this species. It did not seem logical on the basis of topography that a dorsomedial nucleus immediately adjacent to the stria terminalis should receive retinal input. Only an experimental study was capable of demonstrating to which subsystem the so-called nucleus of the stria terminalis actually belonged. Its position in the diencephalon suggests that it is part of the lateral geniculate nucleus rather than some other visual nucleus found in another brain subdivision.

There is no *one* feature that will be universally useful for determining homologies, but as many lines of evidence as possible should be utilized. Surely fiber connections will usually be important. Topographic and topologic position of the nuclear group, morphology, and staining characteristics of individual neurons in the nucleus, histochemical properties, embryologic origin, and in some instances relation to reliably occurring sulci may also be very useful. In regard to diencephalic sulci, it is possible that the relationship of nuclear groups to certain sulci may be a reliable item of evidence for inferring homologies within a group, for example the salamanders, but may be unreliable for inferring homologies between salamanders and fishes. The reliability of characters considered for this purpose must be examined in relation to the particular case. This again points up a need for more intragroup comparisons using experimental methods than has been customary. Only in this way can the range of variation in nervous system characters be determined. Studies of this sort have been somewhat inhibited by a general lack of interest in comparative neurology on the part of the neurobiological establishment, as well as by the view that similar studies of many different animals are pedestrian.

IV. The Nervous System and the Assessment of Primate Relationships

With the exception of Clark's (1971 and elsewhere) use of central nervous system characters in his arguments for classification of the tupaiids within the order Primates, this organ system has seldom been used in primate taxonomy for the assessment of phylogenetic relationships. He contrasted the brain of living tupaiids with the brains of extant insectivores on the one hand and primates on the other. No comparisons were made with other mammalian brains. He noted (p. 241) that inferences concerning the brain of the fossil *Plesiadapis* drawn from study of the skull suggest that this fossil primate had a brain more primitive than *Ptilocercus*. *Ptilocercus* is an extant tupaiid which lacks all of the advanced primate-like characters found in *Tupaia*. This, of itself, suggests that these primate-like features are most likely to be independent acquisitions, a result of either convergence or parallelism. At any rate, it implies that the ancient prosimians were probably not derived from ancestors with brains closely resembling the modern *Tupaia*.

As shown previously (Campbell, 1966b, 1974), the list of characters described by

Clark (1971) in which the brain of *Tupaia* resembles that of primates consists essentially of characters related to the visual system. When examined carefully it is found that some of these characters are possessed by nonprimate mammals with well-developed visual systems and some others only superficially resemble primates. For example, it is true that tupaiids possess a laminated dorsal lateral geniculate nucleus as do primates (as well as a number of other mammals); however, the pattern of input from the retina of the eye to the nucleus is unlike that found in prosimians or anthropoid primates.

Clark (1971) argued that the latter character, as well as the position of the tupaiid pyramidal tracts in the dorsal funiculus of the spinal cord, which contrasts with that of primates in the lateral funiculus, are examples of tupaiid uniqueness. It was his view that these "aberrant specializations" did not argue against primate affinity. In order for the central nervous system to contribute valid evidence of primate affinity for the tupaiids, one would have to be able to demonstrate shared derived character states in this system.

In regard to the pyramidal tract, this structure composed of both corticobulbar and corticospinal fibers has been studied with experimental methods in representatives of 11 mammalian orders. These include the Monotremata, Marsupialia, Insectivora, Primates, Artiodactyla, Perissodactyla, Lagomorpha, Rodentia, Edentata, Carnivora, and Hyracoidea. Additionally, there is nonexperimental data available for the Proboscidea, Cetacea, and Chiroptera. To my knowledge the Dermoptera, Tubulidentata, Sirenia, and Macroscelidea are unknown insofar as the position and extent of the pyramidal tract is concerned.

Corticospinal fibers in mammals typically form a major, usually crossed, tract in either the dorsal, lateral, or ventral funiculus of the spinal cord. Other corticospinal fibers may be found in smaller numbers in other funiculi where they may or may not form conspicuous tracts. In most mammalian orders the major pyramidal tract decussates in the brain stem at its junction with the spinal cord. Decussations at more rostral levels of the neuraxis have been described in some mammals, e.g., in the echidna (Goldby, 1939) and in bats (Fuse, 1926). In the Insectivora and Hyracoidea, corticospinal fibers do not decussate in the brainstem at any one conspicuous location, but cross to the opposite spinal gray at the level of their termination in the spinal cord.

The position of the major tract has been used as a taxonomic character because of its apparent consistency within orders. Strominger (1969) compared edentates of two different infraorders. Both the sloth *Choloepus* and the armadillo *Dasypus* have crossed corticospinal fibers which descend to thoracic levels in the lateral and ventral funiculi. In *Dasypus* the ventral tract is emphasized, while in *Choloepus* the lateral tract is the major one. This then is the only known exception to consistency of this character within orders. It should be remembered that with the exception of the order Primates where 14 species have been examined, most orders are represented by samples of no more than 4 species.

Noback and Shriver (1966) thought that forebrain efferents to the brainstem appeared in nonmammalian vertebrates ancestral to mammals. They suggested that the original corticospinal tracts were "terminal collaterals" of these corticoreticular fibers which came to extend into the spinal cord. They further suggest that the corticospinal tracts evolved independently in the several orders of mammals and that once the funicular position of the main tract was established in the ancestor of the order, that location was permanently established for all subsequent members of that order.

This hypothesis is more plausible than the suggestion often made in the literature that one funicular position, usually the dorsal one, is the "primitive" position. This is based on the fact that "primitive" mammals, such as marsupials and rodents, have primarily dorsal tracts. The implication is made that funicular position changes to the lateral one when the "advanced" orders, e.g., carnivores and primates, evolved. This latter view ignores the fact that most mammalian orders diverged from the ancient insectivore-like stock in the Cretaceous period and have been evolving independently since that time.

It is now known that the pyramidal system is not an exclusively mammalian invention as Noback and Shriver thought. Corticobulbar pathways similar to those of mammals, including those dorsolateral fibers corresponding to the bundle of Bagley described first in the sheep, as well as the medial fibers traveling in the pyramids, have been described in an owl, *Speotyto* (Karten, 1971; Karten and Dubbeldam, 1973). In addition, some of these fibers enter the dorsal funiculus of the rostral spinal cord. This is as well-developed a pyramidal system as is found in some mammalian taxa. It is apparently not so well developed in all birds, as the pigeon *Columba livia* has pyramidal-tract projections which cannot be traced caudally beyond the prerubral field in the rostral midbrain (Karten, personal communication).

There is as yet no evidence that closely similar pathways exist in extant reptiles; however, such studies are only now being attempted. It is possible that the avian pyramidal system has arisen by convergence in certain birds and that the ancient reptiles did not possess a pyramidal system. More experimental data from birds and reptiles are needed to enable one to make a reasonable inference in this regard.

Both the visual and motor pathways of primates are significantly different in organization from most other mammals which have been comparably examined. They are also quite different from the extant insectivores which have been examined (Campbell *et al.*, 1967). The fact that tupaiids do not resemble either group insofar as these derived character states are concerned is strongly suggestive that they have taken a course divergent from that of primates. It is not yet possible to make dogmatic statements about this issue because our experimental evidence is still limited to less than half of the extant mammalian orders. Further, only a few species, and in some cases a single one, have been examined thus far in each order represented.

The sampling problem is a highly significant one. Until recently experimental studies of primate brain organization were essentially confined to *Macaca*. Of course there has been a certain amount of random, gross, descriptive study of the cytoarchitecture and/or myeloarchitecture of various primate brains. In recent years, with the burgeoning interest in primate biology, there has been an increase both in descriptive and experimental studies of a number of species. Experimentally derived data are now available on various aspects of the nervous systems of some prosimians including *Nycticebus* and *Galago*, New World monkeys such as *Saimiri* and *Aotus*, and Old World primates such as *Macaca* and *Pan*.

With the advent of detailed intragroup comparisons using experimental methods, some interesting intragroup differences may be appearing. For example, Spatz and Tigges (1972) described a difference in the organization of the cortical visual areas between *Macaca mulatta* and *Saimiri sciureus*. In *Macaca*, area 17 (the primary visual area or striate cortex) projects topographically to two extrastriate visual areas, area 18 and area 19. In *Saimiri* an area 19 was not identifiable, but a topographic projection from area 17 to area 18 was present. It is not yet clear whether or not this is characteristic only of *Saimiri*, is present in some or all New World monkeys, or is more widely distributed in primates. Only more extensive sampling will determine the true situation.

One of the issues again of interest in primate taxonomy concerns the relationships of the tarsiers. Are they more closely related to anthropoids or to the lemurs and lorises? Nervous system characters could potentially be useful evidence in this regard. For example, there has been a consistent dichotomy between these two major groups of primates insofar as the organization of their central visual system is concerned. The lateral geniculate nucleus of almost all primates has been generally described as being composed of 6 major cell layers or laminae. Clark (1971) has emphasized the difference in orientation of these laminae between anthropoids on the one hand and prosimians on the other.

Clark described the anthropoid condition as "everted" and the prosimian condition as "inverted." In an early publication (Clark, 1930) he described the lateral geniculate nucleus of *Tarsius* as conforming to the anthropoid configuration. More recently, however, Clark (1971) grouped *Tarsius* with the lemurs and lorises in possessing "inverted" lateral geniculate nuclei. In anthropoids the neurons of laminae 1, 4, and 6 receive retinal axons from the contralateral eye, while axons from the ipsilateral eye terminate upon cells in laminae 2, 3, and 5. In prosimians the contralateral eye projects to laminae 1, 5, and 6 and the ipsilateral eye to laminae 2, 3, and 4 (also see Noback, this volume).

Tarsius has been described as possessing either 3 (Clark, 1930) or 4 (Hassler, 1966) laminae when studied with Nissl stains. It is possible that neither view is correct. More than one lamina was found to be hidden in the parvocellular cell mass of the lateral geniculate nucleus of *Saimiri* and *Ateles* when experimental methods of study were utilized. A modern study of the visual system of *Tarsius*, including use of experimental techniques, would determine the configuration of the lateral geniculate nucleus, whether of the prosimian or anthropoid type, and the pattern of retinal input to the nucleus. The accessory optic system is not known to be divergent in anthropoids and prosimians. All primates examined thus far lack a retinal projection to a medial terminal nucleus of the accessory optic system and possess a lateral terminal nucleus which receives retinal axons. Projection to a dorsal terminal nucleus has not always been described (e.g., Campos-Ortega and Glees, 1967; Tigges and Tigges, 1969); however, it has been noted by some observers in both prosimians and anthropoids (Campbell, 1969; Giolli and Tigges, 1970).

Even more important than the problems inherent in sampling sufficiently extensively have been the relatively unsophisticated approaches which have been adopted for inferring nervous system phylogeny in the past. The now infamous rat–cat–monkey–man comparisons which led to highly unreliable phylogenetic inferences are all too familiar. Although seldom seen in the more recent literature, many clinicians, neurophysiologists, and others persist in the view that this is a valid approach.

In recent years there has been some emphasis on making comparisons within lineages, e.g., examining brain characters in a series of insectivores, prosimian primates, anthropoid primates, and man, if one were interested in the phylogeny of the human brain. That this might be an important advance over the more common rat–cat–monkey–man comparisons was made clear to me by Dr. John A. Jane in the early 1960s. We have used this scheme for comparison in drawing inferences concerning the evolutionary history of the human corticospinal motor system (Jane *et al.*, 1969), and its use has been advocated elsewhere (Hodos and Campbell, 1969). I still consider it an advance; however, the methods used for judging which characters are ancestral and which are derived within a lineage must also be included in one's approach if valid phylogenetic inferences are to be drawn (Campbell, 1975). This will usually necessitate the examination of more than 3 or 4 species if anything more specific than gross generalizations are to result from such a study.

A common plan has been to make comparisons between the hedgehog, the tree shrew, and a primate such as *Galago*. If whatever nervous system feature is studied forms a morphological sequence of increasing complexity, then this is inferred to be the historical course of events in the primate lineage. It is sometimes forgotten that characters do not always progress over the course of time from simplicity to complexity in organization. The use of the tree shrew as an "intermediate" has been a particularly vexing problem. Since it is unlikely that the tree shrews are primates or primate ancestors, a number of problems arise in attempting to use them as intermediates between what is considered to be the insectivore condition on the one hand and the primate condition on the other. Extant tree shrews, especially *Ptilocercus*, surely retain many ancestral features. The central nervous system characters which are usually considered intermediate to the hedgehog and primate conditions are, however, usually derived from the visual system. This system is almost surely homoplastic in regard to its primate similarities. The primate lineage may never have passed through a stage with a visual system resembling that of extant tree shrews.

I believe that the central nervous system can be useful in inferring phylogeny and that we can learn something about the evolutionary history of primate and other nervous systems by studying extant animals. However, valid inferences cannot be drawn unless comparative neurologists make themselves aware of the complexities involved in making such inferences. They must become familiar with the methods used in evolutionary biology for performing these analyses. A great deal of work will be required, much of it seemingly pedestrian and repetitive.

This is an exciting era in comparative neurology. After a prolonged period of relative inactivity in this field, new investigative tools are revealing unimagined errors and inadequacies in our understanding of nervous system organization and evolution. Primate neurology will certainly share in this renaissance. It should be remembered that the concepts of primate brain evolution found in most textbooks and popular accounts are largely derived from primarily descriptive studies in the eras of Grafton Elliot Smith and the pioneering experimental studies of Wilfrid Le Gros Clark in the 1930s and 1940s. There is a great deal to learn, and we now have powerful new tools to aid us in this endeavor. How well these tools are used will depend on how carefully the principles of evolutionary biology are heeded and allowed to give direction to future work.

V. References

ARIËNS KAPPERS, C. U., G. C. HUBER, and E. C. CROSBY. 1936. *The Comparative Anatomy of the Nervous System of Vertebrates, Including Man.* MacMillan, New York.

BERGQUIST, H. 1932. Morphologie des Zwischenhirns bei niederen Wirbeltieren. *Acta Zool.* **13**:57–304.

BOCK, W. J. 1969. Discussion: The concept of homology. *In* J. M. Petras and C. R. Noback, eds., *Comparative and Evolutionary Aspects of the Vertebrate Central Nervous System. Ann. N.Y. Acad. Sci.* **167**:71–73.

BUXTON, D. F., and D. C. GOODMAN. 1967. Motor function and the corticospinal tracts in the dog and raccoon. *J. Comp. Neurol.* **129**:341–360.

CAMPBELL, C. B. G. 1966a. Taxonomic status of tree shrews. *Science* **153**:436.

CAMPBELL, C. B. G. 1966b. The relationships of the tree shrews: The evidence of the nervous system. *Evolution* **20**:276–281.

CAMPBELL, C. B. G. 1969. The visual system of insectivores and primates. *Ann. N.Y. Acad. Sci.* **167**:388–403.

CAMPBELL, C. B. G. 1972. Evolutionary patterns in mammalian diencephalic visual nuclei and their fiber connections. *Brain Behav. Evol.* **6**:218–236.

CAMPBELL, C. B. G. 1974. On the phyletic relationships of the tree shrews. *Mammal Rev.* **4**(4): 125–143.

CAMPBELL, C. B. G. 1975 (in press). What animals should we compare? *In* B. Masterton *et al.*, eds., *Evolution, Brain, and Behavior: Persistent Problems.* Erlbaum and Associates, Hillsdale, N.J.

CAMPBELL, C. B. G., and W. R. HAYHOW. 1972. Primary optic pathways in the duckbill platypus, *Ornithorhynchus anatinus*: An experimental degeneration study. *J. Comp. Neurol.* **145**:195–208.

CAMPBELL, C. B. G., and W. HODOS. 1970. The concept of homology and the evolution of the nervous system. *Brain Behav. Evol.* **3**:353–367.

CAMPBELL, C. B. G., J. A. JANE, and D. YASHON. 1967. The retinal projections of the tree shrew and hedgehog. *Brain Res.* **5**:406–418.

CAMPBELL, C. B. G., D. YASHON, and J. A. JANE. 1966. The origin, course and termination of corticospinal fibers in the slow loris, *Nycticebus coucang* (Boddaert). *J. Comp. Neurol.* **127**:101–112.

CAMPOS-ORTEGA, J. A., and GLEES, P. 1967. The subcortical distribution of optic fibers in *Saimiri sciureus* (Squirrel Monkey). *J. Comp. Neurol.* **131**:131–142.

CHAMBERS, W. W., and C. N. LIU. 1957. Corticospinal tract of the cat. An attempt to correlate the pattern of degeneration with deficits in reflex activity following neocortical lesions. *J. Comp. Neurol.* **108**:23–55.

CLARK, W. E. LE GROS. 1930. The thalamus of *Tarsius*. *J. Anat.* **64**:371–414.

CLARK, W. E. LE GROS. 1971. *The Antecedents of Man*, 3rd ed. Quadrangle, Chicago.

COWAN, W. M., D. I. GOTTLIEB, A. E. HENDRICKSON, J. L. PRICE, and T. A. WOOLSEY. 1972. The autoradiographic demonstration of axonal connections in the central nervous system. *Brain Res.* **37**:21–51.

CROWLE, P. K. 1974. Experimental Investigation of Retinofugal Connections to the Diencephalon and Midbrain of Chiroptera. Thesis, Indiana University.

EBBESSON, S. O. E. 1967. Ascending axon degeneration following hemisection of the spinal cord in the Tegu lizard (*Tupinambis nigropunctatus*). *Brain Res.* **5**:178–206.

EBBESSON, S. O. E., 1972. New insights into the organization of the shark brain. Proc. Elasmobranch Biol. Symp. 1971. *Comp. Biochem. Physiol.* **42**:121–129.

EBBESSON, S. O. E., and L. HEIMER. 1970. Projections of the olfactory tract fibers in the nurse shark (*Gingylymostoma cirratum*). *Brain Res.* **17**:47–55.

EBBESSON, S. O. E., and D. M. SCHROEDER. 1971. Connections of the nurse shark's telencephalon. *Science* **173**: 254–256.

EBBESSON, S. O. E., J. A. JANE, and D. M. SCHROEDER. 1972. A general overview of major interspecific variations in thalamic organization. *Brain Behav. Evol.* **6**:92–130.

ELLIOT SMITH, G. 1924. *Essays on the Evolution of Man.* Oxford University Press, London.

FINK, R. P., and L. HEIMER. 1967. Two methods for selective silver impregnation of degenerating axons and their synaptic endings in the central nervous system. *Brain Res.* **4**:369–374.

FUSE, G. 1926. Vergleichend-anatomische Beiträge zur Kenntnis über die sog. obere, zweite oder proximale Pyramidenkreuzung bei Edentaten, sowie bei einigen fliegenden Saügern. *Arb. Anat. Inst. Sendai* **12**:47–92.

GHISELIN, M. T. 1966a. An application of the theory of definitions to systematic principles. *Syst. Zool.* **15**:127–130.

GHISELIN, M. T. 1966b. On psychologism in the logic of taxonomic controversies. *Syst. Zool.* **15**:207–215.

GIOLLI, R. A., and J. TIGGES. 1970. The primary optic pathways and nuclei in primates, pp. 29–54. *In* C. R. Noback and W. Montagna, eds., *Advances in Primatology*, Vol. 1. Appleton-Century-Crofts, New York.

GOLDBY, F. 1939. An experimental investigation of the motor cortex and pyramidal tract of *Echidna aculeata*. *J. Anat.* **73**:509–524.

GUILLERY, R. W. 1970. The laminar distribution of retinal fibers in the dorsal lateral geniculate nucleus of the cat. A new interpretation. *J. Comp. Neurol.* **138**:339–368.

HALL, W. C., and F. F. EBNER. 1970a. Parallels in the visual afferent projections of the thalamus in the hedgehog (*Paraechinus hypomelas*) and the turtle (*Pseudemys scripta*). *Brain Behav. Evol.* **3**:135–154.

HALL, W. C., and F. F. EBNER. 1970b. Thalamotelencephalic projections in the turtle. (*Pseudemys scripta*) *J. Comp. Neurol.* **140**:101–122.

HASSLER, R. 1966. Comparative anatomy of the central visual systems in day- and night-active primates, pp. 419–434. *In* R. Hassler and H. Stephan, eds., *Evolution of the Forebrain.* Georg Thieme Verlag, Stuttgart.

HAYHOW, W. R. 1967. The lateral geniculate nucleus of the marsupial phalanger, *Trichosurus vulpecula*. An experimental study in relation to the intranuclear optic nerve projection fields. *J. Comp. Neurol.* **131**:571–604.

HERRICK, C. J. 1933. The amphibian forebrain. VI. Necturus. *J. Comp. Neurol.* **58**:1–288.

HINES, M. 1929. The brain of *Ornithorhynchus anatinus*. *Philos. Trans. Roy. Soc. London, Ser. B* **217**:155–287.

HODOS, W., and C. B. G. CAMPBELL. 1969. *Scala Naturae*: Why there is no theory in comparative psychology. *Psychol. Rev.* **76**:337–350.

JANE, J. A., C. B. G. CAMPBELL, and D. YASHON. 1969. The origin of the corticospinal tract of the tree shrew (*Tupaia glis*) with observations on its brain stem and spinal terminations. *Brain Behav. Evol.* **2**:160–182.

JOHNSON, J. I., and M. P. MARSH. 1969. Laminated lateral geniculate in the nocturnal marsupial *Petaurus breviceps* (sugar glider). *Brain Res.* **15**:250–254.

JONES, RUTH MCCLUNG. 1950. *McClung's Handbook of Microscopical Technique.* Hoeber, New York.

KÄLLEN, B. 1951. Embryological studies on the nuclei and their homologization in the vertebrate forebrain. *K. Fisiogr. Saellsk. Lund Handl. N.F.* **62**:34.

KARTEN, H. J. 1967. The organization of the ascending auditory pathway in the pigeon (*Columba livia*). I. Diencephalic projections of the inferior colliculus (nucleus mesencephali lateralis, pars dorsalis). *Brain Res.* **6**:409–427.

KARTEN, H. J. 1968. The ascending auditory pathway in the pigeon (*Columba livia*). II. Telencephalic projections of the nucleus ovoidalis thalami. *Brain Res.* **11**:134–153.

KARTEN, H. J. 1971. Efferent projections of the wulst of the owl. *Anat. Rec.* **169**:353 (abstract).

KARTEN, H. J., and J. DUBBELDAM. 1973. The organization and projections of the paleostriatal complex in the pigeon (*Columba livia*). *J. Comp. Neurol.* **148**:61–90.

KARTEN, H. J., and W. HODOS. 1970. Telencephalic projections of the nucleus rotundus in the pigeon (*Columba livia*). *J. Comp. Neurol.* **140**:35–51.

KARTEN, H. J., and A. M. REVSIN. 1966. The afferent connections of the nucleus rotundus in the pigeon. *Brain Res.* **2**:368–377.

KARTEN, H. J., W. HODOS, W. J. H. NAUTA, and A. M. REVSIN. 1973. Neural connections of the visual "Wulst" of the avian telencephalon. Experimental studies in the pigeon (*Columba livia*) and owl (*Speotyto cunicularia*). *J. Comp Neurol.* **150**:253–278.

KUHLENBECK, H. 1929. Uber die Grundbestandteile des Zwischenhirnbauplans der Anamnier. *Morphol. J.* **63**:50–95.

KUYPERS, H. G. J. M. 1958. Pericentral cortical projections to motor and sensory nuclei. *Science* **128**:662–663.

KUYPERS, H. G. J. M. 1960a. Central cortical projections to motor and somatosensory cell groups. *Brain* **83**:161–184.

KUYPERS, H. G. J. M. 1960b. Central cortical projections to motor somatsosensory and reticular cell groups, pp. 138–143. *In* D. B. Tower and J. P. Schadé, eds., *Structure and Functions of the Cerebral Cortex.* Elsevier, Amsterdam.

KUYPERS, H. G. J. M. 1964. The descending pathways to the spinal cord, their anatomy and function, pp. 178–202. *In* J. C. Eccles and J. P. Schadé, eds., *Progress in Brain Research*, Vol. 2. Elsevier, Amsterdam.

LANKESTER, E. R. 1870. On the use of the term homology in modern zoology, and the distinction between homogenetic and homoplastic agreements. *Ann. Mag. Nat. Hist.* **6**:34–43.

MARCHI, V., and ALGERI, G. 1897. *Riv. Sper. Freniatr. Med. Leg.* **12**:3.

NAUTA, W. J. H., and P. A. GYGAX. 1954. Silver impregnation of degenerating axons in the central nervous system: A modified technique. *Stain Technol.* **29**:91–93.

NAUTA, W. J. H., and L. F. RYAN. 1952. Selective silver impregnation of degenerating axons in the central nervous system. *Stain Technol.* **27**:175–179.

NIEUWENHUYS, R., and T. S. BODENHEIMER. 1966. The diencephalon of the primitive bony fish Polypterus in the light of the problem of homology. *J. Morphol.* **188**:415–450.

NOBACK, C. R., and J. E. SHRIVER. 1966. Phylogenetic and ontogenetic aspects of the lemniscal systems and the pyramidal system, pp. 316–325. *In* R. Hassler and H. Stephan, eds., *Evolution of the Forebrain*. Georg Thieme Verlag, Stuttgart.

PETRAS, J. M., and R. A. W. LEHMAN. 1966. Corticospinal fibers in the raccoon. *Brain Res.* **3**:195–197.

RIOCH, D. McK. 1929. Studies on the diencephalon of carnivora. I. The nuclear configuration of the thalamus, epithalamus, and hypothalamus of the dog and cat. *J. Comp. Neurol.* **41**:1–120.

SIMPSON, G. G. 1961. *Principles of Animal Taxonomy*. Columbia, New York.

SPATZ, W. B., and J. TIGGES. 1972. Species difference between Old World and New World monkeys in the organization of the striate–prestriate association. *Brain Res.* **43**:591–594.

STROMINGER, N. L. 1969. A comparison of the pyramidal tracts in two species of edentate. *Brain Res.* **15**:259–262.

TIGGES, J., and M. TIGGES. 1969. The accessory optic system and other optic fibers of the squirrel monkey. *Folia Primat.* **10**:245–262.

9

The Visual System of Primates in Phylogenetic Studies*

CHARLES R. NOBACK

I. Introduction

A well-developed visual sense coupled with different degrees of manual and digital dexterity are two dominant expressions which characterize the order Primates. Thus, an understanding of the neurobiology of the optic and motor systems of the members of this order of mammals can shed light on several problems of phylogenetic significance.

A comprehensive understanding of the evolution of the visual system during primate phylogeny is possible only if comparisons can be made of the visual systems of the common ancestors of the primates and a representative series of subsequent phyletic stages. Unfortunately such a panoramic view cannot be attained, because the crucial stages are lost in the geologic past. The endocasts of primate fossil brains do present some idea as to the size of the "visual cortex" during primate evolution from the contours of the occipital lobe (Radinsky, 1972).

II. Classic Neural Pathways of the Visual System

Environmental stimuli can evoke activity in a sensor (e.g., rods and cones of the eye, tactile corpuscle of the skin). If adequately stimulated, sensors produce neural influences

* Supported by NIH Grants NS-03473 and NS-12436.

CHARLES R. NOBACK · Department of Anatomy, College of Physicians and Surgeons, Columbia University, New York, New York.

which are conveyed through pathways of the nervous system. Each pathway consists of sequences of processing stations (called nuclei, ganglia, bodies, or cortex) and bundles of axons (nerves, tracts, or radiations).

In the classic literature, the visual system of mammals is considered to convey neural codes from the eye to the brain via two basic pathways. The pathway associated primarily with the conscious appreciation of wavelengths in the visual electromagnetic spectrum is the retinogeniculostriate pathway from the eye to the visual cortex (Fig. 1). In this pathway, the retina, the lateral geniculate body of the thalamus, and the visual cortex are linked into a hierarchically organized sequence of neural processing stations. The other pathway is associated primarily with optic reflexes; it is the retinotectal pathway to optic tectum (superior colliculus and pretectum) of the midbrain. In addition to its important role in visuomotor control, the optic tectum also has a functional correlate in visual perception as well.

The following presentation concentrates primarily on the retinogeniculostriate pathway. Although the basic elements of this pathway system are probably similar in all primates and *Tupaia*, important differences do exist among the various groups of these mammals. Some of these differences will be noted and their possible phylogenetic significance will be presented.

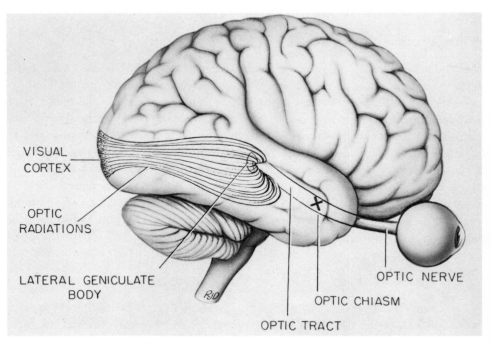

FIG. 1. The retinogeniculostriate pathway conveys influences viewed by the eye from the retina to the lateral geniculate body (via the optic nerve, chiasm, and tract) to the primary visual cortex (area 17, striate cortex) via the optic radiation.

III. Light Gathering, Visual Acuity, and Color Vision

201

THE VISUAL
SYSTEM
IN PRIMATE
PHYLOGENY

The primates have evolved efficient eyes and optic systems. Most species of this order do not possess as efficient a light-gathering eye as many carnivores or as exquisite a visual acuity as some birds. However, the eyes and visual pathways of primates do combine a generally effective light-gathering capacity, visual acuity, and color sense in a remarkable optic system (Elliot Smith, 1928; Duke-Elder, 1958; Prince, 1956). The light gathering is performed by the pupillary apparatus (iris diaphram) and the rod and cone cells of the retina. The former regulates the amount of light entering the eye and the latter are sensitive photoreceptors in the retina. High visual acuity is associated with the presence of a fovea within the area centralis of the eye of "higher primates," accompanied by the parallel and concomitant development of a foveal system (described below). The cones are color photoreceptors which initiate the neural activity in the linkage of certain neurons of the retinogeniculostriate pathway. In addition, there is an enlargement of the processing stations and tracts associated with this system, including the visual cortex of the forebrain (Fig. 2).

A. Area Centralis (Macula Lutea) and Fovea

Near the posterior pole of the eye of many mammals, including primates and man, is a localized modification of the retina, within which is a funnel-shaped depression or pit. This modification is called the area centralis; the depression is called the fovea. The area centralis is defined "as a circumscribed retinal area within which the retina is so constructed as to afford a marked local increase in resolving power" (Walls, 1942). In higher primates and man, the area centralis is often called the macula lutea (yellow spot) because of the yellowish pigmentation within it. An area centralis can occur without the presence of a fovea, but a fovea can only exist as a specialization within an area centralis. A fovea contains extremely slender, closely packed cones; this partially explains the high visual acuity and color sense which are perceived by the foveal system. This system is composed of the foveal cones and their neuronal linkages within the retinogeniculostriate system. (The foveal system is discussed further in Section IV C.) Although some rods are present, the photoreceptors in the nonfoveal area centralis are primarily cones.

IV. The Visual Systems

Until a few years ago, visual perception in mammals had been considered to be subserved exclusively by the retinogeniculostriate pathway and the cerebral cortex. The basic structures of this pathway system comprise, in order, the eye, optic nerve, optic chiasma, optic tract, lateral geniculate body, optic radiation, and visual cortex (Figs. 1 and 2). The visual cortex is generally subdivided into a striate cortex (area 17, primary visual cortex) and a circumstriate cortex (areas 18 and 19). The circumstriate cortex of each hemisphere surrounds the striate cortex. It is connected with its counterpart of the contralateral hemisphere by nerve fibers which pass through the corpus collosum. The

Charles R.
Noback

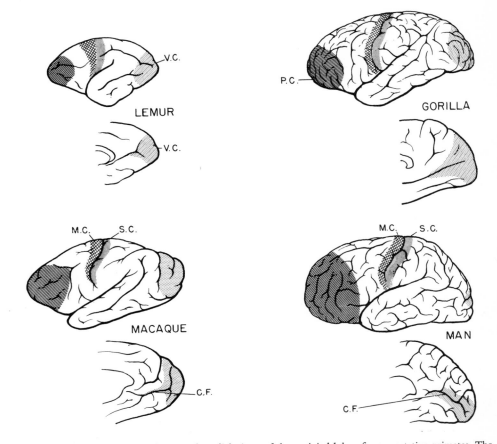

Fig. 2. Lateral views of the cerebrum and medial views of the occipital lobe of representative primates. The primary visual cortex (V.C.) is actually larger in the higher primates, but is relatively smaller in proportion to the total amount of cerebral cortex. The association areas of the occipital, parietal, and temporal lobes (white) are both absolutely and relatively larger in apes and man. Figures are drawn in the following scales: generalized lemur (1×); macaque (1/2×); gorilla (1/3×); and man (1/4×). C.F. = calcarine sulcus; M.C. = primary motor cortex (area 4); P.C. = prefrontal lobe; S.C. = primary sensory cortex (areas 3, 1, 2); V.C. = primary visual cortex (area 17, striate cortex). (Courtesy of Academic Press; from Noback, C. R. and N. Moskowitz, 1963.)

striate cortex of one hemisphere does not have direct connections through the corpus callosum with the striate cortex of the other hemisphere.

The eye contains the retina which is composed mainly of photoreceptor cells (rods and cones) and an intricate organized complex of interacting neurons (horizontal cells, bipolar cells, amacrine cells, and ganglion cells). The retina is actually a distal extension of the central nervous system (forebrain). In fact it is the only truly mobile portion of the brain—it moves as the eye moves. The rods and cones are the photoreceptive cells where transduction and genesis of receptor (generator) potentials occur. The ganglion cells are the output neurons whose axons form the optic nerve, chiasma, and optic tracts. These axons are the channels conveying the coded information from the retina to the lateral geniculate body or to the optic tectum. In a sense the retina is structurally and functionally organized as a mosaic, with each ganglion cell and its axon acting as a channel conveying neural influences from each unit of the mosaic. The lateral geniculate body is a processing station located in the thalamus. The geniculate body projects its output to the visual cortex via the fibers of the optic radiation (Fig. 1).

The role of the geniculate body may be (1) to recode the coded input received from the retina, (2) to discard certain extraneous activity (called noise), or (3) to modify the input by compressing it into fewer channels. Simply stated, the processing nuclei are presumed to suppress noise and, equally important, to enhance the signals of the code.

The environment viewed by the eye is called the field of vision. Each point in the field of vision is represented in a single receptor spot in the retina of one or both eyes, one small group of neurons in one lateral geniculate body, and in a small group of neurons of the striate cortex on one side. Furthermore, each spot in the visual field viewed by both eyes is conveyed via two separate channels (one channel from each eye) via the retinogeniculostriate pathway. These two channels remain separate through the entire pathway until they reach the striate cortex. The act of binocular fusion of the neural influences conveyed via the two channels from one spot in the visual field occurs in the striate cortex.

Recently, other systems subserving visual perception have been proposed. These various concepts of the visual system(s), all of which may be valid, indicate that a definitive understanding of the functional aspects of many levels of visual perception is not possible at the present time.

According to Diamond and Hall (1969), Mishkin (1972), and Snyder (1973) two visual systems are present in mammals. They are the retinogeniculostriate system, and the retinotectal system. The latter comprises the pathway from retina to tectum of the midbrain (superior colliculus) to pulvinar of the thalamus to a portion of neocortex of the temporal lobe. The retinogeniculostriate system is presumed to subserve the more complex levels of visual analysis, while the retinotectal system is presumed to be involved in learning based on complex visual information.

According to Schneider (1969), the cortex of the retinogeniculostriate system subserves the analyses of visual patterns, whereas the superior colliculus of the retinotectal pathway subserves the analysis of spatial orientation. In this context, the striate and circumstriate cortices are involved with answering the question "What is it?" and the superior colliculus (actually this structure is a cortex) with answering the question "Where is it?" in terms of visuomotor responses. In primates, the midbrain does participate in motion perception and acts in motor detection.

Ingle (1973), on the basis of studies of vision in nonmammalian vertebrates, suggests that there are two visual mechanisms (or systems) for the analysis of movements. One system is involved with the detection of moving objects. The other system is concerned with the recognition of the background of movement related to the movement of the eyes of the animal.

Trevarthen (1968) suggests that, on the basis of experiments on rhesus monkeys, the "two visual systems" comprise a system for "ambient vision," and a system for "focal vision." Ambient vision is concerned with the analysis of movement of the subject commanding visual attention versus the environment. Thus vision is served primarily through the visual pathways commencing with the rods and/or the peripheral portions of the retina. Focal vision is involved with the detailed visual analysis of the objects in the environment through the visual pathways commencing with the fovea (the foveal system).

A. Eye

1. Rods and Cones. The photoreceptors of the retina are the rods (rod cells) and cones (cone cells). These cells are sensitive to the narrow band of electromagnetic waves called the visual spectrum. The actual sensors are the photosensitive pigments located within the membranes of the disks in the outer segment of each rod and cone. These pigments are all similar biochemically; however, the slight differences among these pigments are critical to their differential responses to the visual spectrum. They are composed of a specific type of protein, called an opsin, which is bound to a chromatophore with a special configuration (Wald, 1968). The chromatophore is called retinaldehyde (retinal, retinene). Rhodopsin is the photopigment in the rods of primates. It is sensitive to the entire visual spectrum, with a maximum scotopic sensitivity at about 500 nm. The cones of man and many primates contain three photopigments, with each cone containing one of three pigments with peak absorptions at approximately 435 nm (blue), 535 nm (yellow/green), and 565 nm (red). These cones are called blue cones, yellow/green cones, and red cones, respectively. All four photopigments in the rods and cones possess the same chromatophore (11-cis retinaldehyde) and are linked to one of four different opsins (Wald, 1968). It is within the outer segment of each rod and cone that transduction and genesis of generator (receptor) potential takes place. Once light waves trigger photochemical activity, the role of light is finished. Further processing of the neural influences takes place in the neuroretina and the various nuclear stations of the visual pathways. Color coded neurons have been demonstrated in the retina, lateral geniculate body, and striate cortex of several Old and New World monkeys (Anderson *et al.*, 1962; Hubel and Wiesel, 1968).

The retinas of most primates contain both rods and cones. Rods are the low-intensity receptors for nocturnal or dim-light perception. Cones are high-intensity receptors for daylight vision and color perception. Nocturnal primates have few, if any, cones.

2. Color Vision in Nonprimate Vertebrates. The ability to perceive color (hue discrimination) is said not to be common among vertebrates, but it is much more widely distributed than has been previously thought (Autrum and Thomas, 1973). These authors present a comprehensive review of the current status of knowledge of color vision in vertebrates. Information is still incomplete because many species have not been studied utilizing modern investigative methods. For example, recent studies demonstrate that the cat can react to colors. This is based on behavioral and electrophysiological experiments (see Autrum and Thomas, 1973 for the literature).

In general, color vision is well developed in highly visually-oriented diurnal vertebrates with a cone-rich retina possessing a fovea. Color vision has probably evolved independently a number of times during vertebrate phylogeny. It is definitely present in some species in such unrelated vertebrates as teleost fish, amphibians, reptiles, birds, and mammals. The color vision of many fish and their ability to distinguish colors are very similar to those of man (Buddenbrock, 1958). A number of teleost fish have 2–3 different types of cones as determined by the spectral sensitivity of the receptor pigments and microspectrophometric studies of single cones (summary in Autrum and Thomas, 1973). Many species of amphibia in such genera as *Triturus, Bombina, Bufo,* and *Rana* have a trichromatic color sense (reviewed in detail by Autrum and Thomas, 1973).

Many reptiles and birds are capable of seeing colors. These include turtles (*Chrysemys, Testudo,* and *Clemmys*) and lizards (*Anolis* and *Lacerta*). Some reptiles are color blind. All birds investigated (e.g. hummingbirds, pigeon, and fowl) possess color vision similar to that of man. Many species of mammals are probably color-blind. Species in several orders of nonprimate mammals have been demonstrated to be capable of seeing colors. These include the insectivore *Erinaceus europaeus*; rodents such as *Sciurus vulgaris, Citellus citellus,* and *Clethrionomys glareolus*; carnivores such as viverrids, the wolf, dog, and cat; and perissodactylans such as the horse and zebra. The ground squirrel *Citellus citellus* and the gray squirrel *Sciurus carolinensis* have a pure cone retina. For many details and the literature relating to the above account, refer to Autrum and Thomas (1973).

3. Color Vision in Primates. In general, there is a correlation between the photo-receptors and the behavioral patterns of a species. Those primates with a high proportion of cones are predominantly diurnal in habit, while those with retinas composed primarily of rods are essentially nocturnal in habit. Thus, color vision is associated with diurnal primates and tupaiids, while weak or possibly absent color vision is associated with nocturnal primates (Wolin and Massopust, 1970; Autrum and Thomas, 1973). Among the Primates, all the Haplorhini are diurnal, with the exception of *Tarsius* and the cebid *Aotus.* More variation exists among the Strepsirhini; most are nocturnal (Lepilemurinae, Cheirogaleinae, Daubentoniidae, Galagidae, and Lorisidae), except for the diurnal members of the family Indriidae (excluding *Avahi*) and the subfamily Lemurinae (*Hapalemur* and *Lemur*). This subject is reviewed by Pariente (1970) and Charles-Dominique (this volume). The tupaiids are diurnal.

Apparently all species of Ceboidea, Cercopithecoidea, and Hominoidea have retinas with both rods and cones. These primates presumably all have color vision, although only a few species have been tested with experimental techniques for color vision. *Aotus,* the nocturnal member of the Ceboidea, has an almost pure rod retina (Murray *et al.,* 1973). *Tarsius* with its rod retina is a nocturnal species which is said to be color-blind.

The Lemuroidea have basically rod retinas. Because some members of this group may have some cones, it is possible that some diurnal species such as *Indri indri* and *Propithecus verreauxi* may have some color vision. Conelike cells have been described to be present in the retinas of these species. *Lemur mongoz* is said to be probably color-blind (Bierens Dehaan and Frima, 1930).

The *Lorisoidea* have essentially pure rod retinas with few, if any, cones. They are nocturnal animals which are possibly color-blind.

Tupaia has a retina probably with the three cone types. This diurnal mammal with a predominately cone retina has good color vision in red, yellow/green, and blue (Tigges, 1963*a*; Shriver and Noback, 1967). A few rods may be present in the retinas of this genus.

On the basis of behavioral and physiological studies of macaques and the squirrel monkey, De Valois and Jacobs (1968) have analyzed the status of color vision in these species. The macaques (*Macaca irus* and *Macaca nemestrina*) and normal humans have essentially the same trichromatic vision. On the other hand, squirrel monkeys (*Saimiri*) have poorer color vision that is qualitatively different from normal human trichromats but similar to protanomalous humans (weakness in the red system of trichromatic color vision). This difference in color vision of man and Old World monkeys (*Macaca*) from that of New World monkeys (*Saimiri* and *Cebus*) is the primary basis for the suggestion by De Valois and Jacobs that basic differences in color vision may exist between these two major branches of Anthropoidea. The marmoset (*Callithrix*) can behaviorally see color throughout the visual spectrum (Miles, 1958). The chimpanzee, orangutan, and gibbon have excellent color vision (Grether, 1940; Tigges, 1963*b*).

In summary, the analysis of color vision in primates and other vertebrates is incomplete. In addition to behavioral studies of primates, more detailed anatomical and physiological analyses of their retinas and neural pathways are needed to clarify the precise status of color vision in primates.

4. Ocular Fundus. The ocular fundus is the posterior portion of the interior of the eye. It can be examined in the living animal with an ophthalmoscope. The ocular fundi of the Ceboidea (except *Aotus*), Cercopithecoidea, and Hominoidea are all basically similar to the human pattern. These primates exhibit a characteristic pattern of arcuate blood vessels (Wolin and Massopust, 1970). The Lemuroidea, Lorisoidea, *Tarsius*, and *Aotus* have similar vascular patterns. The fundus pattern of *Tupaia* is unlike that in any primate. In *Tupaia*, the central retinal area is present far posterior and lateral to the nerve head. The blood vessels radiate like the spokes of a wheel from the optic papilla (Wolin and Massopust, 1970). The central retinal area in *Tupaia* is somewhat removed from the nerve head; this is consistent with the fact that the eyes of *Tupaia* are laterally oriented. In primates, the eyes are generally directed forward.

5. Tapetum Lucidum. The tapetum lucidum is that portion of the choroid layer of the eye which reflects light (eye shine). It is generally well developed in many nocturnal terrestrial vertebrates and some fish. Light which has passed through the retina is reflected back to the retina by the tapetum lucidum; this reflected light stimulates the rods and cones. The vision of these animals is rendered more sensitive, especially when light is scarce. However, sharpness of focus is sacrificed and the image is blurred or even multiplied.

The tapetum lucidum is apparently absent in all Haplorhini including *Tarsius* and *Aotus*. *Aotus* has an eye shine which is probably the result of myelinated fibers within the retina (Jones, 1965).

Most, if not all, Strepsirhini have a tapetum lucidum, but information concerning the tapetum in these primates is incomplete. The following account is based on the review of Wolin and Massopust (1970), who cite the relevant literature. Each of the lorisids and galagids examined is said to have a tapetum. *Galago crassicaudatus* has a tapetum cellularis composed primarily of cells, whereas *Perodicticus potto* has a tapetum fibrosum composed of collagenous connective tissue fibers. Depending upon the investigator, the thin tapeta of *Loris tardigradus*, *Nycticebus coucang*, and *Galago senagalensis* are either of the cellular or fibrous type. Of the lemuriforms, some have tapeta whereas others apparently lack a tapeta. A tapetum cellularis is present in the nocturnal *Microcebus murinus*, the diurnal *Propithecus verreauxi*, and the diurnal and crepuscular *Lemur catta*. The diurnal *Lemur macaco fulvus* may have a tapetum cellularis, but this has not been definitely established. Other species of the genus *Lemur* are reported to lack a tapetum (*Lemur macaco* and *Lemur rufifrons*, Kolmer, 1930). Among the indriids, a tapetum fibrosum has been described in the nocturnal *Avahi laniger* and the diurnal *Indri indri*. Clearly, more evidence is needed concerning the presence and histological structure of the tapeta in strepsirhines.

In a study of reflected light (eye shine) into the fundus of the eyes of 13 species of Malagasy lemuriforms, Pariente (1970) concluded that the tapetum lucidum is present in (1) the diurnal *Propithecus verreauxi*, (2) the diurnal and slightly crepuscular *Lemur variegatus*, *Lemur rubriventer*, *Lemur macaco*, *Lemur catta*, *Lemur mongoz*, and *Hapalemur griseus*, and (3) the nocturnal *Microcebus murinus*, *Microcebus coquereli*, *Cheirogaleus medius*, *Cheirogaleus major*, *Phaner furcifer*, and *Daubentonia madagascariensis*.

B. Law of Newton–Müller–Gudden

1. General Statement. In nonmammalian vertebrates, the retinofugal fibers project from the retina, cross the midline (decussate) through the optic chiasma, and terminate in the contralateral side of the brain. This complete decussation means that, in these vertebrates, the influences generated by stimuli from the visual field viewed by one eye are conveyed to processing stations of the opposite side of the brain.

Mammals differ from other vertebrates in that some retinofugal fibers from each eye cross the midline and project to the contralateral (opposite) side of the brain, while other fibers do not cross the midline but project to the ipsilateral (same) side of the brain (Figs. 3–5). The presence of both decussating and nondecussating retinofugal fibers in primates as well as all other mammals means that the influences from each eye are projected to both the contralateral and ipsilateral halves of the brain. In primates, this neuroanatomical feature is significant in several functional activities subserved by the visual system. These activities include the ability to judge distances accurately while jumping and leaping and the skill to make fine discriminating movements during the search and manipulation of food and other objects.

The relative number of crossed and uncrossed retinofugal fibers is roughly proportional to the degree to which the eye is directed. Those mammals, including *Tupaia*, with laterally directed eyes have a high proportion of decussating retinofugal fibers relative to the number of nondecussating fibers (Fig. 5). Those mammals, including primates, with frontally directed eyes, have about an equal number of decussating and nondecussating fibers (Fig. 4). Thus, the more the eye is laterally directed, the greater the proportion of decussating fibers, whereas the more the eye is directed frontally, the closer the approach to an equal number of decussating and nondecussating fibers. These relationships are known as the Law of Newton–Müller–Gudden (Walls, 1942).

2. Topological (Retinotopic) Organization within the Visual System. Each focal site in the visual field (environment viewed by the eyes) is projected to a specific focal site in the retina (Figs. 3–5). The lens of the eye reverses the environmental image upon the retina; thus, the visual field of each eye is represented on the retina in a precisely organized

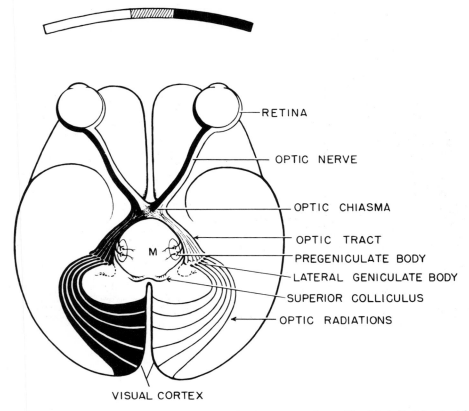

FIG. 3. Diagram of the visual field and the retinogeniculostriate cortex pathway. The right half of the visual field (black) is projected to the left side of the brain; the left half of visual field (white) is projected to the right side of the brain. The portion of the visual field viewed by the area centralis (macula lutea) is indicated by the diagonally hatched block. The fibers projecting from the lateral (temporal) halves of the retinae do not decussate and those projecting from medial (nasal) halves of the retinae decussate in the optic chiasma.

topological (retinotopic) manner. In turn, retinotopic organization of the retina is precisely projected through the retinogeniculostriate pathway and the retinotectal pathway. As a result the lateral geniculate body (pars dorsalis), striate visual cortex, and superior colliculus are retinotopically organized.

In addition, the decussating and nondecussating retinofugal fibers are also retinotopically projected via the optic nerve, chiasm, and optic tract in such a way that the information from one half of the visual fields viewed by both eyes is conveyed to the contralateral lateral geniculate body and striate visual cortex (Figs. 3–5). In other words, in all mammals one half (right or left) of the total visual field is represented in the contralateral lateral geniculate body and visual cortex. An examination of Figs. 4 and 5 indicates the relation of the visual field projected from each eye to the laminae of the lateral geniculate body. Lamination is a basic feature of the mammalian lateral geniculate body; however, the details of the organization of these laminae differ in different groups of mammals. Some laminae represent projections from the contralateral retina, while other laminae represent projections from the ipsilateral retina. (For further details, see Giolli and Tigges, 1970; Noback and Laemle, 1970; Kaas *et al.*, 1972.)

3. *Frontally and Laterally Directed Eyes.* In general, each eye of a mammal views approximately the same amount of the visual field (Figs. 4 and 5); i.e., the angle sub-tended by each uniocular field is about 170° (Duke–Elder, 1958). As a result, mammals with laterally directed eyes such as the rabbit or *Tupaia* can see a large field on one side

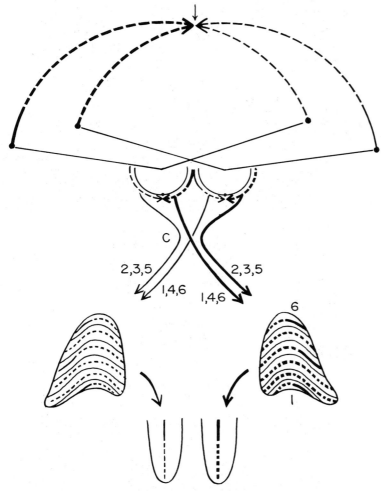

FIG. 4. The retinogeniculostriate pathway of a haplorhine primate with frontally directed eyes. The broken lines represent the portion of the visual fields, its retinal projection, and pathway to the striate cortex from the binocular field (viewed by both eyes). The solid lines represent the portion of the visual fields, its retinal pro-jection, and pathway to the striate cortex from the monocular field (viewed by only one eye). Recall that the pathways are topographically organized. The left visual field (thicker lines) is projected to the right half of the brain. The right visual field (thinner lines) is projected to the left half of the brain. The arrow indicates the site in the visual field which is projected to the area centralis. The laminae of the lateral geniculate body are indicated with their topographic representation; laminae 2, 3, and 5 receive input from the ipsilateral eye (contralateral visual field), and laminae 1, 4, and 6 receive input from the contralateral eye (contralateral visual field). The lower figures represent the striate cortex; the binocular fields are projected to the most occipital regions of the striate cortex. C = optic chiasm.

Charles R.
Noback

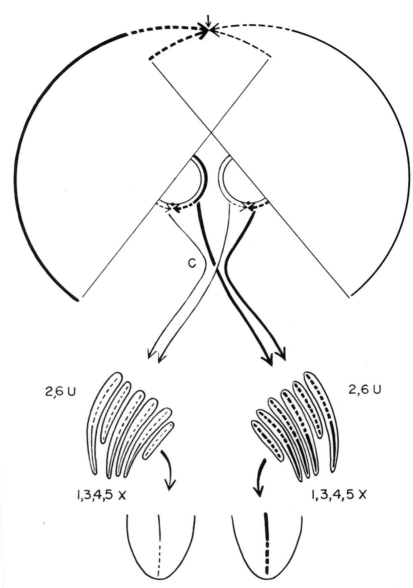

FIG. 5. The retinogeniculostriate pathway of a mammal (*Tupaia*) with laterally directed eyes. The broken lines represent that portion of the visual field viewed by both eyes, and the solid lines represent that portion of the visual field viewed only by one eye. The left visual field (thicker lines) is projected to the right half of the brain. In *Tupaia*, 2, 6 U are the laminae receiving input via uncrossed fibers from the ipsilateral retina and 1, 3, 4, and 5X are the laminae receiving input via crossed fibers from the contralateral retina. The lower figures represent the striate cortex; the binocular fields are projected to the most occipital regions of the striate cortex. C = optic chiasm; U = uncrossed; X = crossed.

of their head with one eye. In addition, they can see only a small area in front simultaneously with both eyes. In such an animal, the binocular field is small and the two monocular fields are large (Fig. 5). On the other hand mammals with frontally directed eyes, such as primates, have large binocular fields and small uniocular fields. The latter are located laterally (Fig. 4). Evidence indicates that frontally directed eyes are the derived state; laterally directed eyes are the general or primitive condition. The evolution of frontally directed eyes from laterally directed eyes has occurred independently in different classes, including fish, birds, and several orders of mammals.

The degree to which a group of vertebrates retains uniocularity or evolves binocularity is related, in part, to its biological value. Uniocularity offers a distinct advantage to prey animals. The resultant wide panoramic visual field facilitates the detection of enemies in many directions. Binocularity is favored by animals in which spatial localization is of paramount importance. This feature is useful to (1) predatory animals in the pursuit of prey, (2) some arboreal animals in their visually guided maneuvers through the branches, and (3) digitally dextrous animals such as primates and raccoons in accurately correlating eye–finger manipulations.

4. Frontally Directed Eyes. The decussating and nondecussating retinofugal projections in those mammals with frontally directed eyes are so organized as to convey neural influences from the right half of the visual field viewed by both eyes to the left half of the brain and from the left half of the visual field to the right half of the brain (Figs. 3 and 4). This is because the right half of the visual field is projected to the nasal (medial) half of the right eye (lens of the eye reverses the image) and to the temporal (lateral) half of the left eye. In turn, the retinofugal fibers from the nasal half of the right eye decussate in the optic chiasma and terminate in the left side of the brain, and those fibers from the temporal half of the left eye do not decussate and therefore terminate in the left side of the brain. These retinofugal fibers of the retinogeniculostriate system terminate in the lateral geniculate body. The fibers from the right and left eyes terminate in different laminae of the lateral geniculate body. The projections from the lateral geniculate body course through the optic radiation to their termination in the striate cortex (primary visual cortex, area 17).

It is important to realize that the pathway from each eye is separated, both structurally and functionally, from that of the other eye during its course from the retina to its termination in the striate cortex. The act of fusion of the information conveyed from the same fields of vision of both eyes occurs initially in the striate cortex. Further processing of the information received by the primary visual cortex occurs in the circumstriate cortex (areas 18 and 19) and other regions of the cerebrum.

5. Laterally Directed Eyes. Mammals (e.g., *Tupaia*) with laterally directed eyes have a proportionately large number of decussating retinofugal fibers and a small number of nondecussating retinofugal fibers (Fig. 5). All the uncrossed fibers originate from a narrow crescent-like area on the temporal side of the retina of each eye. These temporal crescents receive environmental stimuli from the same portion of a narrow band of the visual field. This band viewed by both temporal crescents simultaneously is located directly in front of the animal.

6. Phylogenetic Aspects of the Law of Newton–Müller–Gudden. In mammals, one half of the visual field (right or left) stimulates photoreceptors in both eyes, and the resulting neural activity is conveyed to the contralateral half of the brain. As noted above, this phenomenon is an exquisite example of the precise visuotopic organization of the decussating and nondecussating projections within the visual pathways. This applies to all mammals, regardless of whether the animal's eyes are oriented laterally or frontally.

The differences in the location of the retinal origin and the relative quantities of nondecussating versus decussating fibers between these two groups of mammals pose questions with phylogenetic implications. (1) Have mammals with frontally directed eyes evolved from ancestors with laterally directed eyes and nondecussating retinofugal fibers? If so, this shift from laterally directed to frontally directed eyes must have been accompanied by an intricate reorganization of the components of the retinogeniculostriate pathways. (2) Did the directionality of the eye in these two groups evolve independently before the introduction of the nondecussating retinofugal fibers? If so, this nondecussating component may have appeared independently within the retinogeniculostriate pathways in these two groups.

C. The Foveal System

The foveal system is structurally and functionally specialized for visual acuity and color vision. Undoubtedly, this system exerted a significant role during the evolution of the primates because these visual modalities are critical to many of their behavioral patterns. As noted previously, the fovea and its projections within the retinogeniculostriate pathways are found in all Haplorhini with the exception of *Aotus*. Within the entire retina, the fovea and the surrounding macula form a small circular region. This small region projects to a large portion of the lateral geniculate body and primary visual cortex. An indication of the relation between visual acuity and size of a structure can be gauged from the following figures. In the rhesus monkey, the segment of the visual field viewed by an angle of 1/6 of a degree in the fovea projects to a 1-mm wide sector of the striate cortex. In contrast, the segment of the visual field viewed by an angle of 6° in the peripheral retina also projects to a 1-mm wide sector of the striate cortex. The difference in visual acuity between the fovea and peripheral retina is approximately 36 times.

The neural circuits involved with the perception of color evolved phylogenetically within the retinogeniculostriate pathway. This probably occurred in conjunction with the presumed development of cones from rods. Although the foveal system has a significant role in color vision, the color circuits also commence from cones located in the rest of the retina.

Primates habitually orient themselves visually in space through the foveal system, since it is the retinal site of acute vision.

1. Duplex Retina. A duplex retina is one in which the receptors are both rods and cones and the central visual area contains a macular area with a fovea. The fovea is the small depression in the macula resulting from displacement of the internal layers of the retina. It is both rod-free and avascular. The degree of visual acuity in primates with a fovea is improved for several reasons. An optical advantage is offered because light reaches the cones by passing through fewer retinal layers in the fovea than in other areas of the retina. The lack of vascularity in the fovea is significant because an internal blood supply would interfere with visual acuity (Weale, 1966). This functional sequence of cones, associated foveal neurons, and the pathways from the fovea to the *visual cortex* is called the foveal system. The foveal system is of primary importance because it apparently represents the highest order of neuronal evolution in the visual system. Within Primates, only the Ceboidea (except *Aotus*), Cercopithecoidea, and Hominoidea have duplex retinas (Wolin and Massopust, 1967, 1970). An excellent review of the morphology of the primate retina is presented by Wolin and Massopust (1970).

In diurnal Haplorhini, the retinal arterioles and venules stop short of the fovea. *Aotus* and the Strepsirhini possess a vascular-free central retinal area without a fovea. The nocturnal eye of *Tarsius* has a vascular rod-free fovea which is believed to lack cones (see Wolin and Massopust, 1970). The vascular pattern of the retina of *Tupaia* differs markedly from that of all primates (Wolin and Massopust, 1970).

2. *Depth Perception.* Depth perception is enhanced in that region of the visual fields viewed by both eyes simultaneously. Each spot binocularly viewed is actually seen by a corresponding spot of each retina. Thus, a spot binocularly viewed stimulates the sensors in a spot in one eye and the sensors in a corresponding spot in the other eye. The projection from the one spot relays its influences to a column of cells in the striate cortex, and the projection from the corresponding spot of the other eye relays its influences to the same column of cells in the striate cortex. This means that a group of neurons in the striate cortex receives input from corresponding spots in the retinas of the two eyes. Studies by several investigators demonstrate that some neurons within such a column receive more input from one eye than the other eye, whereas other neurons receive an equal amount of input from each eye. This equality and inequality of the input from both eyes is utilized by the brain as one basis for evaluating depth.

D. The Lateral Geniculate Body

The lateral geniculate body (pars dorsalis) is a major nucleus in the retinogeniculostriate pathway. In many mammals it is composed of relay laminae, which receive input from the retina and relay their output via the optic radiation to the striate cortex. There is no known explanation or functional significance for the laminar pattern. Certain laminae receive projections from the contralateral eye (Fig. 6). Each lamina generally receives projections from only one eye (the exceptions to this are not pertinent to this discussion). The lateral geniculate body of the primates and *Tupaia* are considered to be composed of 6 laminae. Those reports indicating otherwise can be interpreted as fitting into or derived from the 6-laminar pattern. Contrary to the reports stating that *Saimiri* has a 4-layered lateral geniculate body (Hassler, 1966), recent experimental studies indicate that *Saimiri* and *Ateles* have a lateral geniculate body with 6 (or even 7) laminae (Doty *et al.*, 1966). Four layers are presumably present in *Aotus*, *Callithrix*, and *Tarsius* (Jones, 1966a; Hassler, 1966; Polyak, 1957). In any case, these exceptions do not alter the basic concept of the phylogenetic significance of these laminar patterns.

With respect to the projections to the lateral geniculate body, there are apparently 3 basic laminary patterns (Fig. 6) in the primates and *Tupaia*: (1) haplorhine pattern, (2) strepsirhine pattern, and (3) tupaiid pattern. In the haplorhines, laminae 1, 4, and 6 receive projections from the contralateral eye, and laminae 2, 3, and 5 receive projections from the ipsilateral eye. In the Strepsirhines, laminae 1, 5, and 6 receive projections from

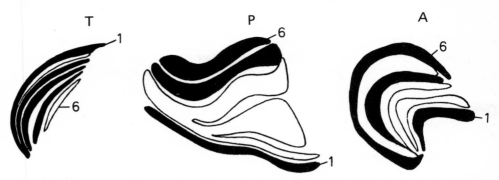

FIG. 6. Laminar patterns of the lateral geniculate body (pars dorsalis) in *Tupaia* (T), the prosimian *Galago* (P) and the anthropoid *Saimiri* (A). Laminae receive projections from the contralateral eye (black) and from the ipsilateral eye (unshaded). Drawn as viewed in cross section. The optic tract is located adjacent to lamina 1.

the contralateral eye and laminae 2, 3, and 4 receive projections from the ipsilateral eye. In *Tupaia*, laminae 1, 3, 4, and 5 receive projections from the contralateral eye, and laminae 2 and 6 receive projections from the ipsilateral eye (Campbell *et al.*, 1967; Laemle, 1968; Campbell, 1972).

Because of the intricacy of the neural connections in the retinogeniculostriate pathway, it is difficult to conceive of one of these laminar patterns evolving into another laminar pattern without a drastic reorganization of the entire system. In this formulation, the lateral geniculate body should not be considered as an independent structure, but rather as a unit integrated into a functional retinogeniculostriate system.

It is probable that these three patterns evolved independently in the early stages of eutherian evolution at the time the uncrossed retinofugal projections appeared (Noback *et al.*, 1969). It is theorized that the retinogeniculocortical pathway evolved in conjunction with the initial stages of the development of both contralateral and ipsilateral retinofugal projections to the lateral geniculate body. In these initial stages, not only did these contralateral and ipsilateral projections make their synaptic connections within the lateral geniculate body, but the three basic laminar patterns were established.

Variations from these three geniculate laminary patterns (e.g., those of *Aotus*, *Callithrix*, and *Tarsius*) may be explained as having evolved from one of these basic patterns. Experimental studies utilizing modern neuroanatomical and electrophysiological methods can produce evidence relevant to the question as to whether the lateral geniculate bodies of *Aotus*, *Callithrix*, and *Tarsius* are actually quadrilaminar in the structural functional sense. Except for the report by Jones (1966a) on one specimen of *Aotus*, no such experimental studies have been published. It is possible that one (or two) of the laminae may receive input from both eyes, and that the laminae may actually be a derived structure composed of the close approximation of two laminae. It is also possible that two laminae may have disappeared or become vestigial during phylogeny. According to Hassler (1966), geniculate laminar layers 5 and 6 are new layers which have been added to the 4 original layers.

In summary, the three geniculate laminary patterns in the haplorhines, strepsirhines, and tupaiids probably become established independently at an early stage of mammalian evolution and persisted in subsequently evolved species.

E. *Aotus* and *Tarsius*

215

THE VISUAL
SYSTEM
IN PRIMATE
PHYLOGENY

The precise phylogenetic relations between *Aotus* and *Tarsius* and to other groups of primates have concerned a number of primatologists. The following brief account presents some known features of the eyes of these two primates which may be of relevance to these investigators.

Aotus and *Tarsius* are nocturnal and/or crepuscular primates, whose eyes retain certain features characteristic of diurnal animals. In general, these features support the concept that these two primates evolved secondarily from diurnality to nocturnality (Walls, 1942; Polyak, 1957; Ogden, 1974). In recent years, a number of investigations on the eye of *Aotus* have been based on use of sophisticated electron microscopic and electrophysiological methods; however, similar studies have not been reported on the eye of *Tarsius*.

Aotus has a duplex retina (Hamasaki, 1967) in which the rods predominate, although cones are present (Jones, 1966*b*; Murray *et al.*, 1973). There is little specialization of a vascularized area centralis (Ferraz de Oliveira and Ripps, 1968), and it seems to lack a fovea (Wolin and Massopust, 1967). Thus, it seems that *Aotus* has a typical peripheral retina, with such a retina largely replacing the central specializations, such as a highly developed area centralis with a fovea (Ogden, 1974). However, there is sufficient specialization to permit histological identification of an area centralis (Jones, 1965; Ogden, 1974). The eye shine of *Aotus* is probably the result of myelinated fibers in the retina, and not of a tapetum lucidum (Jones, 1965). Several adaptations for nocturnal vision are present in *Aotus* (Detwiler, 1941) in order to capture as much light as possible. These include a posterior displacement of the iris diaphragm toward the center of the eye (i.e., the plane with largest diameter). By this adaptation, the iris is able to obtain an increased pupillary diameter which permits more light to reach the retina. The lens has a more spherical shape than in diurnal forms; this more powerful lens refracts the light sufficiently in order to focus the image on the retina.

Tarsius has eyes which are adapted for night vision in conformity with its predominately nocturnal habits. In order to capture as much light as possible in the dark, *Tarsius* has an immense eyeball with a large cornea, lens, and iris diaphragm. Apparently this primate has the largest eye in proportion to body size of any extant mammal (Polyak, 1957). Unlike many nocturnal primates and other mammals, *Tarsius* lacks a tapetum. The structural features of the tarsier retina have been used to help resolve the still-debated question as to whether *Tarsius* should be classified in Prosimii, Haplorhini, or as a transitional form between these two groups. The current status of the morphology of the tarsier's retina is discussed by Rohen (1966), Rohen and Castenholz (1967), and Wolin and Massopust (1970). Apparently the only photoreceptors of the retina are rods; cones are absent (Rohen, 1966). Electrophysiological experiments have not been employed to determine if any spectrally sensitive "cones" are present.

The flat trough-shaped fovea has two features which may have phylogenetic significance (Rohen, 1966): it lacks cones, and in addition, its ratio of rods to ganglion cells is relatively high (about 8:1). The latter indicates that the visual acuity of *Tarsius* is definitely less than that of Haplorhini with lower ratios. (The lower ratio means that higher resolution is possible because more smaller receptive spots are available per unit area of fovea.) The retina of *Tarsius* may be said to be similar to that of other Haplorhini in that it has a fovea within the area centralis. However, this fovea does not resemble those of other haplorhines in the apparent lack of cones and the high ratio of rods to ganglion cells.

F. Visual Cortex

Recent studies indicate that fundamental differences may exist between Old World and New World monkeys with regard to the neural connections of the visual cortex (area 17, striate cortex) and the association visual belts of cortex (areas 18 and 19). This is suggested by evidence reported on the neural connections between these visual cortical areas in *Macaca mulatta* (Cragg and Ainsworth, 1969; Zeki, 1969, 1971) and in *Saimiri sciureus* and *Aotus* (Spatz *et al.*, 1970; Allman and Kaas, 1971; Spatz and Tigges, 1972). In area 17 of *Macaca*, there are neurons which project to visual association areas 18 and 19, whereas in *Saimiri* and *Aotus*, neurons of area 17 project to area 18 but not to area 19. Spatz and Tigges (1972) suggest that on the basis of this evidence the visual cortical association areas in the Old World and New World monkeys have developed independently.

V. Summary

During the adaptive radiation of the primates, the visual system has evolved many morphological characters and physiological expressions which have had functional value in various ecological niches. Attention is directed to some of these features among different groups of primates and tupaiids. Among those discussed are the photoreceptors (rods and cones) of the eye, color vision, foveal system, adaptations to nocturnal and diurnal habits, the Law of Newton–Müller–Gudden, and the retinogeniculostriate visual pathway.

VI. References

ALLMAN, J. J., and J. H. KAAS. 1971. Representation of the visual field in striate and adjoining cortex of the owl monkey (*Aotus trivirgatus*). *Brain Res.* **35**:89–106.

ANDERSON, V. O., B. BUCHMANN, and M. A. LENNOX-BUCHTHAL. 1962. Single cortical units with narrow spectral sensitivity in monkey (*Cercocebus torquatus atys*). *Vision Res.* **2**:295–307.

AUTRUM, H., and I. THOMAS., 1973. Comparative physiology of colour vision in animals, pp. 661–692. *In* R. Jung, ed., *Handbook of Sensory Physiology*, Vol. 7 (No. 3A). Springer-Verlag, Heidelberg.

BIERENS DEHAAN, J. A., and M. J. FRIMA. 1930. Versuche über den Farbensinn der Lemuren. *Z. Vergl. Physiol.* **12**:603–631.

BUDDENBROCK, W. VON. 1958. *The Senses.* University of Michigan Press, Ann Arbor.

CAMPBELL, C. B. G. 1972. Evolutionary patterns in mammalian diencephalic visual nuclei and their fiber connections. *Brain Behav. Evol.* **6**:218–236.

CAMPBELL, C. B. G., J. A. JANE, and D. YASHON. 1967. The retinal projection of the tree shrew and hedgehog. *Brain Res.* **5**:406–418.

CRAGG, B. G., and A. AINSWORTH. 1969. The topography of the afferent projections in the circumstriate visual cortex of the monkey studied by the Nauta method. *Vision Res.* **9**:733–747.

DETWILER, S. R. 1941. The eye of the owl monkey (*Nyctipithecus*). *Anat. Rec.* **80**:233–239.

DE VALOIS, R. L., and G. H. JACOBS. 1968. Primate color vision. *Science* **162**:533–540.

DIAMOND, I. T., and W. C. HALL. 1969. Evolution of the neocortex. *Science* **164**:251–262.

DOTY, R. W., M. GLICKSTEIN, and W. M. CALVIN. 1966. Lamination of the lateral geniculate nucleus in the squirrel monkey, *Saimiri sciureus. J. Comp. Neurol.* **127**:335–340.

DUKE-ELDER, S. 1958. The eye in evolution, pp. 1–843. *In* S. Duke-Elder, ed., *System of Ophthalmology*. Henry Limpton, London.

ELLIOT SMITH, G. 1928. The new-vision. Bowman Lecture. *Trans. Ophthal. Soc. U.K.* **48**:64–85.

FERRAZ DE OLIVEIRA, L., and H. RIPPS. 1968. The "area centralis" of the owl monkey (*Aotes trivirgatus*). *Vision Res.* **13**:219–230.

GIOLLI, R. A., and J. TIGGES. 1970. The primary optic pathways and nuclei of primates, pp. 29–54. *In* C. R. Noback and W. Montagna, eds., *The Primate Brain, Advances in Primatology*. Appleton-Century-Crofts, New York.

GRETHER, W. F. 1940. Color vision and color blindness. *Comp. Psychol. Monogr.* **15**:1–38.

HAMASAKI, D. I. 1967. An anatomical and electrophysiological study of the retina of the owl monkey, *Aotes trivirgatus. J. Comp. Neurol.* **130**:163–174.

HASSLER, R. 1966. Comparative anatomy of the central visual systems in day- and night-active primates, pp. 419–434. *In* R. Hassler and H. Stephan, eds., *Evolution of the Forebrain*. G. Thieme, Stuttgart.

HUBEL, D. H., and T. N. WIESEL. 1968. Receptive fields and functional architecture of the monkey striate cortex. *J. Physiol.* **195**:215–243.

INGLE, D. 1973. Evolutionary perspectives on the function of the optic tectum. *Brain Behav. Evol.* **8**:211–237.

JONES, A. E. 1965. The retinal structure of the owl monkey (*Aotes trivirgatus*). *J. Comp. Neurol.* **125**:19–28.

JONES, A. E. 1966a. The lateral geniculate complex of the owl monkey *Aotes trivirgatus. J. Comp. Neurol.* **126**:171–180.

JONES, A. E. 1966b. Wavelength and intensity effects on the response of single lateral geniculate nucleus units in the owl monkey. *J. Neurophysiol.* **29**:125–138.

KAAS, J. H., R. W. GUILLERY, and J. M. ALLMAN. 1972. Principles of organization in the dorsal lateral geniculate body. *Brain Behav. Evol.* **6**:253–299.

KOLMER, W. 1930. Zur Kenntnis des Auges der Primaten. *Z. Anat. Entwg.* **93**:679–722.

LAEMLE, L. K. 1968. Retinal projections of *Tupaia glis. Brain Behav. Evol.* **1**:473–499.

MILES, R. C. 1958. Color vision in the squirrel monkey. *J. Comp. Physiol. Psychol.* **51**:328–331.

MISHKIN, M. 1972. Cortical visual areas and their interactions, pp. 187–208. *In* A. G. Karczmar and J. C. Eccles, eds., *Brain and Human Behavior*. Springer-Verlag, Berlin and New York.

MURRAY, R. G., A. E. JONES, and A. MURRAY. 1973. Fine structure of photoreceptors in the owl monkey. *Anat. Rec.* **175**:673–695.

NOBACK, C. R., and L. K. LAEMLE. 1970. Structural and functional aspects of the visual pathway of primates, pp. 55–81. *In* C. R. Noback and W. Montagna, eds., *The Primate Brain, Advances in Primatology*. Appleton-Century-Crofts, New York.

NOBACK, C. R., and N. MOSKOWITZ. 1963. The primate nervous system: Functional and structural aspects in phylogeny, pp. 131–177. *In* J. Buettner-Janusch, ed., *Evolutionary and Genetic Biology of Primates*, Vol. 1. Academic Press, New York.

NOBACK, C. R., M. BERGER, L. K. LAEMLE, and J. F. SHRIVER. 1969. Phylogenetic aspects of the visual systems in primates and *Tupaia. Proc. 2nd Int. Congr. Primatol.* **3**:49–54.

OGDEN, T. E. 1974. The morphology of retinal neurons of the owl monkey *Aotes. J. Comp. Neurol.* **153**:399–428.

PARIENTE, G. 1970. Rétinographies comparées des Lémuriens malgaches. *C.R. Acad. Sci., Ser. D.* **270**:1404–1407.

POLYAK, S. 1957. *The Vertebrate Visual System*. University of Chicago Press, Chicago.

PRINCE, J. H. 1956. *Comparative Anatomy of the Eye*. Charles C Thomas, Springfield, Illinois.

RADINSKY, L. 1972. Endocasts and studies of primate brain evolution, pp. 175–185. *In* R. Tuttle, ed., *The Functional and Evolutionary Biology of Primates*. Aldine-Atherton, Chicago.

ROHEN, J. W. 1966. Zur Histologie des Tarsiusauges. *Graefes Arch. Ophthal.* **169**:299–317.

ROHEN, J. W., and A. CASTENHOLZ. 1967. Uber die Zentralisation der Retina bei Primaten. *Folia Primatol.* **5**:92–147.

SCHNEIDER, G. E. 1969. Two visual systems. *Science* **163**:895–902.

SHRIVER, J. E., and C. R. NOBACK. 1967. Color vision in the tree shrew (*Tupaia glis*). *Folia Primatol.* **6**:161–169.

Snyder, M. 1973. The evolution of mammalian visual mechanisms, pp. 693–712. *In* R. Jung, ed., *Handbook of Sensory Physiology*, Vol. 7 (No. 3A). Springer-Verlag, Heidelberg.

Spatz, W. B., and J. Tigges. 1972. Species difference between Old World and New World monkeys in the organization of the striate—prestriate association. *Brain Res.* **43**:591–593.

Spatz, W. B., J. Tigges, and M. Tigges. 1970. Subcortical projections, cortical association, and some intrinsic interlaminar connections of the striate cortex in the squirrel monkey (*Saimiri*). *J. Comp. Neurol.* **140**:155–174.

Tigges, J. 1963a. Untersuchungen über den Farbensinn von *Tupaia glis* Diard 1820. *Z. Anthropal. Morphol.* **53**:109–123.

Tigges, J. 1963b. On color vision in gibbon and orang-utan. *Folia Primatol.* **1**:188–198.

Trevarthen, C. B. 1968. Two mechanisms of vision in primates. *Psychol. Forsch.* **31**:299–337.

Wald, G. 1968. Molecular basis of visual excitation. *Science* **127**:230–239.

Walls, G. L. 1942. *The Vertebrate Eye and Its Adaptive Radiation*. The Cranbrook Institute of Science, Bloomfield Hills, Michigan, 785 pp. Reprinted 1963. Hafner Publishing Co., New York.

Weale, R. A. 1966. Why does the human retina possess a fovea? *Nature* **212**:255–256.

Wolin, L. R., and L. C. Massopust. 1967. Characteristics of the ocular fundus in primates. *J. Anat.* **101**:693–699.

Wolin, L. R., and L. C. Massopust. 1970. Morphology of the primate retina, pp. 1–27. *In* C. R. Noback and W. Montagna, eds., *The Primate Brain, Advances in Primatology*. Appleton-Century-Crofts, New York.

Zeki, S. M. 1969. Representation of central visual fields in prestriate cortex of monkey. *Brain Res.* **14**:271–291.

Zeki, S. M. 1971. Cortical projections from two prestriate areas in the monkey. *Brain Res.* **34**:19–35.

10

Protein Sequence and Immunological Specificity

Their Role in Phylogenetic Studies of Primates

MORRIS GOODMAN

"In regard to classification and all the endless disputes about the 'Natural System', which no two authors define in the same way, I believe it ought, in accordance to my heterodox notions, to be simply genealogical. But as we have no written pedigrees you will, perhaps, say this will not help much; but I think it ultimately will, whenever heterodoxy becomes orthodox, for it will clear away an immense amount of rubbish about the value of characters, and will make the difference between analogy and homology clear. The time will come, I believe, though I shall not live to see it, when we shall have very fairly true genealogical trees of each great kingdom of Nature." [Charles Darwin, letter★ to Thomas H. Huxley.]

"On the view of characters being of real importance for classification, only in so far as they reveal descent, we can clearly understand why analogical or adaptive characters, although of the utmost importance to the welfare of the being, are almost valueless to the systematist. For animals, belonging to two most distinct lines of descent, may have

---

★ Preserved by Francis Darwin (1903, p. 104) quoted by Ghiselin and Jaffe (1973).

MORRIS GOODMAN · Department of Anatomy, Wayne State University, School of Medicine, 540 E. Canfield Ave., Detroit, Michigan.

219

become adapted to similar conditions, and thus have assumed a close external resemblance; but such resemblances will not reveal—will rather tend to conceal their blood-relationship." [Charles Darwin, *The Origin of Species* (p. 420).]

I. Introduction

The classic study of Nuttall (1904), *Blood Immunity and Blood Relationship*, published long before the molecular basis of inheritance had been determined, demonstrated that the proteins of blood can reveal much about the phylogeny of primates and other vertebrates. That the term blood relationship can serve as a synonym for genealogical relationship, i.e., relationship by descent from a common ancestor, is now justified by the findings of molecular biology. Proteins (and blood contains a rich collection readily accessible for study) are especially well suited for genealogical mapping. Consecutive nucleotide base triplets or codons in the segments of DNA called structural genes code via matching RNA codons for correspondingly positioned amino acids in the polypeptide chains of proteins. An amino acid sequence chain of typical length, such as each polypeptide chain in hemoglobin, contains close to 150 amino acids of 20 different kinds. Considering that the theoretical permutations in sequence structure for a chain of this length are astronomical, we can begin to appreciate why the actual permutations produced by evolution provide a key to unravelling the phylogenetic relationships of organisms.

This chapter briefly describes how I arrive at phylogenetic inferences from protein data; phylogenetic conclusions are then presented on the Primates solely in terms of such data. Moreover, these conclusions are represented in a taxonomic classification which attempts to be simply a written description of the genealogical tree of extant primates. In this effort I follow the view of Charles Darwin (1859) and also Hennig (1965, 1966) that taxonomy should describe primarily the genealogical rather than morphological relationships of organic beings.

II. Evolutionary Hypothesis

Three hypotheses on the course of molecular evolution have proven helpful to me in arriving at phylogenetic inferences from protein data: (1) the hypothesis of divergent evolution, (2) the hypothesis of additive evolution, and (3) the hypothesis of parsimonious evolution. The first hypothesis simply states that the more ancient is the common ancestor for a pair of species the greater is the genetic distance for that pair of species. The divergence procedure for depicting phylogeny can be criticized because matrices of genetic distances (such as numbers of amino acid sequence differences among homologous protein chains in different species) reflect patristic similarities (features inherited unchanged from the common ancestor of the whole set of species) and convergent similarities (those acquired by parallel or back changes in different lines of descent) as well as derived similarities (features not present in the common ancestor of the whole set of species, but acquired by forward changes fixed in a more recent common ancestor of only some of the species in the set). Hennig (1965, 1966) correctly emphasizes that phylogenetic relationships can be best determined from shared derived features. Nevertheless, when phenograms are constructed by the divergence procedure from an adequate sample of genetic distances these phenograms are likely to be close to the true cladogeny, since derived similarities should account for most of the groupings. Ideally, rates of molecular evolution should be uniform in all lines of descent. Then, as Moore (1971) demonstrated in a rigorous mathematical treatment of the divergence hypothesis, several computer algorithms are guaranteed to produce the correct genealogical tree from the matrix of genetic distances. The unweighted pair group method (UWPGM) of Sokal and Michener (1958) is the most robust of the divergence algorithms in that it can tolerate appreciable deviations from the uniform rate condition and still produce a correct cladogeny (Moore, 1971).

The second hypothesis, that of additive evolution, is especially helpful when evolutionary rates vary markedly from one lineage to the next. The main condition of the additive hypothesis (Cavalli-Sforza and Edwards, 1967; Moore et al., 1973a) is that the genetic distance observed between any two species in a comparison matrix be proportional to the sum of the changes (such as nucleotide, amino acid, or antigenetic determinant replacements depending on the distance units employed) which actually accumulated in the two lines of descent separating the two species from their most recent common ancestor. Let us say the lineage to species A evolved much more rapidly than the lineage to species B. This would tend to be reflected in the comparison matrix by A showing a larger distance value than B from any compared species which cladistically was as closely related to A as to B. Thus the phylogenetic validity of proposed branching arrangements, such as those produced by the UWPGM divergence procedure, can be examined, and modified if need be, in terms of expectations of the additive hypothesis. The limitations of the hypothesis, however, should be mentioned. Gross inequities in the distribution of convergent similarities among the species in the comparison matrix will violate the additive condition. Also there is often a proportionately larger underestimation of the true genetic distance between more anciently separated species than between more recently separated ones because multiple or superimposed nucleotide replacements at single nucleotide sites increase with evolutionary time, yet cannot be detected from pairwise comparisons of contemporary species.

The third hypothesis, that of parsimonious evolution (a network of descent with the fewest number of changes) is the most powerful of the three for phylogenetic inference. When applied to amino acid sequence data, the parsimony procedure (Moore et al., 1973b; Goodman et al., 1974b) conforms closely to the principles of Hennig (1965, 1966) in that patristic and convergent similarities are distinguished from shared derived similarities, and the latter (the forward mutations fixed in more recent common ancestors) determine the cladistic or genealogical relationships of the species in the parsimony tree. This will become apparent when the parsimony procedure and the results obtained with it are discussed later in this chapter.

III. Homologous Protein Sequence Chains

Polypeptide chains in different species are defined as homologous chains in the strict sense of genetic homology if they are coded for by genes descended from a single ancestral gene which existed in the most recent common ancestor of the species being compared. Polypeptide chains are also homologous chains in the broader meaning of genetic homology as long as they are coded for by genes descended from a single ancestral gene. In this broader definition separate gene lines would not always have to arise from species divergence but could sometimes arise from the ancestral gene duplicating to produce two or more nonallelic genes. Fitch (1970) uses the terms orthologous and paralogous to distinguish between the strict and looser form of genetic homology. As will be seen later, the gene phylogenetic trees constructed by the maximum parsimony method can help identify the strictly homologous or orthologous protein sequences which may rightly be compared in determining the cladistic relationships of the species in which they occur. The evidence for genetic homology among presumed homologous protein sequences in different species is indirect, but nevertheless compelling if genetic homology is taken in its broader meaning. The evidence is simply the similarity observed among the presumed protein homologs with respect to structure and function. For example, there are not only the same number of amino acid residues, 141, in each polypeptide chain (called an alpha chain) of rhesus monkey and human hemoglobins, but the sequence of amino acids in these rhesus and human alpha chains differ at only 4 of their 141 residue positions. Considering the sequence permutations possible for 20 kinds of amino acids at 141 positions (20^{141}), it is virtually certain that genetic homology is responsible for the high degree of similarity in sequence structure between the rhesus and human chains. From what is known about the structure and function of proteins it does not seem possible for any *extensive* portion of two unrelated or nonhomologous polypeptide chains to become similar, i.e., converge over evolutionary time, in sequence structure. The fact that the rhesus and human alpha chains vary at only 4 residue positions comes to light when the sequential order of amino acids in the two chains are compared side by side, i.e., aligned against each other. A simple count then shows that only 4 of the 141 residue positions have differing amino acids in the two sequences.

IV. Amino Acid Sequence Distances among Primates

Table 1 lists polypeptide chains compared in at least two or more primates for amino acid sequence differences. The chain lengths or numbers of amino acid residue positions compared in these protein substances are 141 for alpha hemoglobin chain (α-Hb), 146 for beta hemoglobin chain (β-Hb), 146 for delta hemoglobin chain (δ-Hb), 146 for the gamma hemoglobin chain with glycine at residue position 136 ($^G\gamma$-Hb), 146 for the nonallelic gamma hemoglobin chain with alanine at residue position 136 ($^A\gamma$-Hb), 153 for myoglobin (Myo), 30 for fibrinopeptides A and B (Fib A and Fib B), 104 for cytochrome c (Cyt c), 115 for portions of carbonic anhydrase I (Ca I), and 115 for portions of carbonic anhydrase II (Ca II). Within each of these collections of homologous chains from different primates, the amino acid differences were counted for each pair of species. The combined results of these counts are summarized in Table 2. The upper half of the matrix in this table gives the total number of differing amino acids per number of shared amino acid residue positions for each pair of primates compared, and the lower half of the matrix shows these fractions as percent of differing amino acids or amino acid distance (AAD) values. These matrix values, however, are limited to pairs of primates in which at least three protein chains and 300 amino acid residue positions were compared within each pair. This compensates for the fact that the same protein substances were not compared uniformly in all primate species and tends to ensure that the number of amino acid residue positions compared was not too grossly deficient in any pair of primates for which results are shown in Table 2.

It can be seen that the distances (AAD values) for the human–chimpanzee, chimpanzee–gorilla, and human–gorilla comparisons range from only 0.27 to 0.65 as compared to distances ranging from 2.17 to 2.84 in the comparisons of these species with the Asiatic apes (orangutan and gibbon). Clearly the African apes (*Pan* and *Gorilla*) and *Homo* have diverged much less from one another than from the Asiatic apes (*Pongo* and *Hylobates*). This means, according to the divergent evolution hypothesis, that after the ancestral separation of *Hylobates* and *Pongo* lineages from the remaining hominoids, a common ancestor still existed for *Gorilla*, *Homo*, and *Pan*. It can also be deduced from the distance data in Table 2 that for this particular collection of primates the ancestral separation from the *Homo*–African ape complex increases in the following order: Asiatic apes, cerco-

TABLE I

PEPTIDE CHAINS USED TO CALCULATE AAD VALUES BETWEEN PRIMATES

	α-Hb	β-Hb	δ-Hb	δ-Hb	Gγ-Hb	Aγ-Hb	Myo	Fib A-B	Cyt c	Ca I	Ca II
Man	α-Hb	β-Hb	δ-Hb[a]		Gγ-Hb	Aγ-Hb	Myo	Fib A-B	Cyt c	Ca I	Ca II
Chimpanzee	α-Hb[a]	β-Hb[a]	δ-Hb[a]		Gγ-Hb[a]	Aγ-Hb[a]	Myo[a]	Fib A-B	Cyt c	Ca I[a]	
Gorilla	α-Hb[a]	β-Hb[a]	δ-Hb[a]					Fib A-B			
Orangutan	α-Hb[b]	β-Hb[b]						Fib A-B		Ca I[a]	
Gibbon	pα-Hb[c]	β-Hb[a]	δ-Hb[a]				Myo[a]	Fib A-B			
Macaque	α-Hb	β-Hb					Myo[a]	Fib A-B	Cyt c	Ca I[a]	Ca II[a]
Cercopithecus	α-Hb	β-Hb						Fib A-B		Ca I[a]	Ca II[a]
Squirrel monkey	pα-Hb[c]	β-Hb[a]		δ-Hb[a]			Myo[a]				
Spider monkey	α-Hb	β-Hb[a]		δ-Hb[a]				Fib A-B			
Capuchin monkey	α-Hb	β-Hb						Fib A-B			
Slow loris	α-Hb	β-Hb						Fib A-B			

NOTE: Since the delta hemoglobin chains of ceboids and hominoids are thought to have originated from separate beta gene duplications (Goodman *et al.*, 1974*b*), they are treated not as corresponding chains or homologs of each other in the matrix calculations of Table 2. The sequences for many of the above protein substances are given in Dayhoff (1972, 1973); those which are not are the alpha and beta chains of *Cercopithecus* (Matsuda *et al.*, 1973*b*), spider monkey (Matsuda *et al.*, 1973*d*), capuchin monkey (Matsuda *et al.*, 1973*c*), and slow loris (Matsuda *et al.*, 1973*f*); macaque, and squirrel monkey myoglobins (Romero–Herrera and Lehmann, 1972, 1973*a*); the fibrino-peptides A and B of spider and capuchin monkeys and slow loris (Wooding and Doolittle, 1972); and the primate carbonic anhydrases I and II (Tashian *et al.*, 1972).

[a] Sequence largely inferred from amino acid composition of peptide fragments by homology with known sequences.
[b] Amino acid composition (Buettner-Janusch *et al.*, 1969).
[c] Sequence on the first 31 positions from the *N*-terminal end of the α-hemoglobin chain (Boyer *et al.*, 1972).

TABLE 2

THE AMINO ACID DIFFERENCE MATRIX FOR PRIMATES

	Man	Chi	Gor	Ora[b]	Gib	Mac	Cer	Squ	Spi	Cap	Slo
Man		3/1127	3/463	12/432	12/505	31/799	20/547	29/330	20/317	24/317	36/317
Chimpanzee	0.27		2/463	11/432	12/505	31/684	19/432	30/330	20/317	24/317	36/317
Gorilla	0.65	0.43		7/322	10/352	16/312	16/317	X[c]	20/317	24/317	36/317
Orangutan[b]	2.78	2.54	2.17		X	X	X	X	X	X	X
Gibbon	2.38	2.38	2.84	X		23/354	X	28/330	X	X	X
Macaque	3.89	4.54	5.13	X	6.51		12/542	25/330	23/312	26/312	37/312
Cercopithecus	3.65	4.39	5.06	X	X	2.21		X	22/317	25/317	39/317
Squirrel monkey	8.78	9.08	X	X	8.49	7.58	X		11/323	X	X
Spider monkey	6.31	6.31	6.31	X	X	7.38	6.95	3.39		14/317	32/317
Capuchin monkey	7.56	7.56	7.56	X	X	8.35	7.86	X	4.41		35/317
Slow loris	11.36	11.36	11.36	X	X	11.85	12.28	X	10.08	11.04	

[a] Upper half of matrix: number of differing amino acids per number of shared amino acid positions; lower half of matrix: percent of differing amino acids, the amino acid distance (AAD) values.

[b] The amino acid composition data of orangutan α-Hb and β-Hb chains were used in the estimations of amino acid differences in the comparisons of the orangutan to other hominoids but not in those of orangutan to phylogenetically more distant primates, since amino acid composition data could only be expected to yield correct or nearly correct AAD values between closely related species.

[c] X = comparisons not done because of insufficient data.

pithecoids (*Macaca* and *Cercopithecus*), ceboids (squirrel, spider, and capuchin monkeys), and Lorisoidea (slow loris). Another point of interest is that the distance between *Macaca* and *Cercopithecus* within the Old World monkey subfamily Cercopithecinae is 2.21, which is more than the African ape–*Homo* distance.

V. Immunologically Measured Distances among Primates

In the immunological approach, a vertebrate host such as the rabbit is injected with protein (the antigen) from a donor species. The surface configurations on a donor protein which are different from any configurations on the host proteins stimulate the host's immunological system to produce antibodies. The antibody produced to each such foreign molecular configuration, i.e., to an antigenic determinant, has a combining site which is specifically complementary to and reactive with the antigenic determinant. Because of this, in proper hands, antibodies are superb tools for detecting differences in surface topographies between proteins. Since during evolutionary time amino acid replacements progressively alter the surface topographies of proteins, antigenic differences between proteins should correlate in a general way with amino acid sequence differences. The observations of Arnheim *et al.* (1969) and Prager and Wilson (1971*a,b*) support this expectation.

In my laboratory a large body of data on antigenetic distances among primates has been gathered by the immunodiffusion technique employing Ouchterlony trefoil plates (Goodman, 1963*a*; 1967, 1973; Goodman and Moore, 1971; Goodman *et al.*, 1974*a*; Darga *et al.*, 1973; Baba *et al.*, 1975; Dene *et al.*, 1975). When two related proteins are compared in such a trefoil plate with antisera to one of them (the homologous antigen), the length of the spur of this homologous antigen against the other (the heterologous antigen) will be longer, the greater the dissimilarity between the two proteins in their surface configurations of amino acid groups. If two heterologous antigens are compared, the one which bears the closer relationship to the homologous antigen will yield the longer spur against the other, i.e., a positive net spur size. On the basis of set-theory logic (Moore and Goodman, 1968; Goodman and Moore, 1971) a computer program translates these spur sizes in a series of trefoil Ouchterlony plate comparisons into a table of antigenic distances. When the plates in these tests show multiple precipitin lines (as will happen if the antiserum is directed against a mixture of proteins), each line is counted as a separate reaction in the calculation of average distances from antisera to antigens of either species. An *adjustment procedure* (developed by G. W. Moore and executed by another computer program) is also employed to normalize the antigenic distance values in the different immunodiffusion tables and thereby compensate for the fact that some immunodiffusion tables are better able to capture the full extent of divergence than others. Such tables before adjustment show nonreciprocal antigenic distances between species pairs, i.e., the antiserum to A may show a larger antigenic distance between B and A than the antiserum to B shows. Thus the strategy followed is to assign an *adjustment factor* to each immunodiffusion table and then to multiply the immunodiffusion table by that factor. One table is arbitrarily assigned a factor of 1, while the other tables are assigned factors such that nonreciprocity in the adjusted or normalized tables is minimized.

Spur sizes from over 4800 Ouchterlony trefoil plate comparisons developed with rabbit antisera to whole serum or to specific serum proteins of 20 primate, 3 tree shrew, 1 flying lemur, and 2 elephant shrew species were converted first into tables of normalized antigenic distances and then into a divergence tree (Fig. 1) by the UWPGM algorithm (Dene *et al.*, 1975). The average antigenic distances between branches in this UWPGM tree are given in Table 3, where the taxa formed by the branches are designated by the scheme used in the genealogical classification of Table 4 and depicted in the divergence tree of Fig. 1. *Pan* and *Gorilla* are closest to *Homo*, followed in succession by *Pongo*, gibbons (hylobatines), cercopithecoids (Old World monkeys), ceboids (New World monkeys and marmosets), tarsier, strepsirhines (lemuroids and lorisoids), tupaiids (tree shrews), dermopterans (flying lemur), other eutherian mammals, and marsupials. This order of relationship of *Homo* to other primate lineages was confirmed in over 3000 additional Ouchterlony trefoil plate comparisons using cercopithecoid (*Cercopithecus* and *Macaca*) and ceboid (spider, woolly, and capuchin monkey) antisera to human and chimpanzee serum proteins and using chicken antisera to serum proteins of various primate species. Moreover the albumin antigenic distance values among species of Anthropoidea obtained by Sarich (1970) using the microcomplement fixation technique also depict the close relationship between the African apes and *Homo*, the hominoid affinities of the gibbons, and the subdivision of Anthropoidea into Platyrrhini (Ceboidea) and Catarrhini (Hominoidea and Cercopithecoidea). The very small degree of antigenic distance shown in Table 3 between *Homo* and the African apes (0.73), suggests that *Homo* should be grouped with *Pan* and *Gorilla* in the same subfamily as in Table 4, rather than separated at the family level as in traditional morphologically based taxonomy.

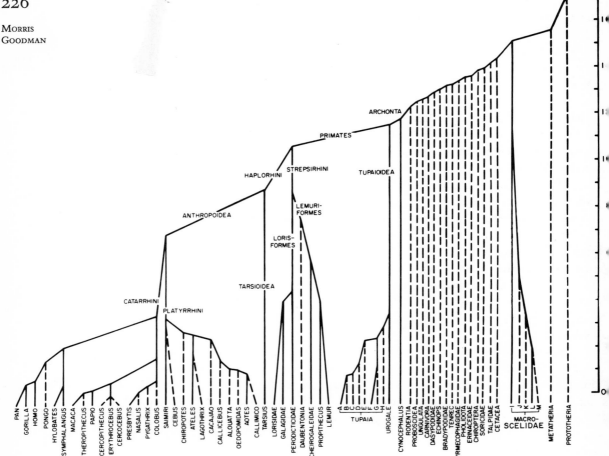

Fig. 1. Heavy lines descend to taxa used as homologous species; dashed lines descend to taxa used only as hetero-logous species. The *Tupaia* species are (A) *T. chinensis*; (B) *T. longipes*, (C) *T. glis*; (D) *T. minor*; (E) *T. palawanensis*; (F) *T. montana*; (G) *T. tana*; and (H) *T. belangeri*. *T. belangeri* at the time this divergence tree was constructed had been tested only with antisera to *T. montana*. Thus its placement reflects only the magnitude of its divergence from *T. montana*. It has now been tested with antisera to *T. chinensis* and has been found to diverge much less from this species than from *T. montana* (Dene, unpublished data). Thus, the next time the divergence tree is constructed, *T. belangeri* is likely to fall closer to *T. chinensis*. The Macroscelidea are (I) *Rhynchocyon*; (J) *Petro-dromus*; (K) *Nasilio*; (L) *Elephantulus myurus*; and (M) *E. intufi*. The ordinate indicates the average antigenetic distance between branches of the tree.

A. A Classification from the Immunodiffusion Data

Tarsius and the Anthropoidea with an antigenic distance of 8.84 between them are slightly closer to each other than to other primates, and similarly Lorisiformes and Lemuriformes are slightly closer to each other (antigenic distance 8.83) than to other primates. This supports the haplorhine–strepsirhine division of the order Primates. The photos in Fig. 2 of a few of the many immunodiffusion plate comparisons developed

TABLE 3

DISTANCES BETWEEN BRANCHES FROM THE UWPGM TREE CONSTRUCTED FROM THE NORMALIZED IMMUNODIFFUSION TABLES OF RABBIT ANTISERA TO THE SERUM PROTEINS OF 26 PRIMATE, TREE SHREW, FLYING LEMUR, AND ELEPHANT SHREW SPECIES (DENE ET AL., 1975).

		Average antigenic distance	Taxonomic designation of the group[a]
Branches within Haplorhini			
Homo	vs. *Pan, Gorilla*	0.73	Homininae
Homininae	vs. *Pongo*	1.56	Hominidae
Hylobates	vs. *Symphalangus*	0.54	Hylobatinae
Hominidae	vs. Hylobatidae	2.14	Hominoidea
Colobus	vs. other Colobinae	0.71	Colobinae
Macaca, Papio, Theropithecus	vs. other Cercopithecinae	0.52	Cercopithecinae
Colobinae	vs. Cercopithecinae	1.50	Cercopithecidae
Hominoidea	vs. Cercopithecoidea	3.23	Catarrhini
Saimiri, Cebus	vs. other Ceboidea	3.34	Ceboidea
Catarrhini	vs. Platyrrhini	6.93	Anthropoidea
Anthropoidea	vs. *Tarsius*	8.84	Haplorhini
Branches within Strepsirhini			
Galago	vs. *Galagoides*	0.57	Galaginae
Nycticebus	vs. *Loris*	2.21	Lorisidae
Perodicticus	vs. *Arctocebus*	0.33	Perodicticinae
Galagidae	vs. Lorisidae	4.00	
Galagidae, Lorisidae	vs. Perodicticidae	4.57	Lorisoidea
Lemur variegatus	vs. other *Lemur*	0.89	*Lemur*
Lemur	vs. *Propithecus*	3.93	Lemuroidea
Microcebus	vs. *Cheirogaleus*	3.63	Cheirogaleidae
Lemuroidea, Cheirogaleoidea	vs. *Daubentonia*	7.63	Lemuriformes
Lorisiformes	vs. Lemuriformes	8.83	Strepsirhini
Branches within Archonta			
Haplorhini	vs. Strepsirhini	10.70	Primates
Tupaia	vs. *Urogale*	3.67	Tupaiidae
Primates	vs. Tupaiidae	11.46	
Primates, Tupaioidea	vs. *Cynocephalus*	11.71	Archonta
Branches within Macroscelidea			
Nasilio	vs. *Elephantulus*	3.34	
Nasilio, Elephantulus	vs. *Petrodromus*	5.06	Macroscelidoidea
Macroscelidoidea	vs. *Rhynchocyon*	11.14	Macroscelidea
Branches within Mammalia			
Archonta	vs. most Eutheria	12.23–14.28	
Most Eutheria	vs. Macroscelidea	15.12	Eutheria
Eutheria	vs. Marsupialia	15.61	Theria
Eutheria	vs. *Tachyglossus*	17.06	Mammalia

[a] Taxa names are from those used in the classification of Table 4.

TABLE 4

A CLASSIFICATION OF EXTANT PRIMATES AND SEVERAL OTHER EUTHERIAN TAXA FROM IMMUNODIFFUSION
DATA (ONLY GENERA TESTED ARE INCLUDED)

Superorder: Archonta
  Order: Dermoptera
          Cynocephalus (flying lemur)
  Order: Tupaioidea
      Family: Tupaiidae
          Tupaia (tree shrew)
          Urogale (tree shrew)
  Order: Primates
    Semiorder: Strepsirhini
      Suborder: Lemuriformes
        Superfamily: Lemuroidea
          Family: Lemuridae
              Lemur (lemur)
          Family: Indriidae
              Propithecus (sifaka)
        Superfamily: Cheirogaleoidea
          Family: Cheirogaleidae
              Microcebus (mouse lemur)
              Cheirogaleus (dwarf lemur)
        Superfamily: Daubentonioidea
              Daubentonia (aye-aye)
      Suborder: Lorisiformes
        Superfamily: Lorisoidea
          Family: Perodicticidae
            Subfamily: Perodicticinae
              Perodicticus (potto)
              Arctocebus (angwantibo)
          Family: Galagidae
            Subfamily: Galaginae
              Galago (bushbaby)
              Galagoides (pigmy galago)
          Family: Lorisidae
              Loris (slender loris)
              Nycticebus (slow loris)
    Semiorder: Haplorhini
      Suborder: Tarsioidea
              Tarsius (tarsier)
      Suborder: Anthropoidea
        Infraorder: Platyrrhini
          Superfamily: Ceboidea
              Saguinus (marmoset)
              Aotus (owl monkey)
              Callicebus (titis)
              Cacajao (uakaris)
              Chiropotes (sakis)
              Cebus (capuchin)
              Saimiri (squirrel monkey)
              Alouatta (howler monkey)
              Ateles (spider monkey)
              Lagothrix (woolly monkey)
        Infraorder: Catarrhini
          Superfamily: Cercopithecoidea

          Family: Cercopithecidae
            Subfamily: Colobinae
              Colobus (guerezas)
              Presbytis (langur)
              Pygathrix (douc langur)
              Nasalis (proboscis monkey)
            Subfamily: Cercopithecinae
              Macaca (macaque)
              Papio (baboon)
              Theropithecus (gelada baboon)
              Cercocebus (mangabey)
              Erythrocebus (patas monkey)
              Cercopithecus (vervet)
        Superfamily: Hominoidea
          Family: Hylobatidae
            Subfamily: Hylobatinae
              Hylobates (gibbon)
              Symphalangus (siamang)
          Family: Hominidae
            Subfamily: Ponginae
              Pongo (orangutan)
            Subfamily: Homininae
              Gorilla (gorilla)
              Pan (Chimpanzee)
              Homo (human)

Other Eutheria
  Order: Macroscelidea
    Superfamily: Macroscelidoidea
      Family: unnamed
              Elephantulus (elephant shrew)
              Nasilio (elephant shrew)
      Family: unnamed
              Petrodromus (elephant shrew)
    Superfamily: Rhynchocyonoidea
      Family: unnamed
              Rhynchocyon (elephant shrew)

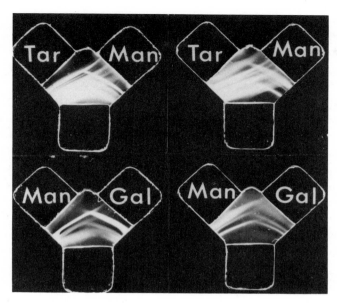

FIG. 2. Trefoil Ouchterlony plate comparisons of rabbit anti-tarsier serum. Reactions in the left plates are against undiluted antigens and in the right plates are against 1/10 dilutions of these antigens. The homologous antigen, *Tarsius* serum, developed 8 distinct precipitin lines, each of which yielded strong spurs against the heterologous antigen, human serum (top two plates). When human serum was compared to another heterologous antigen, *Galago* serum, 3 precipitin lines developed, with the cross-reacting human proteins yielding positive net spurs against the cross-reacting *Galago* proteins.

by rabbit antiserum to tarsier serum are taken from the large body of immunodiffusion results which show why Tarsioidea groups with Anthropoidea to form the taxon Haplorhini. Within the Anthropoidea the divisions generally follow traditional classifications (e.g., Simpson, 1945; Fiedler, 1956; Lasker, 1973), the most notable exception being the grouping of *Pan* and *Gorilla* with *Homo* in Homininae noted above and then the grouping of Ponginae (*Pongo*) with Homininae to form the family Hominidae. As in traditional classifications, hylobatine gibbons show a closer relationship to large-sized apes and humans than to Old World monkeys; thus in the genealogical scheme of Fig. 4 they are placed in the superfamily Hominoidea alongside of Hominidae as the separate family Hylobatidae. Cercopithecoidea (Old World monkeys) contains one extant family, Cercopithecidae, which subdivides as in the traditional scheme into Colobinae and Cercopithecinae. Subdivisions of the Ceboidea are omitted in Tables 3 and 4 because the possibilities for subdividing this superfamily are now being intensively explored by the immunodiffusion technique with antisera to various ceboid species (M. Baba, unpublished data). However, Baba's results are clearly confirming the subfamily grouping of *Ateles* and *Lagothrix* (spider and woolly monkey) into Atelinae, as well as demonstrating a closer relationship of *Alouatta* (howler monkey) to Atelinae than to other ceboids. A quite close relationship between *Chiropotes* (sakis) and *Cacajao* (uakaris) is also evident. Otherwise the various cebid genera seem sharply separated from one another. Moreover, no evidence is provided from the immunodiffusion results for separating marmosets from cebids at the family level (see Fig. 1). All extant Platyrrhini might very well be members of just one genealogical family, the Cebidae.

The branching arrangements within the Strepsirhini are similar to those depicted by traditional views, but there are some exceptions (Dene et al., 1975). Within the lorisiform branch, Asian lorises (Nycticebus and Loris) and African pottos (Perodicticus and Arctocebus) are each closer in immunodiffusion tests to African bushbabies (Galago and Galogoides) than to each other. The divergence tree first groups the African bushbabies and Asian lorises together and shows slightly less antigenic distance between these two groups than between them and the African pottos (4.00 compared to 4.57). However, this grouping might simply reflect strong patristic resemblances (i.e., primitive retentions) between Nycticebus and Galago. Immunodiffusion tables of antisera to nonlorisoid primates show that Nycticebus and Galago cross-react more strongly (show less antigenic distances from the homologous species) than other lorisoids. This suggests, if we reason from the standpoint of the additive hypothesis, that African pottos and African bushbabies might actually be genealogically closer to each other than either is to Asian lorises. Additional molecular data, especially of the type which can be analyzed by the parsimony method, are needed to resolve the question. In the interim, the solution used in Table 4 is subdivision of Lorisoidea into three coequal families: Perodicticidae, Galagidae, and Lorisidae.

In traditional classifications of Lemuriformes, mouse and dwarf lemurs (Microcebus and Cheirogaleus) are placed in Cheirogaleinae and grouped with Lemurinae into the family Lemuridae which then groups with Indriidae into Lemuroidea. Recent morphological studies have challenged this traditional arrangement and suggested that cheirogaleids are cladistically a branch of Lorisiformes rather than of Lemuriformes (see chapters by Tattersall, Szalay, and Cartmill, this volume). In immunodiffusion tests (Dene et al., 1975) Propithecus (Indriidae) and Lemur (Lemuridae) are closer to each other than to cheirogaleids and closer to cheirogaleids than to Daubentonia (Fig. 1). In turn, cheirogaleids are closer (so is Daubentonia) to Propithecus and Lemur than to lorisoids. Moreover, antisera to tree shrews, flying lemur, and various nonstrepsirhine primates show no more divergence for lorisoids (in fact on the average slightly less for Galago and Nycticebus) than for cheirogaleids and lemuroids. Thus the grouping of cheirogaleids with lemuroids rather than with lorisoids in immunodiffusion tests cannot be attributed solely to primitive strepsirhine resemblances but probably reflects a true genealogical relationship. A closer kinship of cheirogaleids to lemuroids than to lorisoids has also been found by Cronin et al., (1974) in microcomplement fixation tests with antisera to strepsirhine and other primate albumins and transferrins and with Phaner representing cheirogaleids. In terms of both divergence and additive hypotheses of antigenic evolution, Cheirogaleoidea belong with Lemuroidea (Lemur and Propithecus) in Lemuriformes, as depicted in Table 4.

The aggregate of immunodiffusion data depicts tree shrews as closer to Primates than other mammals. However, no special relationship to Tarsius and Cynocephalus (flying lemur), as suggested by the initial complement fixation data of Sarich and Cronin

(1974), is apparent. Immunodiffusion tests (Dene *et al.*, 1975) do confirm the close relationship of flying lemur to primates and tree shrews, but show this to be a relationship to Primates as a whole rather than specifically to *Tarsius* and Tupaiidae. On the basis of further tests with antisera to albumins and transferrins, Sarich and Cronin (personal communication) no longer group flying lemur, tree shrews, and *Tarsius* together. However, their tests, unlike ours, do not seem to place *Tarsius* closer to the Anthropoidea.

Antigenic distances from immunodiffusion data (Table 3 and Fig. 1) place the split between strepsirhines and tree shrews at an earlier point in time than the split between strepsirhines and haplorhines. This arrangement, while not inconsistent with the view (e.g., Simpson, 1945) that tree shrews are an early branch of Primates, opposes their inclusion in any taxonomic group including Strepsirhini but not Haplorhini. The taxonomic solution used in Table 4 is to place the separate orders Tupaioidea (tree shrews) and Dermoptera (flying lemurs) with Primates in the superorder Archonta (Fig. 1). This is essentially the arrangement of Gregory (1910), Butler (1956), and McKenna (this volume). Our superorder Archonta (Dene *et al.*, 1975) differs from Butler's in that elephant shrews are excluded, and from McKenna's in that bats are excluded.

Immunodiffusion results, especially those obtained with rabbit antisera to the elephant shrews *Nasilio* and *Elephantulus*, depict the Macroscelidea as an anciently separated branch of Eutheria, with no special relationship to tupaiids, primates, or to any of the lineages currently retained in the Insectivora (Fig. 1). This is essentially what Patterson (1965) found on studying fossil elephant shrews. Moreover the division by Patterson of extant elephant shrews into two major branches, one for *Rhynchocyon* and the other for the remaining genera, is supported by the immunodiffusion results. However, because the average antigenic distance of *Rhynchocyon* from the other genera is so large (11.14, see Table 3), it would seem appropriate to treat the two branches as different superfamilies (Rhynchocyonoidea and Macroscelidoidea) rather than follow Patterson (1965) who treated them as subfamilies in one family. The use of the name Macroscelidoidea in Tables 3 and 4 for the non-Rhynchocyon superfamily is not directly based on immunodiffusion evidence, since *Macroscelides* has not yet been obtained for inclusion in the immunodiffusion tests. However, because the immunodiffusion results, as far as they go, correlate so well with Patterson's findings, it would seem appropriate to follow Patterson's treatment which retains *Macroscelides* not only as the type genus of the non-*Rhynchocyon* extant elephant shrews but also for the whole order.

McKenna (this volume) proposes that Macroscelidea should be grouped with Lagomorpha and Rodentia into the magnorder Ernotheria. The present immunodiffusion results can neither confirm nor oppose this grouping since rabbit antisera cannot be used in a meaningful way to investigate the systematic position of Lagomorpha within the Mammalia. However, if elephant shrews and rabbits were more closely related to each other than to other orders of Eutheria, it would help explain why Macroscelidea is the most divergent eutherian branch in Table 3 and Fig. 1. Since the rabbit would not produce antibodies to any configurations on foreign proteins that are present on its own proteins, a mammal that was genealogically closer to the rabbit than were other mammals would diverge more from an homologous species than would the other heterologous mammalian species (except for rabbit, the immunized host species, if it happened to have been used as a heterologous species in one or another comparison series). Other types of molecular data, such as amino acid sequences on a spectrum of proteins, are needed to investigate interordinal relationships within the Mammalia. As will be evident in the following sections of this chapter, the present sequence data are still too small to yield reliable conclusions on interordinal relationships, but are sufficiently extensive to supplement the immunological data with respect to depicting the genealogy of primate lineages.

VI. Maximum Parsimony Estimation of Evolutionary Change in Proteins

The maximum parsimony principle (Edwards and Cavalli-Sforza, 1963; Estabrook, 1968; Farris, 1970; Fitch, 1971; Moore *et al.*, 1973*b*; Hartigan, 1973), when applied to protein sequences and their underlying codon sequences, states that the best genealogy reconstruction is that which requires the fewest possible nucleotide replacements to account for the descent of the sequences. This principle tends to attribute the maximum number of structural similarities among aligned protein sequences to common ancestry, and thus tends to minimize the number of parallel and back mutations. In the computer programs developed by G. W. Moore, the maximum parsimony analysis is carried out in such a way that often different codons for the same amino acid can be distinguished. Considering the many permutations in amino acids and even more so in codons which are possible owing to mutations, the following probabilistic argument can be used to justify a procedure for reconstructing phylogeny based on the parsimony principle.

If homologous sequences in two lineages have already diverged from each other at an aligned position, the chances are that it would take millions of years of evolution before the same codon could again occupy this sequence position in some of the descendant species. By the time such a parallel or back change occurred in these species, new mutations fixed at other aligned positions would in all likelihood have increased the divergence of their homologous sequences. While such forward mutations would sharply distinguish the anciently separated species, they would also serve (like the derived similarities in Hennig's method of phylogenetic inference) to identify the more recently separated species, In other words, among homologous sequences of diverging species, the codon identities which are the result of common inheritance are likely to be much more frequent than those resulting from parallel or back changes. Thus in reconstructing the phylogeny of living species from their protein sequences, the maximum parsimony method should be used because it seeks that genealogical tree which maximizes among related species the number of codon identities inherited without change from common ancestors and minimizes the number from parallel and reverse changes.

A mathematically rigorous procedure for assigning the most parsimonious reconstructed ancestors to a given genealogy is used (Moore *et al.*, 1973*b*). Since there is no known method, short of an exhaustive search, for finding the most parsimonious genealogy, an heuristic approach is employed (Goodman et al., 1974b) in which an initial UWPGM tree is iteratively improved by a branch-swapping procedure. The search is stopped when it appears that a further swapping of branches will not reduce the length (i.e., the total number of nucleotide replacements) of the genealogy.

Some nucleotide replacements are undoubtedly overlooked in this reconstruction process owing to missing intermediate species. Regions in the genealogy in which few nodal points span long periods of time are most likely to be underrepresented with respect to nucleotide replacements. Since the maximum parsimony method often yields multiple solutions, the solution which most frequently distributes mutations to underrepresented regions in the genealogy is employed. Nevertheless, nucleotide replacements still tend to be more grossly underestimated in sparse regions than in those with many nodal points. When two sequences are directly compared, there can never be more than one nucleotide replacement detected at any particular nucleotide position. However, when the comparison is made through intervening sequences, further replacements are often revealed. An augmentation procedure (described in Goodman et al., 1974b) was designed by G. W. Moore to compensate for the lost nucleotide replacements in sparse regions of a genealogy. The compensations are based on the distribution of nucleotide replacements in the dense regions. The results obtained by this algorithm correlated well with results obtained by the stochastic method based on the Poisson distribution (Holmquist et al., 1972; Jukes and Holmquist, 1972) for estimating hidden mutational change in codons (unpublished data of Moore, Goodman, and Holmquist). Even when the unaugmented nucleotide replacement lengths of a parsimony tree are employed, they reveal much greater divergence, provided the genealogy is relatively dense, between anciently separated sequences than the divergence shown by either amino acid differences or minimum mutation distances obtained from direct comparison of pairs of these sequences. The relatively trivial divergence between African ape and human branches in the parsimony trees described below compared to the large divergence between mammalian orders emphasizes that these hominoids should be grouped together at a relatively low taxonomic level as was already done in Table 4 from immunodiffusion data.

A. Evidence on Primate Phylogeny from Parsimony Trees of Protein Sequences

1. Fibrinopeptide A and B Sequences. The parsimony tree (Fig. 3) for primate and mammalian fibrinopeptides A and B (Goodman, 1973) joins slow loris to the Anthropoidea and joins this primate branch to a rat–rabbit branch; this assemblage then joins other mammalian branches. The tree splits the Anthropoidea into platyrrhines and catarrhines. Chimpanzee and gorilla cannot be distinguished from *Homo*, but the human–African ape ancestor diverges by two nucleotide replacements from the orangutan–human–African ape ancestor and by three replacements from the hominoid ancestor from which the gibbons (*Symphalangus* and *Hylobates*) descend with no replacements on the line to the hylobatine ancestor and only one on the line to *Hylobates*. The order of ancestral branching for gorilla, chimpanzee, and *Homo* cannot be distinguished by the parsimony tree because the human–African ape fibrinopeptide ancestor has the same sequence as the chimpanzee–human ancestor. Similarly the hominoid ancestor cannot be distinguished from the catarrhine ancestor; thus the parsimony tree of fibrinopeptides does not reveal if the gibbons are closer genealogically to other hominoids or to Old World monkeys. Their closer kinship to other hominoids is shown, however, in the parsimony trees of globins. The Homininae fibrinopeptides, consisting of identical *Homo*, *Pan*, and *Gorilla* sequences diverge by 5–8 nucleotide replacements from Old World monkeys (Cercopithecoidea), 9–10 replacements from New World monkeys (Ceboidea), 17 replacements from slow loris (Strepsirhini), and 21–37 replacements from other therian mammal fibrinopeptides.

2. *Cytochrome c Sequences.* Primates are represented in the cytochrome *c* parsimony tree by human, chimpanzee, and rhesus monkey sequences. The human and chimpanzee sequences are identical and diverge from the rhesus monkey sequence by just one nucleotide replacement. This replacement occurs between the catarrhine ancestor and the hominine ancestor. In turn the ancestral catarrhine sequence is separated from its nearest common ancestor with nonprimate mammalian sequences by eight nucleotide replacements (Goodman and Callahan, unpublished data).

3. *Carbonic Anhydrase I and II Sequences.* The genes for carbonic anhydrase I and II,

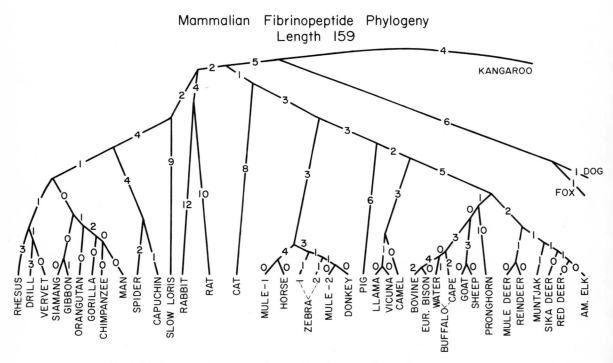

FIG. 3. Maximum parsimony tree of mammalian fibrinopeptide A and B sequences. In the alignment employed, that of Wooding and Doolittle (1972), 27 amino acid residue positions are shared by most species, and the number of nucleotide replacements recorded on the links (the link lengths) are for these 27 positions, except on the terminal links to kangaroo, horse and mule-1, cape buffalo, and rhesus monkey, in which the numbers are for 25, 25, 23, and 25 residue positions, respectively. The amino acid sequences of the 12 primate species are given in Wooding and Doolittle (1972), and the remaining sequences are given in Dayhoff (1972, 1973).

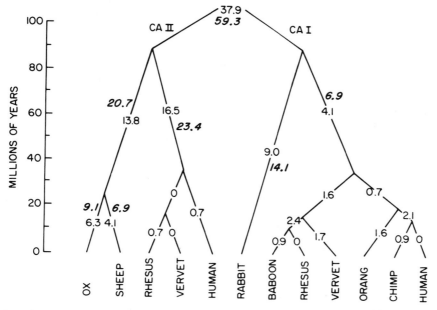

Fig. 4. Maximum parsimony phylogenetic tree of carbonic anhydrase isozymes, CA I and CA II, based on comparisons of partial sequences from 9 mammalian species. The link length numbers are the number of nucleotide replacements per 100 codon residues; augmented link lengths are the italicized numbers. The numbers of nucleotide replacements mentioned in the text are larger than those shown in the figure because for each link they represent the number of replacements for the total number of residue positions shared by the adjacent sequences connected by the link. On most links there were 145 shared residue positions. But on the link to ox (bovine) CA II there were 143; on the links from the catarrhine CA II ancestor to rhesus and vervet CA II, 136; from the eutherian CA I ancestor to rabbit CA I, 78; from the catarrhine CA I ancestor to rhesus CA I, 126; but on the links to baboon and vervet CA I, 117; on the link to orangutan CA I, 126; and on the link to chimpanzee CA I, 117.

like the genes for different globin chain types, arose from a common gene ancestor by the process of gene duplication. A maximum parsimony tree has been constructed (Fig. 4) from sequence data on human, chimpanzee, orangutan, *Cercopithecus*, rhesus, baboon, and rabbit carbonic anhydrase I, and human, *Cercopithecus*, rhesus, sheep, and bovine carbonic anhydrase II (Tashian *et al.*, 1975). In the carbonic anhydrase I region of the tree, one nucleotide replacement separates the hominoid (*Pongo*, *Pan*, and *Homo*) ancestor from the catarrhine ancestor, and another three replacements separate the hominine (chimpanzee–human) ancestor from this hominoid ancestor. While varying from each other by only one nucleotide replacement, the human and chimpanzee sequences are separated from the orangutan sequence by 5–6 replacements, from the Old World monkey sequence by 8–11 replacements, and from the rabbit sequence by 17–18 replacements or by 25–26 replacements when the augmentation algorithm is used to correct for lost replacements. In the carbonic anhydrase II region of the tree, the human, *Cercopithecus*, and rhesus sequences can be originated simultaneously from the catarrhine ancestor, with no nucleotide replacements occurring in the *Cercopithecus* line and one replacement each in the human and rhesus lines. In turn this catarrhine ancestor is separated

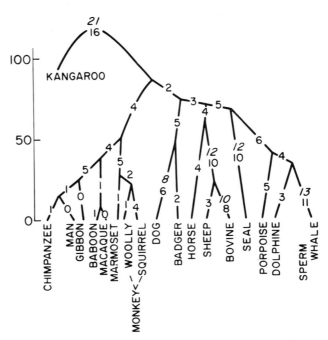

FIG. 5. Parsimony tree of 18 mammalian myoglobin sequences. Link lengths are the numbers of nucleotide replacements between adjacent ancestral and descendant sequences; italicized numbers are the link lengths corrected for superimposed replacements by the augmentation algorithm. The ordinate scale, in millions of years, is inferred from fossil evidence on the ancestral splitting times of the organisms represented by the sequences. Most of these sequences are given in Dayhoff (1972, 1973). Those which are not are from baboon and macaque (Romero-Herrera and Lehmann, 1972), marmoset, spider monkey, and woolly monkey (Romero-Herrera and Lehmann, 1973a), dog (Dumur et al., 1972), and badger (Tetaert et al., 1974).

from the eutherian carbonic anhydrase II ancestor by 24 replacements or 34 replacements when the augmentation algorithm is used.

    4. Myoglobin Sequences. The primates are represented in the phylogenetic tree of myoglobin sequences shown in Fig. 5 by 3 hominoids (*Pan, Homo,* and *Hylobates*), 2 cercopithecoids (baboon and macaque), and 3 ceboids (marmoset, woolly monkey, and squirrel monkey). Figure 5 shows an order of ancestral branching among these primate sequences which corresponds to the traditional taxonomic subdivision of Anthropoidea into Platyrrhini and Catarrhini with the latter splitting into Cercopithecoidea and Hominoidea. The gibbon sequence descends from the hominoid ancestor. Moreover, 5 nucleotide replacements separate this hominoid ancestor from the catarrhine ancestor. Thus clear evidence is provided for the hominoid affinities of the gibbon. The myoglobin parsimony tree also provides evidence for a closer kinship of the chimpanzee to ourselves than to the gibbon, since one replacement separates the hominine (human–chimpanzee) ancestor from the hominoid ancestor. Conversely, joining the gibbon first with either the human or chimpanzee sequence results in a tree with one more replacement than the maximum parsimony tree shown in Fig. 5.

    It is worth noting that the parsimony tree of myoglobins, constructed before the ceboid myoglobin sequences were available, did not require a prior union of human and chimpanzee sequences before attachment of the gibbon sequence. It would have been just as parsimonious for the gibbon to have joined either the human or chimpanzee sequence first. Thus the principle is suggested that the more adequately a phylogenetic group is represented by sequence data (in this case the Anthropoidea by the addition of the ceboid sequences), the more accurately or decisively the branching arrangements within the group can be depicted by the maximum parsimony method.

However, the addition of more sequence data may also raise new questions. This can be illustrated by my latest experiences searching anew for the myoglobin parsimony tree after adding to the data set the recently published *Galago*, *Lepilemur*, and *Tupaia* sequences (Romero-Herrera and Lehmann, 1973b, 1974) and the sea lion sequence (Vigna *et al.*, 1974). The UWPGM tree had a length of 194 nucleotide replacements. It correctly put the catarrhine sequences together, the ceboid sequences, the canoid sequences, the pinniped sequences, the cetacean sequences, and the bovid sequences; it also had the kangaroo descend from the most ancestral branch point but otherwise violated well-substantiated concepts of primate and mammalian phylogeny. Tree shrew joined the catarrhines, then ceboids joined. Horse joined *Lepilemur*; the canoids joined, followed by the pinnipeds. This mixed-up horse, lemurine, carnivore branch then joined the mixed-up tree shrew–Anthropoidea branch. Next *Galago* joined, followed in succession by the cetacean, bovid, and kangaroo branches. Branching topologies that were closer to current ideas on mammalian phylogeny reduced the number of nucleotide replacements needed to account for the descent of these myoglobin sequences. In the tree requiring 183 replacements, the Anthropoidea is correctly arranged, just as in Fig. 5. Strepsirhines (i.e., a *Galago* + *Lepilemur* branch) join the Anthropoidea. Horse joins bovids, and then ungulates join the cetaceans. After a union of this ungulate + cetacean branch with the Primates, tree shrews join the assemblage. Finally the carnivore (canoid + pinniped) branch, followed by the kangaroo, descend from the most ancestral positions on the tree. This topology agrees with some opinions on the phylogenetic origin of Tupaioidea and Carnivora, but disagrees with others. Overall it represents a quite plausible eutherian phylogeny. It can be reduced to 182 replacements by joining tree shrew to *Galago* in the Primates and by moving the carnivore branch next to the ungulate + cetacean branch, leaving the primate (including tree shrew) lineage to descend from the most ancestral eutherian branch point. *Tupaia* has to be joined directly to *Galago* otherwise the length of the tree will increase. The change in the carnivore branch agrees more than the former arrangement with certain prevailing opinions on mammalian phylogeny (however, see McKenna, this volume), but the change in the *Tupaia* branch agrees less well, although it highlights the possible primate affinities of Tupaioidea.

The most parsimonious tree of myoglobin sequences discovered in the present search has a length of 179 nucleotide replacements, only 3–4 replacements less than the two trees with phylogenetically reasonable topologies. This most parsimonious tree appears to me much better in depicting phylogeny than the UWPGM tree but not as good as the two less parsimonious trees described. The several alternative maximum parsimony topologies, which were also found, involved small changes, such as switching of porpoise and dolphin branches, and this would not alter the following description of the relation of the major branches. The Anthropoidea branch remains as in Fig. 5. So does the pinniped + cetacean branch, except that now the pinniped part of the branch consists of sea lion joined to harbor seal. Also as in Fig. 5 the Anthropoidea ancestor connects directly to the inclusive eutherian ancestor. The line from the eutherian to the pinniped–cetacean ancestor gives off first a *Galago* + *Tupaia* branch, then in succession canoid (dog + badger), ungulate (horse + bovid), and *Lepilemur* branches. Although possible primate affinities for *Tupaia* (by way of *Galago*) are suggested, this most parsimonious topology by myoglobin sequences does not accord well with the aggregate of antigenic distance data on serum proteins (Table 3) in which lemuroids and lorisoids are, first, each slightly closer to the other than to Haplorhini (tarsier and Anthropoidea) and, second, when grouped as Strepsirhini are slightly closer to Haplorhini than to tree shrews and nonprimate mammals.

It appears that in the case of myoglobin sequences the maximum parsimony method is depicting an incorrect species phylogeny; it is not, however, depicting an *impossible* gene phylogeny. There might have been, e.g., more than one myoglobin locus in the basal eutherians; if so, after the different orders of eutherian mammals emerged, selection or chance might have activated or maintained in the early lemuroids a myoglobin locus which was closer in descent to the locus utilized by early carnivores and condylarths than to the locus utilized by early haplorhine lineages. It seems more likely, however, that the gene (and species) phylogeny is incorrect; i.e., *Lepilemur* codons which parallel derived codons common to ungulates, carnivores, and cetaceans happen to be numerous enough to be treated by the maximum parsimony method as the same derived codons rather than as parallel ones. (*Lepilemur* diverges by only 12 amino acids from horse, 17 from seal, and 18 from porpoise, but 21 from *Galago* and 17–23 from the ceboid and catarrhine myoglobins.) Considering that the early members of different eutherian orders and suborders separated out of a basal eutherian stock in a relatively short period of evolutionary time, it is easy to see why the present meager myoglobin sequence data, representing these major eutherian lineages, are not likely to depict by parsimony or any other principle a very accurate cladogeny. When the myoglobins of a wider range of lorisoids and lemuroids are sequenced, the maximum parsimony method will be able to reconstruct an ancestral myoglobin sequence for the early lorisoids and another for the early lemuroids. Then a much more accurate positioning of lorisoid and lemuroid branches in the myoglobin parsimony tree can be expected. This may soon be accomplished for the lorisoid branch, because the sequence determination of *Perodicticus* myoglobin has now been completed (Romero-Herrera and Lehmann, 1975) as well as that of *Nycticebus* (Romero-Herrera, personal communication).

Whether or not the *Lepilemur* sequence is paralogously rather than orthologously related to either *Galago* or Anthropoidea sequences should, as indicated, become apparent after myoglobins of additional lemuroids and lorisoids are sequenced and included in the search for the myoglobin parsimony tree. In the absence of such additional data, I feel it is heuristically more useful to assume that the presently sequenced myoglobins are all orthologously related. Then myoglobin can be combined with other sets of orthologous protein sequences to construct a species phylogeny, the rationale being that as more codon positions are examined the likelihood increases that species are grouped in a parsimony

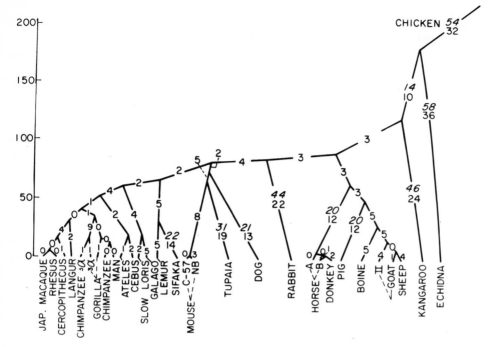

FIG. 6. Parsimony tree of 31 amniote (chicken and mammalian) α-hemoglobin chain sequences. Augmented link lengths are the italicized numbers. The ordinate scale is in millions of years. Sequences not given in Dayhoff (1972, 1973) are from echidna (Whittaker *et al.*, 1973), *Tupaia*, sifaka, and lemur (Hill, 1968), slow loris (Matsuda *et al.*, 1973*f*), *Cebus* (Matsuda *et al.*, 1973*c*), *Ateles* (Matsuda *et al.*, 1973*d*), langur (Matsuda *et al.*, 1973*g*), *Cercopithecus* (Matsuda *et al.*, 1973*b*), Japanese macaque (Matsuda *et al.*, 1973*e*), and chimpanzee and gorilla ³α chains (Boyer *et al.*, 1973).

tree because of true common ancestry rather than convergent resemblances. Such a tree from combined myoglobin, α-hemoglobin, and β-hemoglobin chain sequences is shown in Fig. 8. But let us first consider the parsimony results on alpha chains (Fig. 6) and then on beta chains (Fig. 7) before turning to the tree from the combined sequences.

5. *α-Hemoglobin Chain Sequences.* In addition to the Anthropoidea branch, lorisoid, lemuroid, and tree shrew branches are found in the parsimony phylogenetic tree of amniote α-hemoglobin chain sequences (Fig. 6). The tree shrew is not depicted as a primate in this tree but rather forms with mouse and dog a branch which attaches to the Primates. Nor are lorisoids (slow loris and galago) and lemuroids (lemur and sifaka) closer to each other than to other primates. They both show, however, a closer kinship with the Anthropoidea than with any nonprimate branch. The Anthropoidea separates into platyrrhines and catarrhines. Among the catarrhines, the cercopithecoids separate from the hominoids and then subdivide into colobines (langur) and cercopithecines (*Macaca* and *Cercopithecus*). On the hominoid side, the hominoid ancestor diverges from the catarrhine ancestor by only one nucleotide replacement. The genes at the phenotypically active hominoid alpha locus then descend from the hominoid ancestor without change in the human and chimpanzee lines, while undergoing only one mutation in the

gorilla line. In turn, the catarrhine ancestor is separated from the inclusive eutherian ancestor by 16 nucleotide replacements.

The conservative evolution of the active hominoid locus contrasts with the rapid evolution of the normally silent hominoid $^3\alpha$ locus. Among gorillas and chimpanzees, an occasional animal has a trace hemoglobin in which the alpha chain is of different type from the alpha chain of the predominant hemoglobin. Boyer *et al.* (1973) have sequenced the alpha chains of such trace homeglobins and called them $^3\alpha$ chains. Figure 6 shows that the $^3\alpha$ gene line, after originating from the hominoid alpha ancestor, accumulated 9 nucleotide replacements in descent to the chimpanzee–gorilla ancestor and then one additional mutation each in chimpanzee and gorilla lineages. It has been suggested (Goodman *et al.*, 1974*b*) that the $^3\alpha$ line arose from a duplicated gene linked to a pheno-typically active alpha allele which was increasing in frequency in an early hominoid population. A transcription-blocking mutation then silenced the duplicated locus. This removed it from the constraints of natural selection and permitted it to rapidly accumu-late additional mutations. So far trace hemoglobins with $^3\alpha$-type chains have not been found in human populations. However, if they are discovered and their $^3\alpha$-type chains sequenced, the results might help answer the question of whether or not each of the two African apes shows more recency of common ancestry with each other than with *Homo*. With respect to the principal α-hemoglobin chains, the parsimony criterion does not distinguish the order of ancestral branching among human, chimpanzee, and gorilla sequences, since the separate lines to these three sequences could have originated simultan-eously from the hominoid alpha ancestor without adding any extra mutations to the tree.

6. β-type Hemoglobin Chain Sequences. A phylogenetic tree for tetrapod β-type hemoglobin chain sequences is shown in Fig. 7. Although this is a maximum parsimony tree, several alternative arrangements of the major branches of primates to that depicted in the figure are just as parsimonious. In all such parsimonious arrangements, the Anthropoidea sequences appear as a monophyletic assemblage, and in the majority of these arrangements, they show a closer kinship to lorisoid and lemuroid sequences than to any nonprimate ones. In some of the alternate maximum parsimony trees, the Anthro-poidea are closer to lorisoids than to lemuroids, in others, lorisoids first join lemuroids and this strepsirhine branch unites with the Anthropoidea. In one of the alternative arrangements, cercopithecoid beta branch joins hominoid delta branch first rather than hominoid beta. Maximum parsimony was never achieved by joining hominoid delta first to the ceboid delta branch. Thus these two delta gene lines might have originated from separate duplications of beta genes in the early Anthropoidea. Delta chains replace beta chains in the minor or A_2 adult hemoglobins of ceboids and hominoids, whereas cerco-pithecids lack a minor A_2-type hemoglobin. The detection of a minor hemoglobin component with A_2-like electrophoretic mobility in *Tarsius* (Barnicot and Hewett-Emmett, 1974), and its absence in all strepsirhines examined, suggests the possibility that an original beta–protodelta duplication may have occurred in a primitive haplorhine stock ancestral to both *Tarsius* and Anthropoidea. This could have set the stage for further gene duplications by the mechanism of unequal, paralogous crossing over between beta and protodelta loci, resulting in the secondary origins of delta genes in platyrrhines and independently in catarrhines.

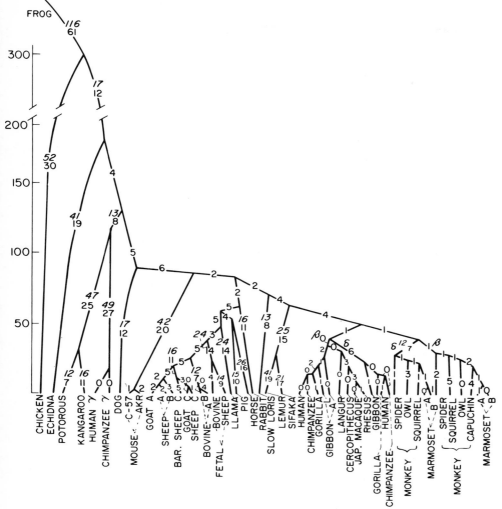

FIG. 7. Parsimony tree of 51 tetrapod (frog, chicken, and mammalian) β-type hemoglobin chain sequences. Augmented link lengths are the italicized (upper) numbers. The ordinate scale is in millions of years. Sequences not given in Dayhoff (1972, 1973) are from chicken (Matsuda *et al.*, 1973a), echidna (Whittaker *et al.*, 1972), lemur, sifaka (Hill, 1968), slow loris (Matsuda *et al.*, 1973f), capuchin monkey (Matsuda *et al.*, 1973c), langur (Matsuda *et al.*, 1973g), and *Cercopithecus* (Matsuda *et al.*, 1973b).

The conclusions on genealogical relationships within the Hominoidea, which can be deduced from the tree in Fig. 7, are supported by all alternative maximum parsimony trees. Several regions of the tree in Fig. 7 depict the phylogeny of hominoid species and in particular demonstrate the close kinship of the chimpanzee to *Homo*. In the gamma region, chimpanzee and human chains are identical. Indeed, this region could have been subdivided into human and chimpanzee $^{G}\gamma$, in which glycine occurs at position 136, and human and chimpanzee $^{A}\gamma$, in which alanine occurs at this position (De Jong, 1971). The two types of gamma chains are encoded by genes which apparently descended from a duplication which preceded the chimpanzee–human divergence (gamma chains occur in other catarrhines, but have not yet been sequenced). In the hominoid delta region of the tree, the human delta chain differs from identical chimpanzee and gorilla delta chains by only one nucleotide replacement, and from gibbon delta chain by two replacements. However, this portion of the tree, while providing evidence for the hominoid affinities of the gibbon, as did the myoglobin maximum parsimony tree, does not distinguish the order of ancestral branching among the hominoids, since the hominoid delta ancestor is identical to the human–African ape and chimpanzee–gorilla delta ancestors. The hominoid beta region of the tree does depict a closer kinship between African apes and *Homo* than between these hominoids and gibbon in that 2–3 nucleotide replacements separate the human–African ape ancestor from earlier hominoid ancestors from which the gibbon lines to alleles in *Hylobates lar* and *Hylobates agilis* descend. The gorilla beta sequence differs from the human–African ape ancestor by only one mutation, and the human and chimpanzee beta sequences are identical to this ancestor. Thus it would be just as parsimonious if the three separate lines to gorilla, human, and chimpanzee beta sequences originated simultaneously from the same common ancestor, i.e., again the parsimony method does not distinguish the order of ancestral branching among human, chimpanzee, and gorilla lineages.

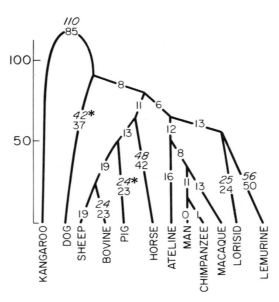

FIG. 8. Parsimony tree of 12 mammalian species, each represented by its combined myoglobin and α- and β-hemoglobin chain sequences. Augmented link lengths are the italicized numbers. The ordinate scale is in millions of years. The "hybrid species" ateline, lorisid, and lemurine, are described in the text. Except for dog and pig, each species is represented by 440 residue positions (153 + 146 + 141 for myoglobin and the two hemoglobin chain sequences). *Because of two incomplete myoglobin sequences (Dumur *et al.*, 1972; Floch *et al.*, 1973), dog is represented by 409 residue positions (122 + 146 + 141) and pig by 401 residue sequences (114 + 146 + 141). Thus, the link lengths of dog and pig are probably slightly less than they would have been if complete myoglobin sequences had been available.

B. Combined Myoglobin and α- and β-Hemoglobin Chain Sequences

In order to produce the cladogram shown in Fig. 8, "hybrid taxa" had to be created. For example ateline is a hybrid of spider monkey (alpha and beta chains) and woolly monkey (myoglobin), lorisid is a hybrid of slow loris (alpha and beta chains) and galago (myoglobin), and lemurine is a hybrid of *Lemur* (alpha and beta chains) and *Lepilemur* (myoglobin). The genealogy shown is the most parsimonious topology found after an extensive search by the branch-swapping procedure. However, two other topologies were equally parsimonious. In one, pig and horse are first joined before they join the bovids, otherwise the branching arrangement is the same as in Fig. 8. In the other alternative the only change is that the strepsirhine branch (lorisoid + lemurine) unites with the ungulates rather than with the Anthropoidea. If the changes from both equivalently parsimonious trees are incorporated into one topology, a less parsimonious tree is obtained. Thus two out of three times the primate placement of the strepsirhine (lorisid + lemurine) branch is confirmed by maximum parsimony using combined sequence data. Moreover maximum parsimony had only once united lorisoid and lemuroid sequences into a strepsirhine branch in the individual gene phylogenies (this happened in one of the alternative arrangements for β-type chains described in the preceding section), whereas with these combined sequences encompassing many more codon positions in the alignment employed, the lorisoid lineage and lemuroid lineage were consistently grouped as one strepsirhine branch in the most parsimonious topologies.

The phylogeny in Fig. 8 from this combined sequence data agrees with the individual gene phylogenies in showing that the genetic divergence between branches of Homininae (from the combined data on *Homo* and *Pan*) is negligible compared to the large divergence between species in different eutherian orders. Only one nucleotide replacement separates human and chimpanzee in this data set, but 143 nucleotide replacements separate lemurine and bovine. Such findings highlight the possibility that the ancestral human–African ape separation may be much more recent than traditional morphological evidence would indicate, as Sarich and Wilson have maintained (Sarich and Wilson, 1967; Wilson and Sarich, 1969; Sarich, 1970), but also support my proposal that molecular evolution decelerated in the hominoids (Goodman, 1961, 1963b, 1965; Goodman *et al.*, 1971, 1972), owing in part to a considerable increase in the length of the generation time in hominoids.

VII. The Genealogy of Primates in Broader Molecular Perspective

Following the view of Hennig (1965, 1966) that a taxonomic classification can serve as a written description of the genealogical or cladistic relationships of organisms (see, however, Simpson, this volume, for an opposing view), I have attempted in Table 4 to present a classification of Primates based on accumulated immunodiffusion evidence, which I believe reflects genealogical relationships on the whole. However, the evidence from immunodiffusion data is based solely on the divergence and additive hypotheses of molecular evolution, since such data are not suited to analysis by the parsimony method.

We can now examine the relationships described in this classification from the fuller body of the molecular data including that in which parsimony analysis is utilized. No molecular data, other than immunodiffusion, exist for the flying lemur, but tree shrews are better represented in that myoglobin and α-hemoglobin chain amino acid sequences have been inferred for *Tupaia*. Parsimony analysis of these sequence data tend more to exclude tree shrews (especially the α-chain results) from the order Primates but also highlight affinities to the Primates (myoglobin results). Thus the tentative solution adopted from immunodiffusion evidence of grouping Primates and Tupaioidea (also Dermoptera) in the superorder Archonta would seem a valid compromise (also see McKenna, this volume). Within the Primates parsimony analysis of the available amino acid sequence data fully supports the separation of Anthropoidea (Platyrrhini + Catarrhini) as a monophyletic branch distinct from Lorisiformes and Lemuriformes and provides some support (Fig. 8) for the grouping (from immunodiffusion data) of Lorisiformes and Lemuriformes into Strepsirhini.★ No amino acid sequence data have been reported on *Tarsius*; although the sequence of its hemoglobin is now being determined (Barnicot, unpublished data). Thus the phylogenetic validity of Haplorhini (Tarsioidea + Anthropoidea) suggested by immunodiffsuion divergence data can not yet be examined by parsimony analysis of sequence data. There are some rather fragmentary DNA data, however, which support the cladistic positions assigned to tarsier and tree shrews in Table 4. These data (Hoyer and Roberts, 1967), from the standpoint of the divergent evolution model, show Anthropoidea to be closer in repeated DNA sequences to tarsier than to slow loris, galago, lemur, tree shrew, mouse, hedgehog, and chicken (the other non-Anthropoidea species included in the comparison series) and also show Anthropoidea to be closer to tree shrew than to mouse, hedgehog, and chicken.

The divisions within the Anthropoidea depicted in Table 4 are well supported by parsimony analysis of the available amino acid sequence data, by divergence analysis of the data on repeated DNA sequences (Hoyer *et al.*, 1964, 1965; Hoyer and Roberts, 1967; Martin and Hoyer, 1967), and on single-copy DNA (Kohne *et al.*, 1972; Hoyer *et al.*, 1972; Kohne, this volume). To recapitulate, Anthropoidea divides into infraorders Platyrrhini and Catarrhini. The infraorder Platyrrhini had only one extant family, Cebidae (with marmosets included with New World monkeys in this family), but Catarrhini subdivides into the superfamilies Cercopithecoidea and Hominoidea. Cercopithecoidea contains one extant family, Cercopithecidae which splits into subfamilies Colobinae and Cercopithecinae (unpublished DNA data of Hoyer, and hemoglobin sequence data agree with the immunodiffusion data on the splitting of the Cercopithecidae). Hominoidea splits into Hylobatidae and Hominidae. The subfamily Hylobatinae (*Hylobates* + *Symphalangus*) is the only extant group in Hylobatidae. The family Hominidae divides into the subfamilies Ponginae for *Pongo* and Homininae for *Pan*, *Homo*, and *Gorilla*. As yet molecular data do not reveal which two of these three hominines share the most recent common ancestor.

★ The new taxonomic category of *semiorder* is utilized in this classification for Strepsirhini and Haplorhini, in order to preserve the suggested sister group relationships between the suborders Tarsioidea and Anthropoidea and between the infraorders Platyrrhini and Catarrhini. The category semiorder is equivalent to Hill's (1953) *grade* but is preferable to the latter because the concept of grade is generally used to reflect a level of anagenetic advancement, not a taxonomic category.

VIII. References

ARNHEIN, N., E. M. PRAGER, and A. C. WILSON. 1969. Immunological prediction of sequence differences among proteins. *J. Biol. Chem.* **244**:2085–2094.

BABA, M. L., M. GOODMAN, H. DENE, and G. W. MOORE. 1975. Origins of the Ceboidea viewed from an immunological perspective. *J. Hum. Evol.* **4**:89–102.

BARNICOT, N. A., and D. HEWETT-EMMETT. 1974. Electrophoretic studies on prosimian blood proteins, pp. 891–902. *In* R. D. Martin, G. A. Doyle, and A. C. Walker, eds., *Prosimian Biology*. Duckworth, London.

BOYER, S. H., A. N. NOYES, G. F. TIMMONS, and R. A. YOUNG. 1972. Primate hemoglobins: polymorphisms and evolutionary patterns. *J. Hum. Evol.* **1**:515–543.

BOYER, S. H., A. N. NOYES, M. L. BOYER, and K. MAN. 1973. Hemoglobin ³α-chains in apes. *J. Biol. Chem.* **248**:992–1003.

BUETTNER-JANUSCH, J., V. BUETTNER-JANUSCH, and G. A. MASON. 1969. Amino acid compositions and amino-terminal end groups of alpha and beta chains from polymorphic hemoglobins of *Pongo pygmaeus*. *Arch. Biochem. Biophys.* **133**:164–170.

BUTLER, P. M. 1956. The skull of *Ictops* and the classification of the Insectivora. *Proc. Zool. Soc. London* **126**:453–481.

CAVALLI-SFORZA, L. L., and A. W. F. EDWARDS. 1967. Phylogenetic analysis: models and estimation procedures. *Evolution* **21**:550–570.

CRONIN, J. E., V. M. SARICH, and Y. RUMPLER. 1974. Albumin and transferrin evolution among the Lemuriformes. *Am. J. Phys. Anthropol.* **41**:473–474.

DARGA, L. L., M. GOODMAN, and M. L. WEISS. 1973. Molecular evidence on the cladistic relationships of the Hylobatidae, pp. 149–162. *In* D. M. Rumbaugh, ed., *Gibbon and Siamang*, Vol. 2. S. Karger, Basel.

DARWIN, C. 1859. *The origin of species by means of natural selection, or the preservation of favoured races in the struggle for life*. Doubleday, Garden City. 517 pp. Reprinted ed.

DARWIN, F. 1903. *More letters of Charles Darwin: a record of his work in a series of hitherto unpublished letters*, Vol. 1. Appleton, New York. 494 pp.

DAYHOFF, M. O. 1972. *Atlas of Protein Sequence and Structure*, Vol. 5. National Biomedical Research Foundation, Silver Spring. 542 pp.

DAYHOFF, M. O. 1973. *Atlas of Protein Sequence and Structure*, Vol. 5. Supplement 1. National Biomedical Reserach Foundation, Silver Spring. 114 pp.

DE JONG, W. W. W. 1971. Chimpanzee foetal haemoglobin: structure and heterogeneity of the γ-chain. *Biochim. Biophys. Acta* **251**:217–226.

DENE, H., M. GOODMAN, W. PRYCHODKO, and G. W. MOORE. 1975 (in press). Immunodiffusion systematics of the Primates. III. The Strepsirhini. *Folia Primatol.*

DUMUR, V., M. DAUTREVAUX, and K. HAN. 1972. The partial amino acid sequence of dog myoglobin. *FEBS Lett.* **26**:241–244.

EDWARDS, A. W. F., and L. L. CAVALLI-SFORZA. 1963. The reconstruction of evolution. *Ann. Human Genet.* **27**:104.

ESTABROOK, G. F. 1968. A general solution in partial orders for the Camin-Sokal model in phylogeny. *J. Theor. Biol.* **21**:421–438.

FARRIS, J. S. 1970. Methods for computing Wagner trees. *Syst. Zool.* **19**:83–92.

FIEDLER, W. 1956. Übersicht über das System der Primaten, pp. 1–226. *In* H. Hofer, A. H. Schultz, and D. Stark, eds., *Primatologia*, Vol. 1. S. Karger, Basel.

FITCH, W. M. 1970. Distinguishing homologous and analogous proteins. *Syst. Zool.* **19**:99–113.

FITCH, W. M. 1971. Toward defining the course of evolution: minimum change for a specific tree topology. *Syst. Zool.* **20**:406–416.

FLOĆH, R., M. DAUTREVAUX, and K. K. HAN. 1973. Sequence partielle des acides amines de la myoglobine de porc. *Biochimie* **55**:95–98.

GHISELIN, M. T., and L. JAFFE. 1973. Phylogenetic classification in Darwin's monograph on the sub-class Cirripedia. *Syst. Zool.* **22**:132–140.

GOODMAN, M. 1961. The role of immunochemical differences in the phyletic development of human behavior. *Hum. Biol.* **33**:131–162.

GOODMAN, M. 1963a. Serological analysis of the systematics of recent hominoids. *Hum. Biol.* **35**:377–436.

GOODMAN, M. 1963b. Man's place in the phylogeny of the primates as reflected in serum proteins, pp. 204–234. *In* S. L. Washburn, ed., *Classification and Human Evolution*. Aldine, Chicago.

GOODMAN, M. 1965. The specificity of proteins and the process of primate evolution, pp. 70–86. *In* H. Peters, ed., *Protides of the Biological Fluids*—1964. Elsevier, Amsterdam.

GOODMAN, M. 1967. Deciphering primate phylogeny from macromolecular specificities. *Am. J. Phys. Anthropol.* **26**:255–275.

GOODMAN, M. 1973. The chronicle of primate phylogeny contained in proteins. *Symp. Zool. Soc. London* **33**:339–375.

GOODMAN, M., and G. W. MOORE. 1971. Immunodiffusion systematics of the primates. I. The Catarrhini. *Syst. Zool.* **20**:19–62.

GOODMAN, M., J. BARNABAS, G. MATSUDA, and G. W. MOORE. 1971. Molecular evolution in the descent of man. *Nature* **233**:604–613.

GOODMAN, M. J. BARNABAS, and G. W. MOORE. 1972. Man, the conservative and revolutionary mammal: Molecular findings on this paradox. *J. Hum. Evol.* **1**:663–686.

GOODMAN, M. W. FARRIS, Jr., G. W. MOORE, W. PRYCHODKO, E. POULIK, and M. W. SORENSON. 1974a. Immunodiffusion systematics of the primates. II. Findings on *Tarsius*, Lorisidae and Tupaiidae, pp. 881–890. *In* R. D. Martin, G. A. Doyle, and A. C. Walker, eds., Duckworth, London.

GOODMAN, M., G. W. MOORE, J. BARNABAS, and G. MATSUDA. 1974b. The phylogeny of human globin genes investigated by the maximum method. *J. Mol. Evol.* **3**:1–48.

GREGORY, W. K. 1910. The orders of mammals. *Bull. Am. Mus. Nat. Hist.* **27**:1–524.

HARTIGAN, J. A. 1973. Minimum mutation fits to a given tree. *Biometrics* **29**:53–65.

HENNIG, W. 1965. Phylogenetic systematics. *Ann. Rev. Entomol.* **10**:97–116.

HENNIG, W. 1966. *Phylogenetic Systematics*. University of Illinois Press, Urbana. 263 pp.

HILL, W. C. O., 1953. *Primates, Vol. I. Strepsirhini*. Edinburgh Univ. Press, Edinburgh.

HILL, R. L. 1968. Unpublished results, pp. 1–174. *In* H. A. Sober, ed., *Handbook of Biochemistry*. Chemical Rubber Co., Chicago.

HOLMQUIST, R., L. R. CANTOR, and T. H. JUKES. 1972. Improved procedures for comparing homologous sequences in molecules of proteins and nucleic acids. *J. Mol. Biol.* **64**:145–161.

HOYER, B. H., and R. B. ROBERTS. 1967. Studies of nucleic acid interactions using DNA-agar, pp. 425–479. *In* H. Taylor, ed., *Molecular Genetics Part II*. Academic Press, New York.

HOYER, B. H., B. J. McCARTHY, and E. T. BOLTON. 1964. A molecular approach in the systematics of higher organisms. *Science* **144**:959–967.

HOYER, B. H., E. T. BOLTON, B. J. McCARTHY, and R. B. ROBERTS. 1965. The evolution of polynucleotides, pp. 581–590. *In* V. Bryson and H. J. Vogel, eds., *Evolving Genes and Proteins*. Academic Press, New York.

HOYER, B. H., N. W. VAN DE VELDE, M. GOODMAN, and R. B. ROBERTS. 1972. Examination of hominid evolution by DNA sequence homology. *J. Hum. Evol.* **1**:645–649.

JUKES, T. H., and R. HOLMQUIST. 1972. Estimation of evolutionary changes in certain homologous polypeptide chains. *J. Mol. Biol.* **64**:163–179.

KOHNE, D. E., J. A. CHISCON, and B. H. HOYER. 1972. Evolution of primate DNA sequences. *J. Hum. Evol.* **1**:627–644.

LASKER, G. W. 1973. *Physical Anthropology*. Holt, Reinhart, and Winston, New York. 424 pp.

MARTIN, M. A., and B. H. HOYER. 1967. Adenine plus thymine and guanine plus cytosine enriched fractions of animal DNA as indicators of polynucleotide homologies. *J. Mol. Biol.* **27**:113–129.

MATSUDA, G., T. MAITA, K. MIZUNO, and H. OTA. 1973a. Amino acid sequence of a β-chain of AII component of adult chicken haemoglobin. *Nature* **244**:244.

MATSUDA, G., T. MAITA, B. WATANABE, A. ARAYA, K. MOROKUMA, M. GOODMAN, and W. PRYCHODKO. 1973b. The amino acid sequences of the α and β polypeptide chains of adult hemoglobin of the savannah monkey (*Cercopithecus aethiops*). *Hoppe-Seyler's Z. Physiol. Chem.* **354**:1153–1155.

MATSUDA, G., T. MAITA, B. WATANABE, A. ARAYA, K. MOROKUMA, Y. OTA, M. GOODMAN, J. BARNABAS, and W. PRYCHODKO. 1973c. The amino acid sequences of the α and β polypeptide chains of adult hemoglobin of the capuchin monkey (*Cebus apella*). *Hoppe-Seyler's Z. Physiol. Chem.* **354**:1513–1516.

MATSUDA, G., T. MAITA, Y. SUZUYAMA, M. SETOGUCHI, Y. OTA, A. ARAYA, M. GOODMAN, J. BARNABAS, and W. PRYCHODKO. 1973d. Studies of the primary structures of α and β polypeptide chains of adult hemoglobin of the spider monkey (*Ateles geoffroyi*). *Hoppe-Seyler's Z. Physiol. Chem.* **354**:1517–1520.

MATSUDA, G., T. MAITA H. OTA A. ARAYA, Y. NAKASHIMA, V. ISHII, and M. NAKASHIMA. 1973e. The primary structures of α and β chains of adult hemoglobin of the Japanese monkey (*Macaca fuscata fuscata*). *Int. J. Pept. Protein Res.* **5**:405–418.

MATSUDA, G., T. MAITA, B. WATANABE, H. OTA, A. ARAYA, M. GOODMAN, and W. PRYCHODKO. 1973f. The primary structures of the α and β polypeptide chains of adult hemoglobin of the slow loris (*Nycticebus coucang*). *Int. J. Pept. Protein Res.* **5**:419–421.

MATSUDA, G., T. MAITA, Y. NAKASHIMA, J. BARNABAS, P. K. RANJEKAR, and N. S. GANDHI. 1973g. The primary structures of the α and β polypeptide chains of adult hemoglobin of the human langur (*Presbytis entellus*). *Int. J. Pept. Protein Res.* **5**:423–425.

MOORE, G. W. 1971. A Mathematical Model for the Construction of Cladograms. Ph.D. Thesis. North Carolina State University, Raleigh. 262 pp.

MOORE, G. W., and M. GOODMAN. 1968. A set theoretical approach to immunotaxonomy: analysis of species comparisons in modified Ouchterlony plates. *Bull. Math. Biophys.* **30**:279–289.

MOORE, G. W., M. GOODMAN, and J. BARNABAS. 1973a. An iterative approach from the standpoint of the additive hypothesis to the dendrogram problem posed by molecular data sets. *J. Theor. Biol.* **38**:423–457.

MOORE, G. W., J. BARNABAS, and M. GOODMAN. 1973b. A method for constructing maximum parsimony ancestral amino acid sequences on a given network. *J. Theor. Biol.* **38**:459–485.

NUTTALL, G. H. F. 1904. *Blood Immunity and Blood Relationship.* Cambridge University Press, Cambridge. 444 pp.

PATTERSON, B. 1965. The fossil elephant shrews (family Macroscelididae). *Bull. Mus. Comp. Zool., Harvard Univ.* **133**(6):295–335.

PRAGER, E. M., and A. C. WILSON. 1971a. The dependence of immunological cross-reactivity upon sequence resemblance among lysozymes. I. Micro-complement fixation studies. *J. Biol. Chem.* **246**:5978–5989.

PRAGER, E. M., and A. C. WILSON. 1971b. The dependence of immunological cross-reactivity upon sequence resemblance among lysozymes. II. Comparison of precipitin and micro-complement fixation studies. *J. Biol. Chem.* **246**:7010–7017.

ROMERO-HERRERA, A. E., and H. LEHMANN. 1972. The myoglobin of primates. III. Cercopithecidae (old world monkeys): *Papio anubis* (olive baboon) and *Macaca fascicularis* (= irus, crab eating monkey). *Biochim. Biophys. Acta* **278**:465–481.

ROMERO-HERRERA, A. E., and H. LEHMANN. 1973a. The myoglobin of primates. IV. New world monkeys: Cebidae: (1) *Saimiri scuireus* (squirrel monkey); (2) *Lagothrix lagothricha* (Humboldt's woolly monkey); Callitrichidae: *Callithrix jacchus* (common marmoset). *Biochim. Biophys. Acta* **317**:65–84.

ROMERO-HERRERA, A. E., and H. LEHMANN. 1973b. The myoglobin of primates. V. Prisomians: *Galago crassicaudatus* (thick-tailed galago) and *Lepilemur mustelinus* (sportive lemur). *Biochim. Biophys. Acta* **322**:10–22.

ROMERO-HERRERA, A. E., and LEHMANN, H. 1974. The myoglobin of primates. VI. *Tupaia glis belangeri* (common tree shrews). *Biochim. Biophys. Acta* **359**:236–241.

ROMERO-HERRERA, A. E., and H. LEHMANN. 1975. The myoglobin of primates. VII. *Perodicticus potto edwarsi* (potto). *Biochim. Biophys. Acta* **393**:205–214.

SARICH, V. M. 1970. Primate systematics with special reference to Old World monkeys—A protein perspective, pp. 175–226. In J. R. Napier and P. H. Napier, eds., *Old World Monkeys, Evolution, Systematics, and Behavior.* Academic Press, New York.

SARICH, V. M., and J. E. CRONIN. 1974. Primate evolution at higher taxon levels: A molecular view. Paper presented at the 43rd Annual Meeting of the American Association of Physical Anthropologists, Amherst, Massachusetts.

SARICH, V. M., and A. C. WILSON. 1967. Immunological time scale for hominoid evolution. *Science* **158**:1200–1203.

SIMPSON, G. G. 1945. The principles of classification and a classification of mammals. *Bull. Am. Mus. Nat. Hist.* **85**:1–350.

SOKAL, R. R., and C. D. MICHENER. 1958. A statistical method for evaluating systematic relationships. *Kans. Univ. Sci. Bull.* **38**:1409–1438.

TASHIAN, R. E., R. J. TANIS, R. E. FERRELL, S. K. STROUP, and M. GOODMAN. 1972. Differential rates of evolution in carbonic anhydrase isozymes of catarrhine primates. *J. Hum. Evol.* **1**:545–552.

TASHIAN, R. E., M. GOODMAN, R. J. TANIS, R. E. FERRELL, and W. R. A. OSBORNE. 1975, pp. 207–223. Evolution of the carbonic anhydrase isozymes. In C. L. Markert, ed., *Isozymes IV—Genetics and Evolution.* Academic Press, New York.

TETAERT, D., K. K. HAN, M. T. PLAUCOT, M. DAUTREVAUX, S. DUCASTAING, L. HOMBRADOS, and E. NEUZIL. 1974. The primary sequence of badger myoglobin. *Biochim. Biophys. Acta* **351**:317–324.

VIGNA, R. A., L. J. GURD, and F. R. N. GURD. 1974. California sea lion myoglobin: complete covalent structure of the polypeptide chain. *J. Biol. Chem.* **249**:4144–4148.

WHITTAKER, R. G., W. O. FISHER, and E. O. P. THOMPSON. 1972. Studies on monotreme proteins. I. Amino acid sequence of the β-chain of haemoglobin from the echidna, *Tachyglossus aculeatus aculeatus*. *Aust. J. Biol. Sci.* **25**:989–1004.

WHITTAKER R. G., W. O. FISHER, and E. O. P. THOMPSON. 1973. Studies on monotreme proteins. II. Amino acid sequence of the α-chain of haemoglobin from the echidna, *Tachyglossus aculeatus aculeatus*. *Aust. J. Biol. Sci.* **26**:877–888.

WILSON, A. C., and V. M. SARICH. 1969. A molecular time scale for human evolution. *Proc. Natl. Acad. Sci. Wash.* **63**:1088–1093.

WOODING, G. L., and R. F. DOOLITTLE. 1972. Primate fibrinopeptides: evolutionary significance. *J. Hum. Evol.* **1**:553–563.

I I

DNA Evolution Data and Its Relevance to Mammalian Phylogeny

DAVID E. KOHNE

I. Introduction

What are the molecular events which result in the evolutionary changes which we see as the appearance of new species? The current view of the role of deoxyribonucleic acid (DNA) indicates that evolutionary changes at the organism level must have been preceded by some quantitative or qualitative change in DNA. Our interest was to determine the pattern of DNA changes with evolution in the hope that the evidence obtained might give some insights into evolutionary mechanisms.

It is obvious that DNA has changed during the evolution of mammals. This paper will primarily be concerned with how helpful these type of data actually are in studying phylogeny. In order to put this study in its proper perspective it is worthwhile to discuss (1) what the DNA in the cell does, (2) how DNA can change and the possible consequences of these changes, (3) what DNA changes can be studied and the significance of the data obtained, and (4) what other types of DNA changes should be studied.

DAVID E. KOHNE · Department of Experimental Pathology, Scripps Clinic and Research Foundation, La Jolla, California.

II. The Function of DNA

What is the function of DNA in a mammalian cell? Some unknown fraction of DNA is necessary to code for the structural elements of the cell. This primarily involves genes coding for a variety of proteins (for the purpose of simplicity enzymes are here included in the class of structural proteins). How much of the DNA is involved in this function is not known. Another class of DNA must be involved in the dynamics of the organization of the cell and the organism. These control-DNA genes could: (1) code for a protein, (2) code for just RNA, (3) be involved in the organization of the DNA and have no gene product, and (4) serve as receptor sites for other control molecules and have no direct gene product. The fraction of DNA in the cell which falls into this class is not known. It may well be that a large fraction of DNA in mammalian cells has no present-day use.

III. Changes in DNA

DNA can change during evolution in a variety of ways. It can change by: (1) single-base mutations or nucleotide substitution, (2) adding new genes or DNA sequences, and (3) rearrangement of existing genes or DNA sequences.

This paper summarizes the results of experiments which measure evolutionary changes in the nucleotide sequences of the bulk DNA of a variety of mammals. These changes have been caused by point mutations (single-nucleotide substitutions) which have occurred in DNA over time. Depending on the function of the particular segment of DNA affected, such a change may have a wide range of possible effects on the organism. It is possible to visualize a class of DNA sequences that has no function whatsoever. Any mutation occurring in this region has no effect on the organism, and all such changes can be passed on to the progeny as "neutral mutations." After a long period of evolution such a DNA sequence could be expected to retain only 25% of its nucleotides in common with its ancestral sequence.

At the other extreme there may be classes of DNA sequences involved in regulation of or in acting as templates for certain classes of RNA wherein any change on the sequence is significant. A large share of these changes can be expected to be deleterious and therefore eventually eliminated from the population. Such a class of DNA sequences can be expected to suffer as many "hits" as any other class, but the changes will not be passed along. The genes for ribosomal RNA may be examples of this class of highly conserved DNA sequences where the differences between ribosomal genes in different animals are minimal, even after long periods of evolution.

Another class of DNA can be defined as that which determines the templates for proteins requiring an unchanged sequence of amino acids. Because of the redundancy of the code, however, some changes in such DNA sequences could be tolerated. One third of the possible mutations will occur in the third position of the codons, and a considerable proportion of these will only change the transfer RNA needed to carry the same amino acid. Thus roughly one fourth of the mutations can produce changes to be propagated, and the resulting difference after a long period of evolution will approach 17%. The histone protein which is carried in common by cow and pea may be an example of the product of this type of DNA. Presumably the DNA coding for this would differ by about 17%.

Finally, the remainder of the DNA sequences can be considered in a class intermediate between these more readily defined extremes. Some changes could be tolerated because of redundancy in the code, or because the replacement of one amino acid by another made no difference, while in adjacent locations any change might be lethal. The proportions of allowed and prohibited changes might vary from one extreme to the other in this class of DNA.

DNA can also be changed in a qualitative and quantitative manner by the addition of new DNA sequences during evolution. It is known that new DNA sequences have been continually added to existing DNA throughout mammalian evolution (Kohne, 1970). Since the divergence of mouse and rat for example, a fraction of DNA, which now comprises 10% of the total mouse DNA, has been added to the mouse genome. The mechanism of addition of these sequences is not known. One way to add new DNA to a cell is by viral infection. A large number of different viruses can insert their nucleic acids into the chromosomal DNA of the host (Gelb et al., 1971; Chattapadhyay et al., 1974). Any viral genes which were inserted into the germ-cell chromosomes could well be preserved through evolution.

The introduction of new DNA sequences into a cell could cause a variety of changes. The new sequences could serve as raw material for the formation of new genes and new combinations of genes. Both the addition of new repeated and viral DNA could alter established patterns of expression and cause great changes. Some viruses contain control elements which enable them to mesh with the functions of the cell. If such a virus were integrated into an unused portion of the DNA, it is conceivable that the formerly unused DNA might become functional and, in a small proportion of cases, actually useful. It is possible that viruses serve as both a selective and a driving force during evolution.

A well-known way of qualitatively altering DNA is by rearrangement of existing sequences. The genetic mechanisms of chromosomal breaks, crossovers, and transloca-tions result in rearrangement. The genetic effects of rearrangement are large and almost certainly play a large role in evolution.

IV. Composition of Mammalian DNA

All higher organisms examined thus far have contained large fractions (40–60%) of the total DNA as families of repeated DNA sequences (Britten and Kohne, 1968) (Table 1). A family of repeated nucleotide sequences is composed of many member sequences, each of which may reassociate with any other member of that particular family. There are many families per cell, and the member sequences of each family are usually not identical to one another in sequence but are similar enough to reassociate. Related species contain at least some of the same families of repeated DNA sequences,

TABLE I
CHARACTERISTICS OF MAMMALIAN DNA

Repeated DNA

1. Repeated DNA is composed of groups or families of DNA sequences. Each member of a family can interact and reassociate with any other member of that family.
2. There are at least several families in each cell. The number of members per family is about 10^5 for mammals.
3. Family members are generally similar but not identical to each other. There are exceptions to this.
4. Repeated DNA reassociates very rapidly because of the high concentrations of each sequence in each cell. It is easily purified.

Nonrepeated DNA

1. Each nonrepeated sequence is present one time per haploid cell.
2. Nonrepeated DNA sequences reassociate very slowly because of their low concentration per cell. They are readily purified.
3. Most of the potential genetic information is contained in this fraction.

since the repeated DNA of one species will reasssociate with that of another species. If two related species contain similar repeated DNA families, it seems probable that they both inherited these families from their most recent common ancestor. It is not known whether the individual members of a related family were identical or only similar at the time of divergence. It is not possible, then, to say whether the interspecies differences seen between these sequences have occurred since the time of divergence. Thus the comparison of DNA in its repeated form will not yield the true extent of nucleotide sequence change when correlated with the time since divergence.

Nonrepeated DNA sequences make up a large fraction of mammalian DNA (40–60%) and each sequence occurs only once per haploid cell (Britten and Kohne, 1968). Reassociated nonrepeated DNA has a thermal stability which indicates that essentially perfect nucleotide pair matching is present. Nonrepeated DNA sequences held in common between species must be the direct descendants of the same common ancestor sequence which was present in the most recent common ancestor. Nonrepeated DNA is thus highly suitable for determining the extent of nucleotide change since the divergence of two species.

V. What Changes Can Be Measured

The addition of DNA to the genome during evolution and the evolutionary changes in nucleotide sequence can be measured by nucleic acid hybridization techniques. The rationale and results of these experiments have been described elsewhere (Kohne, 1970; Kohne et al., 1972). A summary explanation is presented in Fig. 1 and Table 2. The number of nucleotide substitutions which have occurred since the divergence of a variety of mammalian species was measured (Table 3) (Kohne et al., 1972). The pattern of relationships measured by DNA similarity agrees with the classical paleontological view of the relationships between the mammals tested. Mouse DNA, for example, is more closely related to that of rat than to any primate, while that of man is closer to that of chimpanzee than to any other primate measured.

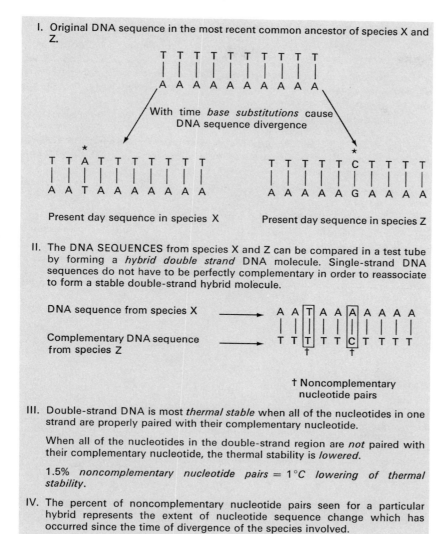

I. Original DNA sequence in the most recent common ancestor of species X and Z.

With time *base substitutions* cause DNA sequence divergence

Present day sequence in species X

Present day sequence in species Z

II. The DNA SEQUENCES from species X and Z can be compared in a test tube by forming a *hybrid double strand* DNA molecule. Single-strand DNA sequences do not have to be perfectly complementary in order to reassociate to form a stable double-strand hybrid molecule.

DNA sequence from species X ⟶

Complementary DNA sequence from species Z ⟶

† Noncomplementary nucleotide pairs

III. Double-strand DNA is most *thermal stable* when all of the nucleotides in one strand are properly paired with their complementary nucleotide.

When all of the nucleotides in the double-strand region are *not* paired with their complementary nucleotide, the thermal stability is *lowered*.

1.5% *noncomplementary nucleotide pairs* = 1°C lowering of thermal stability.

IV. The percent of noncomplementary nucleotide pairs seen for a particular hybrid represents the extent of nucleotide sequence change which has occurred since the time of divergence of the species involved.

FIG. 1. Rationale for DNA comparison. Asterisk (★) indicates nucleotide substitution.

One of the interesting aspects of these studies was the ability to calculate a rate of nucleotide substitutions by correlating the extent of nucleotide substitutions with absolute divergence times for the various species comparisons. Based on the commonly accepted divergence times, the rates of nucleotide substitution for primates and rodents are quite different (Table 4). Rodents have a 10–15 times faster average rate of nucleotide substitution than do primates. Further, the rate of substitution in the primate line appears to have changed greatly during the process of primate evolution with the rate of substitution being high during the early stages of primate evolution and much slower recently (Fig. 2, Table 5). Table 5 presents the rates of change calculated for the different time periods shown in Fig. 2. The absolute values for the rates quoted in Table 5 are probably

TABLE 2

COMPARISON OF HUMAN DNA SEQUENCES WITH CHIMPANZEE DNA SEQUENCES

1. Isolate Man radioactive nonrepeated DNA sequences. There are about 5×10^9 nucleotides per cell in this fraction. All mammals have about the same amount of DNA per cell. DNA used in these experiments has been sheared to 400 nucleotide long pieces.

2. a. Mix a small amount of human radioactive nonrepeated DNA with a large amount of nonradioactive DNA from the chimp. Denature and incubate.

   b. For the control, a small amount of human radioactive nonrepeated DNA is mixed with a large amount of nonradioactive human DNA. Denature and incubate. In this case the radioactive DNA will form perfectly matched nucleotide pair duplexes.

3. Isolate both the hybrid (Man–chimp) and control (Man–Man) double-strand molecules and determine their thermal stabilities (TS):

$$TS = (TS \text{ of Man–Man duplex}) - (TS \text{ of Man–chimp hybrid})$$

4. Percentage of noncomplementary nucleotide pairs (or percentage of nucleotide substitutions since the divergence of Man and chimp) is

$$\Delta TS \text{ (°C)} \times \frac{1.5\% \text{ noncomplementary nucleotide pairs}}{1°C \text{ lowering of } TS}$$

TABLE 3

EXTENT OF NUCLEOTIDE SEQUENCE CHANGES

DNAs compared Radioactive–nonradioactive	Normalized percent reaction of radioactive DNA with nonradioactive DNA	Change in thermal stability (°C)[a]	Percent nucleotide substitutions since divergence
Human–human	100	0	0
Human–chimp	95–100	1.5	2.4
Human–gibbon	95–100	3.5	5.3
Human–green monkey (GM)	95–100	6.3	9.5
Human–capuchin	90–95	10.5	15.8
Human–galago	≈60	≈28	≈42
Human–rat	≈15–20	—	—
Green monkey (GM)–GM	100	0	0
GM–Rhesus	100	2	3
GM–Human	95–100	6.4	9.6
GM–chimp	95–100	6.4	9.6
GM–gibbon	95–100	6.4	9.6
GM–capuchin	90–95	11	16.5
GM–galago	≈60	≈28	42
Mouse–mouse	100	0	0
Mouse–rat	≈80–85	18	27
Cow–cow[b]	100	0	0
Cow–sheep[b]	95–100	6	9
Cow–pig[b]	≈80	14	21

[a] See Table 2.
[b] Laird et al. (1969).

TABLE 4

RATES OF NUCLEOTIDE SUBSTITUTION CALCULATED IN TERMS OF ABSOLUTE TIME

DNAs compared	Millions of years since divergence of species 10^6	Nucleotide substitutions since divergence (%)	Changes per 10^6 years (%)	Nucleotide pair changes per year[a]
Man–Man	0	0	0	0
Man–chimp	15	2.4	0.08	1.6
Man–gibbon	30	5.3	0.09	1.8
Man–green monkey (GM)	45	9.5	0.1	2
Man–capuchin	65	15.8	0.12	2.4
Man–galago	80	42	0.26	5.2
GM–GM	0	0	0	0
GM–rhesus	15	3.5	0.12	2.4
GM–Man	45	9.6	0.11	2.2
GM–chimp	45	9.6	0.11	2.2
GM–gibbon	45	9.6	0.11	2.2
GM–capuchin	65	16.5	0.12	2.4
GM–galago	80	42	0.25	5.0
Mouse–mouse	0	0	0	0
Mouse–rat	10	27	1.3	26
Cow–sheep[b]	25	9	0.18	3.6
Cow–pig[b]	55	21	0.19	3.8

[a] The nonrepeated DNA genome size is 2×10^9 nucleotide pairs per haploid cell.
[b] Laird et al. (1969).

not very accurate. Much better data is needed to generate accurate values by the method of analysis used. The data appear sufficiently good, however, to support the changing trend seen in Table 5. It seems that the rate of change was much higher during the early part of the primate evolution than more recently.

When generation time is taken into account, however, the rodent and primate rates of nucleotide substitution (the rate is recalculated as nucleotide changes per generation) are essentially the same (Table 6). This would suggest that the generation time length of a species affects the rate of nucleotide substitutions which occur. The generation time used for the calculation was in each case the shortest generation time of the two species

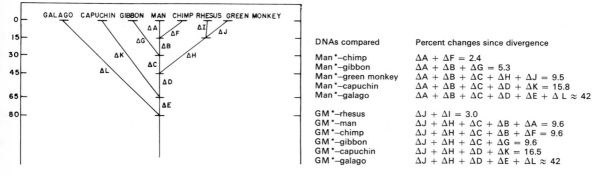

DNAs compared	Percent changes since divergence
Man*–chimp	$\Delta A + \Delta F = 2.4$
Man*–gibbon	$\Delta A + \Delta B + \Delta G = 5.3$
Man*–green monkey	$\Delta A + \Delta B + \Delta C + \Delta H + \Delta J = 9.5$
Man*–capuchin	$\Delta A + \Delta B + \Delta C + \Delta D + \Delta K = 15.8$
Man*–galago	$\Delta A + \Delta B + \Delta C + \Delta D + \Delta E + \Delta L \approx 42$
GM*–rhesus	$\Delta J + \Delta I = 3.0$
GM*–man	$\Delta J + \Delta H + \Delta C + \Delta B + \Delta A = 9.6$
GM*–chimp	$\Delta J + \Delta H + \Delta C + \Delta B + \Delta F = 9.6$
GM*–gibbon	$\Delta J + \Delta H + \Delta C + \Delta G = 9.6$
GM*–capuchin	$\Delta J + \Delta H + \Delta D + \Delta K = 16.5$
GM*–galago	$\Delta J + \Delta H + \Delta D + \Delta E + \Delta L \approx 42$

FIG. 2. Primate phylogenetic tree.

Table 5

Pattern of Rate of Change (see Fig. 2)

Nucleotide change during period (%)	Length of period × 10⁶ years	Nucleotide change per 10⁶ years during period (%)
$\Delta A = 1.2$	15	0.08
$\Delta B = 1.5$	15	0.1
$\Delta C = 1.7$	15	0.11
$\Delta D = 3.4^a$	20	0.17
$\Delta E = 12.6^a$	15	0.85
$\Delta F = 1.2$	15	0.08
$\Delta G = 2.6$	30	0.086
$\Delta H + \Delta J = 5.2$	45	0.11
$\Delta K = 8.1^a$	65	0.12
$\Delta L = 21^a$	80	0.26

[a] Estimated by assuming that an equal number of changes has occurred in each line since the time of divergence. It is interesting to note that if this assumption is not valid, the constant-rate-of-change hypothesis cannot be true for nucleotide changes.

compared. In calculating the values for Table 6 it is assumed that the present-day generation times provide a reasonable approximation of the generation time history of the various species lines. The rates calculated in this manner are surprisingly similar for the primate, rodent, and bovine lines.

The changes in the absolute-time-based rates of nucleotide change seen in Table 5 could be explained by considering the probable generation time history of the species lines. If the generation times have increased in the development of the present-day primates, the slowing of the absolute-time-based rate of change would be the result of a decrease in the number of generations per time period as the primates evolved through time. It appears that in fact the generation time of at least the higher primates has increased considerably during primate evolution, relative to that of the primitive primates. The ancestral primate was presumably a small rodent-sized creature which probably had a short generation time. There is today a general, but not absolute, relationship between generation time and animal size. In general the larger the animal the longer its generation time. This subject is discussed further by Kohne et al. (1972).

VI. Evolutionary Effect of Observed Nucleotide Sequence Changes

The degree of DNA sequence difference between chimpanzee and man is about 2%. This means that 2 out of every 100 bases have been changed since divergence of these species. There are about 10^{10} nucleotides in both human and chimpanzee DNA. A 2% change means that a total of 2×10^8 nucleotide changes have occurred since the time of divergence. What is the nature of these changes? As mentioned earlier the effect of a nucleotide substitution in an organism will depend upon where the substitution takes place. Nucleotide changes in nonfunctional DNA will have no effect on the organism and will be preserved. A change in DNA which codes for a protein also may not have any effect on the organism. The DNA sequence which codes for a protein whose amino acid sequence cannot be changed can change without affecting the amino acid sequence. In many nucleotide triplets which code for specific amino acids the third nucleotide can be changed to any other base without changing the amino acid inserted into the protein. Such a DNA sequence could theoretically change by 17% without affecting the amino acid sequence. For example, the nucleotide triplets UCU, UCC, UCA, and UCG all code for the amino acid serine. In many proteins the amino acid sequence can be changed without any apparent effects on the organism. Thus even nucleotide changes which result in an amino acid change may have little or no selective force on them.

It is not known how much of the DNA of any eukaryote is actually used to code for protein. A large fraction of the DNA is probably involved in the "control" aspects of the cell. Another large fraction of the DNA may have no present day function. Nucleotide changes occurring in this latter class would be fixed at the same rate at which they occur since no selective force is acting on them.

The majority of the changes seen in these studies represent nucleotide changes which have been "fixed" and have existed in these species lines for millions of years. Lethal and highly deleterious changes would be quickly eliminated from the species DNA. The majority of changes seen in these studies represent nucleotide changes which have either a positive selection value, a very small negative selection value, or are neutral.

As the above discussion indicates, many nucleotide changes which have occurred could have little or no effect on the phenotype of the organism and would have little selection pressure on them. It seems likely that a large fraction of the nucleotide

TABLE 6

RATE OF NUCLEOTIDE SUBSTITUTION CALCULATED IN TERMS OF GENERATION TIME

DNAs compared	Nucleotide substitutions per year since divergence	Shortest generation time of the pair of species[a]	Nucleotide pair changes per generation	Nucleotide pair changes per germ cell division[b]
Man–chimp	1.6	10	16	0.32
Man–gibbon	1.8	10	18	0.36
Man–green monkey (GM)	2	2–4	4–8	0.12
Man–capuchin	2.4	2–4	4.8–9.6	0.15
Man–galago	5.2	1–2	5–10	0.15
GM–rhesus	2.4	2–4	4.8–9.6	0.15
GM–human	2.2	2–4	4.4–8.8	0.13
GM–chimp	2.2	2–4	4.4–8.8	0.13
GM–gibbon	2.2	2–4	4.4–8.8	0.13
GM–capuchin	2.4	2–4	4.8–9.6	0.15
GM–galago	5.0	1–2	5–10	0.15
Cow–sheep	3.6	1–2	3.6–7.2	0.1
Cow–pig	3.8	1–2	3.8–7.6	0.1
Mouse–rat	26	0.25	6.5	0.13

[a] The generation time is taken as the gestation time plus the time needed to reach a fertile state (Anderson and Jones, 1967).
[b] Assumes a total of 50 cell divisions to go from zygote to zygote.

substitutions measured is subjected to little or no selective pressure. These changes would be fixed in a population at the same rate at which they occur. Thus the *measured rate* of nucleotide change is probably close (within a factor of 2–3) to the *actual rate* at which nucleotide changes have occurred during evolution.

As has been indicated, a major difficulty in interpreting DNA comparison data results from not knowing the function of most of the DNA. It is now possible to isolate and purify specific RNA populations which code for a single protein. Hemoglobin and γ-globulin are examples. Thus a nucleic acid sequence with a known function can be isolated, the evolution of a DNA sequence which codes for a specific function can be studied, and the comparative evolution of a gene and its product examined. Utilizing such RNAs the existence of "neutral" nucleotide changes can be examined. It may also be possible to ascertain whether the rate of nucleotide substitution seen for bulk DNA is an "unselected" rate and therefore representative of the actual nucleotide change rate during evolution.

The data on nucleotide substitution rates suggest that the rate of DNA molecular evolution may have been different for different species. If true, this might be a troublesome observation. An assumption basic to evolutionary studies has been that morphological and molecular similarity of two species is an index of the temporal relationship between those species. In this view species which have diverged most recently would be more similar molecularly and morphologically. This situation is certainly true if the rate of molecular evolution has been the same for all species at any time during evolution. If, however, the rate of molecular and morphological change can be different for different species lines during the same evolutionary period, the situation becomes difficult. In this case it is possible (particularly in the case of closely similar species) that similarity between species will give an erroneous view of the relative times of divergence of the species in question. Therefore it is important to know something about the rate at which change has occurred during the evolution of various species lines, since different rate models give different interpretations for similarities seen between species.

If one assumes that the rate of molecular evolution is constant, then the DNA data indicate that rat and mouse diverged about 90–120 million years ago and the divergence of green monkey and man occurred about 30–40 million years ago. Paleontologists tell us, however, that rat and mouse diverged no more than 10–20 million years ago. If the paleontologists are right, it is clear that at least DNA evolution can be much more rapid in one species line than another. If the rate of divergence is constant, then the paleontologists are far off in their estimate of the divergence time of mouse and rat.

VII. Addition of DNA to the Genome

Data on the addition of repeated DNA to mammalian genomes during evolution also indicate that DNA was added to some species lines at a much faster rate than in other lines (Kohne, 1970). Again these rates of addition were calculated using commonly accepted divergence times. Repeated DNA appears to have been added continuously during evolution. These sequences provide a very specific molecular marker for differentiating closely related species, since each species examined thus far has contained a repeated DNA fraction which is not shared by other species.

In the last decade it has become clear that virus infection of mammalian cells can result in the integration of viral genes into chromosomal DNA. These viral genes can be passed on to the next generation via the germ cells. New genetic material which has specific functions, i.e., viral genes, can be added to the gene pool of the host by viral infection. These viral functions can be activated in some cases with the result that normal cells become cancerous. A common feature of all mammalian viruses is the "defective" virus. A large number of viral progeny are defective in some way. Infection of cells by these defective viruses results in no virus production. Fragmentary evidence indicates that the "defective" virus genes can also be incorporated into the host cell DNA. Another feature of many viruses is that they produce a class of defective viruses which contain genes of the host cells on which the virus is grown. Many viruses grown on one species can infect cells of a different species. It has not yet been demonstrated that the DNA of the previous host, which was contained in those viruses, can be integrated into the DNA of the newly infected host. However, it seems probable that this occurs, although very rarely.

An enormous number of different viruses are present in each organism. Viruses are, simply speaking, very small bags which contain genes: either host or viral specific genes, or both. The virus is very mobile and readily infects members of the same species as its host as well as members of completely different species. The virus–cell interaction certainly results in the integration of virus genes into the host cell and probably occasionally results in the integration of any other DNA contained in the virus. Any normal or defective virus capable of infecting the germ cells has the potential to add new genes, either viral or from another species, to the species line. Evidence exists which suggests that many virus-like genes are present in "normal" cells and that these viral genes are passed through the germ line. Infection of germ cells by defective particles could result in the integration of viral genes which were "inactive." Because of their inactivity they could easily be fixed in the population and be available for future evolutionary experiments at the molecular level.

The effect of virus–host interactions on evolution has not been thoroughly studied. The author believes that these interactions have been and will continue to be an incredibly important evolutionary force. In this view an understanding of the role viruses play in evolution may provide insights into a major evolutionary mechanism for change.

VIII. DNA Evolution vs. Fossil Evolution

Comparing molecular changes with organismal changes is always a difficult task. Comparison of the pattern and character of these "DNA changes" with the patterns derived from the classical fossil record should, however, provide new insight into the nature of evolutionary forces. The evidence available thus far allows a *very* tentative correlation to be made. During the period where the classical fossil record indicates that extensive speciation occurred in the primates, it appears that DNA was added to the presumptive human genome line at a much faster rate than during other periods for which data are available. It also appears that the rate of nucleotide sequence divergence was much higher relative to other periods. It is not known whether these increased rates are the result or the cause of the extensive speciation seen during this period. The correlation of nucleotide sequence change with generation time suggests that the generation time of the early primate ancestors of man that lived during this period was relatively short. The rate of mixing of genetic material to provide new genetic combinations (e.g., translocations, inversion, etc.) should also be dependent on generation time. It does not seem unreasonable to speculate that the rapid addition, divergence, and mixing of DNA may have played a role in the burst of speciation seen during this period. Much more extensive work needs to be done, however, before any strong conclusions can be made.

IX. What We Need to Know

The current view of the role of DNA in evolution indicates that evolutionary changes at the organismal level must be preceded by some qualitative or quantitative change in the DNA. We have described a variety of changes which can occur in DNA. We know a reasonable amount about changes which occur in structural-type genes and their possible effect on the organism. A question that arises is: Has the accumulation of individual base changes in structural genes been the primary "cause" for evolutionary change? The author's response is that it seems unlikely that "point mutations" serve this role. Of course no hard evidence exists which can prove or disprove this.

It is not known how much of the DNA of mammalian cells is involved in "control" aspects of the cell. Control functions would be involved in such things as the mechanism of differential expression of DNA, making sure timing is proper during development; making sure that organs are the right shape, and others. We have very little data on the types of changes which occur in these control genes and the effect of those changes which occur. Changes in this fraction of DNA must be very important in evolution. Rearrangement of control genes through crossover and translocation could have far-reaching effects. An accumulation of rearrangements of structural genes could also have great evolutionary consequences. Subtle changes in the timing of expression of important structural genes could result in evolutionary types of changes. The areas of control genes and DNA rearrangements are at the present time difficult to study at a molecular level. Knowledge in these areas is almost certainly of major importance for understanding the role of DNA changes in evolution.

X. References

ANDERSON, S., and J. K. JONES, Jr. 1967. *Recent Mammals of the World*. Ronald Press, New York.

BRITTEN, R. J., and D. E. KOHNE. 1968. Repeated sequences in DNA. *Science* **161**:529–540.

CHATTAPADHYAY, S. K., D. R. LOWY, N. M. TEICH, A. S. LEVINE, and W. P. ROWE. 1974. Evidence that the AKR murine-leukemia-virus genome is complete in DNA of the high-virus AKR mouse and incomplete in the DNA of the "virus-negative" NIH mouse. *Proc. Natl. Acad. Sci. U.S.A.* **71**:167–171.

Gelb, L. D., D. E. Kohne, and M. A. Martin. 1971. Quantitation of Simian virus 40 sequences in African green monkey, mouse and virus-transformed cell genomes. *J. Mol. Biol.* **57**:129–145.

Kohne, D. E. 1970. Evolution of higher-organism DNA. *Quart. Rev. Biophys.* **3**:(3)327–375.

Kohne, D. E., J. A. Chiscon, and B. H. Hoyer. 1972. Evolution of primate DNA sequences. *J. Hum. Evol.* **1**:627–644.

Laird, C., B. L. McConaughy, and B. J. McCarthy. 1969. Rate of fixation of nucleotides in evolution. *Nature* **224**:149–154.

STREPSIRHINE
PHYLOGENY

12

The Bearing of Reproductive Behavior and Ontogeny on Strepsirhine Phylogeny

R. D. MARTIN

I. Rationale

Any attempt to reconstruct evolutionary relationships between species is obviously heavily dependent upon speculation, mitigated only in rare cases where direct reference may be made to appropriate, well-documented fossil evidence. It is a fundamental feature of all phylogenetic reconstructions involving living forms that—whether or not relevant fossil evidence is available—the most suitable first step is the formulation of a set of hypotheses about ancestors linking the living forms. Recognition and allocation of fossil evidence is generally dependent upon prior development of explicit or implicit hypotheses about ancestral stocks relating certain living forms. The major, unique advantage of fossil material—if correctly interpreted—is that it permits the addition of a *time scale* to hypothetical phylogenetic trees.

It is unfortunate that hypotheses about common ancestors are generally implied rather than stated in the literature. This is associated with widespread confusion between

R. D. MARTIN · Wellcome Institute of Comparative Physiology, The Zoological Society of London, Regent's Park, London N.W.1, U.K.

the two essentially distinct activities of *classification* and *phylogenetic reconstruction*, unfortunately encouraged by Hennig's (1950) "phylogenetic systematics." Although classification is now usually guided by evolutionary hypotheses and although it is useful to have a preliminary outline classification when attempting phylogenetic reconstructions, two quite different approaches are really required. Implicit hypotheses about common ancestors are perhaps adequate for the establishment of useful classifications, and a certain degree of vagueness about ancestral relationships is probably commendable. All that is necessary is that any given classification should be *compatible* with the apparent phylogenetic relationships (see also Simpson, this volume). A classification need not be interpreted as an unequivocal statement of a set of phylogenetic relationships; in fact the most successful (i.e., widely adopted and long-lived) classifications would seem to be those compatible with a variety of phylogenetic trees. Classifications are arbitrary constructs following a certain set of rules and serving a number of different utilitarian functions (Mayr, 1974). A published "phylogenetic tree," on the other hand, represents an attempt to portray the reality of evolutionary history, and it is in this area that Hennig (1950) has had a major beneficial influence on recent studies. Reconstructions of phylogenetic history may change rapidly, while a well-known classification (e.g., Simpson's 1945 classification of mammals)—which is the basis for communication among scientists —should only be modified if it is eventually in direct conflict with the accumulated consensus of informed opinion about probable evolutionary relationships.

That said, it should be made clear at once that this paper is primarily concerned with the attempt to reconstruct actual evolutionary relationships, and that it is concerned with classification only insofar as an economical, hierarchical set of names is necessary to refer to the animals discussed (Table 1). The following discussion is dependent at the outset on the assumptions underlying this set of names, thus illustrating the importance of classifications to those interested in phylogenetic reconstructions.

The primary question is: Given a group of living (and perhaps fossil) species for comparison, how can one develop reliable hypotheses about ancestral relationships? Discussion of *factual evidence* utilized for hypotheses must be clearly distinguished from discussion of *methodology* inherent in interpretation of the facts. The attempt is therefore made here to distinguish (1) the distribution of various reproductive and ontogenetic characters in primates, and (2) the evolutionary interpretation of each set of characters. Special attention must be given to methodology, since this governs both the marshalling of facts for evolutionary interpretation and the manner in which hypotheses are developed. A set of rules must be generated according to which the various characters of living forms may be objectively assessed in order to produce reliable hypotheses about ancestral relationships.

It is usually accepted that morphological characters typical of a species are based on inherited characteristics. This assumption is difficult to test, but it is justifiable on theoretical grounds. The same assumption can be used about the transmission of certain species-typical behavior patterns, although more care is necessary because much behavior develops plastically within the lifetime of the individual and because much less research has been carried out on the inheritance of behavioral differences. For this reason, it is often preferable to concentrate on behavioral characters linked to morphology (viz.,

TABLE I

PROVISIONAL CLUSTERS OF LIVING PRIMATES[a]

Small Clusters	Inclusive Clusters		
Mouse lemur group (Cheirogaleinae) True lemur group (Lemurinae) Indri group (Indriidae) Aye-ayes (Daubentoniidae)	Malagasy lemurs (Lemuriformes)	Prosimians (Prosimii)	Strepsirhines (Strepsirhini)
Bushbabies (Galaginae) Lorises (Lorisinae)	Loris group (Lorisiformes)		
Tarsiers (Tarsiidae)	Tarsier group (Tarsiiformes)		
Marmosets (Callitrichidae) True N.W. monkeys (Cebidae)	New World monkeys (Ceboidea)	Simians (Anthropoidea)	Haplorhines (Haplorhini)
Cercopithecine monkeys (Cercopithecinae) Colobine monkeys (Colobinae)	Old World monkeys (Cercopithecoidea)		
Gibbons and siamangs (Hylobatidae) Great apes (Pongidae) Humans (Hominidae)	Old World hominoids (Hominoidea)		
Tree shrews: Diurnal tree shrews (Tupaiinae) Nocturnal tree shrews (Ptilocercinae)	Tree shrews (Tupaiidae)		

[a] The formal classificatory terminology (parentheses) is largely based on Simpson (1945). However, the tree shrews have here been excluded from the order Primates, while the gibbons, siamangs, and the great apes are allotted to 2 separate families, since it is now widely accepted (e.g., see Simons, 1972) that the Hylobatidae are really very distinct from the Pongidae. The terms Strepsirhini and Haplorhini have been adopted from Osman Hill's monographic work on the Primates (see Hill, 1953).

functional morphology) or to physiological functions (Martin, 1972*a*). Nevertheless, one need not exclude the application of the comparative method to features of behavior which occur in a fair range of living species and which would appear to behave like "conservative" morphological characters, even if there is no obvious link with a morphological base. Thus, in the following discussion, the same method will be applied to both morphological and behavioral characters, as long as the latter are both species typical and common to a number of species in a cluster.

The first step, then, is the compilation of a list of character similarities and differences among the species compared. It is generally accepted, however, that construction of phylogenetic trees purely on the basis of similarities and differences is totally inadequate, even if very large numbers of characters are included in the comparison. The most obvious reason given for this is that the phenomenon of convergence (including parallelism) is widespread, giving rise to similarities between species not actually derived from a common ancestor possessing the characters concerned. This introduces a distinction between homologous characters (retained in unbroken sequence from a common ancestor) and convergent characters (independently acquired), and much evolutionary research is concerned with distinguishing the two classes of character.

However, it is also necessary to define any common ancestors postulated and to define which ancestors are presumed to have possessed the characters involved (Hennig, 1950; Martin, 1968a). In comparing, for example, a cluster of 3 living species (A, B, C), it is insufficient to state that B and C are closely related to one another and less closely related to A because the former two species share a greater number of apparently homologous characters (derived from a common ancestor). This statement implies that the characters exclusively shared by B and C were present in a more recent common ancestor giving rise to these two species, but not present in the early ancestral stock giving rise to A, B and C. If B and C are actually more closely related to each other than either is to A, then it is likely that they will share more homologous characters; *but this need not necessarily be the case.* One must exclude the possibility that the homologous characters shared by some of the living forms were already present in the early common ancestor, and that the forms lacking such characters have in fact lost or modified them through subsequent evolution (Fig. 1). One might, for example, encounter a situation where, in a cluster of 3 living species, A and B share the greatest number of homologous characters, yet where B and C are actually the most closely related, having diverged from a later ancestral stock. For example, it has been argued elsewhere (Martin, 1968a,b, 1969) that many of the apparently homologous morphological characters shared by tree shrews and (some) primates have been retained by both living groups from the eutherian mammal ancestral stock and not from a later, distinct ancestral primate stock.

The same point is made by the distinction between *cladistic* and *patristic* relationship (Harrison and Weiner, 1964, p. 19). This distinction emphasizes the phenomenon of *differential rates of evolution* rather than the individual character states involved. In Fig. 1, the 3 species A, B, and C are all derived from an early ancestral stock, while B and C are divergent offshoots of a specific later ancestral stock. The common retention of 50% early ancestral, or plesiomorph, homologous characters by A and B (patristic relationship) might obscure the later divergence of B and C from a subsequent common ancestor (cladistic relationship). Because C has, in this case, evolved very rapidly since its divergence from B, while both A and B have evolved relatively slowly, B and C ultimately share only 35% homologous characters. There is very little common retention of early ancestral (plesiomorph) homologous characters (10%), but a highly significant retention of several later ancestral (apomorph) characters (25%) developed in the specific stock giving rise to B and C. This situation has probably occurred repeatedly in the evolution of the primates.

One might therefore ask whether our present phylogenetic trees for primate evolution would be at all acceptable or reliable in the absence of fossil evidence. In short, given only information on the living species, would there be any objective means of distinguishing patristic from cladistic relationships in the reconstruction of a phylogenetic tree? If there is no reliable procedure, interpretation of reproductive characters (for instance) must always remain an entirely speculative exercise. However, it is the author's belief that such a reliable procedure can be developed using objective criteria.

The primary theoretical requirement in tree-building is the formulation of hypo-

MORPHOLOGICAL DIVERGENCE

269

STREPSIRHINE
REPRODUCTIVE
BEHAVIOR
AND ONTOGENY

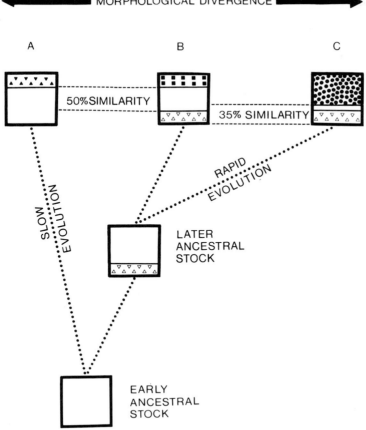

FIG. 1. The numbers of homologous characters shared between living species (A, B, C) cannot be used as a reliable basis for tree-building if rates of evolution differ. Primitive features of the "early ancestral stock" ("plesiomorph characters," Hennig, 1950) must be distinguished from distinctive features of the "later ancestral stock"
("apomorph characters").

theses about characters present in a network of "ancestral stocks" (Martin, 1973) and
about the relationships of those "stocks" to one another as well as to the living forms
compared. The term "ancestral stock" needs definition. One requirement of evolutionary
theory, recently underlined with great clarity by Hoffstetter (1973), is that if the order
Primates is genuinely monophyletic, all living primates must be derived from a single
species population which probably existed in the Upper Cretaceous. Similarly, all
monophyletic subgroups within the Primates must be derived from later, individual
species populations. Theoretically, the ultimate aim of phylogenetic tree-building is to
identify such ancestral species populations in terms of (at least some of) the characters
they possessed and the times at which they existed. But techniques of phylogenetic
reconstruction will probably never be that accurate, and the term "ancestral stock"
(Martin, 1973) can be used to accept this inherent degree of inaccuracy. An "ancestral
stock" may be taken as a time bracket including a number of closely related, divergent

species including that which actually gave rise to the descendant living forms of that "stock." Accordingly, a list of characters of an ancestral stock (rather than an ancestral species) will represent an average for the various species included in the time bracket. Even if we were to find the fossil remains of an actual ancestral species (a highly unlikely event), it is almost certain that its identification as such could not be conclusive because of the limited resolving power of available phylogenetic techniques.

One widespread approach, which has its uses but should be avoided in the latter stages of tree-building, is that of treating the living primates as an evolutionary series or as a model thereof. In the past, evolutionary hypotheses have often been confused with this concept of the living primates as forming a "phylogenetic scale" (*Scala naturae*) with man at the summit (Martin, 1973). In the field of primate reproduction, a most explicit example of this procedure is to be found in Hill's (1932) monograph on placentation. It is sometimes revealing, but ultimately inadequate, to discuss primate evolution in terms of an ascending stepladder passing through the following living species as the stages: hedgehog, tree shrew, lemur, tarsier, monkey, ape, man. This approach must necessarily be eliminated if one is to attempt the reconstruction (however speculative) of actual ancestral stages. A living lemur cannot be simply equated with the ancestral primate stock; the characters of the latter must be established according to a methodology which recognizes the fact that *all* living primates have undergone evolutionary modification since their initial divergence from the ancestral stock.

II. Practical Application

The first task, given a preliminary clustering of species (Table 1), would seem to be the exclusion of convergent characters from evaluation of the species clusters; but this introduces an element of circularity. Convergent similarities are customarily defined as those not retained from a common ancestral stock, yet the aim of excluding convergent characters from consideration is precisely that of identifying and describing such ancestral stocks! One practical means of sidestepping this obstacle is to use two rules of thumb:

1. With complex characters, it may be possible to define convergence as similarity which, when examined in detail, proves to be only superficial. That is, it may be expected with complex characters that similarity acquired by convergence may be underlain by obvious differences in more fundamental aspects. This criterion would apply, for example, to the "tooth-comb" found in tree shrews, most lemuriforms, and all lorisiforms. In tree shrews, the comb consists of 6 lower incisors, whereas in most lemuriform species (Lemuridae, *sensu* Simpson, 1945) and all lorisiforms, it consists of 2 lower canines and 4 lower incisors. Superficial resemblance is exhibited in the possession of a 6-tooth comb in all 3 groups, whereas detailed study shows that the comb is only really *homologous* in lemuriforms and lorisiforms.

2. For simple characters, no such distinction may apply. One can utilize the further criterion that the occurrence of simple similarities between diverse groups which are very different in all or most other characters considered is likely to be a result of convergence, although it could also be the result of sporadic retention of primitive (plesio-

morph) characters. This argument would, for example, be applicable to the convergent formation of postorbital bars in the skulls of marsupials, carnivores, ungulates, hyracoids, tree shrews, and primates (Martin, 1968a), particularly since the appearance of this character is sporadic within groups other than the order Primates.

Thus, it is of the utmost importance to examine all characters in depth, and greater value should be placed on complex characters (e.g., placentation) than on simple characters (e.g., litter size) in reconstructing evolutionary history. It is also apparent that accuracy is likely to increase with the numbers and diversity of characters and species considered. To be sure, there is always the theoretical possibility (referred to as parallelism) that two descendent species may have developed profound similarities independently because of parallel adaptation after their divergence, from a common ancestor lacking the characters concerned but possessing some kind of basis canalizing their development. There is an entire spectrum from "obvious" convergence to "unidentifiable" parallelism, but one can at least advance some way towards a reliable phylogenetic tree by excluding all cases of convergence identifiable by thorough investigation.

This leaves the really difficult problem of recognizing differential evolutionary rates, which cannot be confidently resolved by simply assuming that some species (e.g., living prosimians) have generally evolved more slowly than others (e.g., simians) for some reason. In fact, the problem of differential evolutionary rates is one which is central to all attempts to reconstruct evolutionary trees—including those based on rigorously defined biochemical data (e.g., see Romero-Herrera et al., 1973, for discussion).

Biochemists must also eventually come to grips with the patristic/cladistic problem in their attempts to produce reliable phylogenetic trees (cf. Cook and Hewett-Emmett, 1974). A partial answer to this problem is in fact provided by the tentative reconstruction of actual ancestral stages and determination of the DNA nucleotide base changes necessary to give rise to the observed amino acid sequence changes (Cook and Hewett-Emmett, 1974; Romero-Herrera et al., 1973). Reconstructed "ancestral proteins" must also be functionally plausible as well as structurally conceivable. Further, the various reconstructed ancestral proteins for a given ancestral stock must be functionally compatible with one another. Inclusion of the functional aspect yields an additional source of information which may possibly be exploited to varying extents for all evolutionary reconstructions, biochemical or otherwise, In sum, any reconstructed common ancestor must be a functionally plausible entity; all of its postulated characters must be compatible. This gives yet another reason for considering as many characters as possible in any evolutionary reconstruction.

To return to comparative anatomy and behavior, the central problems of convergence and potential confusion between patristic and cladistic similarity must be resolved in any objective attempt to assess the relative merits of various hypothetical trees. As an interim solution, for the purposes of discussing reproductive evolution, it is suggested that the following pragmatic rules may be used in narrowing the range of theoretical possibilities and formulating a preliminary phylogenetic tree.

1. One must consider as many characters as possible and examine each character as carefully as possible in order to distinguish superficial or simple functional similarity

(probably resulting from convergence) from fundamental, detailed similarity (probable homology). Only fundamental similarities will be used for tree-building.

2. Fundamental similarities common to all (or virtually all) the species compared may provisionally be regarded as homologous characters present in the ancestral stock giving rise to all species in the cluster. For example, since the possession of a grasping hallux on the foot is a universal character among living primates (excluding man), it is likely that this character was present in the ancestral primate stock. By considering species clusters in hierarchical order (Fig. 1; Table 1), provisional branching points (ancestral stocks) can be described in terms of universal characters present in one or more of the clusters of living species.

3. Clustering on the basis of fundamental similarities must subsequently be subjected to further refinement to distinguish cladistic from patristic clustering.

4. In the overall formulation of the tree, the principle of parsimony may be used to exclude suggested evolutionary relationships which involve considerable duplication of evolutionary changes. That is to say, the parsimony principle may be applied in extreme cases where the phylogenetic scheme suggested in highly uneconomical, but not to distinguish between trees which differ only mildly in the total amount of evolutionary change suggested.

5. The principle of the logical antecedent may in many cases be used to select between alternative hypotheses in an evolutionary tree. [This principle corresponds to Hennig's (1950) concept of "ontogenetic character precedence."] For example, on the basis of numerical considerations alone, it would not be possible to decide whether the bicornuate uterus (typical of all living prosimians) or the simplex uterus (typical of all living simians) represents the condition present in the ancestral primate stock. However, detailed analysis of embryological development and of the condition found in many other mammals, in birds, reptiles, amphibians, and fish indicates with a high degree of probability that the bicornuate condition is more primitive than the simplex condition. All the evidence indicates that a bilaterally symmetrical, separate system logically precedes the high degree of fusion represented by the simplex uterus. Thus, it is very likely that the ancestral primate had a bicornuate uterus. It is evident that application of the logical antecedent requires a broad zoological knowledge and also benefits greatly from a consideration of embryology and ontogeny. In practice, use of the logical antecedent requires: (a) consideration of characters found in representatives of clusters larger than the main group investigated (e.g., study of the primates requires consideration of mammals and even of other terrestrial vertebrates), and (b) detailed analysis of individual characters to include developmental aspects.

6. Having revised the preliminary tree in the light of these considerations, the overall reliability of the hypotheses included may be assessed in terms of functional feasibility and functional compatibility in the list of characters postulated for each ancestral stock. The longer the list of characters for each stock, the easier it will be to examine them in such functional terms. In this context, it is very useful to have available empirical rules which have been found to apply generally to the living species concerned and which can be utilized for interpretations of functional aspects of postulated ancestors. For example, it is probably valid to apply to a hypothetical ancestor the observation that a relatively

large brain size in an adult mammal is usually correlated with a relatively long gestation period and a long life span.

7. Any conclusions relevant to durable morphological features may be checked against available fossil evidence.

III. Reproductive Characters

Reference must be made to several categories of information in evaluating the evolution of reproductive characters in the strepsirhines (Table 1):

1. Basic reproductive features of the terrestrial vertebrates.
2. The distribution of reproductive characters in eutherian mammals generally, with particular emphasis on likely characters of the ancestral eutherian stock.
3. The reproductive characteristics of the Tupaiidae, with reference to the possible relationship of the tree shrews to the ancestral primate stock.
4. Reproductive characters typifying all or most primates and distinguishing them from other eutherian mammals.
5. Reproductive characters distinguishing the strepsirhines from the haplorhines (tarsiers, monkeys, apes, and man), indicating a possible division of the early primates into two stocks.
6. Adaptive radiation within the strepsirhines in terms of reproductive characters.

Compilation and interpretation of reproductive data within all of these 6 categories are necessary in order to examine reliably the evolution of the strepsirhines in terms of a succession of ancestral stocks.

Despite the fact that the living therian mammals (marsupials and eutherians) are broadly distinguished from other living vertebrates by a network of characters involved in viviparous reproduction and suckling, little concerted attempt has so far been made to reconstruct the likely reproductive characters of the ancestral eutherian mammal stock. Yet it is impossible to assess (for example) the likely placentation of the ancestral primate stock without a prior hypothesis concerning the placentation of ancestral eutherian mammals. It is not sufficient to discuss primate placentation purely in terms of evolution within the order Primates, a fact which Luckett (1974a, and this volume) has aptly demonstrated by reference to monotremes, marsupials, and eutherian mammals.

Thus, some kind of base-line must be provided by hypothetical reconstruction of the ancestral eutherian stock. Although this can only be considered briefly here, it is an indispensable preliminary to the discussion of primate evolution.

Extrapolation backwards from universal characters of living eutherians indicates that the ancestral eutherians were probably viviparous, with well-developed placentation, and that there were well-established adaptations for suckling behavior. Nourishment of the embryo(s) within the maternal uterus would probably have been ensured jointly by diffusion between maternal and fetal blood vessels and by secretion of "uterine milk" by uterine glands. Placentation would have involved (probably successively) vascularization of the chorion by both the yolk sac (choriovitelline placentation) and the allantois (chorioallantoic placentation). The ovaries of the female would have been contained in a bursa, and mature Graafian follicles would probably have been produced on a regular cyclical basis (estrous cycle) except during gestation or nonbreeding seasons. The likely time of ovulation for each group of maturing follicles would have been associated with a relatively restricted phase of receptivity for mating (estrus). Following internal fertilization the developing zygote would have implanted in the uterine wall at the bilaminar blastocyst stage. Hormonal maintenance of embryonic development, involving a functional corpus luteum, and the timing of birth would have followed the basic mammalian pattern (Perry, 1971). After birth the infants would have been fed with milk produced by a bilaterally symmetrical system of two longitudinal bands of subcutaneous mammary tissue leading to paired teats, the number of which would have been correlated with the litter size (Schultz, 1948). The milk would have contained fat, sugar, protein, and small quantities of other essential components (vitamins, trace elements, etc.). The composition of the milk would have been correlated with the pattern of maternal suckling behavior (Ben Shaul, 1962).

With many other reproductive characteristics there is no universal pattern among living mammals, and some kind of evolutionary interpretation is required. The principle of the logical antecedent, as explained previously, suggests that a bilaterally symmetrical, paired arrangement of the female urogenital tract (Müllerian ducts) preceded any condition involving midline fusion. If this principle is combined with recognition of the universal *minimal* condition found in living eutherians, one can conclude that a bipartite or bicornuate uterus was present in ancestral eutherians, but that some degree of midline fusion had taken place to produce a unified central vagina with a long urogenital sinus linking this and the urethra to the vulva. This confluency of urinary and genital systems into the urogenital sinus would have been a residual reflection of the embryonic association between urinary and genital systems.

A. Placentation

The principle of the logical antecedent might also be applied to the interpretation of early eutherian mammal placental membrane relationships on the basis of Grosser's (1909) system of classification. Modern eutherian mammals exhibit a widely scattered distribution of maternal–fetal membrane relationships in the definitive placenta, ranging from the least invasive (epitheliochorial) to the most invasive (hemochorial), and one cannot therefore apply the principle of universal characters. However, the principle of the logical antecedent dictates that highly invasive placentation systems must have been preceded by noninvasive placentation systems, since there was doubtless a stage in the evolution of mammalian placentation where retention of the fertilized egg and subsequent embryonic development took place without an intimate placental attachment (advanced ovoviviparous stage). Hence, it is highly unlikely that the most invasive form of placentation found among living eutherian mammals (hemochorial placenta type) was present in the ancestral eutherian mammal. It might, therefore, be argued that the least invasive form of placentation among living eutherian mammals (epitheliochorial type) was the ancestral eutherian condition. This point of view has, for example, been supported by

Hill (1932) and by Luckett (1969, 1974a,b, and this volume). However, it is possible that the evolution of placentation began long before the appearance of the ancestral eutherian stock, and the logical noninvasive predecessor of eutherian placentation may well have occurred far back among the mammal-like reptiles (premammalian stage). This possibility was suggested very early on in the development of the science of embryology by Hubrecht (1908). In addition, some characteristics of various modern epitheliochorial placentae, such as the presence of chorionic vesicles associated with the outlets of the uterine glands, indicate that such placentae may well represent a specialized adaptation towards the intensive exploitation of "uterine milk" as a major source of embryonic nutrition (Martin, 1969). Uterine milk seems to be an important, but not exclusive, source of nutrition in mammalian embryogeny generally. It is therefore equally reasonable to assume, as did Mossman (1937) on the basis of broad morphological comparison of placentation in the mammals, that the ancestral eutherian mammal may have possessed a blastocyst with an active, invasive trophoblast and a moderately invasive (zonary labyrinthine endotheliochorial) placenta exploiting both uterine milk and diffusion between maternal and fetal blood vessels. In the adaptive radiation of the mammals, subsequent specialization may have led frequently to enhanced exploitation of uterine milk (epitheliochorial placentation, possibly derived by paedomorphosis), or to intensification of diffusion between maternal and fetal blood vessels (hemochorial placentation). Such divergent specialization might be expected to lead to a bimodal frequency distribution of placental types along the axis of increased invasiveness (epitheliochorial: –syndesmochorial: –endotheliochorial: –hemochorial). In fact actual, definite epitheliochorial or syndesmochorial types of placentation (the less invasive types) seem to be far less common than endotheliochorial or hemochorial types of placentation, in terms of the numbers of major groups of mammals. Further, the epitheliochorial type is more common than the syndesmochorial, while the endotheliochorial is apparently less common than the hemochorial. Indeed, the syndesmochorial type has rarely been reported (Starck, 1965) and may not occur at all (Luckett, personal communication). This is surprising, considering that it would be a logical transitional stage between a postulated epitheliochorial ancestral type and an endotheliochorial or hemochorial type (see Amoroso, 1952, pp. 148–149; and Starck, 1965, p. 271 for tabulation). Thus, both epitheliochorial and hemochorial placenta types are more frequent than would be expected from a hypothesis of general radiation from an epitheliochorial type, and the resulting bimodal pattern is nowhere more apparent than within the order Primates itself (see later).

It has been cogently argued by Luckett (1974a, and this volume) that the epitheliochorial type of placentation is, in fact, the primitive condition for eutherian mammals. The most significant evidence in favor of this interpretation is the fact that the ontogeny of fetal membrane relationships and detailed fine structure of the definitive epitheliochorial placenta are essentially the same in diverse groups of mammals with this type of placenta (strepsirhine Primates, Artiodactyla, Perissodactyla, Cetacea, Pholidota). Originally, Luckett (1969) adopted a modified form of Mossman's (1937) interpretation that the endotheliochorial type of placenta was primitive for eutherian mammals. Although the ancestral eutherian condition was even then postulated as epitheliochorial, Luckett (1969, p. 426) suggested a basic dichotomy between mammal groups with endotheliochorial or hemochorial placentation and those with "retained" epitheliochorial placentation. This led to the somewhat unlikely hypothesis that strepsirhines and haplorhines were separately derived from the "basal eutherian stock." As Luckett has now shown (this volume), such a conclusion is actually unnecessary if epitheliochorial placentation is regarded as a primitive (plesiomorph) character for eutherian mammals generally.

The detailed similarities between the epitheliochorial placentae of diverse mammal groups now exhibiting this type of placentation can be explained either by retention of primitive eutherian characters, or as a result of convergence for similar functions, or as a combination of both effects. Since the other placentation characters postulated as primitive by Luckett (1974a) are also found in some mammals with endotheliochorial placentation (large yolk sac in early embryogeny, temporary choriovitelline placenta, large allantoic vesicle, amniogenesis by folding), the only question to be resolved is whether the ancestral mammal necessarily exhibited a noninvasive, epitheliochorial placenta.

Apart from the considerations already outlined above, there are two major factors of importance here. Firstly, it is perhaps misleading to compare the localized invasive placenta with the diffuse noninvasive placenta. A more meaningful comparison might be between the smooth, noninvasive areas of the chorion of species with an invasive placenta and the diffuse epitheliochorial placenta. The "free" areas of the chorion in species with endotheliochorial and hemochorial zonary or discoid placentae are doubtless involved in absorption of substances contained in uterine milk and are not greatly dissimilar in some forms (e.g., tree shrews) from nonvillous areas of chorion involved in fetal–maternal tissue apposition in forms with an epitheliochorial placenta. Secondly, Kihlström (1972) has recently shown that, in mammals generally, long gestation periods (relative to body size) are found with species having either "villous hemochorial placentae" or epitheliochorial placentae, whereas mammals with short gestation periods (relative to body size) have "labyrinthine hemochorial" or endotheliochorial placentae with an intermediate degree of invasiveness. Although objections may be raised with respect to Kihlström's classification of placental types (since individual species are not identified), the fact remains that living mammals with epitheliochorial placentas typically have long gestation periods. Since it is likely (see below) that ancestral eutherian mammals had small brains and short gestation periods (relative to body size), they may well have had placentae with an intermediate degree of invasiveness. Hence, as suggested previously (Mossman, 1937; Martin, 1969), modern villous hemochorial or epitheliochorial placentae may logically be regarded as products of dichotomous specialization away from an ancestral, moderately invasive placenta type and associated with the development of relatively large brains. (However, see Luckett, this volume, for a different interpretation.)

B. Testicular Descent

Similar problems arise with the question of the descent of the testes in male mammals. Many (but not all) living mammals exhibit descent of the testes to a posteroventral location, but there are some in which the testes remain within the general abdominal cavity, in rare cases lying in the "reptilian position" close to the kidneys. Since embryonic evidence and evidence from amphibians, reptiles, birds, and monotremes indicate that the abdominal, infrarenal position is the logical antecedent to the descended position, it seems highly unlikely that the ancestral eutherian mammals exhibited full, postpenial descent of the kind found in the Primates (excluding tree shrews) and some other modern eutherian mammals. However, descent of the testes to an extra-abdominal site is so widespread among both eutherian and marsupial mammals that it seems fairly certain that the ancestral therian mammal must have differed markedly from the reptilian condition in some way. This dilemma may be resolved through the observation (Glover and Sale, 1968) that in male rock hyraxes, where the testes themselves remain in the "reptilian" infrarenal position, the epididymis is in fact descended in that the cauda epididymis lies close to the ventral body wall. This indicates, on the one hand, that testis descent may be related to sperm storage rather than to sperm production in a body zone to some extent isolated from the elevated core body temperature of mammals, and, on the other hand, that the relatively few mammals which nowadays do not exhibit testicular descent may have actually undergone secondary specialization because of certain features in their evolution. It has already been pointed out (Martin, 1968b) that burrowing habits, aquatic life and/or very large body size are commonly correlated with the absence of testicular descent in eutherian mammals. One might therefore conclude that the ancestral eutherian mammal did exhibit some degree of testicular descent (or at least descent of the cauda epididymis), perhaps associated with the formation of a ventral, prepenial cremaster sac. It is assumed that the ancestral eutherian mammal would have exhibited an elevated core body temperature and moderately developed homeothermy (a universal minimal condition in living eutherians), and that such descent of the testis (or cauda epididymis) was necessary for effective sperm storage. Since the descending testis must always pass backwards along the external face of the pubic bones (symphyseal area) when a true scrotum is formed, the prepenial position must logically precede the parapenial and postpenial positions in mammalian evolution.

C. Estrous Cycle

Returning to the female, there is the question of the estrous cycle and the manner in which ovulation is provoked. Strictly speaking, an "estrous cycle" as such is probably largely a product of captive conditions, since in most cases under natural conditions one would undoubtedly be dealing with a "pregnancy cycle," probably alternating with a nonbreeding season and perhaps preceded by a small number of infertile cycles just after puberty and/or at the beginning of each such season. Hence, it seems logical that (at least during the breeding season) cyclical ovulation was originally simply a reflection of the time taken for maturation of a new set of mature Graafian follicles (follicular phase) following any estrus not succeeded by fertilization (i.e., the equivalent of a short-phase estrous cycle; see Fig. 2). Another major question is that of the actual elicitation of ovulation. Some authors (e.g., Conaway, 1971) have proposed that induced ovulation is more primitive than spontaneous ovulation, while others (e.g., Weir and Rowlands, 1973) regard spontaneous ovulation as the primitive condition. There is also the view (Asdell, 1966) that any attempt at evolutionary interpretation is fruitless. Very little

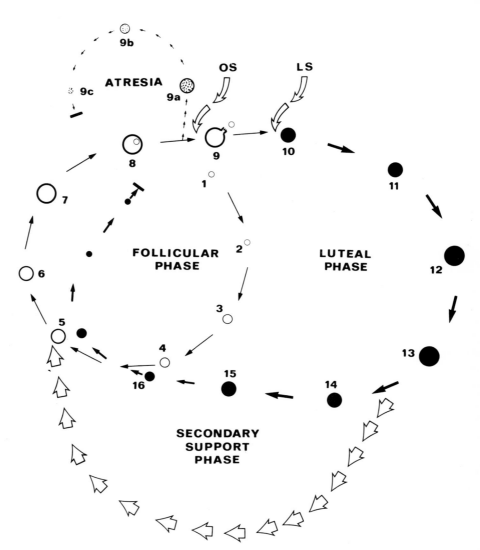

FIG. 2. Schematic representation of female reproductive cycles in mammals. The simplest estrous cycle is produced by periodic maturation of groups of Graafian follicles (follicular phase, 1–9). If an ovulatory stimulus (OS) and a luteinization stimulus (LS) occur (spontaneously or following mating) a luteal phase (10–16) is added to the follicular phase (long estrous cycle). Primates generally seem to have a long estrous cycle whether or not mating occurs. If pregnancy follows mating, the luteal phase is usually extended and with long gestation periods secondary support mechanisms may operate (e.g., production of accessory corpora lutea; replacement of luteal function by the placenta). Follicles are shown as white circles; black circles indicate the corpus luteum following ovulation. Follicles which fail to develop fully or do not erupt because of the lack of an ovulatory stimulus subsequently degenerate (atresia).

comprehensive information is available for mammals generally because of the extensive work involved in clearly distinguishing induced from spontaneous ovulation in any one species. In fact, work on laboratory mammals shows that the situation is more complex than implied by a two-way distinction. There are, probably, three basic categories (Perry, 1971) involved in mammalian female estrous cycles, and they doubtless grade to some extent into a continuum (Conaway, 1971).

1. Ovulation is induced by mating; development of a functional corpus luteum follows automatically.
2. Ovulation is spontaneous; development of a functional corpus luteum (luteinization) is induced by mating.
3. Both ovulation and subsequent development of a functional corpus luteum occur spontaneously.

The first two categories are effectively similar in that there will only be a short estrous cycle (nonovulatory in the first case) if mating does not take place. With the last category, on the other hand, the estrous cycle is long whether or not mating occurs. Of course, given a natural situation in which fertilization *always* occurs whenever a female comes into estrus, the only practical difference between any of the three categories of estrous cycles under natural conditions would be dependent on the follicular phase. However, in a natural situation involving polyestrous females where fertilization does not always occur at estrus, categories (1) and (2) would generally increase the fertilization rate as compared to category (3). A high fertilization rate would be further assisted by occurrence of postpartum estrus.

As has been previously stated (Conaway, 1971; Weir and Rowlands, 1973), mating systems involving induction of either ovulation or formation of a functional corpus luteum (luteinization) generally typify small-bodied, short-lived staple prey species which produce relatively large litters (e.g., many insectivores, many small-bodied rodents). Systems involving spontaneous ovulation and corpus luteum development, on the other hand, generally occur in medium- to large-bodied, long-lived herbivores or carnivores which are not subject to a high predation rate and which produce relatively few infants per year (e.g., many hystricomorph rodents, primates, some carnivores). It should be added that species with a very restricted reproductive season each year would not be under any pronounced selection pressure to either accelerate or decelerate their estrous cycles.

Given the presence of mature Graafian follicles, the chances of successful fertilization would probably be increased if the time of ovulation or luteinization itself were to be fairly flexible, depending on the arrival of a suitable male and the act of mating (induced ovulation). With spontaneous ovulation and luteinization, initiation of the full ovarian cycle whether or not mating has taken place implies more sophisticated coordination of male and female behavior (social organization, but not necessarily a gregarious habit), and induced ovulation or luteinization would accordingly be the logical antecedent. It is likely that social organization—hence behavioral coordination of males and females— was rudimentary in ancestral eutherian mammals and that induced ovulation or luteinization would have increased the chances of successful fertilization. This likelihood is increased by the fact that ancestral eutherians were small-bodied and probably produced large litters (see later). Indeed, the appearance of spontaneous ovulation followed by corpus luteum development in some lines of mammalian evolution would require establishment of definite links between males and females, since any reduction in reproductive success would block the necessary evolutionary transition. Such a transition also amounts to a shift from male to female determination of the time of fertilization. Thus, one would expect to find spontaneous ovulation and luteinization only in those living mammal species which have well-established patterns of social contact between males and females or where there are special reasons for female determination of the time of fertilization. Given a short-phase estrous cycle with periodic maturation of batches of Graafian follicles (Fig. 2), the simplest and most effective system would be that in which ovulation or corpus luteum development is provoked by the courtship and mating activities of the male, asssuming some initial behavioral indicator that the female is in the appropriate phase of follicular ripening ("behavioral estrus"). With a rudimentary system of social organization, it is possible that a female would not be fertilized at all times of receptivity because of the occasional absence of a suitable male, but that the frequent maturation of follicles and dependency of ovulation or luteinization upon copulation would maximize the chances of successful fertilization whenever a suitable male was encountered. The induced mode seems to be associated with short follicular cycles in eutherian mammals generally.

Given the suggestion that the ancestral eutherian mammal was nocturnal (Young, 1962; Romer, 1966), and almost certainly adapted predominantly for olfactory perception, it is likely that signals transmitted between males and females in association with reproduction were predominantly olfactory in the ancestral eutherian stock. The estrous state of the female was probably indicated by vulval/vaginal secretions and by products excreted in the urine, while the male may have possessed olfactory characteristics signalling his suitability for mating. Olfactory indicators of female readiness for mating would seem to be virtually universal among living mammals, including diurnal species.

D. Gestation Periods

Fertilization and implantation of the blastocyst is followed by a characteristic gestation period which differs from species to species primarily according to body weight, litter size, relative brain size, and the developmental state of the young at birth. For comparative purposes, it is preferable to talk in terms of relative length of gestation, since larger mammals tend to have longer gestation periods, whatever other factors may be involved. In considering the likely gestation period of the ancestral eutherian mammal, reference must ideally be made to a standardized body size. Among various living mammals of any selected size there is considerable variation in gestation period length, and one cannot apply the concept of universal characters. The logical antecedent would imply that a short gestation period (simple embryonic development, small relative brain size) would be ancestral to a long gestation period (complex embryonic development, relatively large brain); but there is the limitation that the infants must emerge in a state compatible with postnatal survival, given the basic mammalian complement of maternal care.

In a premammalian stage lacking maternal care, any viviparous offspring would necessarily have been born in a state permitting them to survive independently. However, it has already been established (above) that maternal care involving suckling was probably present in the ancestral therian mammal stock, if not earlier. The infants could hence be born in a relatively immature state, although the degree of immaturity would have been limited by the relatively unsophisticated level of maternal care at that stage. On the other hand, the very existence of universal maternal care indicates that all eutherian mammals must have passed through a stage where the infants were heavily dependent upon additional support early in postnatal development, possibly relying entirely upon maternal milk for their nutrients for the first days or weeks after birth. Even in those living species in which the infants are relatively self-sufficient from birth onwards (e.g., in locomotor terms), at least the basic elements of maternal care (such as suckling) are retained, however briefly. Thus, it can be argued that the ancestral eutherian mammal had a relatively short gestation period, with the infants born in a relatively helpless condition (altricial state; see later).

A major factor which must be considered is that of relative brain size, since modern mammal species with relatively large brains typically have relatively long gestation periods (Portmann, 1965). Since the early eutherian mammals undoubtedly had very small brains relative to body size (Martin, 1973), this provides a further argument for postulating a relative gestation period of short duration in the ancestral eutherians, particularly since Sacher and Staffeldt (1974) have shown that the absolute length of the gestation period of placental mammals is closely correlated with the weight of the infant's brain at birth. On this basis, one can suggest three stages in the evolution of mammalian gestation periods:

1. Pretherian mammal stage, with a reduced macrolecithal egg, noninvasive, incipient placentation (ovoviviparous condition) and moderately developed maternal care. The infants were already born in a relatively immature condition, and did not move independently for some time after birth. The relative brain size was very small, and the relative gestation period (given the immature condition of the young at birth) was hence fairly short.

2. Ancestral eutherian stage, with a microlecithal egg, moderately invasive (endo-theliochorial) placentation, and a well-established pattern of maternal care involving suckling and maternal "toilet" behavior. The infants emerged in an early stage of development and were "helpless" for some time after birth. The relative brain size was still quite small and the relative gestation period was possibly somewhat longer, although increase in brain size over stage 1 was offset by somewhat earlier emergence of the infants from the uterus.

3. In the subsequent radiation of the eutherian mammals, placentation (cf. Starck, 1965) became specialized in many lines, either towards more effective utilization of uterine milk (epitheliochorial placentation) or towards more efficient diffusion between maternal and fetal blood (hemochorial placentation). All eutherian mammal groups underwent expansion in relative brain size (Martin, 1973), some more than others, and

the relative gestation period became longer in those species with the greatest development of the central nervous system. Maternal behavior became emphasized in some lines and reduced in others, where extension in gestation periods led to large-brained infants born at a stage of increased physical development. In some lines, gestation periods may actually have become shorter as a device to maximize litter size (e.g., myomorph rodents; see Portmann, 1965).

E. Development of Offspring at Birth

This leads to the question of the state of the infants at birth in the various lines of eutherian evolution. Portmann (1939, 1941, 1965) has drawn a distinction between altricial and precocial offspring among living mammals, and this dichotomy is so pronounced that the state of the offspring at birth is usually a familial and often an ordinal characteristic, although there are a small number of exceptions. We are thus confronted with a "conservative character" of great value in assessing mammalian evolution (see Starck, 1965). The following list of differences between the altricial and precocial types is generally applicable to eutherian mammals:

Altricial type	*Precocial type*
1. Adults usually construct nests, at least when dependent offspring are present. Adult body size typically small.	1. Adults do not normally construct nests at any stage. Adult body size tends to be medium or large.
2. Infants are born naked, and the ears and eyes are closed by a membrane for some time after birth. Initially, the young usually exhibit imperfect homeothermy compared to adults.	2. Infants are born with at least a moderate covering of hair, and the ears and eyes are open at birth or soon afterwards. Homeothermy is typically well-developed at birth compared to the adult condition.
3. The lower jaw is incompletely developed at birth and the middle ear is hence at an early stage of development. The teeth erupt quite late in postnatal development.	3. The lower jaw is well developed at birth and the middle ear is fairly well developed. The teeth erupt quite soon after birth, at least in small-bodied forms.
4. The gestation period is relatively short; litter size and teat count are large.	4. The gestation period is relatively long; litter size and teat count are very small.
5. Infants typically have low mobility at birth.	5. Infants typically have high mobility at birth.
6. The relative brain size of the neonate and of the adult is small, and the brain grows considerably after birth.	6. The relative brain size of the neonate and of the adult is large, and the brain usually grows only moderately after birth.
7. The adults are generally nocturnal in habits.	7. The adult tends to be diurnal in habits, though a fair number of smaller species are nocturnal.

Portmann concluded that the ancestral eutherian mammals gave birth to quite large litters of altricial offspring, thus agreeing with conclusions based on gestation periods, relative brain size, and ubiquitous eutherian maternal behavior. In modern mammals with precocial infants, the "nest phase" has (so to speak) been included in the gestation period, and the state of the offspring at birth is generally comparable with that of altricial young when first emerging from the nest. This is supported by the fact (Portmann, 1959) that in precocial eutherian mammal young the eyelids fuse over the eye *in utero* and then reopen again near the time of birth. Altricial young in the ancestral mammal stock were doubtless nest-living, at least while dependent on their mothers. This is in agreement with the observation that modern nest-living mammals with altricial young are generally of small body size, since it is widely accepted that ancestral mammals were small creatures (Young, 1962; Romer, 1966).

Having established some of the likely features of ancestral eutherian mammals, it is possible to go on to consider the reproductive characters of tree shrews and primates against the general background of mammalian adaptive radiation.

IV. Tupaiid Reproduction

The reproductive characters of tree shrews have already been discussed extensively (Martin, 1968b), and only salient features deserve repetition: A long urogenital sinus is present in all female tree shrews, while in males the descended testes occupy a weakly developed, prepenial scrotum and may be retracted into the abdomen in adults. The female exhibits a short-phase estrous cycle with a phase length in the region of 6–12 days, and there is reason to believe that ovulation or corpus luteum development is induced rather than spontaneous (Conaway and Sorenson, 1966). The apparent gestation period is relatively short (approx. 45–50 days) in all species studied (see Martin, 1968b), and the litter size is moderate (usually 2–4 infants). All tree shrews appear to inhabit nests of some kind, usually lining them with leaves or other debris, and the offspring are definitely of the altricial kind and kept in a nest for a fairly long period of time (approximately 1 month). Placentation is definitely endotheliochorial (Hill, 1965; Luckett, 1968, and this volume), contrary to earlier reports of hemochorial placentation, and there is a peculiar form of bidiscoid attachment to special placentary pads. The placentation is choriovitelline for the first third of pregnancy, with definite vascularization of the yolk sac. Chorioallantoic placentation is established just before the latter half of pregnancy, but the yolk sac nevertheless persists throughout pregnancy (Luckett, 1974a,b).

It has already been reported (Martin, 1966, 1968b) that at least one tree shrew species, *Tupaia belangeri*, exhibits an unusual system of maternal behavior in that the infants are born in a separate nest cavity which the mother visits only once every 48 h for suckling. The female does not groom her infants, and she does not carry them either for retrieval or for transport to another nest. It was suggested (Martin, 1968b) that this restricted pattern of maternal care might be typical of tree shrews at least within the subfamily Tupaiinae, but this suggestion was subsequently questioned by Sorenson (1970). Nevertheless, recent studies of *Tupaia minor* and *Lyonogale tana* (D'Souza and Martin, 1974) provide further indication that the "absentee system" of maternal care and the 48-hr suckling rhythm may well be typical of several or all tree shrew species. It is likely that Sorenson (1970) failed to observe this behavior because of different, apparently stressful laboratory conditions which suppressed the normal suckling rhythm and led to the high mortality rates among his laboratory-reared tree shrew offspring. With *Tupaia belangeri*, von Holst (1969) has elegantly demonstrated a relationship between environmental/social stress and the disruption of reproductive behavior. It is only under comparatively stress-free captive conditions that the absentee system and the 48-hr suckling rhythm are observed, and appropriate observations (D'Souza and Martin, 1974) have now shown that these features are typical of *Tupaia minor* and *Lyonogale tana* as well. This pattern of behavior is extremely unusual among mammals with altricial young, and to date comparable habits have only been found in rabbits.

V. Primate Reproduction

Turning now to the order Primates (excluding tree shrews), it is first of all necessary to establish whether there are any universal characters indicating a distinction of the ancestral primate from the ancestral eutherian mammal.

Although it is highly likely that all primates can be derived from a common ancestor still possessing a bicornuate uterus (retained in prosimians), it is a universal feature of all living primates (as with many other mammals) that the urogenital sinus has been considerably reduced in the female and persists only as a very short vestibule, if present at all. Indeed, in most female primates the urethra opens to the exterior on a prominence quite separate from the vagina. In the males, the testes are always descended into a distinctive scrotum, typically postpenial in position. Testicular descent is notably precocious and in many (perhaps all) cases the testes are descended to the region of the inguinal canal at birth. Once the testes have descended into the scrotum, constriction of the inguinal ring occurs, and the testes are never retracted back into the abdominal cavity (though retraction into the lower end of the inguinal canal may occur, e.g., in the seasonally breeding Madagascar lemurs).

In female primates generally, there is a long-phase estrous cycle (Fig. 2) with a phase length of 25 days or more, although there have been conflicting reports for one New World monkey species (*Saimiri sciureus*) indicating a shorter phase length (see Hafez 1971, for summary). In addition, primates studied to date generally seem to have spontaneous rather than induced ovulation. Estrus is typically marked by swelling and reddening of the vulval area (and perhaps adjacent skin regions), and a specific olfactory signal (pheromone) is commonly produced at such times. In small-bodied primates (galagines; cheirogaleines; *Lepilemur*) the vulva is characteristically closed externally between estrus periods and while gestation is in progress. (It is possible that *internal* sealing of the female tract may occur in larger-bodied prosimians.) If it is true, as is often assumed, that the ancestral primates were small bodied, like the ancestral mammals, then it is possible that external vulval closure was exhibited by the ancestral primates, or at least by ancestral strepsirhines. Periodic opening and closure of the vulva would have necessitated separation of the urethra from the vagina, thus obliterating the antecedent urogenital sinus. Such periodic closure of the vulva is more likely to be associated with monestrous seasonal breeding and/or a long-phase estrous cycle than with a short-phase estrous cycle and/or year-round breeding, since the possible advantages (e.g., increased vaginal hygiene?) of periodic closure would be balanced against the limitations of highly repetitive opening and closure of the vulva. There may also be a relationship with spontaneous ovulation followed by corpus luteum development, since opening of the vulva with long intervening periods of closure (i.e., a long estrous cycle phase length) would in any case require some system of control of ovulation other than the simple process of cyclical maturation of batches of Graafian follicles, and the emphasis on timing of fertilization would be shifted to the female. Loss of the urogenital sinus and periodic vulval closure similarly typifies hystricognath rodents, which also commonly exhibit spontaneous ovulation and corpus luteum development.

As already mentioned, the primates are divided sharply into two groups in terms of placentation: epitheliochorial in strepsirhines and hemochorial in haplorhines. Whatever the type of placentation, however, it is characteristic of all living primates that the yolk sac is reduced in the later stages of pregnancy or lost altogether and that (relative to body size) the gestation period is long, compared with classical nest-living mammals such as most insectivores, most rodents, and the tree shrews. The infants are always born in a condition closely similar to the precocial type defined by Portmann. Hair is always through at birth, and the eyes and ears are open at or soon after birth. There is typically one infant at birth, although some small cheirogaleines and galagines typically have 2 or 3 infants per litter and the Callitrichidae usually have dizygotic twins. The teat count, which is correlated with litter size (Schultz, 1948) is small in all primates (maximum: 3 pairs). The teeth of the infants usually begin to erupt quite early, though in larger-bodied primates (especially simians) emergence tends to be retarded. Tooth-eruption time is, in fact, a complex feature not only because of the body-size factor but also because an extended gestation period is associated (both in primates and in other mammals) with an extended period of dependence on the mother and of general postnatal development. Sexual maturity is usually achieved at a more advanced age, and the adults tend to have much longer life spans. But the fact remains that tooth eruption occurs much earlier in small-bodied primates generally than in small-bodied mammals with altricial offspring.

It is generally true, in accordance with Portmann's classification, that primates do not build or use nests, and the most common pattern is that the infants are carried from birth onwards, throughout the period of dependency, clinging to the fur of the mother or some other adult. However, among the strepsirhines several small-bodied forms (some cheirogaleines, the aye-aye, and galagines) construct leaf nests, and others (some cheirogaleines, galagines and lemurines) make use of tree hollows or bundles of epiphytes as retreats or as nests for offspring (cf. Martin, 1972a). Such nest use is unusual, since most mammal species with young of the precocial type (as defined above) tend to be of large body size and are emancipated from nests by the very fact that the infants are born at an advanced state of development.

The association between long relative gestation period and large relative brain size is confirmed in that all primates have relatively large brains, although this is considerably more marked in simians than in prosimians (Stephan, 1972; Martin, 1973). It is highly likely that the ancestral primates were distinguished from the ancestral eutherian mammals both by markedly larger brains and by longer gestation periods at any given body size.

In direct contrast to tree shrews, all living primates exhibit intensive mother–infant relationships. The infants are typically with their mothers for all or most of the day from birth onwards, and there is frequent suckling throughout the period that the infant is in contact with the mother. (The Callitrichidae exhibit a curious exception in that the father carries the twins for most of the day from birth onwards, at least in captivity, and the lorisids are unusual in exhibiting "baby-parking"; but the criterion of frequent suckling still applies.) The primate mother exhibits grooming of her offspring, and some kind of carriage is shown by all primate species. In species which utilize nests, the infants are typically carried in the mother's mouth, while species which do not use nests typically exhibit carriage on the parent's fur. Despite this close mother–infant contact and frequent suckling, the rate of postnatal development is slow in many primates, in association with the late attainment of physical and sexual maturity and the long life span. However, the rate of growth of nest-young in galagines and cheirogaleines is little different from that found in tree shrews and other nest-living mammals, and it is therefore possible that ancestral primates did not differ significantly from ancestral eutherians in this particular respect. Nevertheless, primates do seem to be rather distinctive generally in possessing a long period of lactation (relative to body size), surpassing that typical of nest-living mammals (see Buss, 1971, for data on primates).

Thus, overall there is strong evidence that the ancestral primates were distinguished from the ancestral eutherians by an increase in the relative length of the estrous cycle and of the gestation period and in the duration of postnatal development, and that this difference went hand-in-hand with an increase in relative brain size and considerable enhancement of the mother–infant bond and associated behavior.

One can now go on to consider in more detail distinctions between the strepsirhines and the haplorhines and to examine the adaptive radiation of the strepsirhines in terms of reproductive characters. It has already been established (e.g., see Luckett, 1974a,b, and this volume) that the strepsirhines and the haplorhines differ markedly in their placentation and in associated features, and it would be interesting to see whether such differences are matched by a dichotomy in reproductive features other than those directly concerned in placentation. If it is accepted that the ancestral strepsirhine was nocturnal (Martin, 1973), it was likely to have been of small body size (cf. Charles-Dominique, this volume) and hence generally comparable to modern cheirogaleines and galagines. On this basis, one could postulate a nest-living ancestral strepsirhine, perhaps capable of constructing spherical leaf nests of the kind now built by *Microcebus murinus*, *Galago demidovii*, and *Daubentonia madagascariensis*. With the haplorhines, by contrast, there is no evidence to suggest a nest-building ancestor, even if of small body size. This difference, if valid, could be correlated with the apparent early dichotomy in the habits of strepsirhines and haplorhines, in that the former were probably predominantly nocturnal, while the latter were probably predominantly diurnal (Martin, 1973). It has already been suggested (Martin, 1972a) that nest-building or use of tree hollows may be of special advantage to nocturnal mammals in that, with respect to visually oriented predators, it is more important for a nocturnal prey species to be effectively concealed from view in the daytime than for a diurnal species to be hidden from view at night. Diurnally active arboreal primates typically sleep out on branches at night; nocturnally active primates usually use leaf nests, tree hollows, or dense foliage as cover during the daytime. There is also a correlation with body size in that nocturnal primates are generally smaller than diurnal primates, and small mammals more often use nests.

The differences in placentation and nest-building between the strepsirhines and the haplorhines may be linked on the basis of three observations:

1. In primates there is a complex relationship between gestation period and maternal weight, but overall there is no distinction between strepsirhines and haplorhines in terms of relative gestation period (Fig. 3A).
2. At any given maternal weight, a haplorhine produces a neonate which is 2–3 times as heavy as that of a strepsirhine (Leutenegger, 1973; Fig. 3B).
3. For placental mammals generally (Sacher and Staffeldt, 1974), the brain weight of the neonate is closely correlated with the gestation period.

From this it follows that in a given gestation period the hemochorial placentation of the haplorhines may be necessary for the production of an infant of greater overall size than would perhaps be possible with the epitheliochorial placentation of the strepsirhines. However, it also follows that the type of placenta is not a limiting factor for fetal brain growth, since in all eutherian mammals (regardless of placenta type and maternal body size), neonatal brain weight is apparently determined predominantly by the gestation period. It is, of course, conceivable that the ancestors of the strepsirhines may have been under selection pressure to produce small offspring relative to other primates (e.g., as a result of living in an environment with low or unpredictable food availability) and that this favored the utilization of epitheliochorial rather than hemochorial placentation.

Since small-bodied strepsirhines produce much smaller infants than do haplorhines of comparable size, it is also understandable that small strepsirhine species may require breeding nests for thermoregulatory reasons. Thermoregulatory problems for the infants would evidently be more pronounced if the mother were nocturnally active and hence absent from the infants during the coldest periods of the day.

Further correlation with nocturnal habits is suggested by the fact that in lemurs and lorises there is apparently heavy dependence upon olfactory signals, even among diurnal forms (some or all Lemurinae, some or all Indriidae). The strepsirhines are, of course, characterized by the retention of the naked rhinarium (logically antecedent to the haplorhine condition), and it has already been suggested (Martin, 1973) that there is a definite link between the naked rhinarium and the olfactory Jacobson's organ, which is apparently active in all adult nocturnal strepsirhines and inactive in adult haplorhines, with the possible exception of *Tarsius* and some ceboids (see Starck, this volume). In association with this predominance of olfaction, both male and female strepsirhines often exhibit glandular developments in the genital region (scrotum; vulva), and in all female strepsirhines there is some physical association between the urethra and the conspicuous clitoris. (There is usually a urinary gutter extending some way along the clitoris, and in some cases, as in the Lorisidae, this is actually enclosed as a tube within the clitoris.) This would be a useful adaptation in increasing control over olfactory marking by means of urine trails left on fine branches (e.g., indicating the female's estrous condition). No such direct association between the urethra and the clitoris is found in the haplorhines, which are predominantly diurnal and depend far more heavily on visual perception, and in which the clitoris is in any case usually inconspicuous.

There are also a number of distinctions between strepsirhines and haplorhines in terms of reproductive morphology. The ovarian bursa typical of most eutherian mammals is still present as a well-developed structure in strepsirhines, but it is considerably reduced or absent in haplorhines (Lange, 1920; Ioannou, 1971). No explanation has yet been offered for the unusual loss of the ovarian bursa in haplorhines, which might (at first sight) be expected to lead to an increase in ovular failure in a group typified by a very small litter size. However, an observation made by Elert (1947) on the human female at the time of ovulation might provide an answer: The oviduct funnel was seen applied to the surface of the ovary and wandering over it, while the ovary itself was seen to move. It is possible that in haplorhines generally an improved system of active uptake of the egg by the oviduct funnel may have evolved to replace the containing function of the ovarian bursa. If this is the case, the reduction or loss of the bursa could represent a complex ancestral haplorhine character of importance comparable to that of various features of placentation. It should perhaps be added here that strepsirhines never exhibit menstruation at the end of the luteal phase, although haplorhines usually do so (barely noticeable in tarsiers, weakly evident in New World simians, and well marked in most Old World simians). But this difference is doubtless closely allied to the differences in placentation. The maximum teat count in living strepsirhines is normally 3 pairs, while in haplorhines it is 2 pairs, and this may indicate that ancestral strepsirhines had larger litters than ancestral haplorhines, with the ancestral simians perhaps undergoing a further reduction to one pair of teats. Thus, there do seem to be several accessory reproductive characters which support the interpretation, initially based on placentation, that there was an early dichotomy between strepsirhines and haplorhines.

The Cheirogaleinae and the Galaginae are extremely similar in many behavioral and morphological features. One interpretation of this is that both subfamilies have retained many primitive characters from a small-bodied, nocturnal strepsirhine ancestor (Charles-Dominique and Martin, 1970; Martin, 1972a). Alternatively, on the basis of a

FIG. 3. Logarithmic plot of the relationship between gestation period (G in graph A) or single neonate weight (N in graph B) and maternal body weight (M). Data from Leutenegger (1973), supplemented by additional information compiled by the author. Dotted lines are orthogonal regression lines.

Strepsirhines (black squares)—(1) *Microcebus murinus*; (2) *Cheirogaleus major*; (3) *Hapalemur griseus*; (4) *Lemur catta*; (5) *Galago demidovii*; (6) *Galago senegalensis*; (7) *Galago crassicaudatus*; (8) *Arctocebus calabarensis*; (9) *Perodicticus potto*; (10) *Nycticebus coucang*. Haplorhines (black circles)—(11) *Tarsius spectrum*; (12) *Callithrix jacchus*; (13) *Saimiri sciureus*; (14) *Cebus capuchinus*; (15) *Allouatta villosa*; (16) *Macaca mulatta*; (17) *Hylobates lar*; (18) *Pongo pygmaeus*; (19) *Pan troglodytes*; (20) *Gorilla gorilla*; (21) *Homo sapiens*.

There is no clear-cut separation between the two groups in the relationship between gestation period and maternal body weight, but haplorhines and strepsirhines are clearly distinct in terms of neonatal weights relative to maternal weight. In the latter case, the separate orthogonal regression lines for haplorhines and strepsirhines have the same basic formula ($\log N = k + 0.71 \log M$), indicating a similar allometric relationship, but with different values for the constant k.

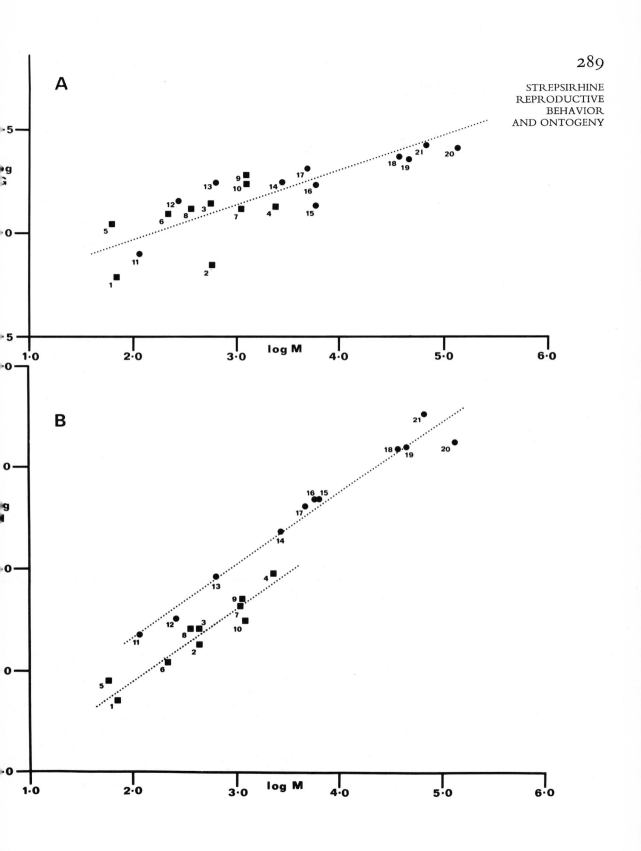

small number of shared basicranial characters interpreted as specialized (apomorph) features, Szalay (Chapter 5, this volume) and Cartmill (this volume) have suggested that the Cheirogaleinae and Galaginae are specialized offshoots from a later stock distinct from a common strepsirhine ancestor which resembled the other Malagasy lemurs in at least some of its basic characteristics. However, there is little indication of any marked divergence between lemurs and lorises in reproductive characters, since virtually all characters can be found in some representatives of both groups. This may simply be a reflection of the fact that complex reproductive characters are very conservative in evolution, and there may have been only minor changes since the lemurs diverged from the lorises. Comparison of the various reproductive characters of *Galago demidovii* (Charles-Dominique, 1972) with those of *Microcebus murinus* (Martin, 1972b) shows that these two species are remarkably alike. The only overall generalization which does seem

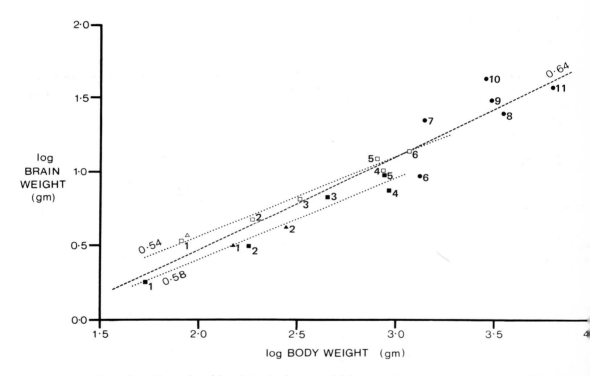

FIG. 4. Logarithmic plot of the relationship between adult brain weight (B) and adult body weight (W) in tree shrews and prosimians. The overall linear regression line (dashed), excluding tree shrews, follows the formula: log B = k + 0.64 log W. However, small-bodied nocturnal lemurs fall on a regression line (dotted) of slope 0.58 which lies below the dotted regression line for the nocturnal lorises, which has a similar slope (0.54). It can be stated that these two groups of nocturnal strepsirhines approximately follow a relationship of the form: log B = k + 0.56 log W, and that for any given body size a lorisid typically has a larger brain than a small-bodied nocturnal lemur (data from Stephan *et al.*, 1970). Small-bodied nocturnal lemurs (black squares)—(1) *Microcebus murinus*; (2) *Cheirogaleus medius*; (3) *Cheirogaleus major*; (4) *Avahi laniger*; (5) *Lepilemur ruficaudatus*. Diurnal lemurs and aye-aye (black circles)—(6) *Hapalemur simus*; (7) *Lemur fulvus*; (8) *Propithecus verreauxi*; (9) *Varecia variegatus*; (10) *Daubentonia madagascariensis*; (11) *Indri indri*. Tree shrews (black triangles)—(1) *Tupaia glis*; (2) *Urogale everetti*. Nocturnal lorisids (open squares)—(1) *Galago demidovii*; (2) *Galago senegalensis*; (3) *Loris tardigradus*; (4) *Galago crassicaudatus*; (5) *Nycticebus coucang*; (6) *Perodicticus potto*. Open triangle—*Tarsius syrichta*.

to be valid for the two groups is that, relative to body size, the gestation period of nocturnal lemurs (Fig. 3) is shorter than that of lorises (all of which are nocturnal), and this agrees with the fact that adult nocturnal lemurs have a smaller relative brain size than adult lorises (Fig. 4). Diurnal lemurs—presumably in association with the shift from nocturnal habits—have gestation periods and relative brain sizes comparable to those of lorises.

As far as reproductive morphology is concerned, it has been suggested that the lorises differ from the lemurs in that the former are characterized by enclosure of the urethra within the clitoris in females, forming a urethral canal running to the tip. However, a condition approximating this is found in cheirogaleines and *Lepilemur*. All strepsirhines seem to possess a baculum, and in male lemurs the os penis is often bifid at the tip in contrast to the lorises. All lorises studied to date apparently show extension of oogenesis into adult life, and this might be considered to be a specific lorisiform feature. The same condition is found in *Daubentonia*, however, and it seems more likely that some characteristic of the ancestral strepsirhine stock itself may be involved.

In terms of behavior, the same diffuse picture emerges. Nest-building, or at least nest use, is typical of the Cheirogaleinae, the aye-aye, and some Lemurinae among the lemurs (Martin, 1972a), and it is also exhibited by the Galaginae among the lorises. It may hence be suggested that some Lemurinae and the Indriidae have abandoned nest use in association with a shift from nocturnal to diurnal habits, while the Lorisinae have abandoned nest use in association with their slow-climbing locomotion, unfavorable for return to a fixed nest site each night. Small-bodied Galaginae carry their infants in the mouth, as do the Cheirogaleinae and some Lemurinae; the remaining lemurs and the Lorisinae always carry their babies on their fur. This difference reflects a distinction between nest use and emancipation from a nest. A further correlation exists between nest use and litter size, in that those species which have a nest often have litters of 2 or 3 young (some Cheirogaleinae, *Varecia variegata*, some Galaginae), while those that do not build nests typically have one infant (some Lemurinae; Indriidae; Lorisinae). Among the Lorisinae and Galaginae, the phenomenon of baby-parking during the night seems to be common, but among the lemurs this has only been reported for *Lepilemur* to date. It is possible that competition and higher predation risks in the habitats occupied by the lorises in Africa and Asia favor concealment of infants at night away from a definite nest area and away from any area where the mother is engaged in feeding.

It has often been suggested (e.g., Le Gros Clark, 1971) that it is typical of the "lower primates," and hence "primitive," for breeding to be seasonal, while breeding among the simians is nonseasonal. This is an over-simplification of a complex situation. It is true that all information to date indicates that all Madagascar lemurs exhibit a strict seasonal pattern of breeding (see Martin, 1972a), even in rain forest areas where seasonal climatic variation is not extreme (Fig. 5). With the lorises, on the other hand, seasonality seems to be to some extent facultative (as in simians), depending on the extent to which local climatic conditions exhibit pronounced seasonal variation. With the two galagine species which inhabit dryer zones outside the Congo/Guinea rain forest block, restricted seasonal breeding is common (*Galago senegalensis, G. crassicaudatus*); but variation can occur even within these two species, according to local conditions. Within the Congo/Guinea

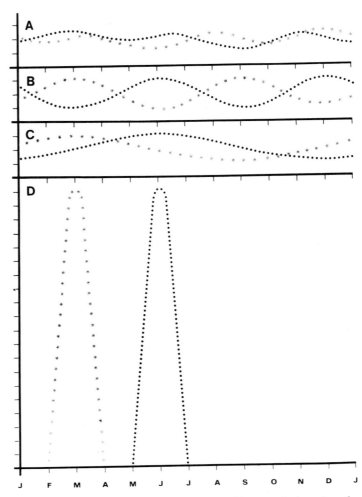

FIG. 5. Schematic illustration of varying degrees of seasonality, taking a standard number of matings (asterisks) leading to a standard number of births (heavy dots) after a gestation period of 3 months: (A) no seasonality, irregular fluctuation in breeding; (B) two annual peaks of births; (C) one annual peak of births; (D) strict seasonality, births confined to a very restricted period of the year. All lemurs exhibit type D seasonality. Among the lorisids, only the potto may exhibit such strict seasonality. Type C is found with most lorisids (in equatorial rain-forest areas), while *Galago senegalensis* and *Loris tardigradus* may exhibit type B.

rain-forest block, there is some evidence for different patterns of seasonal breeding for *G. demidovii*, *G. alleni*, *Euoticus elegantulus*, *Perodicitcus potto*, and *Arctocebus calabarensis*, ranging from strict seasonality in the potto through varying degrees of monthly variation in birth frequency in the other 4 species (Dr. P. Charles-Dominique, personal communication). Overall, there is a general distinction from the strict breeding seasonality typical of the lemurs. The same seems to apply to *Nycticebus coucang* of southeast Asia, but not to *Loris tardigradus* (see Manley, 1966; Hafez, 1971). One can summarize by saying that all lemurs are strictly seasonal in breeding, while lorises generally exhibit some signs of seasonality, the intensity usually depending on local environmental conditions (see Fig. 5). It is important to note that the timing of the breeding season seems to be determined on the basis of the annual variation in length of daylight in some lemur species, if not all. This is feasible under natural conditions because of the considerable distance between Madagascar (even its northernmost tip) and the equator, whereas the Congo/Guinea rain forest block and the region inhabited by *Nycticebus* straddle the equator. Accordingly, most lorises now occur in areas where annual variation in daylight is minimal, and perhaps unsuitable as an indicator for seasonal breeding, and where there are two rainy seasons per year as opposed to the single rainy season in Madagascar. It is possible that, for these two reasons, seasonality in the lorises typically represents a facultative response to local conditions. Most clear-cut cases of seasonality in the lorises (*G. senegalensis*, *G. crassicaudatus*, *Loris tardigradus*) concern species living some distance from the equator; but even here seasonal breeding does not seem to be automatically linked to annual variation in length of daylight.

Taking into account the fact that seasonal breeding can be correlated with vaginal closure and separation of the urethra from the vagina, it may be suggested that the ancestral strepsirhine stock lived during the Early Tertiary in an area distant from the equator and subject to marked seasonal climatic and vegetational variation, with concomitant seasonality in breeding (Fig. 6). Descendants of this stock which reached Madagascar retained marked seasonality even in rain forest areas where facultative seasonal breeding might otherwise not have appeared; the remnants of the strepsirhine stock in Africa and Asia became largely equatorial in distribution, thus becoming exposed to two annual rainy seasons and poorly defined annual variation in length of daylight, with the result that seasonality became a facultative response. (Alternatively, it might be suggested that the lemurs and lorises diverged before strict seasonality of breeding emerged.) Within the seasonal zone inhabited by the ancestral strepsirhines, special adaptations would have been necessary for survival during the dry season. It is possible, for example, that the characteristic tooth-comb of the strepsirhines may have evolved as a device for efficient gum collection in a season of low fruit availability (see Martin, 1972*a*). The suggestion that the tooth-comb may have been present in the ancestral strepsirhine stock is greatly strengthened by the observation (Charles-Dominique and Martin, unpublished data) that the timing and sequence of eruption of the teeth in young *Microcebus murinus* and *Galago demidovii* follows a virtually identical pattern. In both species, the infants are born with the milk incisors and canines through in the upper jaw and the milk tooth-comb and the anterior premolars through in the lower jaw. It is characteristic of both species that the upper and lower middle milk premolars are replaced after the upper and lower posterior milk premolars. Finally, in both species the tooth-comb, lower anterior premolars, upper incisors, upper canines, and upper anterior milk premolars are all replaced before the molars emerge, while the middle and posterior premolars are replaced after the third molar has erupted.

VI. Speculative Review

On the basis of interpretation of the distribution of reproductive characters in living mammals, it is possible to postulate ancestral conditions for the following set of stocks:

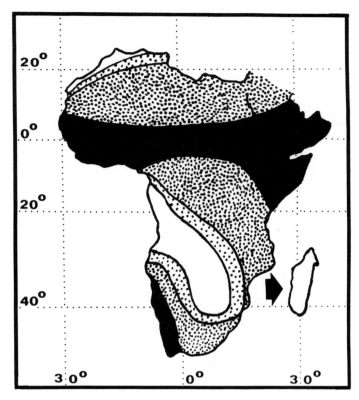

FIG. 6. Idealized climatic subdivision (ignoring surface relief) for Africa in its more southerly position during the Early Tertiary (after Robinson, 1973). The equatorial rain-forest belt (black) was further to the north than in modern times, and the ancestral strepsirhines which invaded Madagascar (arrow) possibly came from an area with humid (heavy stippling), subhumid (light stippling), or arid (white) climatic conditions and probably fairly pronounced seasonality. (Climatic subdivision of Madagascar not shown.)

ancestral eutherian mammal, ancestral primate, ancestral strepsirhine, ancestral haplorhine. Barring parallel evolution in numerous characters, it would seem that the ancestral primate stock *can* be distinguished from the ancestral placental mammal stock in terms of later reproductive adaptations, correlated to some extent with an increase in relative brain size and with a shift from an altricial to a precocial condition of the offspring at birth. By the same token, the tree shrews would not seem to derive from the later, ancestral primate stock since: (1) they have retained certain characters lost in the evolution of the ancestral primate from the ancestral eutherian, and (2) they seem to lack all features which are specific (apomorph) to the ancestral primate stock. Moreover, the diurnal tree shrews (Tupaiinae) investigated to date under relatively nonstressful conditions in captivity (D'Souza and Martin, 1974) all exhibit a specific pattern of maternal care apparently amounting to a specialization towards minimal mother–infant contact. It is difficult to conceive of the tree shrews as derived from the ancestral stock of an order which is consistently characterized—more than any other extant mammal group—by the elaboration of maternal care, juvenile dependence, and enhanced social organization.

Taking into consideration the reproductive features of all living primates, including information on placentation (Luckett, 1974*a,b*), there would seem to be a very strong case for postulating an early dichotomy between the lines leading to the extant primates: the strepsirhines and the haplorhines. In terms of phylogeny, this is an important observation. However, this does not necessarily mean that one should classify all primates into the two groups Strepsirhini and Haplorhini (following Hill, 1953). In the first place, we have no means of assessing the reproductive characters among the major early primate groups (e.g. Adapidae, Microchoerinae, Omomyinae); secondly, it should not be forgotten that classification represents an arbitrary subdivision of the primate phylogenetic tree. It is very useful, in discussing divisions between living primates, to be able to utilize the two terms "strepsirhine" and "haplorhine," as is indicated above; it has yet to be shown that these two terms are the most useful in classifying the order Primates as a whole. [See Mayr (1974) for further discussion of the distinctions between phylogeny and classification.]

As far as the strepsirhines themselves are concerned, the reproductive evidence indicates that they may have originated in an area with marked seasonality and that reproductive characters retained or developed at that stage have generally been conserved by both lemurs and lorises with only minor modifications, largely in relation to diverging habits (nocturnal or diurnal), body size, or morphological specializations (e.g., in locomotion). In several characteristics, the postulated ancestral stock of the living strepsirhines differs from the haplorhine stock through apparent separate specialization, as in the development of a close relationship between the urethra and clitoris in the female, the development of pronounced adaptations for a seasonal environment, the elaboration of nest use and nest construction, and in probable specialization towards the epitheliochorial type of placenta.

For the primates generally, including all strepsirhines, it seems likely that a basic pattern of spontaneous ovulation, vulval closure, and a long-phase estrous cycle was developed at a very early stage. As suggested above, this would imply that the time of copulation was determined predominantly by the female's behavior and that there was a fairly well-developed system of social organization (probably based on an overlapping system of male and female home ranges, with established contacts between specific males and females) which ensured that the male and the female would encounter one another at the appropriate times for copulation (see Martin, 1972*a*). Overall, the primate system of reproduction is based on the "strategy" of a small number of relatively large-brained offspring with a long life span which are well protected from predators, as opposed to the more common "strategy" of mass-production of infants by nest-living (altricial) mammals. This primate "strategy" is doubtless enhanced by elaborate maternal care, which at the same time contributes to effective socialization in preparation for socially organized adult life. Indeed, the precocious descent of the testes in infant male primates may in some way be related to early distinction between males and females in the socialization process, thus increasing the effectiveness of social organization.

This brief conclusion amply illustrates the degree to which speculation is involved in the interpretation of characters exhibited only in living primates; but hopefully there is also some indication that such speculation may be both useful and justifiable.

ACKNOWLEDGMENTS

I am grateful to the following research workers for their valuable help in discussion and/or in criticism of the draft manuscript: Dr. P. Charles-Dominique, Prof. G. A. Doyle, the late Dr. R. F. Ewer, Dr. W. P. Luckett, Dr. J.-J. Petter, Dr. A. Petter-Rousseaux, and Mrs. F. D'Souza.

The information and concepts contained in the paper were developed in the course of 10 years of research work supported at various times by the German Academic Exchange Service, the Science Research Council, the Medical Research Council, and The Royal Society (London).

VII. References

AMOROSO, E. C. 1952. Placentation, pp. 127–311. In A. S. Parkes, ed., Marshall's Physiology of Reproduction, Vol. 2. Longmans, Green & Co., London.

ASDELL, S. A. 1966. Evolutionary trends in physiology of reproduction. Symp. Zool. Soc. London 15:1–13.

BEN SHAUL, D. M. 1962. The composition of the milk of wild animals. Int. Zoo. Yearb. 4:333–342.

BUSS, D. H. 1971. Mammary glands and lactation, pp. 315–333. In E. S. E. Hafez, ed., Comparative Reproduction of Nonhuman Primates. Charles C Thomas, Springfield, Illinois.

CHARLES-DOMINIQUE, P. 1972. Ecologie et vie sociale de Galago demidovii (Fischer 1808; Prosimii). Z. Tierpsychol. Beiheft 9:7–41.

CHARLES-DOMINIQUE, P., and R. D. MARTIN. 1970. Evolution of lorises and lemurs. Nature 227:257–260.

CONAWAY, C. H. 1971. Ecological adaptation and mammalian reproduction. Biol. Reprod. 4:239–247.

CONAWAY, C. H., and M. W. SORENSON. 1966. Reproduction in tree-shrews. Symp. Zool. Soc. London 15:471–492.

COOK, C., and D. HEWETT-EMMETT. 1974. The uses of protein sequence data in systematics, pp. 939–958. In R. D. Martin, G. A. Doyle, and A. C. Walker, eds., Prosimian Biology. Duckworth, London.

D'SOUZA, F., and R. D. MARTIN. 1974. Maternal behaviour and the effects of stress in tree-shrews. Nature 251:309–311.

ELERT, R. 1947. Der Mechanismus der Eiabnahme im Laparoskop, Zbl. Gynäk. 69:38–43.

EVERETT, J. W. 1961. The mammalian female reproductive cycle and its controlling mechanisms, pp. 497–555. In W. C. Young, ed., Sex and Internal Secretions, Williams and Wilkins, Baltimore.

GLOVER, T. D., and J. B. SALE. 1968. The reproductive system of the male rock hyrax (Procavia and Heterohyrax). J. Zool. London 156:351–362.

GROSSER, O. 1909. Vergleichende Anatomie und Entwicklungsgeschichte der Eihäute und der Placenta (mit besonderer Berücksichtigung des Menschen). Wilhelm Braumüller, Vienna and Leipzig.

HAFEZ, E. S. E. 1971. Reproductive cycles, pp. 160–204. In E. S. E. Hafez, ed., Comparative Reproduction of Nonhuman Primates. Charles C Thomas, Springfield, Illinois.

HARRISON, G. A., and J. S. WEINER. 1964. Human evolution. pp. 3–98. In G. A. Harrison, J. S. Weiner, J. M. Tanner, and N. A. Barnicot, eds., Human Biology. Clarendon Press, Oxford.

HENNIG, W. 1950. Grundzüge einer Theorie der phylogenetischen Systematik. Deutscher Zentralverlag, Berlin.

HILL, J. P. 1932. The developmental history of the Primates. Phil. Trans. R. Soc. London, Ser. B. 221:45–178.

HILL, J. P. 1965. On the placentation of Tupaia. J. Zool. London 146:278–304.

HILL, W. C. O. 1953. Primates I: Strepsirhini. Edinburgh University Press, London.

HOFFSTETTER, R. 1973. Origine, compréhension et signification des taxons de rang supérieur: quelques enseignements tirés de l'histoire des mammifères. Ann. Paléontol. 59:137–169.

HUBRECHT, A. A. W. 1908. Early ontogenetic phenomena in mammals and their bearing on our interpretation of the phylogeny of the vertebrates. Q. J. Microsc. Sci. 53:1–181.

IOANNOU, J. M. 1971. Female reproductive organs, pp. 131–159. In E. S. E. Hafez, ed., Comparative Reproduction of Nonhuman Primates. Charles C Thomas, Springfield, Illinois.

KIHLSTRÖM, J. E. 1972. Period of gestation and body weight in some placental mammals. Comp. Biochem. Physiol. 43:673–679.

LANGE, D. DE. 1920. Vorläufige Mitteilung über die Beschaffenheit der Ovarialtasche von Chrysochlorus, Galeopithecus und Tupaja. Bijdr. dierkd. K. Zool. Genootsch. 22:227–232.

Le Gros Clark, W. E. 1971. *The Antecedents of Man*. Edinburgh University Press, London.

Leutenegger, W. 1973. Maternal–fetal weight relationships in primates, *Folia Primatol.* **20**:280–293.

Luckett, W. P. 1968. Morphogenesis of the placenta and fetal membranes of the tree shrews (Family Tupaiidae). *Am. J. Anat.* **123**:385–428.

Luckett, W. P. 1969. Evidence for the phylogenetic relationships of tree shrews (Family Tupaiidae) based on the placenta and foetal membranes. *J. Reprod. Fertil. Suppl.* **6**:419–433.

Luckett, W. P. 1974a. Comparative development and evolution of the placenta in primates. *Contrib. Primatol.* **3**:142–234.

Luckett, W. P. 1974b. The phylogenetic relationships of the prosimian primates: Evidence from the morphogenesis of the placenta and foetal membranes, pp. 475–488. *In* R. D. Martin, G. A. Doyle, and A. C. Walker, eds., *Prosimian Biology*. Duckworth, London.

Manley, G. H. 1966. Reproduction in lorisoid primates. *Symp. Zool. Soc. London* **15**:493–509.

Martin, R. D. 1966. Treeshrews: Unique reproductive mechanism of systematic importance. *Science* **152**:1402–1404.

Martin, R. D. 1968a. Towards a new definition of Primates. *Man* **3**:377–401.

Martin, R. D. 1968b. Reproduction and ontogeny in tree-shrews (*Tupaia belangeri*), with reference to their general behaviour and taxonomic relationships. *Z. Tierpsychol.* **25**:409–532.

Martin, R. D. 1969. The evolution of reproductive mechanisms in primates. *J. Reprod. Fertil. Suppl.* **6**:49–66.

Martin, R. D. 1972a. Adaptive radiation and behaviour of the Malagasy lemurs. *Phil. Trans. R. Soc. London, Ser. B.* **264**:295–352.

Martin, R. D. 1972b. A preliminary field study of the lesser mouse lemur (*Microcebus murinus* J. F. Miller 1777). *Z. Tierpsychol. Beiheft* **9**:43–89.

Martin, R. D. 1973. Comparative anatomy and primate systematics. *Symp. Zool. Soc. London* **33**:301–337.

Mayr, E. 1974. Cladistic analysis or cladistic classification? *Z. Zool. Syst. Evolut.-Forsch.* **12**:94–128.

Mossman, H. W. 1937. Comparative morphogenesis of the fetal membranes and accessory uterine structures. *Contrib. Embryol. Carnegie Inst. Wash.* **26**:129–246.

Perry, J. S. 1971. *The Ovarian Cycle of Mammals*. Oliver and Boyd, Edinburgh.

Portmann, A. 1939. Die Ontogenese der Säugetiere als Evolutionsproblem. *Biomorphol.* **1**:109–126.

Portmann, A. 1941. Die Tragzeit der Primaten und die Dauer der Schwangerschaft beim Menschen; ein Problem der vergleichende Biologie. *Rev. Suisse Zool.* **48**:511–518.

Portmann, A. 1959. *Einführung in die vergleichende Morphologie der Wirbeltiere*. Benno-Schwabe & Co. Verlag, Basel.

Portmann, A. 1965. Über die Evolution der Tragzeit bei Säugetieren. *Rev. Suisse Zool.* **72**:658–666.

Robinson, P. L. 1973. Palaeoclimatology and continental drift, pp. 451–476. *In* D. H. Tarling and S. K. Rumcorn, eds., *Implications of Continental Drift to the Earth Sciences*, Vol. 1. Academic Press, New York.

Romer, A. S. 1966. *Vertebrate Paleontology*, 3rd ed. University of Chicago Press, Chicago.

Romero-Herrera, A. E., H. Lehmann, K. A. Joysey, and A. E. Friday. 1973. Molecular evolution of myoglobin and the fossil record: A phylogenetic synthesis. *Nature* **246**:389–395.

Sacher, G. A., and E. F. Staffeldt. 1974. Relation of gestation time to brain weight for placental mammals: implications for the theory of vertebrate growth. *Am. Nat.* **108**:593–616.

Schultz, A. H. 1948. The number of young at birth and the number of nipples in primates. *Am. J. Phys. Anthropol. (N.S.)* **6**:1–23.

Simons, E. L. 1972. *Primate Evolution: An Introduction to Man's Place in Nature*. Macmillan Co., New York.

Simpson, G. G. 1945. The principles of classification and a classification of mammals. *Bull. Am. Mus. Nat. Hist.* **85**:1–350.

Sorenson, M. W. 1970. Behavior of tree shrews, pp. 141–194. *In* L. A. Rosenblum, ed., *Primate Behavior*, Vol. 1. Academic Press, New York.

Starck, D. 1965. *Embryologie*. Georg Thieme Verlag, Stuttgart.

Stephan, H. 1972. Evolution of primate brains: a comparative anatomical investigation, pp. 155–174. *In* R. Tuttle, ed., *The Functional and Evolutionary Biology of Primates*. Aldine-Atherton, Chicago.

Stephan, H., R. Bauchot, and O. J. Andy. 1970. Data on size of the brain and of various brain parts in insectivores and primates, pp. 289–297. *In* C. R. Noback and W. Montagna, eds., *The Primate Brain*. Appleton-Century-Crofts, New York.

von Holst, D. 1969. Sozialer Stress bei Tupajas (*Tupaia belangeri*). *Z. vgl. Physiol.* **63**:1–58.

Weir, B. J., and I. W. Rowlands. 1973. Reproductive strategies in mammals. *Ann. Rev. Ecol. Syst.* **4**:139–163.

Young, J. Z. 1962. *The Life of the Vertebrates*. Clarendon Press, Oxford.

13

Relationships among the Malagasy Lemurs

The Craniodental Evidence

IAN TATTERSALL AND JEFFREY H. SCHWARTZ

I. Introduction

It has been very pertinently remarked by Charles-Dominique and Martin (1970) that comparisons of the Malagasy prosimians with other primates have generally been restricted to *Lemur*, or, more rarely, to other members of Lemurinae. *Lemur* has, in fact, become the stereotype of the group which, for no better reason than its restriction to Madagascar, has been regarded as the result of a single adaptive radiation; this supposition has, in turn, apparently discouraged the performance of an overall comparative analysis of the lemurs. Even those recent publications (e.g., Charles-Dominique and Martin, 1970; Tattersall, 1973*a,b*), which have hinted that the lemurs might form a less homogeneous group than traditionally supposed, have done little more than merely to make the suggestion.

In this paper the results of a preliminary comparative analysis of the craniodental morphology of the lemurs are presented. Much of the detailed comparative information

IAN TATTERSALL · Department of Anthropology, The American Museum of Natural History, New York, New York.

JEFFREY H. SCHWARTZ · Department of Anthropology, University of Pittsburgh, Pittsburgh, Pennsylvania.

300

IAN
TATTERSALL
AND
JEFFREY H.
SCHWARTZ

necessary for such a study is unavailable at present, so the theory of relationships given in Fig. 5 can be no more than tentative. Considerations of space preclude the presentation of the lengthy comparative descriptions on which this analysis is based; a fuller discussion of cranial morphology will be found in Tattersall and Schwartz (1974).

It should be noted that we take "lemur" to be synonymous with "Malagasy primate"; we use it, therefore, as a combination taxonomic/zoogeographic, rather than as a purely taxonomic, term. In the initial discussion we follow generally accepted terminology, and speak of the Cheirogaleinae (*Cheirogaleus, Microcebus, Phaner*); Lemurinae (*Lemur, Varecia, Hapalemur, Lepilemur*); Indriinae (*Indri, Propithecus, Mesopropithecus, Avahi*); Archaeolemurinae (*Archaeolemur, Hadropithecus*); and Palaeopropithecinae (*Palaeopropithecus, Archaeoindris*).

II. Methods

The methodology primarily followed in formulating the theory of lemur relationships expressed in Fig. 5 is that generally described as "cladistic." Basic to this procedure is the ability to discriminate between primitive and derived character states. In any group of related organisms a spectrum of variation (morphocline), continuous or otherwise, is generally found in the state of any given character. One of these states will usually represent the primitive condition, i.e., that state of the character which was present in the common ancestor of the group as a whole. All deviations from this state may be regarded as derived.

The distinction between primitiveness and derivedness introduces the possibility of forming what Hennig (1966) has called "sister groups," i.e., those pairs of taxa, at any

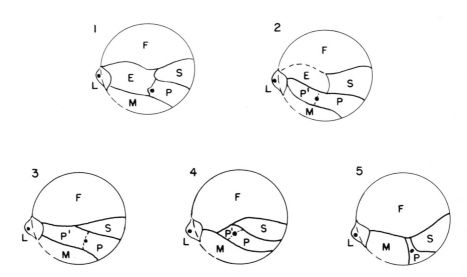

FIG. 1. Schema of the arrangement of the bones in the medial orbital wall of extant lemurs. 1, Lorisidae; 2, *Microcebus* and *Cheirogaleus*; 3, Lemuridae; 4, Indriidae; 5, *Daubentonia*. (1–4 modified from Kollman, 1925.

level of the taxonomic hierarchy, which are most closely related to each other by virtue of recency of common ancestry. Two taxa, A and B, may be held to form a sister group no matter how many taxa more closely related to A than B is are unknown.

Since primitive character states may be shared by organisms which are only remotely related, it follows that sister groups can be recognized only on the basis of uniquely shared derived character states. How, then, do we determine that a given character state is primitive or derived? Or, in other words, how do we infer the polarity of the morphocline in a particular characteristic? Even where (as is only barely the case for the Malagasy lemurs) a pertinent fossil record exists, we cannot turn to it, since there is no *necessary* relationship between the stratigraphic age of an organism and the primitiveness or derivedness of aspects of its morphology. We are, then, confined to comparative or morphogenetic information. Few details of morphogenesis are known for the lemurs, which leaves us with comparative morphology.

The criterion adhered to here lies in the communality of possession of character states. A character state typical of a large number of the taxa within a sister group, and which, in particular, is shared with related taxa of approximately equivalent rank, may reasonably be inferred to be primitive for that group. Among the lemurs a variety of factors occasionally impede definitive employment of this criterion, but in general it may be held to be reliable.

The transition from the analysis of single characters to the determination of sister groups, however, introduces complications. Usually it is found that a taxon appears to be allied with certain taxa in one set of characteristics, yet allied with others when different characters are considered. Such situations result from parallelism, and it is clear that among the lemurs parallelism has been as common an occurrence as among any other group of mammals. The provisional theory of relationships put forward here (Fig. 5) has been arrived at by the assessment of a considerable variety of derived character states and the relative probabilities involved; several alternative hypotheses, many of them more familiar to primatologists than the one proposed, were rejected in the process.

After an initial sister group A has been defined, it is compared as a whole with the other taxa under consideration. Sister group B is thus determined exactly as was A, i.e., on the basis of unique derived characteristics; it may be another species, or, indeed, any or all of the other taxa under consideration. But to say that sister group A, already a combination of taxa, is being compared with other taxa, is really to say that a theoretical construct is being compared. This construct corresponds to the morphotype of the hypothetical common ancestor of the group; the morphotype thus consists of the primitive states for the group in all characters under consideration. These states may, however, be derived with respect to the common ancestor of the next larger grouping. We have chosen here to discuss the phylogenetic relationships of the Malagasy lemurs within the context, where possible, of the construction of ancestral morphotypes for the various sister groups shown in Fig. 5. This has been done largely because it is the most economical way of expressing the distribution among the lemurs of primitive and derived character states. States considered primitive for more inclusive groupings are generally not listed subsequently where they persist in the ancestral morphotypes of subgroups of these larger assemblages.

Ian
Tattersall
and
Jeffrey H.
Schwartz

III. Cranial and Dental Evidence

Recent surveys of prosimian relationships, none of which have treated the question in much detail, have generally focused on temporal arterial circulation as a decisive factor in deciphering the phylogeny of the lower primates. It is generally agreed that "lemurine type" circulation, in which the stapedial artery is large and the anterior carotid missing, is primitive for modern Prosimii as a whole (e.g., Clark, 1971; Szalay and Katz, 1973), and this is certainly reasonable on grounds of communality of possession. Equally, the condition shared by the cheirogaleines and lorisids, in which an enlarged ascending pharyngeal artery is present, has plausibly been viewed as an homologous derived character (Szalay and Katz, 1973; Szalay, Chapter 5, this volume; Cartmill, this volume). The conclusion drawn from this, that the cheirogaleines form a sister group of the lorisids, appears plausible. However, that the remaining Malagasy primates form an homogeneous whole, at least with respect to the former, demands thorough examination. Common possession of a uniform, primitive basicranial morphology by the Malagasy lemurs reveals nothing more than that such a character was present in the common ancestor of the group as a whole. It tells us nothing about phylogenetic relationships within that group. In discussing lemur relationships, then, we must turn to the consideration of other characters.

The common ancestor of the indriines, palaeopropithecines, archaeolemurines, and *Daubentonia* (for want, at this stage, of a better term, the "*Indri* group") possessed a relatively globular braincase with a short facial skeleton tucked somewhat beneath it. The orbits were widely separated and fairly well frontated; in the medial orbital wall the frontal was in broad contact with the maxilla (Fig. 1). This latter bone was robust and relatively deep. The corpus of the mandible was deep and laterally narrow: the unfused symphysis was long and oblique with a well-marked genial fossa and the gonial angle expanded and well rounded-out. A large digastric was present, reflecting itself in an excavated attachment area on the medial aspect of the horizontal ramus and in the presence of a salient paroccipital process. The considerably raised condyle possessed an extended, transversely convex articular surface, and reposed in a deep glenoid fossa. The nasal area was high and relatively broad. An inflated bulla was present: the tympanic ring and carotid circulation were as in the indriines. The dental comb possessed but two teeth, bilaterally; these were perhaps less slender and procumbent than those of the indriines. Three premolars were present in the upper and lower jaws. The molar teeth were not dissimilar to those of *Propithecus*; M^1 and, more so, M^2 were squarish, with a well-marked cusp at each corner and more buccal elaboration than seen in *Indri*; M^3 was considerably reduced. The lower molars possessed 4 main cusps, the lingual pair more salient than the buccal; a small but distinct paraconid was present in addition, while M_3 also possessed a differentiated, if small, hypoconulid. In both their development and their eruption, initial and complete, the premolars appeared in the sequence $P^4_4 \, P^3_3 \, P^2_2$, and the upper canine appeared centrally in the total sequence. Three deciduous, as well as 3 permanent, premolars were present. dP_4 was a 4-cusped molariform tooth, but was triangular in outline because the anterior cusp pair was more closely approximated than the posterior.

The hypothetical morphotype of the common ancestor of the other lemurs and the lorisids ("lemur/lorisid group"), taken together, differs strongly from that just described for the *Indri* group in almost all characters beyond the simple number of nonincisor teeth and the condition of the ear region. This ancestor possessed a relatively (not extremely) long facial skeleton, situated anterior to, rather than beneath, the less-inflated neurocranium. The horizontal ramus of the mandible was shallow and slender, its ovoid symphysis unfused. The condyles, wide transversely but restricted anteroposteriorly and lacking transverse convexity, were raised above the level of the tooth row, but not to a very considerable extent; the glenoid fossa was not excavated. The gracile postglenoid process was not confluent with the bulla. The mandibular angle was hooked, rather than inflated, with very little ventral extension. The nasal region lacked a high profile. The auditory bulla was well inflated, and the tympanic ring lay inside it, angled medially. The carotid foramen was lateral; a large stapedial artery was present, and the anterior carotid was lacking. The paroccipital process was very little pronounced, and the orbits were moderately frontated and convergent. The premaxillary and nasal bones did not project anteriorly beyond the alveolar margin; the posterior palatine foramen was formed as in *Microcebus*. The upper incisor teeth were probably high-crowned and slender, and the canine quite long and robust. The dental comb possessed 6 stout teeth, procumbent but not pronouncedly so. Both upper and lower anterior premolars were robust, somewhat caniniform teeth, noticeably more salient than those posterior to them. P^4 was probably bicuspid, the buccal cusp exceeding the lingual in size. M^1 and M^2 bore 3 principal cusps, and were cingulated lingually: the protocone was symmetrically placed with respect to the higher buccal cusps. M^3 was small and tricuspid. M_1 and M_2 were subequal in size, bearing a small entoconid and a paraconid shelf. A hypoconulid was present in the small M_3. The deciduous teeth resembled their permanent successors, except in the case of dP_4^4, which were molariform. The teeth of the dental comb were the first to appear in both the deciduous and permanent dentitions, although canine eruption was delayed relative to that of the incisors. Development and eruption of the upper canine were closely associated with those of P_2^2, and both permanent and deciduous premolars developed and erupted in the sequence $P_2^2 P_4^4 P_3^3$. The last 3 teeth of the permanent set to develop and erupt were $M_3^3 P_4^4 P_3^3$.

The common ancestor of the cheirogaleines and lorisids possessed an inflated auditory bulla enclosing a free tympanic ring. An anterior carotid foramen was present, reflecting the presence of an anterior carotid artery. The promontory artery was larger than the stapedial branch; the internal carotid penetrated the bulla posteromedially rather than posterolaterally. The postglengoid process was fused medially to the bulla; in the nasal fossa the large first endoturbinal descended to overlap the maxilloturbinal. The dental comb was probably more procumbent and less robust than that of the common ancestor of the lemur/loris group as a whole. P_2 was inclined forwards. The hypocone on M^1 and M^2 was well differentiated, and situated posterolingually; it was accentuated by waisting of the posterior margin of the tooth. The paracone was conical, while the metacone was compressed about an oblique axis. A distinct entoconid was present in the lower molars, and the paraconid shelf was reduced.

It should probably be noted that this hypothetical common ancestor is likely to have

304

IAN
TATTERSALL
AND
JEFFREY H.
SCHWARTZ

been remote in time; thus, although attention has recently been drawn by Charles-Dominique and Martin (1970) to the similarities between *Microcebus murinus* and *Galago demidovii*, their differences are equally striking, despite the fact that the two taxa share similarities in size and life-style.

As a group, the cheirogaleines are distinguished from the lorisids largely by the retention of the primitive state in a variety of characteristics which in the lorisid common ancestor were derived. Such derived states include a somewhat higher vaulting of the neurocranium, mastoid inflation posterior to the bulla, loss of the stapedial artery, reduction of the posterior palatine foramen, loss of the prenasopalatine lamina in the medial orbital wall, fusion of the tympanic ring to the lateral bullar wall, some filling out of the mandibular angle, obliquity of the mesiodistal axis of the upper canine, and the molarization of P^4. Derived character states in the ancestral cheirogaleine include the lowering of cusp relief in the molar teeth, and the enlargement of the occlusal surfaces of the upper incisors, particularly I^1. The tooth-comb was more gracile and procumbent, and M_1^1 were the first teeth of the permanent sequence to erupt (a condition apparently independently derived in the galagines).

Within Cheirogaleinae, *Phaner* is distinct from the common ancestor of *Microcebus* and *Cheirogaleus* (as indeed from both modern genera), in possessing the following derived character states: stapedial artery lacking; mandibular condyles compressed laterally; hypoconulid absent from M_3; mandibular angle less hooked; cranial roof vaulted (Fig. 2); dental comb more slender and procumbent; P_2 more vertical; I^1 enlarged and forwardly projecting; last permanent teeth to develop and erupt: $P_4^4 M_3^3 P_3^3$.

The common ancestor of Lemurinae and *Megaladapis* remained quite close to that of the lemur/loris group. The dental comb was, however, more gracile and procumbent; both upper incisors were reduced in height but broadened mesiodistally; the toothcomb was the earliest of the permanent series in both development and eruption; $\underline{C}$ P_2^2 developed as a unit, but the eruption of $\underline{C}$ was delayed; M_2^2 developed after this unit, but erupted before the canine. The protocone of the upper molars was large and somewhat anteriorly placed; the posterior palatine foramen bounded the entrance of a short bony canal. The lateral pterygoid lamina was fused posteriorly to the bulla.

Somewhat more change is evident in the later evolutionary stages of this group. The common ancestor of *Lemur* and *Varecia* (Fig. 2) possessed an elongate facial skeleton, parallel temporal lines, and ventrally protruding medial pterygoid laminae. P^2 was small; P^3 was the most salient upper premolar. The lingual moieties of the upper molars were shifted anteriorly; in P^4 a small internal cusp was closely approximated to the protoconid, and large anterior and posterior foveae were present. M_2 lacked a hypoconulid. The upper deciduous premolars erupted in the sequence dP^2 dP^3 dP^4.

The ancestral morphotype of *Hapalemur*, *Lepilemur*, and *Megaladapis* (Fig. 2) can be characterized as possessing somewhat downwardly curving nasal bones and a facial skeleton somewhat longer than that found in either of the living genera. Slight mastoid inflation was present, and the temporal lines converged posteriorly. The postglenoid

FIG. 2. Crania of the living Malagasy lemurs in lateral view. A, *Cheirogaleus major* (subadult); B, *Microcebus murinus*; C, *Phaner furcifer*; D, *Lepilemur mustelinus*; E, *Hapalemur griseus*; F, *Lemur rubriventer*; G, *Varecia variegatus*; H, *Propithecus verreauxi*; I, *Avahi laniger*; J, *Indri indri*. Drawings are not to scale.

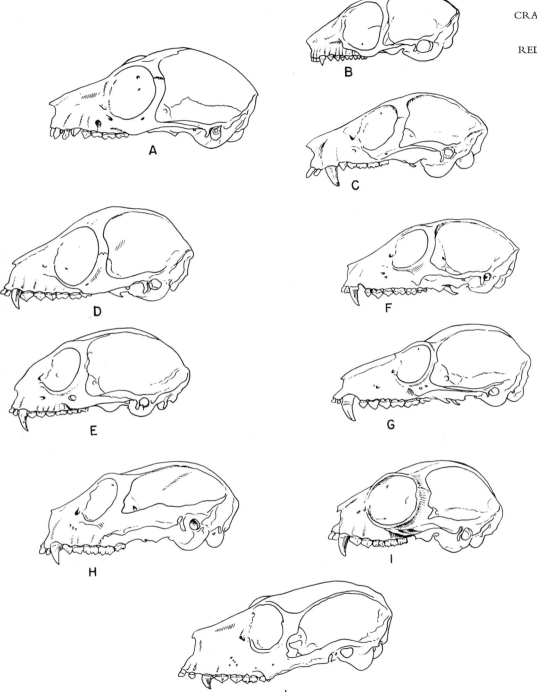

306

IAN
TATTERSALL
AND
JEFFREY H.
SCHWARTZ

process was fused medially to the bulla and was quite long dorsoventrally. The pre-maxilla was short anteroposteriorly; the small I^2 was masked from the side by the canine. In the upper molars, the paracone and metacone were connected by a ridge; the buccal margin of these teeth was quite pronounced. In the lower molars the trigonid basin opened internally. The mandibular angle probably protruded somewhat ventrally and was slightly flexed medially. The premolars developed in the sequence $P^2_2\ P^4_4\ P^3_3$ (i.e., showed the primitive pattern), but erupted in the order $P^4_4\ P^3_3\ P^2_2$.

The common ancestor of *Lepilemur* and *Megaladapis* exhibited a peculiar condition of the temporomandibular joint wherein a secondary articular surface ran perpendicularly down from the posterior aspect of the medial moiety of the condyle (Fig. 3). The stapedial artery was greatly reduced. No upper incisors were present in the permanent dentition. The upper molars were characterized by a posterointernal cingulum, lingual buttressing of the metacone, and the presence of a parastyle. On all lower molars a long, well-defined paracristid and elongated anterior fovea were present but there was no metastylid. M_3 possessed an elongated heel supporting a very substantial hypoconulid. M^1_1 and M^2_2 were the earliest permanent teeth both to develop and to erupt.

Relationships within the *Indri* group are more difficult to discuss within the context of ancestral morphotypes. This is largely because, however these relationships are viewed, the common ancestor of each sister group within this array of related forms remained remarkably primitive in almost all aspects of craniodental morphology. In other words, hardly any of the derived character states to have arisen within the *Indri* group are shared at any level beyond the generic.

The ancestral morphotype of the *Indri* group, it will be recalled, is regarded here as possessing 3 deciduous premolars, replaced by 3 permanent teeth, the latter developing and erupting after M^3_3. In the indriines and palaeopropithecines one of the permanent premolars is lost. In both the indriines and *Daubentonia*, however, the deciduous dentition

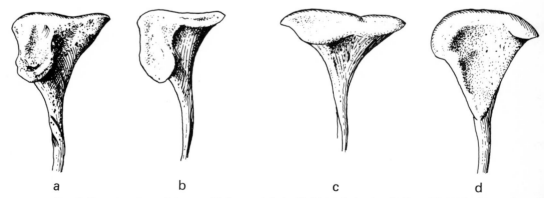

FIG. 3. Posterior view of the mandibular condyle in (a) *Megaladapis edwardsi* (AMNH 3025); (b) *Lepilemur mustelinus* (AMNH 100643); (c) *Lemur fulvus* (AMNH 100528); (d) *Propithecus verreauxi* (AMNH 170473). Right side; not to scale.

shows a highly derived condition, which we consider unlikely to have evolved in parallel within each group. The indriines possess 4 deciduous premolar teeth arranged as in Fig. 4. Conventional reasoning, under which tooth loss is generally regarded as derived, would suggest that 4 premolars represent the primitive condition. However, the recent work of Osborn (1973), on the mechanisms of dental development, supports the suggestion that this state is in fact derived within the *Indri* group.

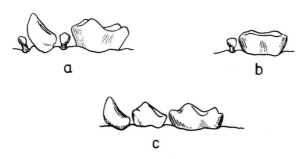

FIG. 4. The lower deciduous premolars in (a) indriines; (b) *Daubentonia*; (c) primitive type with three deciduous premolars.

Tooth budding is initiated by neural crest cells migrating from the neural tube into the jaws. Deciduous premolar budding in mammals is initiated posteroanteriorly, its tempo governed by the rate at which the jaw itself is growing. In addition, the morphology of the deciduous premolars generally decreases in complexity from back to front; this is because the initiator of a tooth family is carried forward with the increase in jaw size and stimulates the development of successively simpler teeth. In the lorisids, lemurids, and archaeolemurines, for example, one sees this morphological gradient in the deciduous premolars in the molariform dP_4, the caniniform dP_2, and the intermediate dP_3.

In the indriines, on the other hand, although dP_4 is molariform, it is not preceded by a merely slightly less complex tooth, but by a markedly reduced, almost rudimentary, dP_3 (Fig. 4). By comparison, dP_2 is much more complex than dP_3 and, like dP_4, is preceded by a small peglike tooth. This is not what one would expect to find if the neural crest cells were invading the developing lower jaw in a primitive single-wave fashion. However, it is what would occur if the first wave were rapidly followed by a second invasion of neural crest cells. The first wave would initiate a molariform dP_4; but the field of inhibition created by the encroaching second wave would cause dP_3 to be diminished in size. The fields of inhibition created by the follicular growth of dP_4 and dP_3 would, in turn, impede the morphogenesis of dP_2; the caniniformity of this tooth would be further enhanced by its incorporation into the canine field (cf. Butler, 1939, 1963). By this time interstitial growth would be minimal, and the total inhibitory effect would demand a diminished dP_1. A mechanism of this type can obviously not be considered primitive when compared to the single-wave-initiated deciduous premolar gradient of other prosimians.

308

I̲a̲n̲
T̲a̲t̲t̲e̲r̲s̲a̲l̲l̲
A̲n̲d̲
J̲e̲f̲f̲r̲e̲y̲ H̲.
S̲c̲h̲w̲a̲r̲t̲z̲

In *Daubentonia* the morphology of the 2 remaining deciduous premolars is extremely close to that of dP_3 and dP_4 in the indriines (Fig. 4). The development of these premolars is evidently initiated by the homolog of the first wave of neural crest cells in the indriines. The effects of the second wave, however, are cancelled by the field of inhibition created by the proximity of the root of the enlarged lower canine, and the appearance of the two anterior lower deciduous premolars is forestalled. It thus appears that the common ancestor of Indriinae and *Daubentonia* already possessed the indriine-type mechanism of lower deciduous premolar development, and that this shared character of the two taxa is derived with respect to all other members of the *Indri* group. A similarly shared derived character lies in the eruption of P^4 before M_3^3.

In almost all cranial characteristics the archaeolemurines (*Archaeolemur majori* especially) are very close to Indriinae (particularly *Propithecus*). Yet in their dentition they are far more divergent from the latter, in both tooth number and morphology, than are the palaeopropithecines. It is possible that the loss of a premolar in each quadrant of the dentition may have taken place independently, and through the operation of different mechanisms, in Indriinae and Palaeopropithecinae. However, in the absence of sufficient information on dental development and eruption in the latter, this cannot be demonstrated; we are left, then, with an obviously derived character, shared by both taxa.

The most striking departure in the palaeopropithecine skull from the indriine condition lies in the structure of the bony ear; however, the sharing between Indriinae and Archaeolemurinae of the same structure of this region is the sharing of a primitive character, and thus at this level is of no taxonomic value. Again, the significance of the reminiscence in form between dP_4 in *Archaeolemur*, *Daubentonia*, and Indriinae is impossible to assess in this context because this tooth is unknown in both *Palaeopropithecus* and *Archaeoindris*.

In details of cranial structure, despite disparities in size, the indriines remain an extraordinarily coherent unit. Nevertheless, *Propithecus*, *Mesopropithecus*, and *Avahi* retain the primitive short face and mandible, while the splanchnocranium and lower jaw of *Indri* are substantially longer (Fig. 2). The corresponding elongation of the neurocranium in *Indri* is a direct result of the functional demands on the masticatory musculature of the longer face. *Indri* also departs from the primitive condition in possessing less complex M^1 and M^2, with distinct reduction of buccal elaboration; in showing better cuspal differentiation on M_3^3; in developing and erupting all the anterior teeth before M_1^1; in having mesiodistally long, relatively low-crowned lateral upper incisors; and (probably) in possessing a fifth endoturbinal. *Avahi* is distinguished from the primitive pattern largely in those features related to the increase in size of its orbits. In addition, however, its nasals do not contact the premaxillae, but jut forward over the nasal aperture; the carotid foramen is less laterally placed; the upper incisors are high crowned but thin, I^2 exceeding I^1 in size; the canine is the last tooth to develop; I^1 develops and erupts before M_1^1; and lateral mastoid inflation is present. *Propithecus*, as far as we have been able to determine, remains very close to the primitive conformation in almost all aspects of cranial and dental morphology.

Since *Indri* and *Avahi* are divergent in distinct ways from a common ancestor morphologically reminiscent of *Propithecus*, it is impossible, on the basis of the evidence used here, to establish unequivocal sister groups within Indriinae, except in the case of *Propithecus* and *Mesopropithecus*.

The dotted lines in Fig. 5 demonstrate the extent to which precise relationships are not determinable, from the evidence considered here, on purely cladistic grounds. Thus we have made no attempt to characterize the common ancestor of the Malagasy primates as a whole, and have left uncertain the relationship shown in Fig. 5 between the *Indri* group and the other forms. This is largely because the two groups are not directly comparable in most of the characters used in the analysis, although the lemur/loris morphotype doubtless represents the primitive condition in most, if not all, of the characters under consideration.

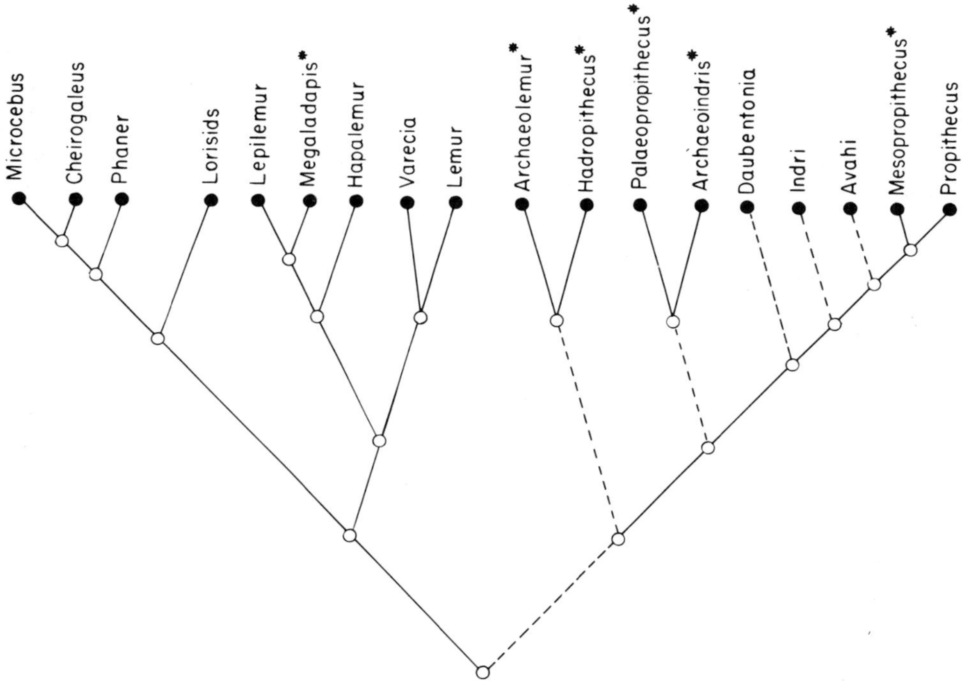

FIG. 5. Provisional theory of relationships among the Malagasy lemurs. Asterisks denote extinct taxa.

The major problem in determining the relationships between the three major groups defined here is that no pair of them is unequivocally united by the sharing of derived characters. What evidence there is, however, suggests that the lemur/loris split occurred subsequent to the divergence of the *Indri* group from the common ancestor. Thus we suggest that the overall common ancestor was characterized by M_2^2 which developed and erupted subsequent to the $\underline{C}$–P_2^2 functional unit, and by premolars which developed and erupted in the order P_2^2 P_4^4 P_3^3. The primitiveness of this latter sequence is suggested not only by comparison within Primates but also by the data presented by Osborn (1970, and references therein), Slaughter *et al.* (1974), and West (1972) on living and extinct insectivores. The lemur/loris ancestor retained the primitive features of dental

310

IAN
TATTERSALL
AND
JEFFREY H.
SCHWARTZ

development and eruption. In the ancestor of the *Indri* group, on the other hand, the coordinated appearance of $\underline{C}$ and $P\frac{2}{2}$ was lost, and the premolars developed and erupted in a posteroanterior sequence.

IV. Classification of the Malagasy Lemurs

In classifying the lemurs we would adhere to the proposition that a classification is *not* a phylogeny. Classifications may be constructed to reflect a view of phylogeny (Fig. 5), but, equally, according to the purposes of the classifier, they may reflect purely patristic affinities, or any combination of the two types of information. In this sense, no classification may absolutely be stated to be right or wrong, at least on the basis of the criteria used. Our preference is for a classification based on phylogeny, but phylogeny more broadly interpreted than in the cladistic sense. Phylogeny, which, quite simply, means evolutionary history, consists of much more than mere branching sequences. We believe that a classification should not be inconsistent with branching sequences, but, rather than being strictly limited to the constructs defined by monophyly, that it should also reflect such phenomena as adaptive divergence. This generally involves raising the status of highly divergent taxa to the rank of the next larger sister group (but not to a higher rank, for this would introduce inconsistency with branching sequences). Given this general approach, there are two ways in which the Malagasy lemurs may be classified on the basis of the evidence available to us. Since evolutionary divergence is taken into account, either arrangement would hold whether or not the *Indri* group is cladistically more distant from the lemur and loris groups than either is from the other.

In order to preserve the conventional taxonomic ranks of the lemur and loris groups, the major divisions in the Malagasy primate fauna would have to be established at the infraordinal level. The resulting classification would be as follows:

Infraorder Lemuriformes Gregory, 1915
   Superfamily Lemuroidea Gill, 1872
      Family Lemuridae Gray, 1821
         Subfamily Lemurinae Mivart, 1864
         Subfamily Lepilemurinae★ Rumpler and Rakotosamimanana, 1971
      †Subfamily Megaladapinae Major, 1894
Infraorder Lorisiformes Gregory, 1915
      Family Cheirogaleidae Gregory, 1915
         Tribe Cheirogaleini‡ Gregory, 1915
         Tribe Phanerini Rumpler and Rakotosamimanana, 1971
Infraorder Indriiformes Tattersall and Schwartz, 1974
      Family Indriidae Burnett, 1828
         Subfamily Indriinae Burnett, 1828
      †Subfamily Palaeopropithecinae Tattersall, 1973
      †Subfamily Archaeolemurinae Standing, 1908
      Family Daubentoniidae Gray, 1870

★ Includes *Lepilemur* and *Hapalemur*.   † Denotes extinct taxon.   ‡ Includes *Cheirogaleus* and *Microcebus*.

The separation at the infraordinal level of Lemuriformes, Lorisiformes, and Indriiformes, while least disturbing the presently accepted classification, also expresses what one may take to be the long separate evolutionary histories of the groups involved. In terms of classificatory balance within the order Primates as a whole, this scheme would best fit with the establishment of a basic strepsirhine/haplorhine dichotomy. But although we have elsewhere expressed a preference for such a classification (Tattersall and Schwartz, 1974), an alternative arrangement, which could better fit the division of living primates into the two suborders Prosimii and Anthropoidea, and which would, further, emphasize the basic unity of the living lower primates, would be equally acceptable:

Infraorder Lemuriformes Gregory, 1915
    Superfamily Lemuroidea Gill, 1872
        Family Lemuridae Gray, 1870
            Subfamily Lemurinae Mivart, 1864
            Subfamily Lepilemurinae★ Rumpler and Rakotosamimanana, 1971
            †Subfamily Megaladapinae Major, 1894
    Superfamily Lorisoidea Gray, 1821
        Family Cheirogaleidae Gregory, 1915
            Tribe Cheirogaleini‡ Gregory, 1915
            Tribe Phanerini Rumpler and Rakotosamimanana, 1971
    Superfamily Indrioidea Burnett, 1828
        Family Indriidae Burnett, 1828
            Subfamily Indriinae Burnett, 1828
            †Subfamily Palaeopropithecinae Tattersall, 1973
            †Subfamily Archaeolemurinae Standing, 1908
        Family Daubentoniidae Gray, 1870

ACKNOWLEDGMENTS

We thank Mr. Nicholas Amorosi, American Museum of Natural History, for preparing the illustrations; the manuscript was typed by Ms. Ginger LoAlbo.

V. References

BUTLER, P. M. 1939. Studies of the mammalian dentition. Differentiation of the post-canine dentition. *Proc. Zool. Soc. London, Ser. B* **109**:1–36.

BUTLER, P. M. 1963. Tooth morphology and primate evolution, pp. 1–14. *In* D. Brothwell, ed., *Dental Anthropology*. Pergamon Press, London.

CHARLES-DOMINIQUE, P., and R. D. MARTIN. 1970. Evolution of lorises and lemurs. *Nature* **22**:257–260.

CLARK, W. E. LEGROS. 1971. *The Antecedents of Man*, 3rd ed. Edinburgh University Press, Edinburgh.

HENNIG, W. 1966. *Phylogenetic Systematics*. University of Illinois Press, Urbana.

KOLLMAN, M. 1925. Etudes sur les lemuriens: la fosse orbito-temporale et l'os planum. *Mem. Soc. Linn. Normandie* (*Zool.*), N.S. **1**:3–20.

★ Includes *Lepilemur* and *Hapalemur*.   † Denotes extinct taxon.   ‡ Includes *Cheirogaleus* and *Microebus*.

312

IAN
TATTERSALL
AND
JEFFREY H.
SCHWARTZ

OSBORN, J. W. 1970. New approach to Zahnreihen. *Nature* **225**:343–346.

OSBORN, J. W. 1973. The evolution of dentitions. *Am. Sci.* **61**:548–559.

SLAUGHTER, B., R. H. PINE, and N. E. PINE. 1974. Eruption of cheek teeth in Insectivora and Carnivora. *J. Mammal.* **55**:115–125.

SZALAY, F. S., and C. KATZ. 1973. Phylogeny of lemurs, galagos and lorises. *Folia Primat.* **19**: 88–103.

TATTERSALL, I. 1973*a*. Cranial anatomy of Archaeolemurinae (Lemuroidea: Primates). *Anthropol. Pap. Am. Mus. Nat. Hist.* **52**(1):1–110.

TATTERSALL, I. 1973*b*. Subfossil lemurs and the "adaptive radiation" of the Malagasy lemurs. *Trans. N.Y. Acad. Sci., Ser. II* **35**:314–324.

TATTERSALL, I., and J. H. SCHWARTZ. 1974. Craniodental morphology and the systematics of the Malagasy lemurs (Primates: Prosimii). *Anthropol. Pap. Am. Mus. Nat. Hist.* **52**:(3):139–192.

WEST, R. M. 1972. Upper deciduous dentition of the Oligocene insectivore *Leptictis* (= *Ictops*) *acutidens*. *Ann. Carnegie Mus.* **44**:25–44.

14

Strepsirhine Basicranial Structures and the Affinities of the Cheirogaleidae

MATT CARTMILL

I. Introduction

With a few exceptions, students of primate evolution have always agreed that the Madagascar lemurs are more closely related to the lorises and galagos than any other living primates. During the 19th century, all these strepsirhine prosimians were usually lumped together as "lemurs," and some workers (e.g., Gray, 1870; Forbes, 1896) erected subfamilies and tribes that mixed Malagasy and African prosimians together in what now seems a haphazard and undiscriminating fashion. Most recent workers have followed Gregory (1915) in distinguishing a loris–galago group (Lorisiformes) from the Madagascar lemurs (Lemuriformes), denying that any member of either group has special affinities with any or all members of the other.

Gregory (1915) listed four principal traits of the Lorisiformes which he felt distinguished them from the Lemuriformes:

1. Stapedial artery reduced.
2. Main branch of internal carotid artery entering braincase in front of the auditory bulla, instead of entering the bulla and running along the promontorium.

MATT CARTMILL · Departments of Anatomy and Anthropology, Duke University, Durham, North Carolina.

3. Zygomatic bone not touching the lacrimal bone in the anterior rim of the orbit.
4. Tympanic bone (the bony frame of the eardrum) forming the lateral wall of the bulla, instead of being suspended within the tympanic cavity.

Later workers have expanded this list considerably. A survey of the literature through 1948 led Hill (1953) to report more than 30 supposed differentiae of the Lorisiformes. The more important and better-documented of these can be briefly listed:

5. Exposure of a flat orbital plate ("os planum") of the ethmoid bone in the medial wall of the orbit.
6. Longitudinal septum inside the bulla, partitioning off a group of air cells medial to the tympanic cavity proper.
7. Cellular pneumatization of the mastoid region (see Szalay, Chapter 5, this volume).
8. Dental enamel showing relatively more interstitial matrix and prisms with wavy (instead of straight) edges (Carter, 1922).

Several of these lorisiform traits, in particular 4 and 5, are also characteristic of tarsiers and some other haplorhines, and so most workers in the field have regarded one or another of them as primitive retentions from the common ancestry of haplorhines and strepsirhines. The Eocene adapids resemble the Madagascar lemurs rather than the lorises in almost all of these features. Therefore, most workers have agreed with Gregory (1915) that the typical adapids already show specializations leading toward the Madagascar lemurs, and that the separation between lorisiforms and lemuriforms must accordingly date back at least to the Eocene. Several paleontologists, including Gregory (1915) and Simons (1962), have proposed that the adapid *Pronycticebus gaudryi* from Eocene or early Oligocene deposits of France may be an ancestral lorisiform, as implied by the name which Grandidier gave this genus. Le Gros Clark (1934) and Szalay (1971*b*) have argued for an even more ancient derivation of the Lorisiformes, suggesting that the lorisiform arrangement of the tympanic region is more primitive than the lemuriform arrangement and thus could not have been derived from the supposedly intermediate condition seen in *Pronycticebus*. (However, see Szalay, Chapter 5, this volume.)

II. Challenges to the Lemur–Loris Dichotomy

Several facts suggest that the lemur–loris separation may be less ancient than Gregory and most subsequent workers have thought. Extant lemuriforms and lorisiforms share at least two specializations not known in adapids or any other fossil primates (although presumably present in the Miocene lorisiforms from East Africa): the clawlike nail on the second toe, and the tooth-comb. The grooming "claw" is an adaptation of a sort that has developed in many other groups of mammals, including marsupials (Jones, 1924) and rodents (Pocock, 1922); it has little more taxonomic value than the shared dental formula of lemurs and lorises, which also distinguishes them from adapids. But, as Martin (1972) and Schwartz (1974) have argued, the detailed similarities between the lemur and loris tooth-combs—incorporation of the canine, development of a lateral flange on the canine, caniniform specializations of the anterior lower premolar, and the sequence of eruption of the lower canine and incisors—are numerous and peculiar enough to suggest that both groups derive from a common ancestor more recent than the Adapidae.

315

STREPSIRHINE
BASICRANIAL
STRUCTURES AND
AFFINITIES OF
CHEIROGALEIDAE

Furthermore, the differences between the two groups are neither profound nor numerous. Many of each groups' supposedly distinctive characteristics, as listed by Hill (1953) and other taxonomists, are either not characteristic or not distinctive. Thus, the cartilaginous tracheal rings of lemuriforms are in fact incomplete dorsally, like those of lorisiforms and other mammals (Straus, 1931; and personal observations). The clitoris is traversed by the urethra in some lemuriforms, just as in all lorisiforms (Hill and Davies, 1954; Hill, 1958; Eckstein, 1958; Petter-Rousseaux, 1964). A coracoid head of the m. biceps brachii, seen in all lemuriforms, is also reported in *Perodicticus* (Miller, 1943) and galagos (Jouffroy, 1962). The first ethmoturbinal covers the maxilloturbinal in *Lepilemur* as well as in the Lorisiformes (Kollmann and Papin, 1925). The infraorbital foramen (or canal) is not "invariably double" (Hill, 1953) in lemuriforms and is ordinarily double or multiple in *Perodicticus* (Le Gros Clark, 1956). A *Loris*-like double supratragus is not characteristic of galagos and is variable in occurrence in lorisines. The zygomatic–lacrimal contact in the anterior orbital margin is not present in all adapids (Major, 1901) and is sometimes seen in galagos (Simons, 1961; Charles-Dominique and Martin, 1970). Examples could be multiplied.

Most of the other traits supposedly distinctive of the Lorisiformes are shared with one or more of the four genera of Madagascar lemurs (*Cheirogaleus, Allocebus, Microcebus, Phaner*) that constitute the family Cheirogaleidae. Like lorisiforms, all the cheirogaleids have a reduced stapedial artery and a large branch of the carotid system bypassing the bulla to enter the braincase at the front end of the petrosal bone (Saban, 1963). *Cheirogaleus* and *Microcebus* (but not *Phaner*) also resemble the loris–galago group in having an orbital exposure of the ethmoid (Kollmann, 1925); cellular pneumatization of the mastoid region is found in *Allocebus* (Petter-Rousseaux and Petter, 1967), but not in other cheirogaleids.

Most of these features that distinguish various cheirogaleids from other lemuriforms and link them to lorisiforms have been known for decades, but until recently they have not been taken very seriously by students of primate phylogeny. Gregory (1915, 1920) interpreted the carotid similarities as convergent specializations. Kollmann (1925) regarded the orbital exposure of the ethmoid as a primitive retention lost in some of the Madagascar lineages. Later systematists and anatomists have similarly regarded the lorisiform–cheirogaleid resemblances either as primitive retentions or convergent specializations, usually mentioning them in footnotes, if at all. Until Rumpler took the step of erecting a separate family Cheirogaleidae in a paper delivered in London in 1972 (Rumpler, 1974), the issue was almost universally obscured by the relegation of the cheirogaleids to subfamily status within the Lemuridae. Charles-Dominique and Martin (1970) had emphasized lorisiform–cheirogaleid resemblances, but they did so to undercut the traditional notion that the lemur–loris separation dates to the Eocene or earlier, not to argue that lorisiforms and cheirogaleids have special affinities. However, they emphasized that the morphological and behavioral resemblances between cheirogaleids and galagos could be explained by positing one or more post-Eocene crossings of the Mozambique Channel in either direction.

Such an explanation has recently been offered by Szalay and Katz (1973), who propose that a basal cheirogaleid stock diverged from the lemurids on Madagascar during the Tertiary, and that Lorisiformes originated from this stock by back-migration across the Mozambique Channel. In this view, the lorisiform–cheirogaleid resemblances are shared derived characters, not primitive retentions from the last common ancestor of the living strepsirhines. This implies that adapid and lemurid traits are primitive for the strepsirhines, and therefore has certain implications for our reconstruction of the phylogeny of the entire order Primates.

To assess the affinities of the cheirogaleids, each special resemblance between the cheirogaleids and lorisiforms must be reexamined to determine whether it represents a uniquely shared derived character (synapomorphy), a primitive retention lost in other strepsirhines, or a convergence produced by parallel selection pressures in the two groups. We will begin with the carotid arteries.

III. The Lorisiform Carotids

In most primates, only two vessels on each side bring significant amounts of blood to the brain: the vertebral artery, which enters the braincase through the foramen magnum, and the promontory branch of the internal carotid, which enters the bulla and traverses the middle-ear cavity. Other arrangements are found in other groups of mammals. In creodonts, and probably in the ancestral placentals, the promontory artery was supplemented by a medial branch of the internal carotid, which entered the braincase via a canal running through the petrooccipital synchondrosis, medial to the bulla (Matthew, 1909). In canoid carnivorans, the promontory artery is wholly supplanted by this medial entocarotid (Matthew, 1909; Story, 1951; Miller et al., 1964). The "internal carotid" of muroid rodents (Guthrie, 1963) is probably also a medial entocarotid (McKenna, 1966). In cats and many ungulates, much of the blood supply to the brain comes via anastomoses between intracranial vessels and one or more branches of the external carotid (Davis and Story, 1943; Daniel et al., 1953).

The carotid pattern seen in lorisiforms and cheirogaleids is unlike that seen in other primates. Although an intrabullar carotid artery is found in these animals, it is tiny and largely supplanted by an artery arising at or near the bifurcation of the common carotid. This artery runs forward medial to the bulla and enters the braincase via a foramen at the anterior end of the petrosal, passing immediately into the cerebral arterial circle ("of Willis"). For the time being, I will refer to this artery by the noncommittal name, "anterior carotid," employed by Saban (1963). There are four interpretations of the homologies of this vessel in the cheirogaleids, the lorisiforms, or both:

1. It represents the internal carotid of *Homo sapiens*—i.e., the promontory artery (Tandler, 1899; Gregory, 1920; Le Gros Clark and Thomas, 1952).
2. It represents the medial entocarotid (Van Valen, 1965; Szalay, 1972).

3. It represents the ascending pharyngeal artery of *Homo sapiens* (Adams, 1957; Saban, 1963; Kanagasuntheram and Krishnamurti, 1965; Krishnamurti, 1968; Bugge, 1972).

4. It is a neomorph, not homologous with any vessel found in other primates or in primitive mammals (Szalay and Katz, 1973; Hoffstetter, 1974).

317

STREPSIRHINE
BASICRANIAL
STRUCTURES AND
AFFINITIES OF
CHEIROGALEIDAE

These interpretations hold vastly different implications for strepsirhine phylogeny. The only primates or close primate relatives of the early Tertiary that retained the medial entocarotid were the Microsyopidae (McKenna, 1966; Szalay, 1969). The Eocene "primates of modern aspect" (Simons, 1972) lack any trace of this vessel, as do all extant and fossil haplorhines and noncheirogaleid lemuriforms. Therefore, if the cheirogaleid or lorisiform anterior carotid represents a persistent medial entocarotid, no known Eocene primate can be directly ancestral or even very closely related to the lorisiforms or cheiro-galeids. If, on the other hand, this artery is a neomorph unique to the cheirogaleids and lorisiforms, then the hypothesis of Szalay and Katz (1973), that the Lorisiformes have been derived from the cheirogaleids by back-migration from Madagascar, becomes more plausible. By the former interpretation, the last common ancestor of the extant strepsi-rhines was probably no later than the Paleocene; by the latter interpretation, all extant strepsirhines may be descended from one ancestral Madagascar lemur species, and the lemuriform–lorisiform divergence could be as recent as the beginning of the Miocene.

A. Human Morphology and Terminology

The structures I wish to talk about have been referred to by a variety of conflicting names, and an explicit definition of some terms is accordingly necessary. These definitions are based on the nomenclatorial assumption that the human *nomina anatomica* take priority where homology is known; that, if an ankle bone in some animal is believed to be homologous with the human talus, it is an error to refer to it as the astragalus.

In a macerated human skull, the foramen lacerum is a large ragged-edged aperture at the anterior tip of the petrous temporal. This hole represents a persistently unossified region of the chondrocranium; in a living human being, the gap in the skull is plugged with cartilage and transmits no large structures. The internal carotid artery does not pass through the foramen lacerum, but merely runs across its internal surface just after exiting from its carotid canal in the petrosal bone. The aperture through which the internal carotid or promontory artery enters the braincase should therefore not be referred to as the foramen lacerum medium (as it often is by those who call the superior orbital fissure the "foramen lacerum anterius"), or as the foramen lacerum anterius (as it is by those who call the superior orbital fissure the "sphenorbital foramen"). This internal carotid aperture has been called the anterior carotid foramen by some authors (Story, 1951; McDowell, 1958), but this term is confusing when "anterior carotid artery" is used *sensu* Saban (1963). Since the Paris *nomina anatomica* include a term for the carotid canal but not for the primitively independent foramina at either end, new names for these foramina are needed.

Several small structures pass through the foramen lacerum in *Homo sapiens*. The most important of these is the nerve of the pterygoid canal ("Vidian nerve"), which runs through a cartilaginous canal that is continuous anteriorly with the pterygoid canal in the sphenoid. The foramen also ordinarily transmits a small accessory meningeal branch of the ascending pharyngeal artery (Romanes, 1964). This accessory meningeal artery supplies the surrounding dura mater and the medial surface of the semilunar ganglion (sensory ganglion of nerve V), where it may anastomose with semilunar branches of the internal carotid (Bergmann, 1942). In mammals with a more perfectly ossified petrosal than our own, the nerve of the pterygoid canal may pass through a bony aperture separate from that traversed by the accessory meningeal branch of the ascending pharyngeal. In such cases, I propose to reserve the term "foramen lacerum" for the arterial foramen, since the term "pterygoid canal" is available and preferable for that transmitting the Vidian nerve and artery. Therefore, the foramen through which the anterior carotid of cheirogaleids or lorisiforms enters the braincase is a foramen lacerum if and only if the anterior carotid is taken to be homologous with the ascending pharyngeal artery.

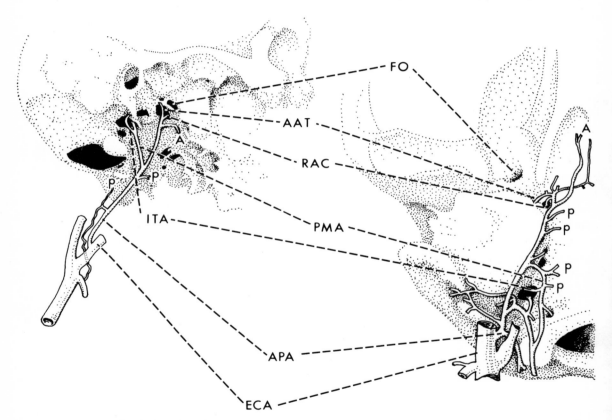

FIG. 1. Ascending pharyngeal arteries. Left, inferolateral view of *Homo sapiens* skull and ascending pharyngeal artery (after various sources). Right, ventral view of *Procyon lotor* cranial base and ascending pharyngeal artery (after Story, 1951). Not to same scale; arterial diameters not to scale. Abbreviations: A, anastomotic palatine branch; AAT, artery to auditory tube; APA, ascending pharyngeal artery; ECA, external carotid artery; FO, foramen ovale; ITA, inferior tympanic artery; P, pharyngeal branches; PMA, posterior meningeal artery; RAC, ramus anastomoticus to internal carotid.

The ascending pharyngeal artery of *Homo* (Fig. 1) arises from the base of the external carotid, or sometimes from the internal carotid or the common carotid bifurcation. From its origin, it ascends on the posterior wall of the pharynx, supplying the pharyngeal musculature and soft palate. Above, it ends in the aforementioned accessory meningeal branch that passes through the foramen lacerum. Posteriorly, it sends off the posterior meningeal artery, which enters the jugular foramen to supply the dura of the posterior cranial fossa. Its inferior tympanic branch accompanies the tympanic branch of cranial nerve IX into the tympanic cavity. Near its termination at the foramen lacerum, the ascending pharyngeal artery gives off a branch to the auditory tube. "For a small artery," as Crafts (1966) remarks, "it certainly has a wide distribution." This peculiar distribution—to jugular foramen, foramen lacerum, tympanic cavity, auditory tube, and pharynx—is sufficiently distinctive to be useful in determining its homologies in other mammals.

B. The Ascending Pharyngeal Artery in Other Mammals

Story's (1951) classic description of the carotid arteries of *Procyon lotor* identifies a medium-sized artery arising from the base of the external carotid as the ascending pharyngeal artery. This artery runs anteriorly between the external carotid and medial entocarotid, passing around the medial side of the bulla and ending in anastomotic palatine branches. In its course, it gives rise to the occipital artery, a posterior meningeal artery that traverses the hypoglossal canal and sends an inferior tympanic branch into the jugular foramen to join the tympanic branch of nerve IX, muscular branches to the sternomastoid and posterior digastric, branches to the walls of the pharynx, and an anterior pharyngeal branch that supplies the auditory tube and sends an anastomotic twig to the internal carotid* through a foramen at the anterior end of the petrosal. This distribution corresponds in detail to that of the corresponding artery in *Homo sapiens* (Fig. 1) and strongly suggests that the two are homologous. The foramen through which the raccoon's ascending pharyngeal artery anastomoses with the internal carotid is therefore a foramen lacerum.

In the other procyonids described by Story, the posterior branches of the ascending pharyngeal artery are more variable in their connections; the occipital artery may arise directly from the external carotid or medial entocarotid, and the posterior meningeal supply tends to be usurped by the occipital. Nevertheless, the anterior branches to the auditory tube and internal carotid remain constant. This description applies equally to *Canis* (Miller *et al.*, 1964) and at least some ursids (Story, 1951; Davis, 1964). The anastomosis with the medial entocarotid is also known in *Mustela* (Davis and Story, 1943), *Vulpes*, and *Martes* (Chapuis, 1966). The ascending pharyngeal–entocarotid anastomosis in dogs is ordinarily of no functional significance (Jewell, 1952; Torre *et al.*, 1959), and this seems to be true of other canoid carnivorans as well. A relatively larger anastomosis is found in *Felis pardus* (Chapuis, 1966). In the domestic cat (Davis and Story, 1943; Daniel

---

* The vessel which receives this anastomotic twig is probably neither the medial entocarotid nor promontory, but a common trunk that would receive blood from both vessels (if they were present) and convey it into the cerebral arterial circle.

319

STREPSIRHINE
BASICRANIAL
STRUCTURES AND
AFFINITIES OF
CHEIROGALEIDAE

et al., 1953), the posterior branches of the ascending pharyngeal artery are taken over by the occipital artery, but the ascending pharyngeal–entocarotid anastomosis is larger and more important than in canoids; the proximal part of the cat's medial entocarotid has become an atrophic cord, and the distal part which enters the cerebral arterial circle is filled via the anastomosis through the foramen lacerum. From that distal part, promontory branches run backward over the promontorium, anastomosing with branches of the inferior tympanic artery accompanying the tympanic branch of nerve IX. In domestic *Felis*, then, the ascending pharyngeal usurps both the medial entocarotid and promontory arterial supplies. An ascending pharyngeal–entocarotid anastomosis is also reported for *Nandinia* (Davis and Story, 1943) and *Herpestes griseus* (Chapuis, 1966). We can infer its presence in the ancestral carnivoran.

The researches of Daniel *et al.* (1953) provide some comparative data on artiodactyls, rabbits, and rats. The ascending pharyngeal–entocarotid anastomosis was not found in *Rattus* or *Oryctolagus*, nor in adult oxen, goats, and sheep. It was found, however, in 5 near-term sheep fetuses. In the domestic pig, this anastomosis persists into adult life, while the entocarotid vessels disappear altogether; the ascending pharyngeal artery provides the major blood supply to the cerebral arterial circle.

Although no other haplorhine primate has so persistent a foramen lacerum as *Homo*, macerated skulls of adult anthropoids usually retain small gaps between the petrosal and alisphenoid of sufficient breadth to transmit small anastomotic vessels. Since an ascending pharyngeal artery of otherwise human type has been observed in *Papio* and *Macaca* (McCoy *et al.*, 1966), it seems probable that ascending pharyngeal–carotid anastomoses could be found in well-prepared specimens of nonhuman catarrhines.

C. The Anterior Carotid in *Microcebus* and Other Strepsirhines

The carotid arteries of an injected specimen of *Microcebus murinus* from the Duke University Primate Facility are depicted semidiagrammatically in Fig. 2. At a point near the posterior edge of the m. cricothyroideus, the common carotid artery divides into the anterior and external carotids. The anterior carotid runs anteriorly and dorsally toward the anterior tip of the petrosal. It is crossed in turn by nerves XII and IX: where it contacts nerve IX, it gives off a minute posterior meningeal artery, which runs with nerve IX upward and backward into the jugular foramen. An extremely fine internal carotid artery arises from the midpoint of the right posterior meningeal artery and runs laterally to a tiny carotid foramen on the posterior surface of the bulla; in the specimen examined, the left internal carotid was absent. The main trunk of the anterior carotid continues on toward the "foramen lacerum medium" in close contact with the dorsal wall of the pharynx; it gives off a few small pharyngeal branches en route and crosses the superior pharyngeal branches of the vagus. Just before reaching the pharyngeal surface of the basicranium, the anterior carotid gives rise to three slender branches: (1) a medial pterygoid branch, which divides into a posteriorly curving branch that accompanies the auditory tube and a somewhat larger artery of supply to the medial pterygoid muscle; (2) a terminal pharyngeal branch that enters an adenoid-like mass of fatty tissue overlying the anterior tip of the petrosal; and (3) a minute meningeal branch.

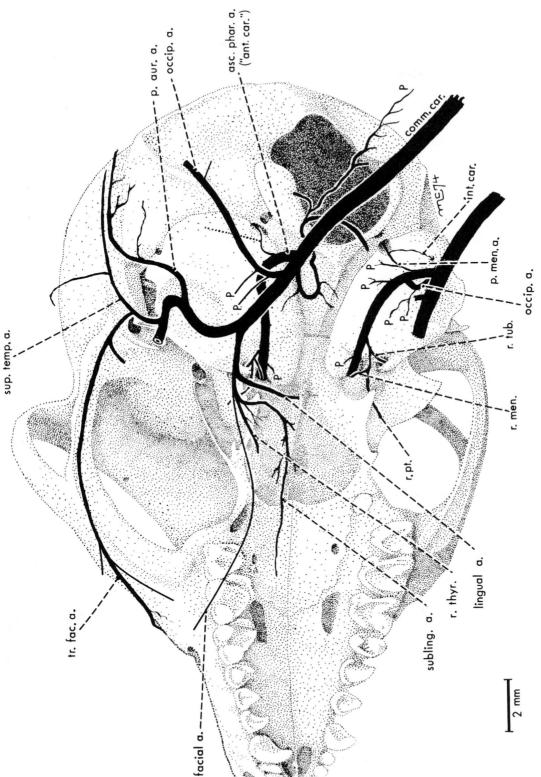

Fig. 2. Carotid arteries of *Microcebus murinus* (semidiagrammatic). The maxillary artery has been truncated just after giving rise to the superficial temporal artery. Abbreviations: asc. phar. a. ("ant. car.") = ascending pharyngeal artery ("anterior carotid"); comm. car. = common carotid; int. car. = internal carotid; occip. 1. = occipital artery; P = pharyngeal branches; p. aur. a. = posterior auricular artery; p. men. a. = posterior meningeal artery; r. men. = ramus meningealis; r. pt. = ramus to pterygoid musculature; r. thyr. = ramus to thyroid gland; r. tub. = ramus tubae auditivae; subling. a. = sublingual artery; sup. temp. a. = superficial temporal artery; tr. fac. a. = transverse facial artery.

This last accompanies the stem of the anterior carotid into the "foramen lacerum medium," which forms a short canal bounded by the petrosal in back and the alisphenoid in front. The meningeal branch supplies the walls of this canal. While still in the canal, the anterior carotid gives off a delicate artery that traverses its own tiny foramen in the roof of the canal, emerging on the inferior surface of the semilunar ganglion, which it supplies. The anterior carotid continues upward and forward, piercing the overlying dura and ending in the cerebral arterial circle (Fig. 3). Just before doing so, it gives rise to four endocranial branches: (1) superior and (2) inferior ophthalmic arteries; (3) a tiny branch that enters the basisphenoid; and (4) a fine artery that runs backward and downward in contact with the wall of the inferior petrosal sinus. No branches of the internal carotid proper could be detected inside the braincase; the anterior carotid was not joined by a promontory artery, and, although the ramus superior of the stapedial artery was present, it had filled with injection mass via its orbital anastomosis with the ophthalmic artery and could not be traced backward into the petrosal.

Published descriptions of the anterior carotid in other strepsirhines are insufficiently detailed to permit more than a schematic comparison. As described by Saban (1963), *Cheirogaleus* differs from *Microcebus murinus* chiefly in that the internal carotid is larger, retains an identifiable anastomosis with the anterior carotid, and arises from the common carotid via an independent entocarotid–posterior meningeal trunk, rather than from the anterior carotid. In adult *Lemur* (Saban, 1963), the internal carotid has a similar independent origin, but posterior meningeal and inferior tympanic branches originate directly from a "*rameau tubaire*," which arises from the internal carotid trunk just distal to the carotid bifurcation and runs forward to the anterior tip of the petrosal, ending there in branches to the walls of the auditory tube. It is highly probable that this *rameau tubaire* is homologous with the cheirogaleid anterior carotid, as Van Valen (1965) first surmised.

In lorisiforms, as in *Microcebus murinus*, the internal carotid is greatly reduced or vestigial; the absence of a functional connection between the proximal and distal extremities of the promontory artery has been reported for *Loris tardigradus*, *Nycticebus coucang*, and *Galago senegalensis* (Kanagasuntheram and Krishnamurti, 1965; Bugge, 1972). The central (or intrabullar) part of the promontory artery lacks a lumen in these animals. The anterior carotid, which functionally replaces the promontory, gives off "branches to the prevertebral muscles, pharynx and soft palate" in *N. coucang* and *G. senegalensis* (Bugge, 1972). Lorisiforms also resemble *Microcebus murinus* in possessing a double ophthalmic artery, which Bugge (1972) found in *Nycticebus* and *Galago senegalensis* but not in *Lemur catta*.

In its known respects, the anterior carotid of lorisiforms differs from that of *Microcebus murinus* chiefly in dividing into a simple extracranial rete just proximal to the "foramen lacerum medium." A rete in this position is reported for *Loris tardigradus* and *Galago crassicaudatus* (Tandler, 1899), and for *Nycticebus coucang* and *Galago senegalensis* (Bugge, 1972). In *G. senegalensis*, *L. tardigradus*, and *N. coucang*, at least, the arterial rete is interlaced with parallel venous channels (Kanagasuntheram and Krishnamurti, 1965), forming a potential counter-current exchange mechanism. The rete is better developed in galagines than in lorisines; in the latter group, it shows great intraspecific variability and may be reduced to a single channel (Kanagasuntheram and Krishnamurti,

1965; Bugge, 1972). No carotid rete was present either intra- or extracranially in the examined specimen of *Microcebus murinus*.

323

STREPSIRHINE
BASICRANIAL
STRUCTURES AND
AFFINITIES OF
CHEIROGALEIDAE

D. Homologies of the Anterior Carotid

From the foregoing description, and from the schematic comparisons presented in Fig. 4, it is evident that the so-called anterior carotid of *Microcebus* is a probable

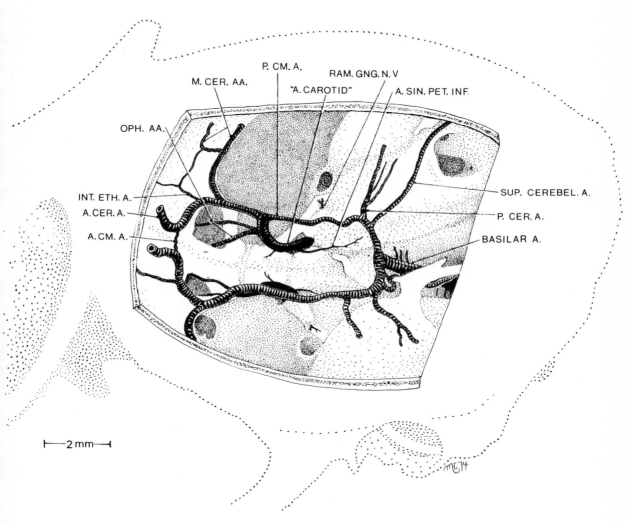

FIG. 3. Arteries at base of brain of *Microcebus murinus*. Abbreviations: "A. CAROTID," "anterior carotid" (ascending pharyngeal artery); A. CER. A., anterior cerebral artery; A. CM. A., anterior communicating artery; A. SIN. PET. INF., arteria sinus petrosi inferioris; INT. ETH. A., internal ethmoidal artery; M. CER. AA., middle cerebral arteries; OPH. AA., ophthalmic arteries; P. CER. A., posterior cerebral artery; P. CM. A., posterior communicating artery; RAM. GNG. N. V., ramus to the semilunar ganglion of nerve V; SUP. CEREBEL. A., superior cerebellar a. The anterior cerebral arteries fuse distally to form a midline vessel, as in other lemuriforms.

homolog of the ascending pharyngeal artery of *Homo*. These vessels resemble each other in that they (1) originate near the bifurcation of the common carotid, often or usually by a common stem with the internal carotid; (2) give off branches to the jugular foramen, the roof of the pharynx, the soft palate, the auditory tube, and the anterior end of the principal entocarotid channel,; and (3) lie in close contact with the upper part of the pharynx, being crossed laterally by nerves XII and IX. The small branch which the anterior carotid of *Microcebus* sends into the semilunar ganglion is probably homologous with the comparable semilunar branches of the ascending pharyngeal in *Homo*: the main anterior carotid channel, which continues past this point into the cerebral arterial circle, derives from an anastomosis between the semilunar and internal carotid arteries, of the sort often found in *Homo*. Even in *Homo*, this anastomosis may in rare cases enlarge to replace the internal carotid, becoming the principal extravertebral channel of blood supply to the brain (Lie, 1968), as in *Microcebus* or the pig.

As far as is known, the anterior carotid of lorisiforms has relationships and branches resembling those of *Microcebus* and *Cheirogaleus*, and it is probable that Adams (1957) was correct in identifying it as "a greatly enlarged ascending pharyngeal artery." The "foramen lacerum medium" of cheirogaleids and lorisiforms can therefore be identified as a true foramen lacerum, homologous with those of *Homo* and *Procyon*.

Adams' conclusion has not gone entirely unquestioned by later writers. Szalay and Katz (1973) mistakenly believed that the ascending pharyngeal artery is not present in adults of any other extant primates and refused to accept Adams' hypothesis for that reason. Kanagasuntheram and Krishnamurti (1965) concurred with Adams, but raised the possibility that the anterior carotid is a neomorph and that a slender parallel vessel seen in *Galago senegalensis* represents the ascending pharyngeal. This parallel vessel appears late in ontogeny in *Galago* and is evidently absent in *Nycticebus* (Davies, 1947). These facts suggest that the smaller vessel is more likely to be a neomorph, or perhaps homologous with pharyngeal branches of the occipital artery, observed in *Microcebus*.

Other suggested homologies of the anterior carotid seem improbable. It is clearly not homologous with the internal carotid, since both vessels are present in *Cheirogaleus*. It is also unlikely to represent the medial entocarotid, since it shows many detailed correspondences to the ascending pharyngeal artery of canoids, which coexists with a medial entocarotid. In dogs, the "ventral petrosal sinus" is paralleled by another venous channel with similar connections, which traverses the same bony canal as the medial entocarotid (Miller *et al.*, 1964). There is presently no evidence as to which of these two sinuses is homologous with the single inferior petrosal sinus of primates; if it is the latter, the small vessel accompanying the sinus in *Microcebus* (Fig. 3, "a. sin. pet. inf.") may represent the persistent distal end of the medial entocarotid.

FIG. 4. Carotid patterns in primates. A, hypothetical ancestral pattern; B, *Homo sapiens*; C. *Lemur* (after Saban, 1963); D, *Microcebus murinus*. Abbreviations: a, ascending pharyngeal; "ac," "anterior carotid" (= ascending pharyngeal); at, branch to auditory tube; ec, external carotid; gV, branch to ganglion of nerve V; ic, internal carotid; it, inferior tympanic; o, occipital; op, ophthalmic; p, pharyngeal branches (hypothetical in *Lemur*); pm, posterior meningeal; pr, promontory; rs, ramus superior of stapedial; "rt," "rameau tubaire" (= ascending pharyngeal); st, stapedial.

325

STREPSIRHINE
BASICRANIAL
STRUCTURES AND
AFFINITIES OF
CHEIROGALEIDAE

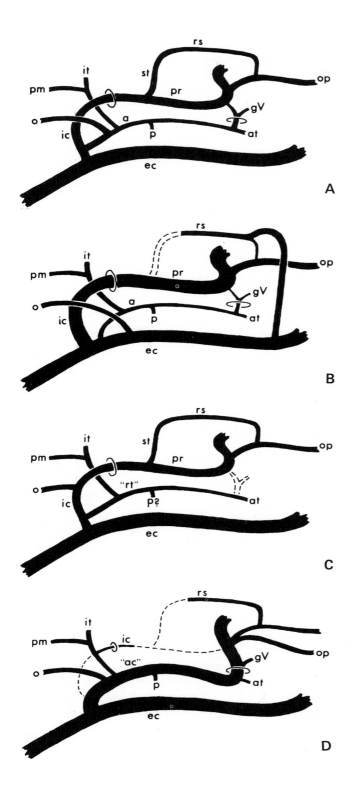

E. Ontogeny and Function of the "Anterior Carotid"

In its primary role as a vessel supplying blood to the muscles of the pharynx, the ascending pharyngeal artery appears to be virtually universal among placental mammals. It is presumably retained from their common ancestor. Its anastomosis with the internal carotid circulation via the foramen lacerum, although widespread, is usually of no functional importance, and its enlargement into a major channel of blood supply to the brain, as in pigs, lorisiforms and cheirogaleids, is uncommon. The wide distribution of this anastomosis, and the observations of Daniel *et al.* on fetal and adult sheep, suggest that a small anastomosis between the internal carotid and the distal part of the ascending pharyngeal artery may be present during early development in all mammals. This embryonic anastomosis ordinarily remains minute during ontogeny and becomes functionless or absent in the adult. Nevertheless, it may persist or become enlarged as a developmental anomaly in almost any species, as in the specimen of *Lemur catta* shown in Fig. 5.

What selective advantage might lead to this anomaly's spread and fixation in a population, as in the cheirogaleids and lorisiforms? Several hypotheses have been advanced. Szalay and Katz (1973), misled by an illustration of Saban's (1963), hypothesize

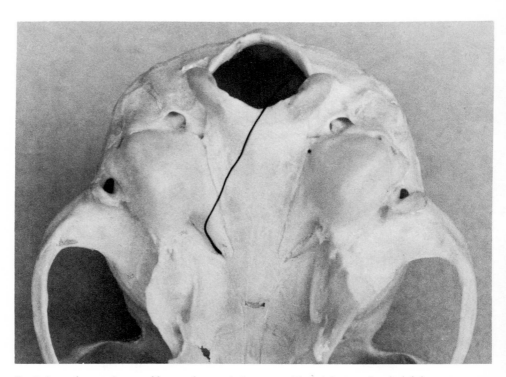

FIG. 5. Anomalous persistence of foramen lacerum in *Lemur catta*. The bristle traversing the left foramen emerges intracranially alongside the hypophyseal fossa. (Specimen courtesy of Dr. P. Charles-Dominique, who detected the anomaly.)

that the "anterior carotid" empties chiefly into the ophthalmic artery and thus probably "arose in response to demands for a greater blood supply to the orbital region." In fact, the ascending pharyngeal of lorisiforms and cheirogaleids feeds into the cerebral arterial circle in the same way as the anthropoid internal carotid does (Fig. 3); indeed, the distal part of the "anterior carotid" is probably homologous with the distal end of the internal carotid. No disproportionate amount of the arterial blood that enters the braincase via the foramen lacerum goes to supply the eyeball, either in *Microcebus* or lorisiforms (Bugge, 1972). Moreover, the promontory artery is perfectly capable of supplying an enlarged eyeball, as it does in *Tarsius* and *Aotus*.

Since an extracranial rete mirabile is developed on the enlarged ascending pharyngeal artery of lorisiforms, determining the function of this rete would be helpful in explaining the enlargement of its parent artery. Ask–Upmark (1953) hypothesized that the rete is needed to protect the tissues of the brain against the possibly traumatic effects of the arterial pulse, undamped in these animals by the walls of a carotid canal. The absence of a rete in *Microcebus murinus* and some individual lorisines refutes this contention. Krishnamurti's (1968) notion, that the lorisiform rete aids venous return by allowing the retial pulse to exert rhythmic pressure on the intertwined venous channels, seems still more implausible.

The experimental work of Baker and Hayward (1967, 1968) demonstrates that the carotid rete has a thermoregulatory function in at least some mammals. In birds, mammals, reptiles, and even fish, the preoptic region of the hypothalamus is an important center for thermoregulation; heating this region results in a drop in metabolic rate and body temperature, while cooling it has the opposite effect (Mills and Heath, 1970). Baker and Hayward demonstrate that in sheep, and probably also in cats, blood temperature in the cerebral arterial circle and in the preoptic hypothalamus varies with that of the nasal mucosa, not with that of the common or external carotid artery. They conclude that in the sheep, venous blood from the nose returns to the cavernous sinus via ophthalmic veins and via anastomoses with the pharyngeal plexus. This cooled blood drains from the cavernous sinus into the venous channels of the intracranial carotid rete. The rete therefore acts as a countercurrent heat exchanger, cooling only the blood entering the brain and thus allowing systemic temperatures to rise temporarily to levels which could not be tolerated if they affected the thermoregulatory centers or other parts of the brain.

This analysis accounts for the fact that carotid retia develop only on arteries which come into close proximity with nasal venous drainage into the cavernous sinus or pharyngeal plexus; no mammal, for instance, develops a rete on the internal carotid. The ascending pharyngeal artery is a favorable site for the development of a rete, since it runs along the back of the pharynx and has an anastomotic terminal branch which parallels the small veins draining out of the cavernous sinus through the foramen lacerum.

The comparative evidence thus suggests that the enlargement of the ascending pharyngeal artery in the lorisiforms has a thermoregulatory function, possibly allowing brief periods of intense activity to produce large temporary increases in systemic temperature without risking thermal injury to the brain or necessitating the dehydration

327

STREPSIRHINE
BASICRANIAL
STRUCTURES AND
AFFINITIES OF
CHEIROGALEIDAE

involved in rapid heat dissipation by panting. This explains the fact that the carotid rete is best developed in galagines and is vestigial or even sometimes absent in the slow-moving lorisines. However, if the enlarged ascending pharyngeal of cheirogaleids serves a similar purpose, the absence of a rete on the ascending pharyngeal artery of *Microcebus murinus* demands explanation. Barnett and Brazenor (1958) observe that the testicular arterial rete characteristic of marsupials is simple in small animals and more complex in large ones. This is a consequence of the square-cube law; in smaller animals the surface-to-volume ratio of each vessel is greater, and a simple rete provides enough surface for exchange of heat. Since the lorisiform "carotid" rete is relatively simple even in larger animals, it may be that mere contact between the ascending pharyngeal artery and nasopharyngeal veins, observed in *Microcebus murinus*, answers the same purpose for this tiny lemur as the retia of galagos. If so, we would expect to see a simpler rete developed in *Galago demidovii* than in larger galagos, and we would also expect to find a true rete in *Phaner furcifer* or other larger cheirogaleids. These predictions remain to be tested.

IV. The Strepsirhine Bulla

The so-called extrabullar tympanic ring is probably the most widely known and accepted diagnostic feature distinguishing the lorisiforms from the lemuriforms. Although it is always possible to distinguish a lorisiform from a Madagascar lemur by examining the bulla, the differences are less profound than Le Gros Clark's (1959, p. 137) much-copied figure suggests, and possibly intermediate conditions are known. In typical Madagascar lemurs, most of the tympanic ring lies internal to the petrosal bulla, articulating with the petrosal and squamosal only at its extremities. Nevertheless, the ring is not, as usually stated, free and intrabullar; the nonarticular central arc of the ring is attached to the bullar meatus by a cartilaginous anular membrane continuous with the cartilage of the external acoustic meatus, and there is a substantial extrabullar exposure of the ring at its posterior articulation, where it contributes to the bullar meatus (Fig. 6A). In older specimens of *Lemur*, the anulus membrane may become ossified as an extension of the petrosal, so that the ring is attached to the bullar meatus by bone instead of cartilage. Van Valen (1965), who noted this phenomenon, felt that it represented a possible transitional stage in the evolution of the lorisiform bulla. In the extinct palaeopropithecines and megaladapines of Madagascar, the entire anulus membrane and much of the external meatal cartilage appears to have been replaced by an ossified extension of the tympanic (Saban, 1963). This is evidently a derived and allometrically conditioned feature (Tattersall, 1973a). In at least some specimens of *Microcebus murinus*, the anulus membrane is replaced almost completely by bony laminae derived laterally from the petrosal and medially from the tympanic ring (Fig. 6B); in a dissected specimen in the Duke University collections, only some 30% of the ring's circumference remains "free" within the bulla where a narrow hiatus still separates the petrosal and tympanic anular laminae. Moreover, as Saban (1963, p. 151) noted, the tympanic ring of *Microcebus murinus* is so closely approximated to the meatus of the bulla that it induces a corresponding ring-shaped deformation of the bullar wall, clearly visible from the exterior. The condition seen in

329

STREPSIRHINE
BASICRANIAL
STRUCTURES AND
AFFINITIES OF
CHEIROGALEIDAE

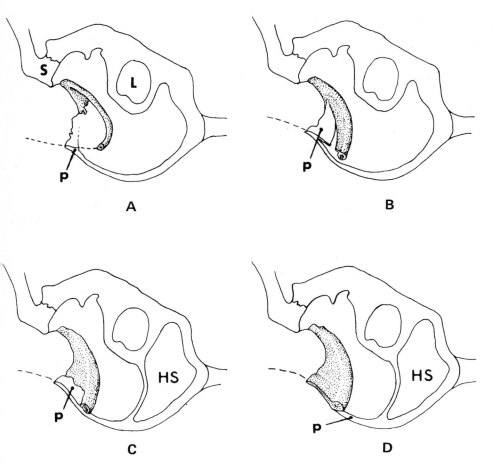

Fɪɢ. 6. Coronal sections through right ear regions of strepsirhine skulls, seen from anterior (diagrammatic). A, *Lemur*; B, *Microcebus murinus*; C, *Loris*; D, typical lorisiform (e.g., *Nycticebus*). The tympanic is stippled; the ectodermal surface of the acoustic meatus and anulus membrane is indicated by a dashed line. Abbreviations: HS, hypotympanic sinus; L, labyrinth of inner ear; p, petrosal; S, squamosal.

Microcebus is thus not far from that found in *Loris* (Fig. 6C), where the petrosal participates in the formation of the inferior margin of the external acoustic meatus, so that the central portion of the tympanic ring (now in direct contact with the inner wall of the bulla) is still enclosed by the petrosal (Saban, 1963). A similar constitution of the bulla is reported for Miocene lorisines (Le Gros Clark, 1956; Szalay, 1972). In typical lorisiforms, the meatal expansion of the tympanic forward from its original posterior contribution to the meatus is complete (Fig. 6D), and the petrosal no longer overlaps the tympanic ring.

The most loris-like bulla found among Madagascar lemurs is that of the rare cheirogaleid *Allocebus trichotis*, known only from four specimens collected between 1874

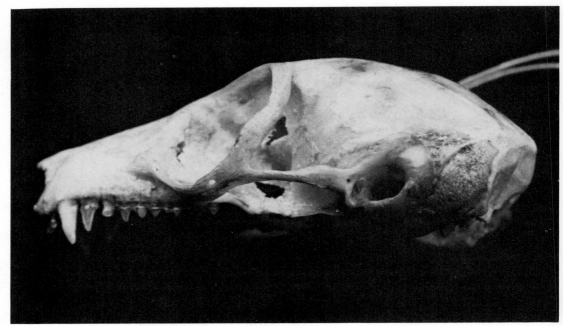

FIG. 7. *Allocebus trichotis*, type skull (BMNH 75.1.29.2), showing cellular pneumatization of mastoid and squamous portions of temporal bone.

and 1965. The type skull (Fig. 7) clearly displays pneumatization of the mastoid region and part of the squamosal, noted by earlier investigators (Major, 1894; Petter-Rousseaux and Petter, 1967). A small damaged area of the right mastoid region shows that the mastoid air sinus is composed of many small cellules separated by delicate bony partitions, as in lorisoids; other cheirogaleids have a simpler, somewhat smaller air sinus in this position. The bulla of *Allocebus* is very weakly inflated, a feature seen in Miocene and Recent lorisines but not in other cheirogaleids. No internal carotid foramina are discernible. As far as I can tell by peering through the external acoustic meatus using a 10× hand lens, the anulus membrane is completely ossified.

A remarkable feature of the bulla of *Allocebus* is that the tympanic cavity is partly subdivided by two low transverse septa springing from the floor of the bulla and running all the way across from the medial edge of the cavity to the tympanic ring. Similar, but more numerous and variable, incomplete transverse septa subdivide the lorisiform hypotympanic sinus (the medial portion of the middle-ear cavity, partitioned off by the principal longitudinal septum; Fig. 6C and D).

Evidently, the systematic differences in bullar construction between the African and Malagasy strepsirhines are not as usually described. A more precise diagnosis would distinguish the Madagascar lemurs by only two characters: the absence of a complete longitudinal intrabullar septum, and the extension of the tympanic air space laterally below the lower edge of the tympanic ring.

A. Ontogeny of the Lemuriform Bulla

331

STREPSIRHINE
BASICRANIAL
STRUCTURES AND
AFFINITIES OF
CHEIROGALEIDAE

The way in which people have thought and written about the mammalian ear region in the past is misleading in some respects, largely because certain ontogenetic facts have not been adequately appreciated. A simplified sketch of the developmental pattern of this region will help to clarify these points.

The middle and outer ears of mammals are derived, respectively, from the first pharyngeal pouch, which is an evagination of the pharynx's endodermal lining, and from the first pharyngeal cleft, an invagination of the ectoderm on the side of the embryonic head. In fish, the pouch and cleft come into contact and open into each other, forming the first gill slit, or spiracle. In mammals (Fig. 8A and B), they come into contact with each other, but do not open; a thin film of mesoderm persists between them. This persisting film of mesoderm, together with the ectoderm and endoderm covering its inner and outer faces, becomes the tympanic membrane (Fig. 8C). The ectoderm-lined invagination from the surface develops into the external acoustic meatus. The evagination from the pharynx becomes the auditory ("Eustachian") tube, and its distal end expands to form the tympanic cavity. Any air-filled space continuous with the tympanic cavity (e.g., the mastoid air cells, or the air space surrounding the middle ear ossicles) is formed by expansion of the first pharyngeal pouch through resorption of the mesodermal derivatives adjoining its endodermal lining.

In all mammals, the floor of the external meatus and tympanic cavity becomes secondarily chondrified. These secondary cartilages may remain cartilaginous, become ossified as extensions of any of the adjoining chondrocranial bones, ossify as extensions of the tympanic, or develop ossification centers of their own. The tympanic bone first appears as a membranous ossification medial to the shaft of Meckel's cartilage, and its posterior crus grows out secondarily as a caudal extension into the mesenchyme surrounding the margin of the eardrum (Goodrich, 1930). Both membrane bones and endochondral bones are preformed in softer tissues enclosing the walls of the cavities; their forms and relations are determined by the shapes which the cavities assume.

If we bear these facts in mind when looking at the bony tissues in the adult animal, it is easy to see that several questions that have preoccupied some earlier investigators are red herrings and that some traditional terminology is misleading. For instance, the lemuriform tympanic ring and eardrum cannot in any functionally meaningful sense be called "intrabullar"; the bulla simply bulges out laterally along the anteroinferior surface of the unossified external acoustic meatus. Similarly, the lorisiform and tarsiid tympanic does not, as Klaauw (1931) and subsequent authors have generally stated, form the lateral wall of the bulla, since the recessus meatus is neither functionally nor developmentally assimilable to the bulla; it is more precise to say that the external acoustic meatus in these groups has a broadly dilated medial end, and ossifies as an extension of the tympanic. Another instance of the misleading effect of the traditional terminology is the dispute between Hürzeler (1948) and Simons (1961) as to whether the middle ear of *Necrolemur* more closely resembles that of lemurids or tarsiids; this debate reduces to the rather moot question of whether *Necrolemur*'s external acoustic meatus ossified as an extension of the tympanic or of the petrosal, since both parties are agreed that *Necrolemur* is like *Tarsius* in

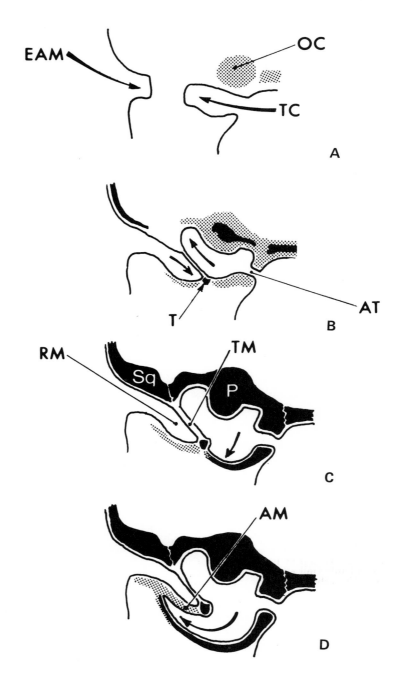

FIG. 8. Development of lemuriform tympanic cavity; diagrammatic coronal sections. A, embryonic stage. No ossification; 1st pouch (TC) and 1st cleft (EAM) separated by thick mesenchymal partition. B, later fetal stage. C, newborn stage. D, postnatal growth of tympanic cavity laterally below eardrum, leaving medial end of auditory meatus surrounded by tympanic cavity. Abbreviations: AM, anulus membrane; AT, auditory tube; EAM, external acoustic meatus; OC, otic capsule; P. petrosal; RM, recessus meatus; Sq, squamosal; T, tympanic ring; TC, tympanic cavity; TM, tympanic membrane. Bone shown in black; cartilage, in stipple.

having an ossified meatus and like *Lemur* in having a tympanic cavity which extends laterally beyond the lower edge of the eardrum.

In early developmental stages of Madagascar lemurs (Major, 1899), the tympanic ring lies in an almost horizontal position beneath the promontorium, and the unossified floor of the bulla is correspondingly very narrow transversely (Fig. 8B). This condition is true of fetal mammals in general. In newborn *Lemur mongoz* (Fig. 9A), and presumably in other lemuriforms, bone growing out from the petrosal into this narrow bullar floor comes almost into direct contact with the entire medial surface of the ring before the ring attains its more vertical adult orientation. This juvenile lemuriform configuration (Fig. 8C) is essentially the same configuration seen in adult ceboids, tarsier embryos (Kampen, 1905), and juvenile *Homo* (Fig. 9B); it differs from that seen in adult tarsiers and catarrhines only in that the tympanic ossification has grown out laterally under the floor of the meatus in the latter animals.

The ceboid bullar configuration in *Lemur mongoz* is lost soon after birth. A narrow, unossified outpocketing of the bullar floor, probably originating from cartilage persisting between the tympanic ring and the bulla, pushes upward and laterally to cover the lateral aspect of the ring's anterior crus (Fig. 9C). Among modern primates, this ontogenetic pattern is limited to, and distinctive of, the Lemuriformes. The medial end of the cartilaginous external acoustic meatus is left projecting into the bullar cavity, forming an anulus membrane attached around the lateral edge of the tympanic ring (Fig. 8D). This anulus membrane, like the eardrum itself, represents a region of persistently unossified contact between the ectodermal meatus and the wall of the tympanic cavity.* The Lemuriformes are, accordingly, the only extant primate group in which part of the lateral wall of the bulla ordinarily remains unossified. This is merely another way of stating Gregory's (1920, p. 211) conclusion that "the 'annulus membrane' is . . . an infolded surface of the bulla where it has grown around the tympanic ring."

B. Phylogeny of the Bulla

Szalay (1972) argued that, since the ectotympanic forms the entire lower margin of the external acoustic meatus in known plesiadapoids, the "intrabullar" tympanic ring of lemuriforms is unlikely to be primitive for the order Primates. He concluded that the lorisiform construction is a primitive retention and that the lemuriform condition is accordingly derived. However, the Szalay–Katz hypothesis, that the lorisiforms are descended from Madagascar cheirogaleids, implies that the lemuriform condition is primitive for the strepsirhines. Szalay's earlier conclusion needs to be reassessed in this light.

As we have seen, the crucial features of the lemuriform bulla that distinguish it from the lorisiform bulla are its lack of complete longitudinal septa and the subtympanic extension of its tympanic cavity. Kampen (1905) considered the lorisiforms to be persistently primitive in the former respect. This view won the concurrence of Gregory (1920), Klaauw (1931), and Saban (1963), all of whom regarded the partial longitudinal

---

* The anulus membrane of *Lemur catta*, and perhaps of other lemurs, contains only skin and mucous membrane, not cartilage.

333

STREPSIRHINE
BASICRANIAL
STRUCTURES AND
AFFINITIES OF
CHEIROGALEIDAE

septum of lemuriforms as in some sense vestigial. This thesis seems not to be supported by the fossil evidence; known plesiadapoids and early primates of modern aspect appear to have lacked a complete septum of the sort seen in lorisiforms.

The second peculiarity of the Lemuriformes, the subtympanic extension of the tympanic cavity, is widespread among Paleogene primates. This point has been obscured in the past by a failure to recognize the ontogenetically secondary character of ossification in the floor of the external acoustic meatus. In *Necrolemur*, an essentially lemuriform configuration of the tympanic cavity must have been achieved in late fetal or early post-natal life. The adult condition as interpreted by Simons (1961) differs from this only in one important respect: the meatal cartilage has been replaced or supplemented by a lateral extension from the ectotympanic bone surrounding the eardrum. Extension of this ossification beyond the lateral edge of the petrosal bulla has produced what the traditional terminology would refer to as an extrabullar ectotympanic, but the fundamental configuration is lemuriform (Fig. 8D) rather than lorisiform (Fig. 8C), and the "extrabullar" situation of a part of the ectotympanic is secondary.

This might be viewed as grounds for serious reconsideration of Hürzeler's (1948) thesis that *Necrolemur* and its relatives are lemuriforms, were it not for the recent discovery of Gingerich (1974) that *Plesiadapis tricuspidens* has a configuration of the tympanic cavity much like that of *Necrolemur* (Fig. 10). Here, too, the "extrabullar" extension of the ectotympanic may be taken as secondary. Thus, a subtympanic extension of the tympanic cavity is known in early representatives of all three major groups derived from protoplesiadapoid stock: Plesiadapoidea (as represented by *Plesiadapis*), Strepsirhini (as represented by the adapids), and Haplorhini (as represented by *Necrolemur*). No other placental group is known to share this feature.* The "intrabullar" eardrum and tympanic ring therefore was probably inherited from the last common ancestor of the three groups named. Accordingly, it cannot be used to argue for lemuriform affinities of *Necrolemur*, which appears from the morphology of its carotids (Szalay, Chapter 5, this volume) and interorbital region to have special relationships with tarsiers and anthropoids.

If the ancestral haplorhine had a subtympanic extension of the tympanic cavity, this trait was lost in anthropoids and in later tarsoids, probably independently. The earliest anthropoids for which we have any evidence of bullar configuration, *Aegyptopithecus* and *Apidium* from the Fayum deposits of Egypt, display the configuration seen in ceboids, lorisiforms, newborn catarrhines, and newborn *Lemur mongoz*. No major alteration of the ontogenetic pattern would have been required to effect this change; it would suffice to arrest development at the stage shown in Fig. 9A.

Gingerich (1973) has proposed that the lemuriform bulla represents the ancestral

---

* The eardrum of tree shrews is "intrabullar" in a different sense; during ontogeny the entotympanic bone, which forms the floor of the tympanic cavity, spreads laterally beyond the tympanic ring into the cartilaginous floor of the external acoustic meatus. However, the tympanic cavity does not extend laterally past the eardrum, and the tympanic ring is in direct contact with the subjacent entotympanic though not co-ossified with it.

FIG. 9. The bulla in juvenile primates. Top, newborn *Lemur mongoz* (uncataloged British Museum specimen), showing direct contact between tympanic ring and lateral edge of petrosal bulla. Middle, newborn *Homo sapiens*, anteroinferior view. As in ceboids or newborn *Lemur*, the nearly horizontal tympanic ring (outlined on right) contacts the narrow floor of the tympanic cavity. FM, foramen magnum; f, foramen lacerum. The petrosal overlaps the posterior edge of the tympanic (asterisk on left, dotted line on right); the significance of this is unclear. Bottom, *Lemur mongoz* (BMNH 05.6.22), stereopair showing initial stages of bullar extension lateral to eardrum.

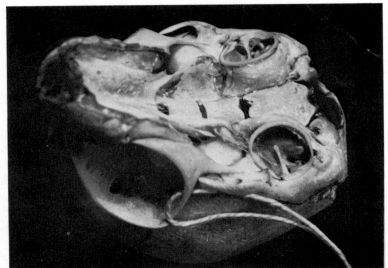

A

335

STREPSIRHINE
BASICRANIAL
STRUCTURES AND
AFFINITIES OF
CHEIROGALEIDAE

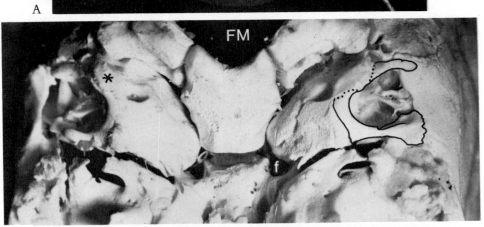

B

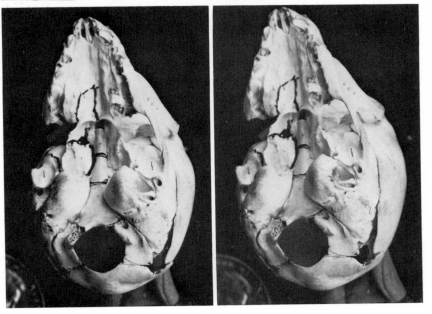

C

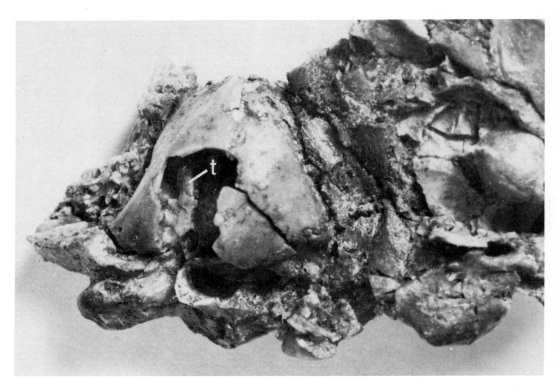

FIG. 10. Ventral view of right bulla of *Plesiadapis tricuspidens* (uncataloged specimen from private collection, Paris). The tympanic cavity extends lateral to the tympanic ring (t), which is attached to the bulla by an ossified anulus membrane and small transverse septa.

condition for anthropoids but not for tarsioids, and that the Anthropoidea may be more closely related to lemurs than to tarsiers. This thesis seems to be contradicted by evidence drawn from biochemistry (see Goodman, this volume), placental morphology and development (see Luckett, this volume), carotid morphology (see Szalay, Chapter 5, this volume), and the anatomy of the orbit and nasal fossa (Cave, 1967; Cartmill, 1972), all of which indicate that anthropoids are more closely related to tarsiers than to lemuriforms living or fossil. Gingerich's hypothesis rests principally on his conclusion that the Oligocene catarrhine *Apidium* had a free, intrabullar tympanic. After examining the specimen and discussing it with Dr. Gingerich, I am persuaded that he has correctly described its morphology, but that the only trait which it displays that is not also seen in adult ceboids is the persistently open articulation between the anterior tympanic crus and the squamosal. Since the tympanic ring must (as Gingerich points out) have been roughly vertical in orientation, any subtympanic extension of the tympanic cavity would have had to extend anterior and at least slightly lateral to the squamosal–tympanic articulation, if it was to have enclosed the anterior crus as it does in lemuriforms. But this area in the fossil is preempted by the postglenoid process and the glenoid fossa. Furthermore, neither a squamosal–petrosal articular surface not a descending bullar wing of the squamosal appears to have been present anterior to the squamosal–tympanic articulation. The unfused squamosal–tympanic articulation, which constitutes the chief similarity between *Apidium* and lemuriforms, persists into early postnatal life in *Homo* (Fig. 9B), and may have persisted into early adult life in the ancestral catarrhines.* If, as suggested above, the lemuriform bullar configuration is primitive for the primates, its persistence in an Oligocene catarrhine would not in any case demonstrate adapid affinities.

It may be objected that it would be equally parsimonious to assume that the lorisiform bullar configuration is primitive and that a subtympanic extension of the tympanic cavity was evolved independently in the lineages leading to *Plesiadapis*, *Necrolemur*, and the adapids and later lemuriforms. Three lineages undergo bullar transformation in this reconstruction. If the lemuriform condition is taken as primitive, the number of lineages in which independent acquisition of the lorisiform–ceboid configuration must be postulated ranges from four (assuming that *Phenacolemur* resembles *Tarsius*) or five (assuming independent acquisition in the Old and New World anthropoids) to two (assuming that *Phenacolemur* resembles *Plesiadapis* and that the living Haplorhini have a common ancestor which is not also ancestral to *Necrolemur*).

In the present state of our knowledge, application of the parsimony principle will not resolve the issue. Our only recourse is to attempt to assess the relative likelihood of evolutionary change in either of the two possible directions. Since the lorisiform–ceboid configuration of the bulla is attained and later lost during the ontogeny of the lemuriform bulla, it would be relatively easy to retain this configuration into adult life. Independent addition of the postnatal lemuriform phase of bullar development, during which the tympanic cavity wraps itself laterally around the bottom of the eardrum, seems correspondingly less likely to have occurred in multiple parallel lineages. Nevertheless, it would be reasonable to postulate such parallelism if it could be shown that this change conferred some adaptive advantage. We must proceed to ask under what circumstances natural selection might tend to favor one bullar configuration over another.

C. Allometry and the Auditory Apparatus

The morphological differences between the bullae of Lemuriformes and Lorisiformes appear to have negligible functional significance; bivariate plots of stimulus frequency against sound pressure at threshold levels are effectively identical for *Lemur catta*, *Nycticebus coucang*, *Perodicticus potto*, and *Galago senegalensis* (Gillette *et al.*, 1973). If one type of bulla has been derived from the other, this change therefore probably was not produced because of any direct selective advantage of the derived type, but occurred merely as a secondary effect of some other change taking place in the transitional lineages.

I suggest that allometric factors might be expected to produce lorisiform morphology in any lemuriform lineage undergoing marked progressive decrease in body size. The mammalian cochlea exhibits pronounced negative allometry; evidently auditory acuity is in some way related to absolute cochlear size, as visual acuity is to retinal area. The components of the middle ear also exhibit negative allometry. Since the required impedance match between the middle and inner ear of mammals is largely effected by the

---

* An unfused squamosal-tympanic joint is found in *Tarsius* (Hershkovitz, 1974), and its presence in *Apidium* is thus no sign of lemur affinities.

337

STREPSIRHINE
BASICRANIAL
STRUCTURES AND
AFFINITIES OF
CHEIROGALEIDAE

difference in surface area between the eardrum and the footplate of the stapes (Littler, 1965), the ratio of eardrum area to area of the fenestra vestibuli must remain relatively constant in mammals of different size; hence, in small mammals with relatively large inner ears, the eardrum is also relatively large. The eardrum of a 90-g *Microcebus*, for example, has absolute linear dimensions almost half those of an adult human being's eardrum.

The air contained in the tympanic cavity is compressed by inward movements of the eardrum and attenuated by outward movements. Since this enforces elastic recoil against any movement of the eardrum, bullar volume represents a mechanical analog of capacitance in an alternating-current circuit. Thus, any decrease in bullar volume relative to eardrum area increases the capacitative reactance, and therefore the input impedance, of the middle-ear linkage. It follows that the transmission characteristics of the middle ear should be improved by opening the bulla, rendering its effective volume infinite. This has been shown to be true of domestic cats (Tonndorf and Khanna, 1968).

Since the ratio of eardrum area to bullar volume would show negative allometry even without allometric variation in its component terms, and since eardrum area in fact shows marked negative allometry, very small mammals are faced with the problem of reducing an augmented impedance at the eardrum. In most diminutive mammals, the problem is handled by increasing bullar volume as much as possible; the mechanical advantage of the ossicular lever system may also be increased (Webster, 1961). In lemuriforms, bullar volume can be increased by a lateral displacement of the tympanic membrane, which will have the secondary effect of reducing the breadth of the anulus membrane. In very small lemuriforms, then, we should expect to find the tympanic ring relatively very large, crowding the floor of the bulla inferiorly, and closely approximated to the lateral bulla wall. This condition is seen in *Microcebus murinus*, the smallest known lemuriform. As shown above, it represents a plausible structural intermediate between *Lemur* and *Loris* (Fig. 6).

If some such process accounts for the disappearance of the subtympanic extension of the tympanic cavity in lorisiforms and modern haplorhines, how can we account for the fixation of this extension in the plesiadapoid ancestry* in the first place? In a diminutive mammal with a relatively huge and horizontally oriented eardrum (e.g., a shrew), a subtympanic extension of the tympanic cavity would significantly and usefully augment the very limited tympanic volume available between the eardrum and the overlying cranial base. It is possible that the lemuriform bullar configuration was acquired for this reason in the insectivoran ancestry of primates; however, many marsupials are also characterized by a lemur-like subtympanic extension of the tympanic cavity, and it has been suggested (Jones and Lambert, 1939) that this may represent a retention from the therian ancestry. If so, we would expect to find an unossified but otherwise lemur-like tympanic cavity in shrews, small didelphids, or other diminutive mammals with relatively large and horizontal tympanic rings.

---

* Examination of the ear region of *Phenacolemur* persuades me that what Szalay called the "petrosal articular surface of the ectotympanic" in 1972, but not in Chapter 5, this volume, is in fact a narrow ossified anulus membrane. This is further support for the notion that primitive primates had a subtympanic extension of the tympanic cavity. (See also Szalay, Chapter 5, Fig. 16.)

V. The Medial Orbital Wall

339

STREPSIRHINE
BASICRANIAL
STRUCTURES AND
AFFINITIES OF
CHEIROGALEIDAE

In most adult mammals, the medial wall of the eye socket is completely ossified. This wall forms as a mosaic of several basicranial and facial bones, usually including the frontal above, the maxilla and palatine below, and the lacrimal at the orbit's anterior edge. The apex of the orbit is always formed by the orbitosphenoid, through which the optic nerve and the ophthalmic artery pass.

In haplorhine primates, the ethmoid element of the cranial base is also exposed in the medial orbital wall, immediately behind the lacrimal. In typical Madagascar lemurs, the ethmoid is not exposed; this area is filled instead by the orbital process of the palatine, which usually extends up to touch the lacrimal and so excludes the frontal from contact with the maxilla. This pattern is also found in *Morganucodon* (Kermack and Kielan-Jaworowska, 1971), polyprotodont marsupials, tree shrews, elephant shrews, leptictids (Butler, 1956), and at least some artiodactyls and carnivorans (Haines, 1950). Many primatologists have accordingly regarded the palatine–lacrimal contact as a primitive retention in Madagascar lemurs.

Major (1901) was the first to describe and figure an ethmoid contribution to the medial orbital wall in strepsirhines. He asserted that such an exposure is characteristic of lorisiforms and is also rather widespread among Madagascar lemurs. As Jones (1917) and Kollmann (1925) demonstrated, Major mistook the anterior portion of the lemuriform palatine for an ethmoid exposure, and most of his observations on Madagascar lemurs are therefore invalid. Jones (1929) mistakenly concluded that Major's observations on lorisiforms were also invalid and proceeded to list the absence of the ethmoid exposure as a diagnostic feature of the Strepsirhini. The true state of affairs was more precisely described by Kollmann (1925), who demonstrated that all lorisiforms and most cheiro-galeids resemble *Tarsius* and higher primates in having an ethmoid contribution to the medial orbital wall.

Since even some experienced morphologists continue to mistake the anterior part of the lemurid palatine for an ethmoid exposure, it is worthwhile to contrast *Microcebus coquereli* with newborn *Lemur mongoz* in this regard (Fig. 11). In the newborn *Lemur*, the postlacrimal element is clearly continuous with the palatine. The center of the orbital part of the palatine bears a large sphenopalatine foramen (nasopalatine foramen of some authors), which transmits branches of the maxillary nerve and artery from the orbital fossa to the nasal fossa. In the adult, the preforaminal part of the palatine becomes inflated and bulges back over the foramen, so that its continuity with the postforaminal part of the bone becomes obscured. In *Microcebus coquereli*, a clear suture separates the pre-foraminal part of the palatine from an ethmoid exposure behind the lacrimal. The identity of this postlacrimal element with the ethmoid is demonstrated by the attachment of the ethmoturbinals to its inner surface and by a suture separating it from the frontal. In *Lemur*, the orbital lamina of the frontal grows down to cover the lateral aspect of the ethmoid; the adult's ethmoturbinals are discernible through this lamina, dorsal to the margins of the preforaminal part of the palatine bone.

If the ethmoid is not exposed in the medial orbital wall, a third mosaic pattern may occur, in which the frontal and maxillary come into contact behind the lacrimal (Fig. 12A). Among Madagascar primates, the frontal–maxillary contact behind the lacrimal is found in the indriines, *Daubentonia*, *Palaeopropithecus* (Hill, 1953), and the archaeolemurines (Piveteau, 1957; Tattersall, 1973*b*). A short frontal–maxillary articulation in this position also occurs more or less commonly in *Phaner* (e.g., AMNH 100831), *Hapalemur* (e.g., AMNH 100534), and *Lepilemur* (Genet-Varcin, 1963). In the archaeolemurines, the frontal–maxillary contact is accompanied by a specialized loss of the preforaminal part of the palatine, yielding a maxillary–orbitosphenoid articulation (Fig. 12C).

Which of these patterns can be taken as primitive? The palatine–lacrimal contact is widespread among nonprimates, and its common occurrence in *Lemur* and *Tupaia* has led some to regard it as the primitive condition. However, an indriid-like frontal–maxillary contact is found in the adapids (Le Gros Clark, 1934; Piveteau, 1957) and in *Plesiadapis* (Russell, 1964). These facts, coupled with the absence of a palatine–lacrimal contact in cheirogaleines and the wide distribution of the adapid pattern among diversely adapted Madagascar lemurs, imply that the frontal–maxillary contact is primitive for the order Primates.

This conclusion is further support for the notion that the primary affinities of the primates are not with the tree shrews and other "menotyphlan" Insectivora, in which the palatine contacts the lacrimal, but with shrews and hedgehogs [Van Valen's (1967) Erinaceota] in which the maxilla contacts the frontal (Butler, 1956; McDowell, 1958). The constant palatine–lacrimal articulation characteristic of the genus *Lemur*, invoked in the past to support inclusion of the tree shrews in the Lemuriformes, appears to be a lemurid peculiarity among primates; its appearance in tree shrews argues *against* their having close ties to the primate ancestry.

Among extant primates, the ethmoid appears in the orbital wall when the orbits are relatively large or closely approximated, or both; the ethmoid exposure can be predicted almost perfectly from skull length and orbital orientation (Cartmill, 1971) or from palate breadth and interorbital breadth. Where orbital size and approximation are not too great relative to the orbit's anteroposterior length, a palatine–frontal contact may persist behind the ethmoid exposure (Fig. 12D)—e.g., in some individuals of *Nycticebus* or *Cheirogaleus medius* (Kollmann, 1925). In more typical cheirogaleids and lorisiforms, the ethmoid extends back to touch the orbitosphenoid (Fig. 12F). Essentially this same configuration is seen in platyrrhines; that seen in catarrhines differs only in having the sphenopalatine foramen located on the orbitosphenoid–palatine suture.

There has been considerable reduction of the palatine in tarsiers, but otherwise their medial orbital wall (Fig. 12E) resembles the transitional pattern seen in some *Cheirogaleus* and *Nycticebus*, with a substantial palatine–frontal contact persisting behind the ethmoid exposure. In adult tarsiers, the sutures in the medial orbital wall usually disappear (Kollmann, 1925). It has been suggested (Cartmill, 1971) that the medial orbital wall of *Necrolemur*, which has a superficially adapid-like configuration (Simons and Russell,

---

FIG. 11. Medial orbital wall of lemuriforms. Above left, newborn *Lemur mongoz* (uncataloged British Museum specimen). The frontal is in broad contact with the palatine, which is divided by a large sphenopalatine foramen (f) into a posterior and anterior half, the latter articulating directly with the lacrimal. Below, stereogram (depth exaggerated) of *Microcebus coquereli* (BMNH 70.1.28.3), showing independent ethmoid element; the prespheno-palatine part of the palatine is small and does not contact the lacrimal. Above right, labeled diagram of right half of stereogram. Abbreviations: eth, ethmoid; fr, frontal; lac, lacrimal; max, maxilla; os, orbitosphenoid; pal, palatine; pt, pterygoid. The arrow in the diagram indicates the sphenopalatine foramen.

341

STREPSIRHINE
BASICRANIAL
STRUCTURES AND
AFFINITIES OF
CHEIROGALEIDAE

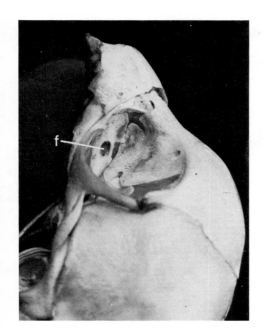

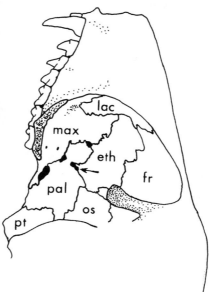

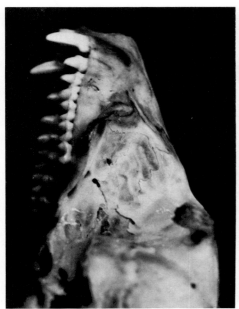

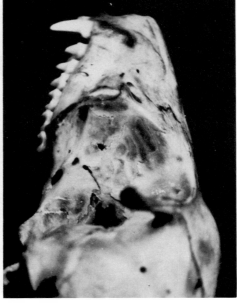

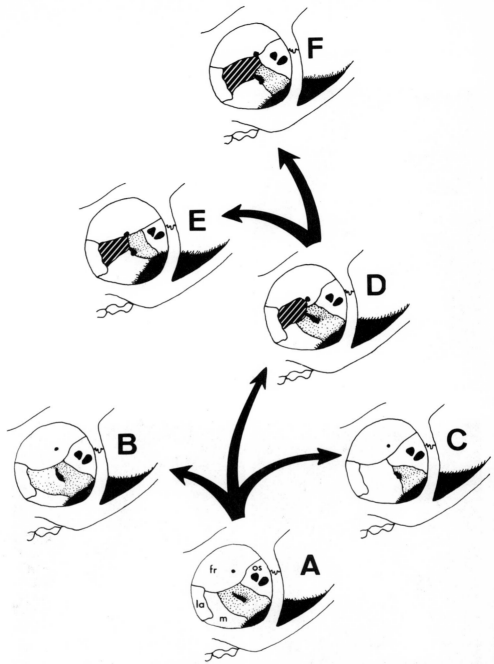

Fig. 12. Medial orbital wall of primates (diagrammatic). Palatine stippled, ethmoid shown by diagonal hachure. A, primitive pattern (adapids, indriids, some *Phaner*); B, typical lemurid pattern, with palatine extending up to lacrimal; C, pattern seen in *Archaeolemur*, with secondary loss of presphenopalatine portion of palatine; D, small ethmoid exposure with persisting frontal–palatine contact (seen in some *Nycticebus* and *Cheirogaleus*); E, pattern seen in *Tarsius*, resembling D but with loss of presphenopalatine portion of palatine; F, typical cheirogaleid, lorisiform, and anthropoid pattern, with ethmoid contacting orbitosphenoid. Arrows indicate probable evolutionary pathways from one pattern to another. Abbreviations: fr, frontal; la, lacrimal; m, maxilla; os, orbitosphenoid.

1960), may have been tarsier-like in early ontogenetic stages, and that the configuration seen in the adult results from a secondary obliteration of the frontal–ethmoid suture, as in adult *Microcebus*. The extreme narrowness of *Necrolemur's* interorbital septum renders it likely that the ethmoid would have been exposed; in some specimens from the Quercy phosphorites (e.g., MNHN 10879 and MNHN 11060), the septum is reduced to a single bony lamina near the orbital apex, as in tarsiers and small anthropoids. *Pseudoloris* probably had a similar arrangement. As noted above, this configuration of the anterior cranial base constitutes an additional reason for positing special affinities between necrolemurines, tarsiines, and anthropoids (Cartmill, 1972).

Since *Pronycticebus*, which is a small adapid with large orbits and clearly demarcated sutures, shows no sign of having had an exposure of the ethmoid in the medial orbital wall, it seems likely that the last common ancestor of the extant primates also lacked such an exposure. The orbital mosaic of the indriids probably preserves the primitive strepsirhine arrangement, from which the patterns seen in *Lemur*, lorisiforms, and the typical cheirogaleids have been derived. Since the appearance of the ethmoid in the medial orbital wall is at least partly conditioned by allometry and orbital approximation, it may have been attained independently in cheirogaleids and lorisiforms (as it was in at least one haplorhine lineage). It is therefore not a conclusive indicator of genetic affinities between the two groups. However, the orbital mosaic of the typical cheirogaleids (Cheirogaleinae) deviates from that of other lemuriforms in a loris-like direction, but retains primitive features (e.g., a substantial preforaminal lamina of the palatine) not seen in lorisiforms. This gradation from primitive to derived traits is suggestive, since it parallels those described above for traits of the carotids and bulla.

VI. Reconstruction of Strepsirhine Phylogeny

We have established that only a few traits distinguish the lorises and galagos from typical Madagascar lemurs, and that most of these traits are either present or prefigured among the cheirogaleids. The phylogenetic correlates of these trait distributions remain to be determined.

Two debatable issues are involved. The first is the phyletic affinities of the European adapids. It has been suggested that independent lorisiform and lemuriform ancestors can be distinguished among the known adapids (Gregory, 1915; Simons, 1962). Others have suggested that the adapids may represent a stage in lemuriform evolution postdating the lemur–loris split (Le Gros Clark, 1934; Szalay, 1971a,b, 1972). A third opinion (Charles-Dominique and Martin, 1970) is that the adapids are a side issue and that the lemurs and lorises derive from a common African ancestor which already possessed a grooming "claw" and dental comb. For our purposes, the first two notions prove to be logically equivalent, and contrast with the third.

The second debatable issue, that of cheirogaleid affinities, is more complex. We can distinguish three significantly different possibilities:

1. The lemur–loris split may antedate the lemur–cheirogaleid split. If so, the

343

STREPSIRHINE
BASICRANIAL
STRUCTURES AND
AFFINITIES OF
CHEIROGALEIDAE

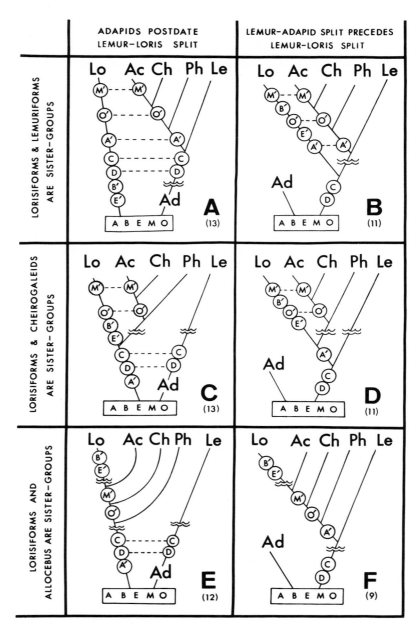

FIG. 13. Models of cheirogaleid relationships. Each of the 6 cladograms (A–F) is the most parsimonious under the assumptions specified for its row and column. Smaller letters signify traits as follows: A, arterial pattern of lemur/adapid type; A', ascending pharyngeal supplants internal carotid; B, bulla extends lateral to eardrum; B', bulla of lorisiform type; C, "claw" on second toe; D, dental comb present; E, enamel of lemur/adapid type; E', enamel of loris type; M, mastoid lacks cellular pneumatization; M', mastoid pneumatized; O, orbital wall lacks ethmoid exposure; O', orbital wall with ethmoid exposure. Abbreviations of taxa (larger letters): Ac, *Allocebus*; Ad, adapids; Ch, *Cheirogaleus* and *Microcebus*; Le, lemuriforms (excluding cheirogaleids); Lo, lorisiforms; Ph, *Phaner*. Numbers in parentheses indicate total number of trait acquisitions (circles) and crossings of Mozambique Channel (wavy lines) required by each phylogeny. Dashed lines show traits acquired in parallel Sequences of events between consecutive branching points are arbitrary.

345

STREPSIRHINE
BASICRANIAL
STRUCTURES AND
AFFINITIES OF
CHEIROGALEIDAE

"sister group" (Hennig, 1965) of the Lorisiformes is the Lemuriformes as a whole.
2. The cheirogaleid–lemur split may antedate a simple cheirogaleid–loris split. If so, the sister group of the Lorisiformes is the Cheirogaleidae.
3. The lorisiforms may represent a relatively recent offshoot from the cheirogaleids. If so, the most likely sister group of Lorisiformes is *Allocebus*.

The possible combinations of these alternatives yield 6 cladograms, diagrammed in Fig. 13. The figure incorporates four additional assumptions: (1) European adapids, like *Notharctus*, lacked a grooming "claw". (2) If adapids are ancestral to any living primates, they are ancestral to the genus *Lemur*. (3) For any trait, the condition found in adapids is the primitive strepsirhine condition, unless the contrary assumption is more parsimonious—which it happens not to be for any of the traits considered here. (4) Derived characters do not revert to the ancestral condition, unless the contrary assumption is more parsimonious—parsimony being evaluated throughout by the number of character transformations and crossings of the Mozambique Channel that a given phylogenetic reconstruction requires.

Given these assumptions, it is easy to see from Fig. 13 that whatever we assume about cheirogaleid affinities, it is always more parsimonious to assume that the specialized dental comb and grooming claw were present in the last common ancestor of the extant strepsirhines—or, in Hennig's (1965, 1966) terminology, that these features are synapomorph for the extant strepsirhines. Similarly, whatever we assume about the adapids, it is always most parsimonious to assume that *Allocebus* has unique affinities with the lorises and galagos. If there is no evidence to the contrary, we cannot do better than to accept these ideas (Fig. 13F) as working hypotheses.

But evidence to the contrary has been proposed and demands to be evaluated. Simons (1962) offered three sorts of evidence in support of his tentative suggestion that an independent lorisiform lineage could be traced into the Adapidae: (1) measurements of the skull of *Pronycticebus*, (2) supposed special resemblances between the dentition of the adapid *Anchomomys quercyi* and that of the Miocene lorisine skull from Rusinga which Walker (1974) has recently reassigned to *Mioeuoticus*, and (3) the bullar morphology of *Pronycticebus*. Most of the cranial measurements used by Simons reflect skull size either directly or allometrically, and the data he presents tell us little beyond the fact that *Pronycticebus* and *Nycticebus* are smaller than *Adapis magnus* and *Lemur*. The supposed dental resemblances of *Mioeuoticus* to *Anchomomys* were discounted by Simpson (1967).

The features of the bulla which Simons considered suggestive of possible lorisiform affinities for *Pronycticebus* are the proximity of the tympanic ring to the bullar meatus and the apparent fusion of the tympanic ring's posterior third to the lateral wall of the bulla. Study of the type specimen (Fig. 14) shows that *Pronycticebus* is less loris-like than *Microcebus murinus* in both these respects. The principal difference between *Pronycticebus* and *Lemur* is that the former's anulus membrane has become partly ossified as an extension of the petrosal. The evidence so far advanced does not warrant tracing the lemur–loris split back into the adapids.

Granted that the morphological evidence favors the Szalay–Katz hypothesis, it remains to be asked whether lorisiforms are derived from Madagascar lemurs by back-

FIG. 14. *Pronycticebus gaudryi*, holotype, MNHN 11056. Stereographic view of right bulla from ventral aspect, showing tympanic ring.

migration, or whether chierogaleids represent a persistently primitive lorisiform population that colonized Madagascar some time after the original colonization by the lemurid ancestors. Two facts favor the former interpretation. The first is that lorisiform resemblances are not uniform among the cheirogaleids; *Phaner* is clearly very much unlike lorises and galagos, whereas a striking suite of morphological and behavioral resemblances links *Microcebus murinus* with some of the smaller galagos (Charles-Dominique and Martin, 1970). Since *Allocebus trichotis* resembles lorisiforms in derived features of the bulla in which it differs both from other Madagascar lemurs and from adapids, it seems unlikely that it represents either an ancestral cheirogaleid stock from which other cheirogaleids have evolved back toward a lemurid-like condition or an ancestral Madagascar lemur stock from which other Madagascar lemurs have evolved back toward an adapid-like condition. Parallel evolution cannot be ruled out, but the anatomical facts suggest that lorisiforms originated from an *Allocebus*-like Madagascar cheirogaleid that managed to recross the Mozambique Channel. The other fact that favors this interpretation is parasitological. Madagascar lemurs are parasitized by several genera of psoroptid mites sufficiently distinctive in features of the cuticle and genitalia to constitute a subfamily Makialginae. A related but peculiarly specialized genus, *Galagalges*, has been recovered from a single specimen of *Galago senegalensis* collected in Zaire; its oddly elongated body and many details of its surface anatomy seemed to Fain (1963a) to warrant erecting a new family Galagalgidae for its reception. Subsequently, Fain (1963b) recovered and described a new psoroptid genus, *Cheirogalalges*. This genus, known to infest both species of *Cheirogaleus*, but not recovered from any other lemurs to date, is intermediate in most respects between the more primitive morphology fround in the mites of typical lemurids and the specialized condition found in the galago parasite. This independent piece of evidence supports the Szalay–Katz hypothesis.

The following reconstruction of strepsirhine phylogeny seems most compatible

347

STREPSIRHINE
BASICRANIAL
STRUCTURES AND
AFFINITIES OF
CHEIROGALEIDAE

with the facts presented above. The last common ancestor of the living strepsirhines colonized Madagascar during the early Tertiary, following the island's separation from the African mainland, or (less probably) during the latest Cretaceous, before separation began. This common ancestor resembled cheirogaleids, indriids, and daubentoniids in lacking a palatine–lacrimal contact in the orbit; resembled lemurids, indriids, and daubentoniids in having neither an orbital exposure of the ethmoid nor a substantial foramen lacerum in the adult; resembled cheirogaleids and lemurids in dental formula; and resembled all extant Madagascar primates in having a subtympanic extension of the tympanic cavity.

The last common ancestor of the extant strepsirhines was probably a small animal, although the evidence for this is tenuous. *Microcebus murinus* and *Cheirogaleus medius* display less divergence of the thumb, both on flat and cylindrical supports, than do various species of the genus *Lemur* (Bishop, 1964; Cartmill, 1974). This is almost certainly a primitive feature lost in larger lemuriforms. It therefore seems likely that a secondary increase in weight has demanded increased specialization of the hand for grasping in the larger lemuriforms, for reasons outlined by Napier (1967).

The ancestral cheirogaleid lineage diverged from the other Madagascar lemurs during the Paleogene. The cheirogaleid stock was distinguished by a persistence and enlargement of the fetal anastomosis between the ascending pharyngeal and internal carotid arteries, but remained otherwise primitive. The phanerine lineage, which must have separated from other cheirogaleids prior to the mid-Oligocene, preserves the ancestral condition in many features, although the only extant phanerine shows several specialized traits (detailed by Tattersall and Schwartz, this volume).

The Cheirogaleinae, regarded here as comprising *Cheirogaleus*, *Microcebus*, and *Allocebus*, are distinguished from phanerines by having an ethmoid exposure in the orbital wall; this may be a convergence, a shared derived trait, or a retention (lost in *Phaner*) from the cheirogaleid ancestry. Within the cheirogaleines, *Microcebus coquereli* appears to be persistently primitive in lacking the physiological and anatomical specializations for seasonal fat storage seen in *M. murinus* and the two *Cheirogaleus* species. If future work discloses other derived features with a similar distribution, we will be justified in concluding that *M. murinus* and *Cheirogaleus* have a more recent common ancestor than either has with *M. coquereli*. This would warrant sinking *Microcebus* into *Cheirogaleus* and resurrecting Gray's (1870) genus *Mirza* for the reception of *M. coquereli*.

Allocebus trichotis clearly deserves separation from the other cheirogaleines at least at the generic level, although the cladistics of this separation are unclear. The cellular mastoid pneumatization which this species shares with lorisiforms is interpreted here as a derived (or synapomorph) feature linking it uniquely to lorisiforms. If so, then the lineage leading to *Allocebus* must have separated from the other cheirogaleine lineages by the early Miocene.

By the middle of the Miocene, the basal lorisiform stock in Africa had already differentiated into lorisines and galagines (Walker, 1974). The divergence of these two groups appears to have been produced by character displacement (Charles-Dominique, 1971). This implies that the ancestral lorisines and galagines were sympatric; therefore, it seems unlikely that the Asian and African lorisines developed their shared specializations independently from a galago-like common ancestry, as Walker (1969) at one time thought possible. Although most of the traits that distinguish lorisines from galagines can be explained in terms of the lorisine locomotor habit (e.g., the vascular bundles in the limbs, reduced tail, decreased orbital frontation, elongated lumbar region, exaggerated grasping specializations of the hands and feet, loss of the ulnocarpal articulation, and perhaps the deflated bulla and simplified carotid rete), some lorisine traits seem unrelated to locomotion (e.g., the venous foramina transversaria of the postcervical vertebrae and the raised temporal lines). The degree of parallel evolution demanded by the hypothesis of lorisine diphyly seems implausible, even granting broadly parallel selection pressures on the various lorisine lineages.

Known features of the Miocene lorisiforms are compatible with the notion that lorisiforms originated from an earliest Miocene cheirogaleid near the *Allocebus* ancestry. Miocene galagines had begun to develop distinctively *Galago*-like femora (Walker, 1970), but their calcanei still retained proportions like those of cheirogaleids (Martin, 1972). The dental combs of the Miocene lorisiforms appear to have been indistinguishable from those of typical modern strepsirhines (Simpson, 1967; Walker, 1969), rather than persistently primitive as Le Gros Clark and Thomas (1952) suggested. Mastoid pneumatization is reported to be less elaborate in the Miocene lorisiform skulls than in those of extant lorisiforms (Le Gros Clark and Thomas, 1952; Le Gros Clark, 1956).

Among extant lorisiforms, *Loris* uniquely resembles *Mioeuoticus* in retaining a subtympanic extension of the petrosal; either this primitive feature has been lost independently in different lorisiform lineages, or *Mioeuoticus* has special affinities with *Loris*. In either event, the persistence of this archaic feature in *Loris* indicates that its ancestral lineage split off from the lorisine ancestry during the early Neogene. Even if Tattersall (1969) is correct in suggesting that *Indraloris lulli* from India may not be an early Pliocene lorisine, the ancestors of *Loris* had probably entered India from Africa by the beginning of the Pliocene.

VII. Taxonomic Implications

The evidence presented here, added to that marshalled by Szalay and Katz (1973), warrants the conclusion that lorises and galagos are more closely related to cheirogaleids, and probably to *Allocebus* in particular, than to other Madagascar lemurs. Those who agree with this conclusion, and who are anxious to make classifications as vertical as possible, will probably favor the transfer of the family Cheirogaleidae to the Lorisiformes. Indeed, this step was taken by some of the other contributors to this volume in the preliminary drafts of their articles. Those who agree with Hennig and his followers that any "paraphyletic" taxon (i.e., one defined on shared traits retained from its common ancestry with "sister" taxa of equivalent rank) is illegitimate, will be required to do this, since the taxon Lemuriformes must be paraphyletic if lemuriforms gave rise to lorisiforms.

I believe that any such move is premature at this time, for several reasons:

1. The derived traits linking cheirogaleids to lorisiforms have a lower phyletic

valence than was thought in the past. The lorisiform ascending pharyngeal–entocarotid anastomosis and "extrabullar" tympanic ring, attained or prefigured in cheirogaleids, represent modified retentions of fetal lemuriform morphology. These retentions could easily have become fixed in parallel lineages for functional reasons suggested above. A trait which appears in the adult by premature cessation of an ontogenetic process cannot be weighted as heavily in judging affinities as a trait whose appearance results from the addition of a new ontogenetic process.

2. The presently available biochemical evidence (Cronin *et al.*, 1974; Goodman, this volume) links cheirogaleids more closely to lemurids and indriids than to lorisiforms.

3. The phylogeny diagramed in Fig. 13F is more parsimonious than its alternatives only if it is assumed that parallel fixation of some new morphological trait in two different populations is about as probable as an animal's managing to cross the Mozambique Channel and reproduce itself on the other side. There is no evidence for this assumption, and the poverty of the mammal fauna of Madagascar suggests that crossings of the Mozambique Channel may be very improbable indeed. If the probability of the parallel appearance of one trait in two lineages is more than three times as great as the probability of a successful crossing of the Mozambique Channel, then the phylogeny suggested by Charles-Dominique and Martin (1970), diagrammed in Fig. 13B, is the most parsimonious. There is no way of assigning a value to these probabilities at present.

4. Even if lorisiforms are descended from early cheirogaleids, the ancestral lorisiform is not likely to have been the first branch to diverge from the cheirogaleid stem. The most probable "sister group" of the lorisiforms is *Allocebus trichotis*. In a strictly vertical system of classification, in which sister groups must have the same taxonomic rank (Hennig, 1965, p. 115), this phyletic interpretation implies that all the lorisiform genera would have to be sunk into a single genus; this would ultimately be grouped together with the cheirogaleine genera into a family Lorisidae, which would have a new family, Phaneridae, as a sister group. The resulting nomenclatorial chaos would far outweigh any gain in the phylogenetic information conveyed by the classification.

Apart from the most enthusiastic phyleticists, all primate systematists would agree in rejecting the foregoing classificatory arrangement, even though it reflects the most probable phylogenetic relationships of the species in question. The simplest reason for rejecting it is that the phylogenetic reconstruction underlying it is based entirely on a few morphological features of extant animals and is actually contradicted by serological data. A few more bits of information might dictate an altogether different reconstruction.

At present, there is virtually no aspect of primate phylogeny on which all competent authorities agree. Even the division of the order into Haplorhini and Strepsirhini, taken for granted by most of the contributors to this volume, is vigorously opposed on reasonable grounds by several morphologists and serologists. Descending to lower levels of the Linnaean hierarchy, various reputable workers have recently proposed phyletic schemes that would group the catarrhines with the lemuriforms, the plesiadapoids with the tarsiers, the tarsiers with the tree shrews, or the gibbons with the cercopithecoids. Clearly, the time is not ripe for attempts to make our classificatory units mirror our ideas about phylogeny. If we wish to go on communicating intelligibly with each other, with our students, and with future generations of primatologists, the best course of

349

STREPSIRHINE
BASICRANIAL
STRUCTURES AND
AFFINITIES OF
CHEIROGALEIDAE

action is to continue using paraphyletic (or "wastebasket") taxa where phylogeny is acknowledged to be uncertain. Because we lack early Tertiary fossil strepsirhines in Madagascar and Africa, the taxon Lemuriformes is one of the most useful wastebaskets we have to work with. Any attempt to split this taxon into vertical phyletic units will almost certainly be prompted by speculative phylogenetic reconstruction, grounded in an absence of relevant fossil data, and productive of unending confusion about nomenclature.

VIII. Summary

The cheirogaleid lemurs of Madagascar resemble the lorises and galagos (Lorisiformes) in having an enlarged ascending pharyngeal ("anterior carotid") artery which provides the major extravertebral blood supply to the brain. Some, but not all, cheirogaleids resemble lorisiforms in having an ethmoid exposure in the medial orbital wall and a reduced subtympanic extension of the tympanic cavity. The cheirogaleid *Allocebus trichotis* is unique among extant Madagascar lemurs in displaying cellular pneumatization of the mastoid and squamous portions of the temporal bone, as in lorisiforms. It is concluded that cheirogaleids probably have closer affinities to lorisiforms than to other Madagascar lemurs. The differential distribution of lorisiform traits among cheirogaleids, together with some parasitological evidence, suggests that lorisiforms arose from a cheirogaleid near *Allocebus*, which invaded Africa from Madagascar sometime around the close of the Oligocene. Nevertheless, since the lorisiform configurations of the bulla and carotids represent modified and functionally significant retentions of fetal lemuriform morphology, and since the orbital exposure of the ethmoid is largely conditioned by factors of allometry and orbital orientation, most of the traits shared by lorisiforms and cheirogaleids might easily have been acquired independently. For this and more theoretical reasons, transfer of the Cheirogaleidae to the Lorisiformes would be premature.

ACKNOWLEDGMENTS

Many people at European and American museums made this study possible by their generosity in admitting me to their collections and finding work space, instruments, and specimens for me. I particularly thank G. B. Corbet and P. H. Napier, at the British Museum (Natural History); D. Goujet, C. Poplin, and D. E. Russell, at the Museum national d'Histoire naturelle in Paris; H. Zapfe, of the Paläontologisches Institut of the University of Vienna; P. D. Gingerich and E. L. Simons, at the Peabody Museum of Natural History; and I. Tattersall at the American Museum of Natural History. I thank Kaye Brown, W. L. Hylander, and R. F. Kay for their generous help and critical comments. I also thank J. Bergeron, P. Charles-Dominique, R. D. Martin, and T. L. Strickler for their kind assistance, and Alice Wheeler for her help in preparing the manuscript. This research was aided by travel funds from the Wenner–Gren Foundation for Anthropological Research, Inc., and by the Anatomy Department and Primate Facility Fund of Duke University.

ADAMS, W. E. 1957. The extracranial carotid rete and carotid fork in *Nycticebus coucang*. *Ann. Zool.* (Agra) **2**:21–38.

ASK-UPMARK, E. 1953. On the entrance of the carotid artery into the cranial cavity in *Stenops gracilis* and *Otolicnus crassicaudatus*. *Acta Anat.* **19**:101–103.

BAKER, M. A., and J. N. HAYWARD. 1967. Carotid rete and brain temperature of cat. *Nature* **216**:139–141.

BAKER, M. A., and J. N. HAYWARD. 1968. The influence of the nasal mucosa and the carotid rete upon hypothalamic temperature in sheep. *J. Physiol.* (*London*) **198**:561–579.

BARNETT, C. H., and C. W. BRAZENOR. 1958. The testicular rete mirabile of marsupials. *Austr. J. Zool.* **6**:27–32.

BERGMANN, L. 1942. Studies on the blood vessels of the human Gasserian ganglion. *Anat. Rec.* **82**:609–630.

BISHOP, A. 1964. Use of the hand in lower primates, pp. 133–225. *In* J. Buettner-Janusch, ed., *Evolutionary and Genetic Biology of Primates*, Vol. 2. Academic Press, New York.

BUGGE, J. 1972. The cephalic arterial system in the insectivores and the primates with special reference to the Macroscelidoidea and Tupaioidea and the insectivore–primate boundary. *Z. Anat. Entwickl.* **135**:279–300.

BUGGE, J. 1974. The cephalic arterial system in insectivores, primates, rodents and lagomorphs, with special reference to the systematic classification. *Acta Anat.* **87**(Suppl. 62):1–160.

BUTLER, P. M. 1956. The skull of *Ictops* and the classification of the Insectivora. *Proc. Zool. Soc. London* **126**:453–481.

CARTER, J. T. 1922. On the structure of the enamel in the primates and some other mammals. *Proc. Zool. Soc. London* **1922**:599–608.

CARTMILL, M. 1971. Ethmoid component in the orbit of primates. *Nature* **232**:566–567.

CARTMILL, M. 1972. Arboreal adaptations and the origin of the order Primates, pp. 97–122. *In* R. Tuttle, ed., *The Functional and Evolutionary Biology of Primates*. Aldine-Atherton, Chicago.

CARTMILL, M. 1974. Pads and claws in arboreal locomotion, pp. 45–83. *In* F. A. Jenkins, J., ed., *Primate Locomotion*. Academic Press, New York.

CAVE, A. J. E. 1967. Observations on the platyrrhine nasal fossa. *Am. J. Phys. Anthropol.* **26**:277–288.

CHAPUIS, G. 1966. Contribution à l'étude de l'artère carotid interne des Carnivores. *Mammalia* **30**:82–96.

CHARLES-DOMINIQUE, P. 1971. Éco-éthologie des Prosimiens du Gabon. *Biol. Gabonica* **7**:121–228.

CHARLES-DOMINIQUE, P., and R. D. MARTIN. 1970. Evolution of lorises and lemurs. *Nature* **227**:257–260.

CRAFTS, R. C. 1966. *A Textbook of Human Anatomy*. Ronald Press, New York.

CRONIN, J. E., V. M. SARICH, and Y. RUMPLER. 1974. Albumin and transferrin evolution among the Lemuriformes. *Am. J. Phys. Anthropol.* **41**: 473–474

DANIEL, P. M., J. D. K. DAWES, and M. M. L. PRICHARD. 1953. Studies of the carotid rete and its associated arteries. *Philos. Trans. R. Soc.* (*London*) Ser. B **237**:173–208.

DAVIES, D. V. 1947. The cardiovascular system of the slow loris. *Proc. Zool. Soc. London* **117**:377–410.

DAVIS, D. D. 1964. The giant panda: a morphological study of evolutionary mechanisms. *Fieldiana* (*Zool. Mem.*) **3**:1–339.

DAVIS, D. D., and H. E. STORY. 1943. The carotid circulation in the domestic cat. *Zool. Ser. Field Mus. Nat. Hist.* **28**:1–47.

ECKSTEIN, P. 1958. Internal reproductive organs. *Primatologia* **3**(1):542–629.

FAIN, A. 1963a. *Galagalges congolensis* g. n., sp. n. Un nouvel Acarien psorique du Galago (Sarcoptiformes). *Rev. Zool. Bot. Afr.* **67**:242–250.

FAIN, A. 1963b. Les Acariens producteurs de gale chez les Lémuriens et les Singes, avec une étude des Psoroptidae (Sarcoptiformes). *Bull. Inst. R. Sci. Nat. Belg.* **39**(32):1–125.

FORBES, H. O. 1896. *A Hand-book to the Primates*. Vol. 1. Edward Lloyd, London.

GAZIN, C. L. 1958. A review of the Middle and Upper Eocene primates of North America. *Smithson. Misc. Coll.* **136**:1–112.

GENET-VARCIN, E. 1963. *Les Singes actuels et fossiles*. N. Boubée et Cie., Paris.

GILLETTE, R. G., R. BROWN, P. HERMAN, S. VERNON, and J. VERNON. 1973. The auditory sensitivity of the lemur. *Am. J. Phys. Anthropol.* **38**:365–370.

GINGERICH, P. D. 1973. Anatomy of the temporal bone in the Oligocene anthropoid *Apidium* and the origin of Anthropoidea. *Folia Primatol.* **19**:329–337.

GINGERICH, P. D. 1974. Cranial Anatomy and Evolution of Early Tertiary Plesiadapidae. Ph.D. Thesis, Yale University, New Haven, Connecticut.

GOODRICH, E. S. 1930. *Studies on the Structure and Development of Vertebrates*. 1958 reprint, Dover, New York.

GRAY, J. E. 1870. *Catalogue of Monkeys, Lemurs, and Fruit-Eating Bats in the Collection of the British Museum*. British Museum, London.

GREGORY, W. K. 1915. On the classification and phylogeny of the Lemuroidea. *Bull. Geol. Soc. Am.* **26**:426–446.

GREGORY, W. K. 1920. On the structure and relationships of *Notharctus*, an American Eocene primate. *Mem. Am. Mus. Nat. Hist.* **3**:49–243.

GUTHRIE, D. A. 1963. The carotid circulation in the Rodentia. *Bull. Mus. Comp. Zool. (Harvard)* **128**:455–481.

HAINES, R. W. 1950. The interorbital septum in mammals. *J. Linn. Soc. London (Zool.)* **41**:585–607.

HENNIG, W. 1965. Phylogenetic systematics. *Ann. Rev. Entomol.* **10**:97–116.

HENNIG, W. 1966. *Phylogenetic Systematics*. University of Illinois Press, Urbana.

HERSHKOVITZ. P. 1974. The ectotympanic bone and origin of higher primates. *Folia Primatol.* **22**:237–242.

HILL, W. C. O. 1953. *Primates: Comparative Anatomy and Taxonomy. I. Strepsirhini*. Edinburgh University Press, Edinburgh.

HILL, W. C. O. 1958. External genitalia. *Primatologia* **3**(1):630–704.

HILL, W. C. O., and D. V. DAVIES. 1954. The reproductive organs in *Hapalemur* and *Lepilemur*. *Proc. R. Soc. Edinburgh, Sect. B* **65**:251–270.

HOFFSTETTER, R. 1974. Phylogeny and geographical deployment of the primates. *J. Hum. Evol.* **3**:327–350.

HÜRZELER, J. 1948. Zur Stammesgeschichte der Necrolemuriden. *Abh. Schweiz. Palaeont. Ges.* **66**(3):1–46.

JEWELL, P. A. 1952. The anastomoses between internal and external carotid circulations in the dog. *J. Anat.* **86**:83–94.

JONES, F. W. 1917. The structure of the orbitotemporal region of the skull of *Lemur*. *Proc. Zool. Soc. London* **1917**:323–329.

JONES, F. W. 1924. *The Mammals of South Australia. Part 2*. British Science Guild, Adelaide, Australia.

JONES, F. W. 1929. *Man's Place Among the Mammals*. Edward Arnold and Co., London.

JONES, F. W., and V. F. LAMBERT. 1939. The occurrence of the lemurine form of the ectotympanic in a primitive marsupial. *J. Anat.* **74**:72–75.

JOUFFROY, F. K. 1962. La musculature des membres chez les Lémuriens de Madagascar. Étude descriptive et comparative. *Mammalia* **26**(Suppl. 2):1–326.

KAMPEN, P. N. VAN. 1905. Die Tympanalgegend des Säugetierschädels. *Gegenbaurs Morphol. Jahrb.* **34**:321–722.

KANAGASUNTHERAM, R., and A. KRISHNAMURTI. 1965. Observations on the carotid rete in the lesser bush baby (*Galago senegalensis senegalensis*). *J. Anat.* **99**:861–875.

KERMACK, K. A., and Z. KIELAN-JAWOROWSKA. 1971. Therian and non-therian mammals. *Zool. J. Linn. Soc. London* **50**(Suppl. 1):103–115.

KLAAUW, C. J. VAN DER. 1931. The auditory bulla in some fossil mammals. *Bull. Am. Mus. Nat. Hist.* **62**:1–352.

KOLLMANN, M. 1925. Études sur les Lémuriens. La fosse orbito-temporale et l'os planum. *Mém. Soc. Linn. Normandie (Caen), Sect. Zool., N.S.* **1**:1–20.

KOLLMANN, M., and L. PAPIN. 1925. Études sur les Lémuriens. Anatomie comparée des fosses nasales et de leurs annexes. *Arch. Morphol.* **22**:1–60.

KRISHNAMURTI, A. 1968. The cerebral arteries of *Nycticebus coucang coucang*. *Folia Primatol.* **8**:159–168.

LE GROS CLARK, W. E. 1934. On the skull structure of *Pronycticebus gaudryi*. *Proc. Zool. Soc. London* **1934**:19–27.

LE GROS CLARK, W. E. 1956. A Miocene lemuroid skull from East Africa. *Brit. Mus. (Nat. Hist.) Fossil Mamm. Afr.* **9**:1–6.

LE GROS CLARK, W. E. 1959. *The Antecedents of Man: An Introduction to the Evolution of the Primates*. Edinburgh University Press, Edinburgh.

LE GROS CLARK, W. E., and D. P. THOMAS. 1952. The Miocene lemuroids of East Africa. *Brit. Mus. (Nat. Hist.) Fossil Mamm. Afr.* **5**:1–20.

LIE, T. A. 1968. *Congenital Anomalies of the Carotid Arteries*. Excerpta Medica Foundation, Amsterdam.

LITTLER, T. S. 1965. *The Physics of the Ear*. Macmillan, New York.

MAJOR, C. I. F. 1894. Über die malagassischen Lemuriden-Gattungen *Microcebus*, *Opolemur*, und *Chirogale*. *Novit. Zool.* **1**:2–39.

MAJOR, C. I. F. 1899. On the skulls of some Malagasy lemurs. *Proc. Zool. Soc. London* **1899**:987–988.

MAJOR, C. I. F. 1901. On some characteristics of the skull in the lemurs and monkeys. *Proc. Zool. Soc. London* **1901**:129–153.

MARTIN, R. D. 1972. Adaptive radiation and behaviour of the Malagasy lemurs. *Philos. Trans. R. Soc. London, Ser. B* **264**:295–352.

MATTHEW, W. D. 1909. The Carnivora and Insectivora of the Bridger Basin, Middle Eocene. *Mem. Am. Mus. Nat. Hist.* **9**:289–567.

353

STREPSIRHINE
BASICRANIAL
STRUCTURES AND
AFFINITIES OF
CHEIROGALEIDAE

McCoy, H. A., D. R. Swindler, and J. W. Albers. 1966. The external carotid artery of the baboon and the rhesus monkey: Branching pattern and distribution, pp. 151–179. *In* H. Vagtborg, ed., *The Baboon in Medical Research*. University of Texas Press, Austin.

McDowell, S. B. 1958. The Greater Antillean insectivores. *Bull. Am. Mus. Nat. Hist.* **115**:113–214.

McKenna, M. C. 1966. Paleontology and the origin of the primates. *Folia Primatol.* **4**:1–25.

Miller, M. E., G. C. Christensen, and H. E. Evans. 1964. *Anatomy of the Dog*. W. B. Saunders, Philadelphia.

Miller, R. A. 1943. Functional and morphological adaptations in the forelimbs of the slow lemurs. *Am. J. Anat.* **73**:153–183.

Mills, S. H., and J. E. Heath, 1970. Thermoresponsiveness of the preoptic region of the brain in house sparrows. *Science* **168**:1008–1009.

Mivart, St. G. 1873. On *Lepilemur* and *Cheirogaleus* and on the zoological rank of the Lemuroidea. *Proc. Zool. Soc. London* **1873**:484–510.

Napier, J. R. 1967. Evolutionary aspects of primate locomotion. *Am. J. Phys. Anthropol.* **27**:333–342.

Petter-Rousseaux, A. 1964. Reproductive physiology and behavior of the Lemuroidea, pp. 91–132. *In* J. Buettner-Janusch, ed., *Evolutionary and Genetic Biology of Primates*, Vol. 2. Academic Press, New York.

Petter-Rousseaux, A., and J.-J. Petter. 1967. Contribution à la systématique des *Cheirogaleinae* (Lémuriens malgaches). *Allocebus*, gen. nov., pour *Cheirogaleus trichotis* Günther, 1875. *Mammalia* **31**:574–582.

Piveteau, J. 1957. *Traité de paléontologie*, Vol. 7. Masson et Cie., Paris.

Pocock, R. I. 1922. On the external characters of the beaver (Castoridae) and some squirrels (Sciuridae). *Proc. Zool. Soc. London* **1922**:1171–1212.

Romanes, G. J., ed. 1964. *Cunningham's textbook of anatomy*. 10th ed. Oxford University Press, London.

Rumpler, Y. 1974. Cytogenetic contributions to a new classification of lemurs, pp. 865–869. *In* R. D. Martin, G. A. Doyle, and A. Walker, eds., *Prosimian Biology*. Duckworth, London.

Russell, D. 1964. Les mammifères Paléocènes d'Europe. *Mém. Mus. Natl. Hist. Nat. (Paris), Sér. C* **13**:1–321.

Saban, R. 1963. Contribution à l'étude de l'os temporal des Primates. *Mém. Mus. Natl. Hist. Nat. (Paris), Sér. A* **29**:1–378.

Schwartz, J. H. 1974. Observations on the dentition of the Indriidae. *Am. J. Phys. Anthropol.* **41**:107–114.

Simons, E. L. 1961. Notes on Eocene tarsioids and a revision of some Necrolemurinae. *Bull. Brit. Mus. (Nat. Hist.) Geol.* **5**:45–69.

Simons, E. L. 1962. A new Eocene primate genus, *Cantius*, and a revision of some allied European lemuroids. *Bull. Brit. Mus. (Nat. Hist.) Geol.* **7**:1–36.

Simons, E. L. 1972. *Primate Evolution: An Introduction to Man's Place in Nature*. Macmillan, New York.

Simons, E. L., and D. E. Russell. 1960. Notes on the cranial anatomy of *Necrolemur*. *Breviora Mus. Comp. Zool. (Harvard)* **127**:1–14.

Simpson, G. G. 1967. The Tertiary lorisiform primates of Africa. *Bull. Mus. Comp. Zool. (Harvard)* **136**:39–62.

Stehlin, H. G. 1912. Die Säugetiere des schweitzerischen Eocaens. VII (1): *Adapis*. *Abh. Schweiz. Palaeont. Ges.* **38**:1165–1298 (not successively paginated).

Story, H. E. 1951. The carotid arteries in the Procyonidae. *Fieldiana (Zool.)* **32**:477–557.

Straus, W. L., Jr. 1931. The form of the tracheal cartilages of primates, with remarks on their supposed taxonomic importance. *J. Mammal.* **12**:281–285.

Szalay, F. S. 1969. Mixodectidae, Microsyopidae, and the insectivore–primate transition. *Bull. Am. Mus. Nat. Hist.* **140**:193–330.

Szalay, F. S. 1971a. Significance of the basicranium of early Tertiary primates for the phylogeny of the order. *Am. J. Phys. Anthropol.* **35**:297.

Szalay, F. S. 1971b. The European adapid primates *Agerina* and *Pronycticebus*. *Am. Mus. Novit.* **2466**:1–19.

Szalay, F. S. 1972. Cranial morphology of the early Tertiary *Phenacolemur* and its bearing on primate phylogeny. *Am. J. Phys. Anthropol.* **36**:59–76.

Szalay, F. S., and C. C. Katz. 1973. Phylogeny of lemurs, galagos and lorises. *Folia Primatol.* **19**:88–103.

Tandler, J. 1899. Zur Vergleichenden Anatomie der Kopfarterien bei den Mammalia. *Denkschr. Kais. Akad. Wiss. (Wien), Math.-Nat. Klasse* **67**:677–784.

Tattersall, I. 1969. More on the ecology of *Ramapithecus*. *Nature* **224**:821–822.

Tattersall, I. 1973a. Subfossil lemuroids and the "adaptive radiation" of the Malagasy lemurs. *Trans. N.Y. Acad. Sci., Ser. 2* **35**:314–324.

Tattersall, I. 1973b. Cranial anatomy of the Archaeolemurinae (Lemuroidea, Primates). *Anthropol. Pap. Am. Mus. Nat. Hist.* **52**:1–110.

Tonndorf, J., and S. M. Khanna. 1968. The quality of impedance matching by the middle ears of cats. *Ann. Otol. Rhinol. Laryngol.* **77**:154–163.

TORRE, E. DE LA, M. G. NETSKY, and I. MESCHAN. 1959. Intracranial and extracranial circulations in the dog: anatomic and angiographic studies. *Am. J. Anat.* **105**:343–382.

VAN VALEN, L. 1965. Tree shrews, primates, and fossils. *Evolution* **19**:137–151.

VAN VALEN, L. 1967. New Paleocene insectivores and insectivore classification. *Bull. Am. Mus. Nat. Hist.* **135**:217–284.

WALKER, A. C. 1969. New evidence from Uganda regarding the dentition of Miocene Lorisidae. *Uganda J. (Kampala)* **33**:90–91.

WALKER, A. C. 1970. Post-cranial remains of the Miocene Lorisidae of East Africa. *Am. J. Phys. Anthropol.* **33**:249–262.

WALKER, A. C. 1974. A review of the Miocene Lorisidae of East Africa, pp. 435–447. *In* R. D. Martin, G. A. Doyle, and A. C. Walker, eds., *Prosimian Biology.* Duckworth, London.

WEBSTER, D. B. 1961. The ear apparatus of the kangaroo rat, *Dipodomys. Am. J. Anat.* **108**:123–148.

HAPLORHINE PHYLOGENY

15

Phylogeny, Adaptations, and Dispersal of the Tarsiiform Primates

FREDERICK S. SZALAY

I. Introduction

The populations of the southeast Asian *Tarsius* are the only living representatives of a group of primates that once flourished in dazzling variety and abundance. The heyday of these small primates, as far as we know, was the Eocene. They were clearly differentiated from a lemuriform ancestry somewhere in the Paleocene and survived into the Miocene. Unlike their contemporaries, the lumuriform adapids which produced species as large as *Leptadapis magnus*, or the several species of *Pelycodus* or *Notharctus*, no known tarsiiform is larger than a South American capuchin or uakari (Fig. 1). What the known tarsiiforms lacked in size, however, they apparently made up in the diversity of their known structural adaptations. This is particularly true of the dentition, hitherto their best known aspect. As inferred from the dental differences, their feeding regimes were probably highly diversified, and, one may suspect, this was also true of the locomotor habits of the various species, although, admittedly, the postcranial morphology is not very well known.

FREDERICK S. SZALAY · Hunter College of the City University of New York, and the American Museum of Natural History, New York, New York.

It is my purpose to summarize here the interrelationships of tarsiiforms, to present some additional evidence for the existence of a haplorhine–strepsirhine dichotomy probably as far back as the Paleocene (see also Szalay, this volume), to suggest some hypotheses for the ancestral tarsiiform adaptations, and, finally, to conclude with some speculations on zoogeographical aspects of tarsiiform evolution along with those of other primates.

II. Phylogeny

A. Historical Review

Views on the relationships of the Omomyidae Trouessart, 1879 (including the Anaptomorphidae Cope, 1883 and Microchoeridae Lydekker 1887) have fluctuated moderately although the creation of these family concepts signified that the group was early recognized as distinct from the Adapidae. A significant early step was that of Schlosser, in 1887, who pointed to special similarities between the North American *Omomys* and the European *Necrolemur*.

In his important 1903 monograph on Eocene primates Wortman presented a major review of primate phylogeny and classification. Unlike most of his paleontologist predecessors studying primate phylogeny, Wortman boldly asserted the closer ties of *Tarsius* and allies to the platyrrhines and catarrhines than to lemuroids, using the suborder Anthropoidea (a concept originally intended, however, for platyrrhines and catarrhines only) for the three former groups. Within the tarsioid primates, which he referred to as Paleopithecini, he recognized the Tarsiidae and the Anaptomorphidae. Astutely, he divided the Anaptomorphidae into the Omomyinae and Anaptomorphinae and he included *Microchoerus* and *Necrolemur* in the latter. Years later, the Tarsiiformes, erected by Gregory (1915), included two families, the Michrochoeridae and the Tarsiidae. The latter encompassed both the Anaptomorphinae and Omomyinae of Wortman. Gregory's Tarsiiformes was of equal rank to the Lemuriformes and Lorisiformes within the suborder Lemuroidea. Consequently, Gregory (1920) again indicated his allocation of the Anaptomorphidae when he referred to skulls of *Tetonius* and *Necrolemur* as those of tarsiiforms.

Simpson, in the primate section of his 1937 milestone faunal study of the Paleocene Fort Union of Crazy Mountain Field, referred to the Eocene nonlemuroids as "acknowledged tarsioids" (p. 145), and considered the Paleocene genera, now grouped under the Paromomyidae, as having tarsioid ties. In years to come he carried this suggestion further (see below) by including the Paromomyinae in the Anaptomorphidae. However, in his study of the skeletal remains of *Hemiacodon* in 1940, Simpson came to doubt the clear-cut differentiation of tarsioid–lemuroid traits in Eocene primates, saying little about the phylogeny *per se* of these groups. Through his study of *Hemiacodon* Simpson essentially cast the most serious doubt on the validity of a tarsiiform–lemuriform dichotomy, a view which he espoused again, with even greater persuasiveness, in his 1955 work on the Phenacolemuridae. Simpson's 1940 and 1945 concept of the Anap-

359

PHYLOGENY,
ADAPTATIONS,
AND DISPERSAL
OF TARSIIFORM
PRIMATES

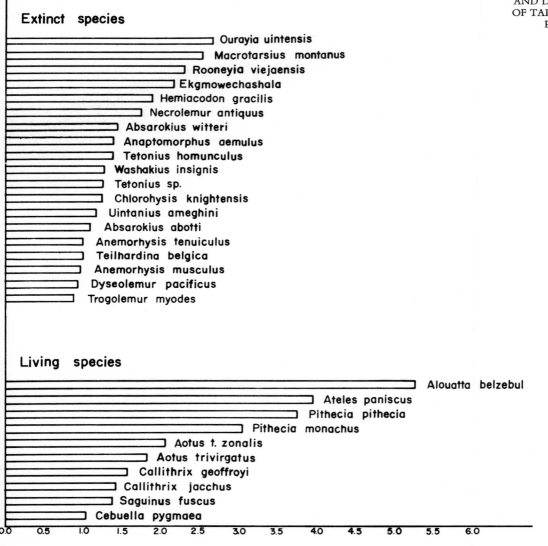

FIG. 1. Estimated lengths of the tooth rows of some better known fossil tarsiiforms compared to those of some extant platyrrhines. The figure gives a very approximate measure of the size range among tarsiiforms.

tomorphidae included the Omomyinae, Paromomyinae, Anaptomorphinae, Necrolemurinae, and Pseudolorisinae. It should be pointed out that, aside from including
several disputed genera and allocating the Paromomyinae to the family, the concept of
the Omomyidae recognized here is essentially the Anaptomorphidae of Simpson (1940,
1945).

Simpson's studies were not the only ones which cast a shadow of doubt on the
phyletic relationships of the omomyids. In 1946 and 1948, Hürzeler published works
on the cranial morphology of *Necrolemur* emphasizing the taxonomic characters of the
ear region. His assessment of the facts compelled him to suggest that anatomical characters allied *Necrolemur* and relatives with the lemuriforms, rather than with the living
Tarsius. Undoubtedly, Simpson (1955) was influenced by Hürzeler, whose work re-
kindled doubts voiced in the former's 1940 studies. In 1955, Simpson considered the
"Anaptomorphidae" not a natural family, and noted that (p. 438),

> . . . there is no convincing evidence that any early primate is more 'tarsioid' than 'lemuroid'
> in natural affinities. The mooted 'tarsioid' characters are some features of the cheek dentition,
> characters for the most part merely primitive for posimians now known (as in the genera just
> named) to be of possible association with 'lemuroid' skulls; enlargement of the orbits (e.g., in
> *Tetonius*,) which does not reach the *Tarsius* extreme or resemble it in detailed anatomy and
> which has certainly occurred independently in many primates, especially those that became
> nocturnal; and in a few cases elongation of the tarsus, which again is not demonstrably like
> *Tarsius* in extent or detail, which also occurs among 'lemuroids' (or, notably, Lorisiformes,
> and also among nominal insectivores), and which in some instances at least (e.g., *Hemiacodon*)
> is different from the truly tarsioid trend and quite surely independent.

Thus, in this rather radical departure from the early, turn of the century, views on the
Eocene primates, Simpson, with the full weight of his authority, again cast serious doubts
on the previously suggested probabilities on "prosimian" relationships. However,
Hürzeler's assertion that the ectotympanic is free in the bulla in *Necrolemur*, contrary to
Stehlin (1916) who demonstrated it to be fused to the rim of the bulla, was reinterpreted
by Simons and Russell (1960) and Simons (1961b) who have suggested that Stehlin's
studies on basicranial morphology were correct.

Gazin (1958) proposed that the Omomyidae should be separated from the Anaptomorphidae on the family level, and made the statement that the genera included in his
concept of the Omomyidae, ". . . are somewhat more primitive or generalized than the
anaptomorphids, and in this way occupy a position away from the anaptomorphids
towards the notharctids" (p. 47). He later remarked in the same work that even after
". . . the removal of omomyids, the anaptomorphids still appear to be a somewhat un-
natural association of genera" (p. 73). In this study, Gazin almost completely refrained
from referring the Omomyidae and Anaptomorphidae to a superfamily, although he
clearly allocated the Eocene notharctines and adapines to the Lemuroidea. His reluctance
may have been partly influenced by the weight of Simpson's positive stance in 1955,
in essence stating that the deciphering of phyletic affinities of Eocene forms is a nearly
hopeless undertaking. McKenna (1960), however, in treating the tarsiiforms of the Four
Mile local fauna, did not follow Gazin, and considered the Anaptomorphinae and
Omomyinae, as well as the Paromomyinae, as of the same family, i.e., presumably a
monophyletic taxon.

361

PHYLOGENY,
ADAPTATIONS,
AND DISPERSAL
OF TARSIIFORM
PRIMATES

In 1961 Simons took the step of including the Microchoerinae in the Tarsiidae (1961*b*). His main reason for this allocation was the alleged homologies of the anterior dentition of *Microchoerus* and *Tarsius*. Simons argued in great detail (pp. 57–61) that the only differences in the dental homologies of *Nannopithex*, *Necrolemur*, *Pseudoloris*, and *Microchoerus* from *Tarsius* were the loss of a single pair of lower incisors in microchoerines compared with *Tarsius* and the possession of a vestigial P_1 in the fossils. Simons has argued that specimens of *Microchoerus edwardsi* show a small alveolus anterior to the enlarged lower tooth, and concluded that this was proof for the latter being the canine. With this arrangement he could postulate that *Tarsius* derived from a primitive microchoerine with a single pair of lower incisors, canines, and 4 premolars. Simons also stated that (p. 67), "Some or all of the forms now ranked in the Omomyidae and Anaptomorphidae may eventually prove to be [tarsioid] (with the advent of new and better specimens) but demonstrating tarsioid characters . . . becomes increasingly difficult with greater antiquity." Simons (1961*a*), in a paper on *Ourayia*, however, simply allocated the Omomyidae to the "?Infraorder Lemuriformes" without any further explanation. Simpson's 1955 statements concerning the Eocene primates were much heeded and undoubtedly influenced students of these forms for several years after.

Simons (1963, p. 90), in a review discussing the Anaptomorphidae (= Anaptomorphinae), noted that, ". . . it is debatable whether anything is gained by ranking this group with the 'tarsioids,' in the absence of sufficient cranial or postcranial remains in this family upon which comparisons could be based." He suggested, however, probable close ties to the Necrolemurinae (= Microchoerinae) which he previously classified within the Tarsiidae. Following Gazin's (1958) remarks on the Omomyidae, *sensu stricto* (= Omomyinae) Simons (1963, p. 94) believed that, "Perhaps the closest relationships of this family are to the notharctines, by way of dental resemblances between the early Eocene English species *Cantius eppsi* and North American *Pelycodus* species, but there are also dental features of similarity to the European necrolemurines, and to the Anaptomorphidae."

In the original description of *Rooneyia*, Wilson (1966), tentatively placing the genus into the Omomyidae, classified the family under the infraorder Lemuriformes; he was perhaps influenced by Simons' (1961*a*) allocation of *Ourayia* to the ?Lemuriformes. One year later, Russell *et al.* (1967) espoused essentially Simpson's 1940 and 1945 concept of the Anaptomorphidae. They included in it the Paromomyinae, Anaptomorphinae, and Omomyinae, but removed the microchoerines into the Tarsiidae, following Simons (1961*b*).

In a "prosimian" classification McKenna (1967) considered the Omomyidae and Anaptomorphidae separate families under the superfamily Omomyoidea (in which he also included the Paromomyidae). However, following Simons (1961*b*), he also treated the Microchoerinae as a subfamily of the Tarsiidae, the latter retained in the superfamily Tarsioidea. His various superfamilies, including his concept of the Omomyoidea, were grouped within the suborder Prosimii, and in the notes after the classification he stated his belief that the omomyids were more closely related to the tarsioids than to the lemuroids.

In the following year, 1968, Robinson took the unusual step of placing the Omomyidae within the Lorisiformes.

B. Relationships within the Tarsiiformes

Figure 2 presents the most plausible phylogeny I could infer for the known tarsiiform genera. Most decisions whether character states were merely omomyid symplesiomorphies or synapomorphies between two or more species of the family were based on the morphology of teeth. This is not an *apologia* for the use of teeth for phyletics but merely a commentary on their use. Tarsiiform dentitions, including cheek teeth, are extremely varied and this quality alone, as in all other mammals, renders them very useful in systematics. Of significance are the extreme complexity of mammalian molar crowns and the fact that relatively minor genetic or developmental changes affecting the dentition can greatly alter the gestalt of a tooth. As the whole tooth changes, however, the resulting new gestalt becomes difficult to analyze character by character.

The detailed arguments for the relationships between species and genera cannot be discussed in this paper and are treated elsewhere (Szalay, in press). In that monographic study I relied heavily on some selected features of the basicrania, dentitions, and postcranial elements of known fossil and living primates, in varying order, to infer phylogeny. I can only briefly touch on a few of these characters here.

1. Cranial Morphology (Figs. 3–6). The figures presented summarize the extent to which fossil tarsiiforms are known by skulls. The more significant aspects of the cranial morphology, those of the basicranium, were briefly discussed (Szalay, this volume) and I will not reiterate the conclusions presented there.

There are a number of outstanding cranial differences between the extant *Tarsius* spp. and *Necrolemur antiquus* which supply strong arguments against classification of these taxa in the same family (see also Szalay, this volume), and therefore they are discussed here. As noted above, the comparison revealed some highly derived conditions of the basicranium in the extant genus but not in the European form, and vice versa. Dorsal and anteromedial to the promontorium of *Tarsius* a diverticulum, an air space, is enormously enlarged, and this is strikingly reflected in the very large anteromedial segment of the bulla. In *Necrolemur* this section of the bulla is undeveloped, although a middle ear diverticulum is present. The skull of *Tetonius* is poorly known, yet there are features on the generotype which suggest a condition not unlike that found in *Necrolemur*. In spite of what is usually stated or implied, petromastoid inflation is minimal or virtually nonexistent in *Tarsius*. There are three major dome-shaped enlargements on the occiput, as in *Necrolemur*. These seem to be contoured to the shape of the two lateral lobes and the vermis cerebelli of the cerebellum. *Necrolemur*, however, has an enormously enlarged patromastoid.

As there is no evidence that this uninflated petromastoid of *Tarsius* is secondary, the condition of *Necrolemur* almost certainly appears to be the more derived character state. The crushed skull of *Tetonius* suggests a petromastoid very much like that of *Necrolemur*, a character of some significance in the assessment of its relative phyletic position. *Rooneyia*, however, shows no excessive inflation of the petromastoid. The petromastoids of *Rooneyia* and *Tarsius* are then either mere symplesiomorphies and therefore should not be used to signify special ties between these two genera, or, as a remote possibility, a convergently derived condition in these two forms. As noted (Szalay, this volume), pronounced inflation of the mastoid was probably present in the anthropoid morphotype, although in the common haplorhine ancestor it was probably uninflated.

363

PHYLOGENY,
ADPATATIONS,
AND DISPERSAL
OF TARSIIFORM
PRIMATES

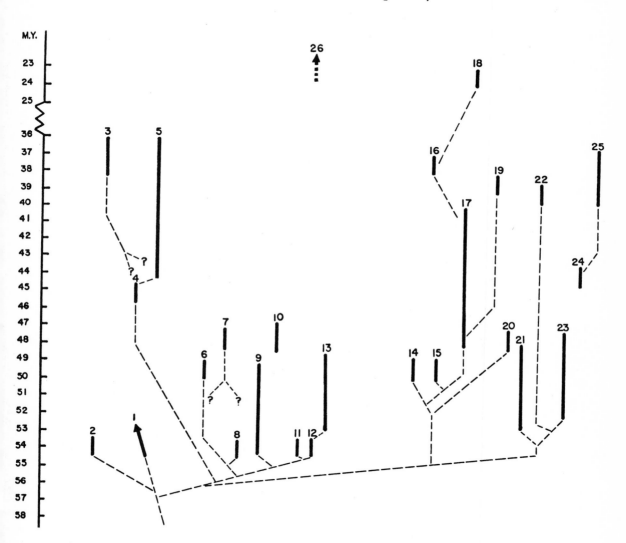

FIG. 2. Known stratigraphic ranges (solid lines) and suggested hypotheses of phylogenetic relationships (broken lines) of taxa, primarily of the Tarsiiformes. 1, segment of the stratigraphic range of the Adapidae; 2, *Donrussellia*, a possible tarsiiform; 3, *Pseudoloris*; 4, *Nannopithex*; 5, *Necrolemur* and *Microchoerus*; 6, *Chlororhysis*; 7, *Anaptomorphus*; 8, *Teilhardina*; 9, *Anemorhysis*; 10, *Trogolemur*; 11, *Mckennamorphus*; 12, *Tetonius*; 13, *Absarokius*; 14, *Loveina*; 15, *Shoshonius*; 16, *Rooneyia*; 17, *Washakius*; 18, *Ekgmowechashala*; 19, *Dyseolemur*; 20, *Hemiacodon*; 21, *Uintanius*; 22, *Chumashius*; 23, *Omomys*; 24, *Ourayia*; 25, *Macrotarsius*; 26, unspecified source of origin for *Tarsius*.

FREDERICK S.
SZALAY

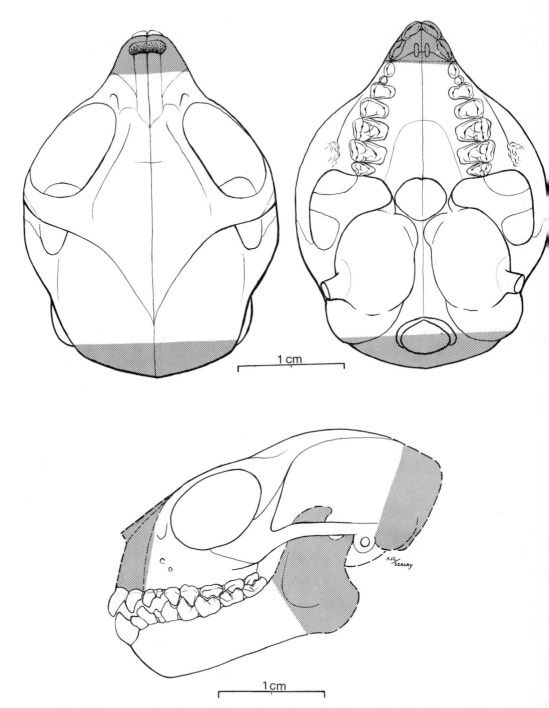

FIG. 3. Reconstructed cranium of *Tetonius homunculus*, early Eocene, Wyoming. The cranial morphology indicates that this taxon is already highly derived by early Eocene times and considerably differentiated from a common tarsiiform ancestor.

365

PHYLOGENY,
ADAPTATIONS,
AND DISPERSAL
OF TARSIIFORM
PRIMATES

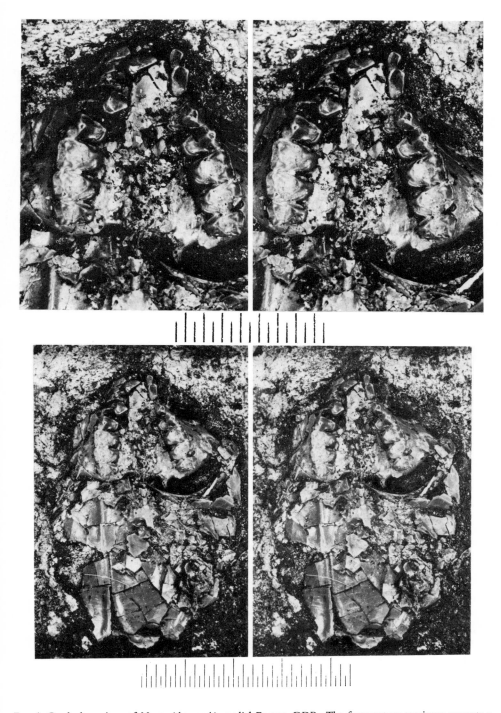

FIG. 4. Crushed cranium of *Nannopithex raabi*, medial Eocene, DDR. The fragmentary specimen suggests a cranial morphology not unlike that of the North American *Tetonius*.

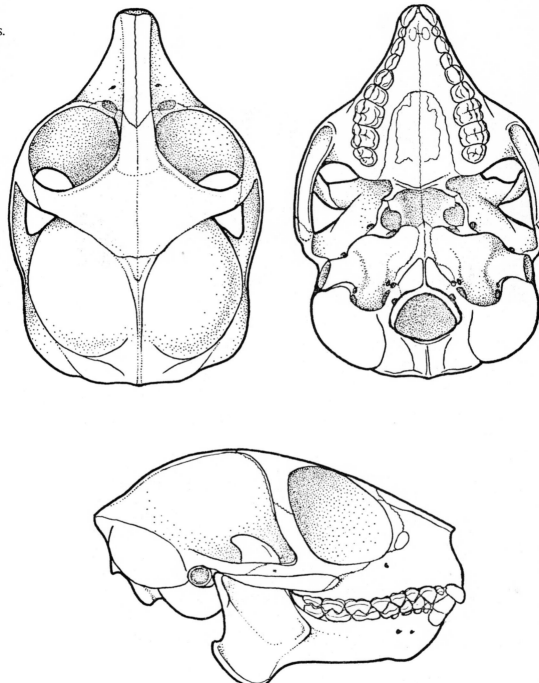

FIG. 5. Reconstructed cranium of *Necrolemur antiquus*, late Eocene, France. Several characters of the cranium of *Necrolemur* suggest special relationships of the microchoerine tarsiiforms with some anaptomorphine omomyids.

367

PHYLOGENY,
ADAPTATIONS,
AND DISPERSAL
OF TARSIIFORM
PRIMATES

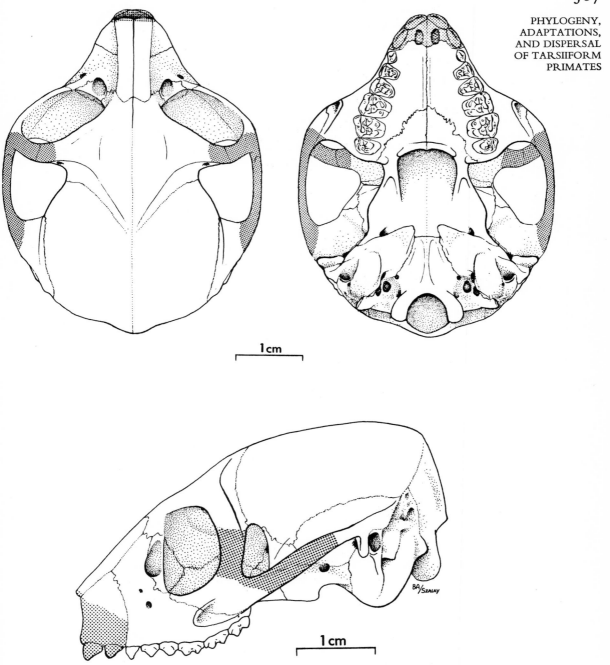

Fɪɢ. 6. Reconstruction of the cranium of *Rooneyia viejaensis*, late Eocene, Texas. Lack of petromastoid inflation in this form hints that a variety of tarsiiforms existed which were distinctly more primitive in basicranial construction than either *Tetonius* or *Necrolemur*.

2. Dentition (Figs. 7–12). The obvious problem posed by genera known only by teeth is that we are usually ignorant of the stage of evolution reached by the remaining anatomy. However, if the dental evidence suggests monophyly of the taxa of the Omomyidae, then the spotty cranial and postcranial evidence in a number of independent lineages perhaps permits some inductive generalizations for the remaining, more poorly known anatomy of the morphotype. In this respect, therefore, the dentitions, in addition to the wealth of information they supply about the genealogy and the feeding mechanism, are also extremely valuable in assessing levels of organization in features other than teeth.

It is difficult to determine what the most primitive omomyid, adapid, or paromomyiform molar morphology was like. The inferred primitive conditions of both omomyid and adapid molars were probably very similar to one another. Morphotypes of both these groups appear to be somewhat more advanced than an inferred primitive paromomyiform, and certainly more advanced in molar morphology than *Purgatorius*. It appears certain that the ancestor of both adapids and omomyids had an I_1^1; $\frac{2}{2}$; C_1^1; P_{1-4}^{1-4}; M_{1-3}^{1-3} dental formula. In addition to these general remarks, I cannot state with any degree of certainty whether the omomyid or adapid morphotypes are more primitive or derived dentally.

Whether the European tarsiiform genera (*Nannopithex, Necrolemur, Microchoerus,* and *Pseudoloris*) are recognized as being sufficiently divergent to represent an independent family or only a subfamily of the Omomyidae is a problem which should be dependent largely on the homologies of the antemolar dentition. If the homologies and the general conformation of these teeth can be shown to be the same as in some omomyid taxa, and the basicranium of *Necrolemur* can be reconciled with those of other omomyids such as *Tetonius*, rather than *Tarsius*, then the Microchoerinae may be placed within the Omomyidae.

The most important first step in deciphering the homologies of the anterior teeth should be the establishment of the premaxillary–maxillary suture. Simons (1961*b*), who has attempted to solve the antemolar homologies of microchoerines and has extensively discussed some aspects of this problem, has bypassed the crucial issue by stating that (p. 58): ". . . the upper dental formulae of all species of both groups (i.e., of microchoreines and *Tarsius*) are apparently the same (2.1.3.3) as are the sizes of the teeth relative to each other . . ." This to me reflects the acceptance of the premaxillary–maxillary suture as originally shown for *Necrolemur antiquus* by Stehlin (1916, p. 1343) and by Simons and Russell (1960). Repeated examinations of all *Necrolemur antiquus* skulls in European and North American collections failed to confirm the suture mesial to the seventh tooth from the back or distal to the second one from the front. One specimen (MNHN 1957-14) may be interpreted in such a way that a suture might have passed between the first and

---

FIG. 7. Composite dentition of *Teilhardina belgica*, early Eocene, Belgium. Note the nearly complete eutherian dentition (one incisor is lost), the large canine, and the distinctly anaptomorphine-like conformation (primarily a reduction in relative size) of the last molars. The cheek teeth of this taxon are exceptionally primitive among the Tarsiiformes, and also compared to most of the lemuriform Adapidae. Based on dental evidence, derivation of this largely primitive tarsiiform from adapid taxa like either *Protoadapis* or *Pelycodus* appears plausible.

369

PHYLOGENY,
ADAPTATIONS,
AND DISPERSAL
OF TARSIIFORM
PRIMATES

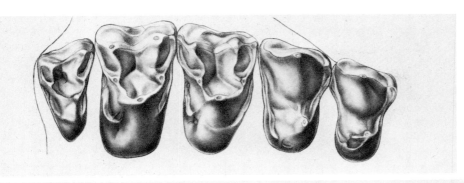

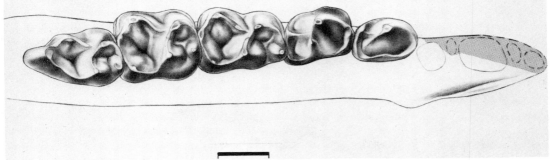

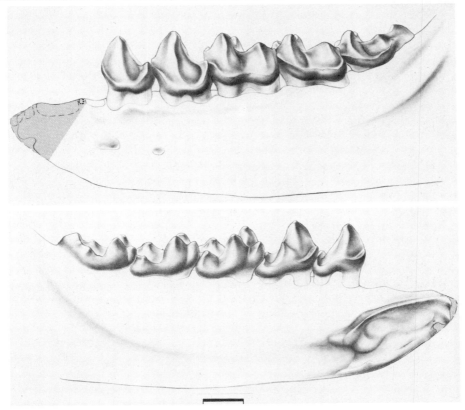

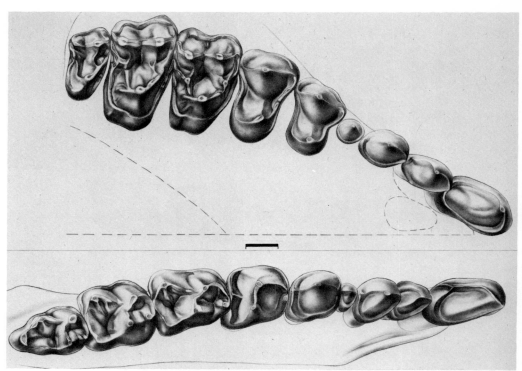

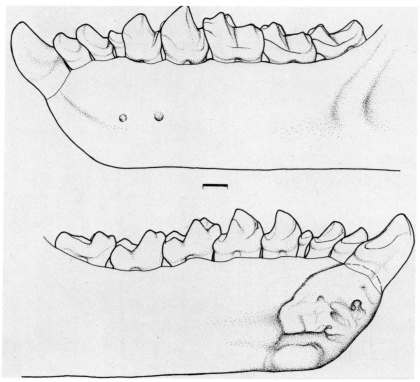

second upper teeth. Fusion apparently occurred at a very early stage in ontogeny between the premaxillary and maxillary bones and this question is still unsettled. I know of no unequivocal evidence which would either support or contradict Stehlin's original designation.

Simons (1961b, pp. 58–61) has argued that the enlarged microchoerine lower incisor is a canine. His evidence consisted of two specimens of *Microchoerus edwardsi* in the British Museum which, according to him, show small alveoli anterior to the enlarged procumbent tooth. He cites additional specimens of the same taxon in the Paris collections. My interpretation of these openings, on specimens which are usually broken, differs from his. I consider the openings as parts of the abundant nutrient canals on *Microchoerus* mandibles. This genus, for adaptive reasons obscure at present, had a great number of nutrient vessels entering and leaving the anterior part of the mandible. This condition is particularly well documented by the numerous, irregularly placed mental foramina on the side of the jaw. Simons' (1961b, p. 57) discussion on occlusion is endorsed, but I believe that strict application of upper and lower tooth occlusal relationships cannot be applied to microchoerine dentitions. There are 9 teeth above and 7 below, although for purposes of discussion the second lower tooth is so small that it takes no part in occlusion, and, therefore, the number of operational lower teeth is better considered as 6.

In order to analyze microchoerine tooth homologies and mechanical function, two important selective factors must be postulated. First, there was selection for either a reduced face, including both the upper and lower jaws, or for reduction of the length of the lower jaw. Second, there was clearly selection for enlargement of the pair of anterior teeth both above and below. I will assume that phylogenetic enlargement of both upper and lower anteriormost teeth occurred simultaneously as their mechanical function and biological role are clearly tied together. As the enlargement of the anterior teeth probably occurred together, and as the upper enlarged tooth is clearly not a canine, it is probable that the enlarged lower tooth is the incisor originally occluding with the enlarged upper one. As this incisive function appears to have been of extreme importance, judged by the size of the teeth in all microchoerines, disturbance of the role of this area by phyletically losing the lower incisor and replacing it with the lower canine was unlikely. Selection for maintaining occlusion of a lower canine with an upper one was unlikely to be of greater importance than the maintenance of the adaptively important occlusion by the incisors. If this was the case, it is likely that following the enlargement of the anterior teeth, tooth reduction occurred distal to them. Clearly, the same number of teeth were not lost above and below. Thus, tooth reduction, a response to either mandibular shortening or disproportionate enlargement of the anterior teeth, occurred in a manner to fill the available space on the mandible with the ontogenetically most developed teeth at the time of eruption.

FIG. 8. Composite dentition of *Tetonius homunculus*, early Eocene, Wyoming and Colorado. Often mistakenly held to be the most primitive of the Anaptomorphinae, this basal Eocene form has already developed impressive specializations of its anterior dentition.

371

PHYLOGENY,
ADAPTATIONS,
AND DISPERSAL
OF TARSIIFORM
PRIMATES

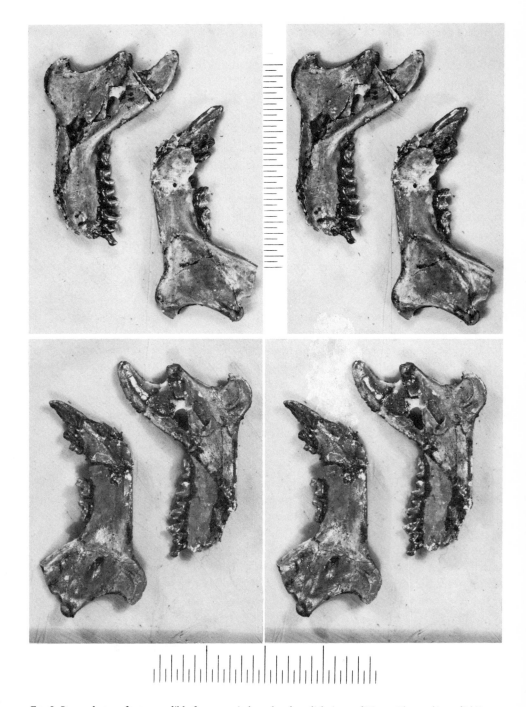

FIG. 9. Stereophotos of two mandible fragments in lateral and medial views of *Nannopithex raabi*, medial Eocene DDR. The most primitive of known microchoerines, this form has already developed the characteristic anterior dentition of this group of tarsiiformes.

373

PHYLOGENY,
ADAPTATIONS,
AND DISPERSAL
OF TARSIIFORM
PRIMATES

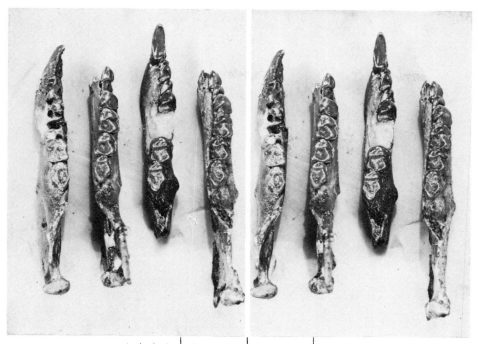

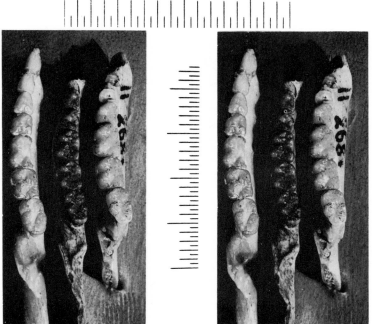

FIG. 10. Stereophotos of dentitions in madibular fragments of *Nannopithex raabi*, early Eocene, DDR (above), and *Necrolemur antiquus*, late Eocene, France (below).

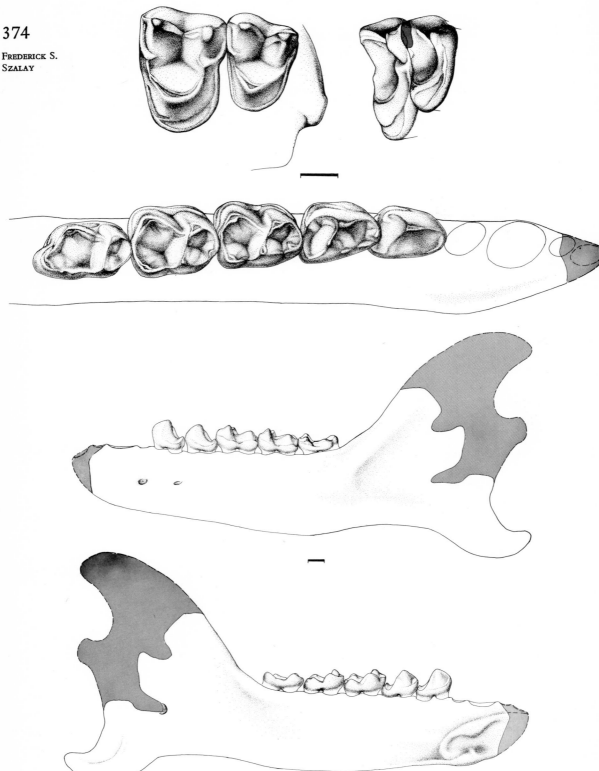

Which of the upper teeth is most likely to be the canine? Lacking the evidence from the suture I cannot decide whether the second or third tooth from the front is the canine. In either case, in my view, the homology of the enlarged anterior teeth do not influence the decision about the upper canine. The two teeth posterior to the enlarged lower incisor are either P_1 and P_2 as suggested by Simons or an incisor and a canine, as I interpret the dental formulae in either *Omomys* or *Tetonius*, for example. If the second upper tooth is the canine, then the derived enlargement of the P^1 could be considered a major distinction of microchoerines from all other tarsiiforms known. Therefore, their separation from the omomyids as a family, the Microchoeridae, could be supported.

Much has been made of the alleged similarities between dentitions of the Quercy phosphorite *Pseudoloris* and *Tarsius* (see in particular Teilhard de Chardin, 1916–1921; Simons, 1961b). Simons has recently placed great emphasis on the resemblances between *Pseudoloris* and *Tarsius* and as noted, he allocated the Microchoerinae to the Tarsiidae. Results of my reexamination of the kinds of similarities between *Pseudoloris* and *Tarsius* differ from those of previous investigators. I found that some of the alleged similarities are not similarities at all, whereas others are convergent. The most important difference between the two genera, previously alleged to be a similarity, is the manner of arrangement and occlusal function of the anterior dentition. These, in turn, strongly suggest and this is borne out by the cheek-tooth morphology, that the premolars and molars are similar as a result of convergence rather than as a result of deriving their conformation from a common ancestor.

The dentition of *Tarsius* and its inferred functional features are so unique that a brief account of these is necessary. Articulating skulls with their mandibles reveals that a distinct incisive stroke is present. During this incisive bite the mesial edges of the lower canine shear against the greatly enlarged central upper incisors and the small lateral pair of incisors. The cutting edges of the two pairs of upper incisors are almost exactly transverse to the long axis of the skull. The molar crests are high and movement of the lower jaw appears to be mostly orthal, producing largely point cutting by the cheek teeth, with a very minimal mesiolingual component. The mandibular fossa is very unusual among primates. Anteroposteriorly long and transversely narrow, it is a trough which permits extensive propalinal but very little transverse movement. This is also reflected by the articular condyle which is long anteroposteriorly but very narrow transversely.

The arrangement of the anterior dentition of *Pseudoloris* is exactly like that of other microchoerines. Unlike the antecanine teeth of *Tarsius*, those of *Pseudoloris* are lined up mesiodistally. In addition to this pronounced difference, the reduced microchoerine lower canine is not the major lower piercing tooth; this role is assumed by the enlarged incisor. Selection for the performance of the same biological roles, piercing and cutting,

375

PHYLOGENY,
ADAPTATIONS,
AND DISPERSAL
OF TARSIIFORM
PRIMATES

FIG. 11. Dental and mandibular remains of *Chumashius balchi*, late Eocene, California. Note the presumably primitive size relationships of the canine to the premolars and molars, the unenlarged condition of the incisors, and the primitive condition of the mandibular angle. The cheek-tooth morphology, perhaps convergently, is very callithricid-like.

Frederick S.
Szalay

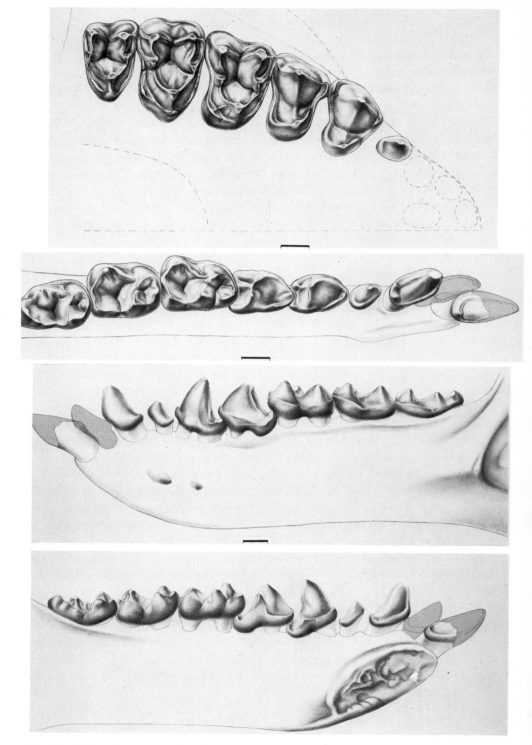

Fig. 12. Dentition of *Omomys carteri*, medial Eocene, Wyoming. Note and contrast the conformation of the last molars in *Omomys*, *Chumashius*, and other omomyines, to that of anaptomorphines, examplified in this paper by *Teilhardina* and *Tetonius*.

by the antecanine dentitions of *Pseudoloris* and *Tarsius* has resulted in highly divergent relative sizes, morphology, and manner of mechanical function, dictated by the morphology of their respective ancestry.

Although the morphology of the postcanine teeth of *Pseudoloris* is similar to that of *Tarsius*, several clues suggest that the similarities are not shared derived character states. Close examination of the molars and premolars reveals that the most important shared similarity between the two genera is the acuity of the cusps and the relatively high crests; there are no other significant similarities beyond these. *Pseudoloris* has a well-defined hypocone suggesting an ancestry not unlike *Nannopithex*.

Until proof is available as to the exact homologies of the upper teeth distal to the incisor, the Microchoerinae are best considered a subfamily of the Omomyidae, but not of the Tarsiidae. As noted (Szalay, this volume), microchoerines do not share some of the characteristic specializations of the tarsiid ear region. Short of a probably shared early tarsiiform specialization of the tarsus and possibly the crus (not as advanced, however, as seen in *Tarsius*), they do not appear to share advanced traits exclusively with the Tarsiidae.

3. Postcranial Morphology (Figs. 13–17). Owing to limits of space and unfinished analysis, I can only touch on a few significant points here. The most complete skeletal remains of any omomyid is still the fragmentary skeleton of *Hemiacodon gracilis*, AMNH 12613, although a fair variety of other omomyids are also known by tarsal and elbow-joint material (Szalay, in press; in prep.). Simpson (1940, pp. 195–197) emphasized that the skeletal remains of *Hemiacodon* were close to his morphological concept of the Lemuroidea. The latter, judged from his comparisons, included such phenetically, temporally, and phyletically widely separated forms as *Notharctus*, *Lemur*, and *Galago*; the last genus was to typify the long-footed lemuroids. He concluded that *Hemiacodon* was skeletally "intermediate" between tarsioids and lemuroids. He also suggested the possibility (p. 197), improbable to him, ". . . either (1) that *Hemiacodon* and its allies are true and fully differentiated tarsioids resembling the lemuroids only by convergence or in pre- or proto-primate characters, or (2) that they are lemuroids convergent toward the tarsioids in some respects."

Perusal of Simpson's discussion of the postcranial anatomy of *Hemiacodon* indicates to me that at that time his overriding concern was the discovery of general similarity which would align the genus with either *Tarsius* or the "lemuroids" (i.e., the category Strepsirhini). His aim was "closeness" of morphology rather than recency of ancestry. He could not find diagnostic features, and it is in this respect that he found the "intermediate" phenetic position of *Hemiacodon* between *Galago* and *Tarsius*. The concept "lemuroid" was used, however, in a rather broad sense in the study, and comparisons of the bones, such as the calcaneum, with various characters of *Galago* rather than with those of an inferred ancestral condition of the "lemuroids" (i.e., strepsirhines) biased his discussion.

Simpson (p. 195) correctly indicated that calcanea of *Hemiacodon* and *Teilhardina* are similar and, although stated differently, he concluded that the similarities indicated close relationships. While in Halle, DDR, in 1970, I attempted to locate the calcaneum and tibia–fibula described and poorly illustrated by Weigelt (1933). Search by Professor

377

PHYLOGENY,
ADAPTATIONS,
AND DISPERSAL
OF TARSIIFORM
PRIMATES

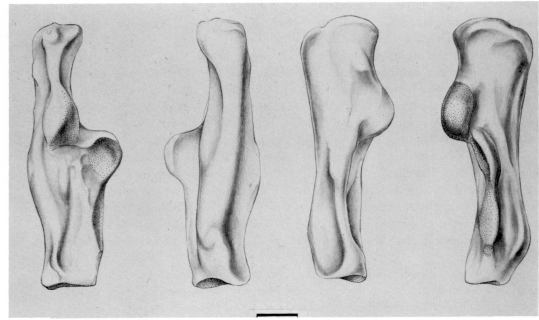

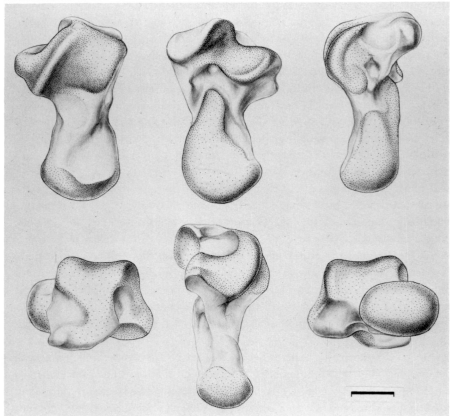

379

PHYLOGENY,
ADAPTATIONS,
AND DISPERSAL
OF TARSIIFORM
PRIMATES

Matthes failed to locate these specimens. As Simpson (p. 196) pointed out, the poor illustrations of the postcranials allocated to *Nannopithex* by Weigelt (1933) show, '. . . considerable resemblance to *Hemiacodon* in the hindlimb." I might further add that, in light of the complex of similarities which tie microchoerines to omomyids and which are clearly derived primate features, there is no reason to suspect that calcaneal elongation in *Nannopithex* and *Necrolemur* are not equally shared primitive omomyid characters.

Broadly based studies on the evolutionary morphology of early Tertiary primate skeletal remains are under way and the results will be published at later dates. Some pertinent analyses have already been communicated (Szalay and Decker, 1974; Decker and Szalay, 1974; Szalay *et al.*, 1975) on the then known foot structure of paromomyiforms and Eocene lemuroids. At this time, however, selected parts of the postcranial anatomy offer some additional information for the assessment of omomyid phylogeny. Because tarsal remains, the calcaneum and astragalus in particular, are known from more early Tertiary species of primates than any other parts of the postcranial anatomy, these elements are excellently suited for a broad comparative study among a whole range of species and supraspecific taxa. Occasional long-bone fragments or pelvic remains are not as well known, and therefore, at present, their value for phylogenetic and functional analyses is considerably less.

Several different kinds of small primate calcanea and astragali are known from various Eocene localities. Many of these tarsals are almost certainly from omomyids, although allocation to genera is uncertain. Probabilities of referral to dentally described taxa vary from locality to locality. Thus, calcanea and an astragalus from Dormael, Belgium are almost certainly those of *Teilhardina belgica*.* Nonadapid and nonparomomyiform calcanea and astragali from the Four Mile localities of Colorado are likely to be remains of one of the two species of *Tetonius* present there, and a genus of omomyids is known by tarsals from the early Eocene Bitter Creek localities recently collected by D. E. Savage and associates. Several types of omomyid calcanea, which are not from *Hemiacodon*, are also known from Bridger beds.

Comparisons have been made with virtually all of the known fossil primate record and with representatives of all genera of primates. Among the living taxa particular emphasis was placed on strepsirhines, *Tarsius*, and platyrrhines. The most significant comparisons have been those with the Paleocene and Eocene fossil record. I will now briefly examine several selected features of the omomyid tarsals, considering the following characters: (1) relative lengths of moment arms of the calcanea, (2) the calcaneocuboid articulation, (3) the peroneal tubercle, (4) the lower ankle joint between the astragalus and calcaneum, (5) the astragalonavicular articulation, and (6) the tibial trochlea, the astragalar canal, and the upper ankle joint, in general.

* The astragalus originally allocated to *Teilhardina* by Teilhard de Chardin (1927) is rodent, and the humerus is anuran.

FIG. 13. Calcaneum and astragalus of *Teilhardina belgica*, early Eocene, Belgium. The moderate elongation of the calcaneum is roughly comparable to that found in the cheirogaleid *Microcebus*. Note the development of the posterior part of the astragalar body, interpreted to be a retention from an adapid condition. This posterior trochlear shelf is one of the conspicuous shared derived characters of almost all known lemuriforms.

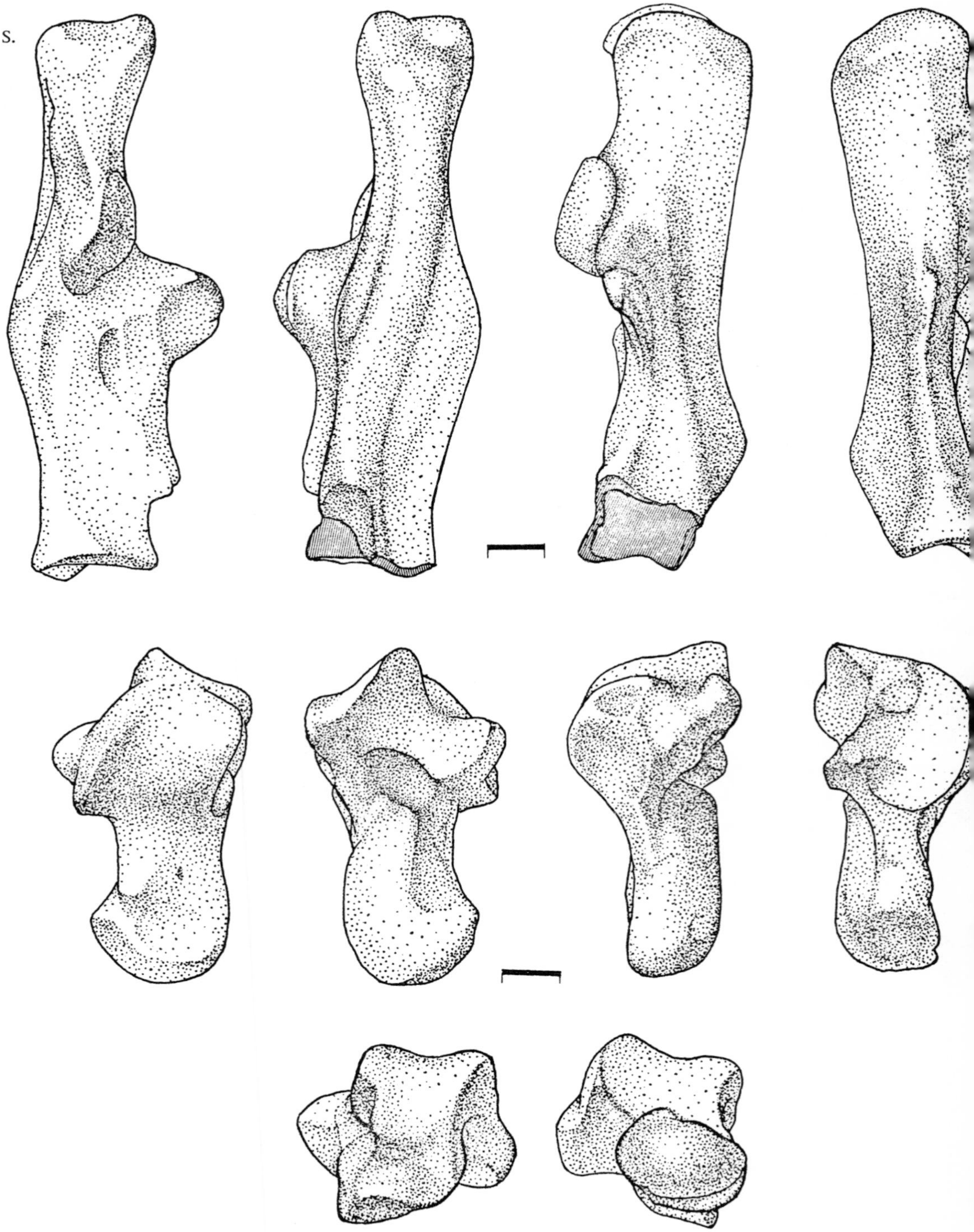

381

PHYLOGENY,
ADAPTATIONS,
AND DISPERSAL
OF TARSIIFORM
PRIMATES

Calcaneal elongation appears to characterize as phyletically distant genera of omomyids as *Hemiacodon*, *Teilhardina*, ?*Tetonius*, the Bitter Creek and Bridger omomyids (with generically uncertain allocation), *Necrolemur*, and *Nannopithex*. Until proven to the contrary, it appears reasonable to assume that calcaneal elongation shown by these taxa is an ancestral omomyid character. This feature, if validly inferred as primitive for the omomyids, separates the family from any known paromomyiforms or the primitive condition inferred for adapids.

Comparisons of omomyid tarsals with adapids reveal that relative elongation of the moment arm is as advanced in *Hemiacodon*, and others, as it is in some cheirogaleids or galagines. Combined characters of any of these taxa, when compared to one another, therefore, indicate that this feature was independently attained by lorisiform cheirogaleids and lorisids on the one hand and the omomyids on the other.

The relevance of the African Miocene lorisiform calcanea recently reported by Walker (1970) must be noted at this point. These bones belong to three taxa which are represented by KNM SGR 2094, SGR 2010, and SGR 2556. The drawings presented by Walker (1970, Fig. 8) are somewhat simplified. In discussing these bones Walker noted (p. 253) that, "The shape of the fossil calcanea, apart from the lesser distal elongation, compares closely with those of modern galagos even to the extent of having a joint surface showing the presence of an anterior calcaneonavicular synovial joint as described by Hall-Crags ('66)." The presence of a well-developed and distally displaced calcaneonavicular joint, as in galagos, cannot be ascertained, and no other shared derived similarities between the calcanea of living galagos and the fossils can be found. Any one of them, on the other hand, could represent the calcaneum of the morphotype, and may even be the actual common ancestor of lorises and galagos, although this is unlikely judged by the cranial specializations attained by lorisids by this time. Rather than sharing special similarities with galagos, the Miocene calcanea are primitive lorisoid and astonishingly similar to those of such cheirogaleids as *Cheirogaleus* and *Phaner* (Decker, personal communication).

Similarity between the Miocene calcanea and those of known omomyids, excluding that of *Necrolemur*, is great. I suspect, however, that this similarity is partly a result of both having aspects of primitive morphology, that of a lemuriform ancestor, and to independent acquisitions, i.e., the distal elongation of the calcaneum. These calcanea are more "cheirogaleid" than "lorisid" and therefore ancestral lorisiform. Basicranial evidence would strongly caution against arguments based on these specimens which may appear substantive in advocating special tarsiiform–lorisiform ties.

The upper ankle joint, as shown by the body of the astragalus, and, at least as indicated by *Necrolemur* and *Nannopithex*, the distally fused tibia and fibula, was significantly different in omomyids compared to adapids and paromomyiforms. Many lemuriforms retain at least the dorsal astragalar foramen leading into the astragalar canal. However, no known omomyid astragali show traces of this canal which, I believe, is primitive for Theria, Eutheria, and, as indicated, was present in the primate morphotype.

The body of the astragalus of *Teilhardina* shows a posterior extension which I

Fig. 14. Composite figures of the calcaneum and astragalus of ?*Tetonius* sp., early Eocene, Colorado.

consider to be homologous to the posterior trochlear shelf of adapids described in Decker and Szalay (1974). As this feature is best developed in *Teilhardina*, but not recognizable unequivocally in *Hemiacodon* or in *?Tetonius*, its moderate expression in the European omomyid might reflect a feature retained from an ancestral lemuriform condition. This posterior trochlear shelf, then, suggested as a character of the lemuriform morphotype, since it is widely present in most subsequent lemuriform taxa, was probably present in the last ancestor common to the Omomyidae.

In summarizing the tarsal evidence, it appears that a number of the character states known for omomyid calcanea and astragali appear to be derived ones when compared to their homologs in the adapid morphotype, yet these bones in the tarsiiform morphotype were similar to the primitive adapid condition. Such features as the screw-type lower ankle joint are primitive for the whole order (Szalay and Decker, 1974).

I should comment here on the phylogenetic significance of nails vs. claws, and vice versa. Nails, at least on several of the cheiridia, were probably primitive in the ancestor of all known strepsirhines and, therefore, probably also in the tarsiiform morphotype. The recent view of Cartmill (1974) concerning the primitive condition of claws in the ancestral strepsirhine and subsequently derived taxa is relevant in this regard as our views differ. The presence of the terminal matrix and deep stratum (the latter usually characteristic of claws) in *Cebus albifrons* does not necessarily, contrary to Cartmill's arguments (1974, pp. 73–74), represent a secondary modification from a marmoset-like condition. It does not mean that by inferring the presence of nails in primitive haplorhines we must hypothesize a lineage leading to *Cebus albifrons* which lost its claws by the Eocene, regained them by the Oligocene, and lost them during the later Tertiary; this scheme is Cartmill's argument of greatest parsimony. To the contrary, presence of histological features of true claws in nailed primates testifies that acquisition of nails did not occur simultaneously with loss of all true claw characters. The presence of such primitive features as the deep stratum and terminal matrix in the few groups of extant fully clawed primates does not bar the more reasonable explanation, at least in the marmosets, that primitive histology was inherited from a nailed ancestor. As the nailed ancestor of extant primates probably possessed this ancestral morphology, independent loss of the two layers of nails in the subsequent lineage is more reasonable an explanation than independent acquisition of nails in four or five lineages of primates that are otherwise divergently adapted. The reorganization of hand and foot morphology in callithricids compared to the more primitive cebid versions would also lend weight to the explanation that marmoset claws are secondary acquisitions.*

---

* In fact the clawed condition of digits II–V contrasted to the nailed and reduced big toe of the foot is one of the most powerful clues about the circumstances of origins of marmosets. The joint morphology between MtI and the entocuneiform leaves little doubt that the decreased range of mobility of that joint was derived from one with a much greater range, as that displayed in cebids like *Saimiri* or *Aotes*. The clawed condition of the digits most functional in locomotion can be tied to the characteristic molar morphology and robust incisors of the marmosets. Marmoset molars bear convergent similarity to those of *Phaner* and *Cheirogaleus* and both of these cheirogaleids are known for their dietary specialization for tree exudates. Some marmosets are also known to feed on resins (*Cebuella*, for example; A. Rosenberger, personal communication). Postulating a gum–resin feeding diet in the callithricid common ancestor would provide the selectional forces for the acquisition of claws. Claw climbing on large tree trunks where sap abounds would be far more efficient in terms of energy expenditure than continuing a grasping climb of the ancestors.

III. Adaptations

383

PHYLOGENY,
ADAPTATIONS,
AND DISPERSAL
OF TARSIIFORM
PRIMATES

This paper emphasizes cladistic aspects of the tarsiiform fossil evidence. I find it necessary, however, to present a brief characterization of the anagenetic advances that we may infer from the fossil record (relying heavily, however, on recent field studies of primates) and contrast this to that of the ancestral lemuriforms.

Whenever complete skulls, dentitions, and skeletons are known (a rare situation, particularly among small vertebrates), analogies with the total morphology of living

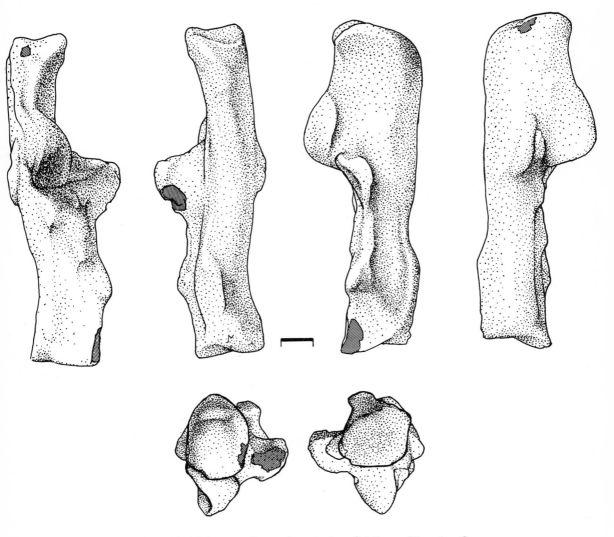

FIG. 15. Omomyid calcaneum (AMNH 88824), gen. and sp. undetermined, medial Eocene, Wyoming. Comparisons among known omomyid tarsal remains hint to a moderate degree of divergence in mechanical function and subsequently differences in their locomotor mode.

forms occasionally leads to recognition of approximate ecological vicars of living animals among well-known fossils. The environment, however, is rarely divided up identically by members of distantly or even closely related radiations. The method of analogy is not only restricted to a few species of fossils but is also quite useless for a number of fossil taxa which probably have no exact or even vaguely similar ecological vicars in the recent faunas.

A two-level conceptual approach, outlined by Bock and von Wahlert (1965), clearly differentiating between mechanical function and biological role, is, in my view, perhaps the most convincing one for understanding organismic morphology, both of living forms and, in a somewhat modified form, of fossils. The phenotype reflects not just the adaptedness to present demands, but it is clearly a compromise between the adaptedness of its ancestors and of its own. It is then of obvious importance that to understand adaptations of various species, one must have some understanding of the adaptations of the inferred or actual common ancestor. Assessments of the ancestral condition, therefore, have important consequences in the evaluation of adaptations in the descendant species.

Finally, diagnostic inferences on the mode of life of a given fossil taxon must necessarily be based on derived character states, as these are the ones differentiating a given form from its ancestor. Primitive character states may only represent adaptations of a precursor, although this clearly does not mean that functions of the primitive morphology and the roles performed by them were not part of the adaptedness of the species analyzed. It does mean, however, that one may make relatively meaningful statements in association with the role of the derived character states only.

I have discussed specific adaptations of the cranium, dentition, and postcranial morphology in a more detailed study. I will restrict myself here to some more general comments.

A. Cranial Morphology (Figs. 3–6)

The orbits on the tiny skull of *Tetonius* were clearly relatively large, a condition allometrically related to the relatively large brain and small absolute size of this taxon. It is impossible to ascertain whether the relative size of the orbits reflects nocturnal adaptations only in *Tetonius homunculus* or also are retentions from a nocturnal ancestor. Judged from the enlarged orbits of *Tetonius* and *Necrolemur*, one might suggest that the ancestral omomyid was probably a nocturnal species.

When studying omomyid maxilla fragments one is struck by the fact that almost all the taxa known by these parts have a base of the zygoma and an orbital shelf that projects a great deal laterally. Lacking the complete skulls for most of the species it is impossible to relate these traits to the relatively large size of the eyes, particularly as all the omomyids are small or tiny animals in which one expects relatively large eyes. Marmosets and tamarins, squirrel monkeys and the talapoin, all diurnal forms, also show the laterally projecting orbital shelf. What all these forms, and probably most of the omomyids, had in common, although the latter to varying degrees, is well-developed orbital convergence. Thus, the greatly projecting orbital shelf may not be indicative of

385

PHYLOGENY,
ADAPTATIONS,
AND DISPERSAL
OF TARSIIFORM
PRIMATES

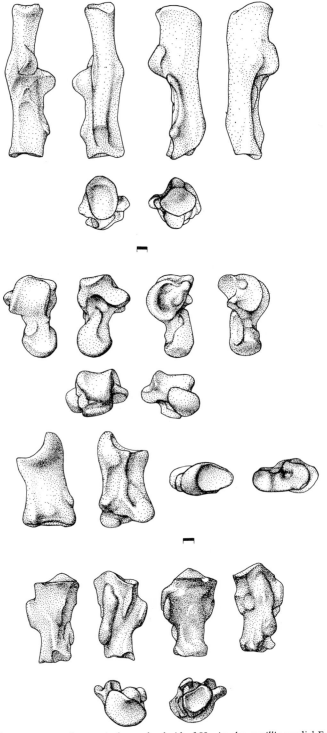

FIG. 16. Calcaneum, astragalus, navicular, and cuboid of *Hemiacodon gracillis*, medial Eocene, Wyoming.

relatively enlarged eyes or adaptations to nocturnality, but perhaps represent the well-developed orbital convergence of the omomyid morphotype. Orbital convergence, however, at least in the case of omomyids, might have been an adaptation enhancing visual acuity in a *rapidly* locomoting ancestor. Whether or not nocturnality was a factor in the development of orbital convergence is a point difficult to argue, but the selective factors brought about by hopping and jumping were likely to have placed great premium on stereoscopy, irrespective of predatory habits (for different but complementary views see Cartmill, 1972).

One of the diagnostic distinctions of haplorhines from strepsirhines, as detailed elsewhere (Szalay, this volume), is the phyletic enlargement of the promontory artery; I can only offer a speculative hypothesis for this condition. Some selective factors in the earliest haplorhines were perhaps the same as in the earliest cheirogaleids. If the haplorhine morphotype was nocturnal, as was likely the first cheirogaleid, selection might have favored an increase of blood supply to the orbits. Although the pathway of blood flow through the ear region is different in the two groups, the selective factors responsible for the existence of these divergent, diagnostic characters might have been the same.

It was suggested by Werner (1960) that failure to enlarge the stapedial artery was a result of the limit set on expansion by the stapes. Perhaps this explains why both the haplorhines, on the one hand, and the cheirogaleids and subsequently lorisids, on the other hand, enlarged and developed the promontory and the ascending pharyngeal arteries, respectively. The enlarged platyrrhine and catarrhine promontory arteries show that the size of this vessel is not limited by basicranial architecture.

The broken basicranium of *Tetonius* shows signs of an enlarged petromastoid area and a relatively very large middle-ear chamber, particularly the hypotympanic sinus essentially as in *Necrolemur*. Petromastoid inflation is often causally tied to bulla inflation and subsequently to the enlargement of the middle-ear cavity. In *Necrolemur*, with an enormous petromastoid, the middle-ear cavity, although large, is not as hypertrophied as that of *Tarsius*. Yet, in *Tarsius*, which has an enlarged hypotympanic sinus, there is no noticeable petromastoid inflation. Among living primates the solitary, nocturnal lorisids have a middle ear with multiple partitions and a greatly inflated petromastoid, chambers which are continuous with the middle ear proper. Charles-Dominique (1971) noted that of the Gabon lorisiforms, galagines feed principally on Orthroptera, Lepidoptera, and small Coleoptera and that (p. 224), "Hearing plays a large part in the detection of these insects—many of which are captured in flight by a rapid extension movement in which the Bush-baby maintains a grasp on the branch with its hind limbs. The ears . . . are particularly mobile, permit detection of insects in flight even through an opaque screen." Are some lorisid modifications of the ear structure related to food-procuring activities? Whatever the specific role of bullar hypertrophy and the possible correlated inflation of the petromastoid to further enlarge the middle-ear cavity, the major selective

FIG. 17. Entocuneiform and metatarsal I of *Hemiacodon gracilis*, medial Eocene, Wyoming. Characters on these bones indicate a powerful, grasping foot.

387

PHYLOGENY,
ADAPTATIONS,
AND DISPERSAL
OF TARSIIFORM
PRIMATES

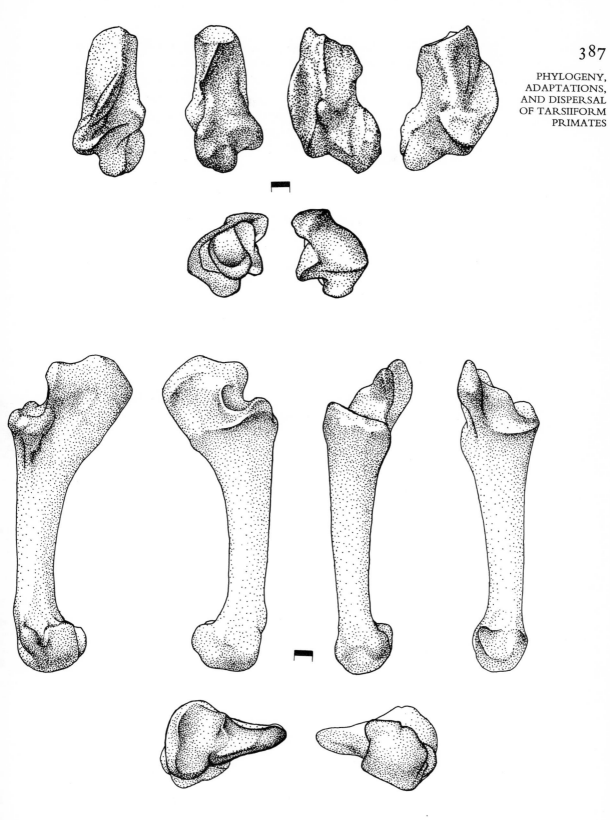

factor was perhaps the improvement of hearing acuity in general. Lay (1972) suggests selection for acuity for a wide range of low-frequency sounds in desert rodents, without any selection for particular frequencies in this range. Lay further convincingly hypothesizes, at least for desert species, that as efficiency of carnivores increases with sparsity of cover where rodents can hide, selection at the same time operates to increase hearing acuity in prey species and perhaps, I suggest, in predators also. Thus perhaps some form of bullar hypertrophy in lorisids, most of which are predatory, and in some omomyids indicates adaptations to both prey and predator detection.

B. Dentition (Figs. 7–12)

The crown morphology of the teeth is as much of a compromise between the influence of the ancestral genotype and current selective factors as any other attributes of an animal. I believe the selective forces derive, following changes in feeding behavior, from a compromise of (1) the texture and hardness of food to be triturated, (2) occlusal requirements rigidly canalized in most mammalian species, and (3) the species-specific emphasis on either the puncturing–crushing, or buccal or lingual phases of the chewing stroke[*] of the masticatory cycle. As (3) can be analyzed from morphology alone (Seligsohn and Szalay, 1975), and (2) is relatively well understood, rigorous studies can yield answers to point (1) for fossil taxa. However, the need to study living species, their specific diets, and related morphology cannot be overemphasized for paleobiology.

In general, the most important finding from numerous field studies of primates has been that sympatric species consistently follow their own dietary regime, subdividing the food resources in a species-specific manner. This obviously suggests that specific selective forces act on the feeding mechanism and this, in turn, suggests that taxon-specific adaptations are causally correlated with feeding adaptations.

As a rule, the biological role of fossil dentitions is usually determined by analogy with dentitions of living species with known diet. This method can be extremely deceptive when only molars are utilized, as some poorly known fossil taxa which share many primitive molar similarities with living forms are often interpreted to have had similar food preferences to the living analogs. Minor changes in living forms which might signify major adaptive shifts from ancestral patterns thus are often overlooked and instead of causal explanations, ill-understood correlations tie the morphology of the crown pattern of living species to their diet. Although still based on analogy, comparison of similar sized complete dentitions, area by area, tooth by tooth, is much more reliable in assessing possible broad feeding preferences of fossil species.

The most thorough and sophisticated surveys of primate diets in the wild (in particular Hladik and Hladik, 1969; Charles-Dominique, 1971; and Petter and Peyrieras, 1970) reveal both species-specific and habitual dietary regimes, particularly in rain-forest environments in which niche width is probably relatively narrow. It has also been

---

[*] These three distinct phases of the power stroke have been recognized and clearly differentiated by Crompton and Hiiemae (1970), and the prevalence of at least the first two appears probable in most therians. See discussion of these modes of chewing in Hiiemae and Kay (1973).

389

PHYLOGENY,
ADAPTATIONS,
AND DISPERSAL
OF TARSIIFORM
PRIMATES

recently established (Hladik and Hladik, 1969; Charles-Dominique, personal communication) that insects supply the three essential dietary components, protein, carbohydrates, and lipids, whereas leaves yield protein and carbohydrates, and fruits contain largely carbohydrates. It follows that frugivores are more dependent on a steady supply of animal food as opposed to folivores, which only need to supplement their diet with occasional insects to make up for the lipid deficiency of leaves. The supplied dietary profiles, researched and analyzed by Hladik and Hladik (1969) and Hladik *et al.* (1971) on several species of platyrrhines, clearly show, as stated by the authors (1969, p. 115), that, "Each species of primate has a specific dietary regime. Each has its own particular preferences. These preferences are not simply 'opportunistic' responses to the food available. Different species feed on different things in the same areas." Charles-Dominique (1971), pointed out that even in the tropical ecosystem (p. 225),

> There is a critical period (dry seasons) during which the availability of food (insects, fruits) reaches a minimum. With some of the prosimian species (for which there is adequate sampling throughout the year), it has been shown that there is a reduction in ingested food and loss in body weight in individual animals during this critical period. Examination of the digestive tract shows that dietary specializations are far more distinct during the peak of the dry season than at any other time during the year. Thus, during the critical period, *E. elegantulus* eats gums and no fruit, whilst *P. potto* eats no gums, whereas these two species will eat both fruit and gums for the rest of the year.

I have surveyed the dental morphology of primate species on which field studies have been conducted and concluded that by a method of analogy comparing the whole dentition of fossil forms to those of similar sized living species it is possible to suggest, at least as a hypothesis to be eventually tested by more rigorous methods (Seligsohn and Szalay, 1975), one of five broad diet categories for fossil taxa. The categories might be called (see also A. Jolly, 1972), from one extreme to the other, folivorous, frugivorous, omnivorous, insectivorous–carnivorous, and gumivorous. Folivorous, insectivorous–carnivorous, and gumivorous specializations are probably recognized with a high degree of probability. Frugivorous and omnivorous feeding regimes, on the other hand, are less well-defined categories, as most species considered frugivorous consume varying amounts of leaves and insects to satisfy protein and lipid requirements.

Dental remains of some Eocene taxa are relatively abundant and a taxon-by-taxon analysis is beyond the scope of this account. Omomyid dentitions are very diversified, presumably as a direct result of their divergent species-specific adaptations. Their diversity is probably also correlated with their absolute size: small mammals, having absolutely less force available for mastication than larger ones, require more exacting occlusal procedures (Walker and Murray, 1975) to deal with the food substances of specific diets.

All four areas of the dentition vary considerably among fossil tarsiiforms. The anterior dentition, however, is only rarely known well. *Tarsius* exhibits an anterior dentition, along with the premolars, in which the very pointed single-cusped morphology of the teeth and their occlusal relations point to maximizing orthal piercing and penetration. The biological role of this anterior dentition, as can be inferred from both laboratory and field accounts, lies in its use as a combination killing and holding device. The prey caught by the hands and feet is killed and probably segmented by the anterior dentition with the help of the hands. What is known of omomyid anterior dentitions has been analyzed in detail elsewhere. These adaptations are tantalizing as they allow a glimpse into an undoubtedly extensive adaptive divergence.

The morphology of omomyid cheek teeth are not stereotyped, and within the same subfamily there appear to be specializations of whole dentitions which may reflect mechanical solutions for extremes ranging between herbivorous feeding (e.g., *Macrotarsius*) and catching and masticating insects and small vertebrate prey (e.g., *Omomys*). In order to show some aspects of the structural and presumably functional diversity of omomyid dentitions I have prepared simple indices of the relative functional areas of each of the cheek teeth from P_3–M_3 (the usual lack of fossils prevented coverage of the anterior dentition) in some of the better known species (Table 1). Relating the loosely

TABLE I

DENTAL MODULES ($m = xA/xA$) OF SOME OMOMYID PRIMATES[a]

Taxon	mP_3	mP_4	mM_1	mM_2	mM_3	mP^3	mP^4	mM^1	mM^2	mM^3
Tetonius homunculus	0.13	0.19	0.25	0.23	0.19	0.19	0.22	0.24	0.24	0.10
Anemorphysis spp. (3 species)	0.11	0.18	0.24	0.23	0.24	0.17	0.19	0.23	0.25	0.15
Absarokius abotti	0.15	0.23	0.23	0.23	0.16	0.20	0.25	0.23	0.23	0.09
Omomys carteri	0.14	0.17	0.23	0.23	0.23	0.14	0.16	0.24	0.25	0.20
Loveina zephyri	0.14	0.18	0.25	0.22	0.20					
Hemiacodon gracilis	0.13	0.15	0.24	0.24	0.23					
Ourayia uintensis	0.12	0.14	0.24	0.24	0.26	0.12	0.14	0.24	0.27	0.22
Washakius insignis	0.11	0.15	0.21	0.22	0.30	0.15	0.18	0.20	0.25	0.22
Shoshonius cooperi						0.15	0.17	0.22	0.25	0.20
Uintanius ameghini	0.18	0.34	0.18	0.16	0.14	0.22	0.22	0.21	0.21	0.13
Chumashius balchi	0.16	0.19	0.23	0.22	0.19					
Dyseolemur pacificus	0.11	0.15	0.22	0.25	0.26					
Ekgmowechashala philotau	0.11	0.21	0.28	0.22	0.18					
Teilhardina belgica	0.12	0.17	0.25	0.27	0.18	0.14	0.20	0.25	0.26	0.14

[a] The estimated functional "area" of a crown is simply the product of the greatest width and length of a tooth.

termed functional area of an upper or lower tooth (the product of length and width) to the sum of the functional areas of the teeth from the third premolar to the last molar, either above or below, allows an informative, if not comprehensive, comparison (Fig. 18). The qualitative details, and their possible reflection of specific biological roles, discussed elsewhere, further underline the great diversity of omomyid feeding adaptations.

The use of either dental proportions (relative functional area patterns, a term coined by Katharine Milton, personal communication) or the mechanical analysis of individual elements of the dentition must be firmly grounded on the best possible assessment of the morphotype for a given taxon. What matters theoretically, I believe, is the evaluation of qualitative change, the adaptive shift in the teeth from inferred common ancestors to descendants in lineages, rather than the differences between many species with ties of varying recency to a higher category. The gist of this type of comparison is that adaptations reflect divergence of one form from another, the ancestor, or between two or more closely related taxa from a common ancestor. These inferred differences should be analyzed for mechanical function of mastication and their significance in terms of a specific diet. However, operational procedures are almost invariably compromised by the available record.

Regardless of the morphology inherited once a lineage has made an adjustment to procurement and processing of its species-specific food resources, dental proportions, coupled with careful qualitative evaluation of crown patterns, may be effective in predicting broad feeding patterns. Theoretically, difficulty in prediction must lie with those taxa which just diverged from the ancestral taxon of an adaptive radiation. They carry with them the attributes of the ancestor, yet, as divergence occurs, these descendant species might show only the rudiments of adaptive responses for the particular new way of life which is divergent from the ancestral one. Operationally, recognition of such problems seems, at least now, almost insurmountable. Antiquity of certain food preferences is as difficult to judge in living taxa as in fossil ones; yet, theoretically, it seems clear that both the relative antiquity of a habitus and, probably much more so, the intensity of selective forces (directed and amplified by competition) must be reflected in the extent of adaptations for a particular feeding regime.

I do not know of any tarsiiform genus which has a fused symphysis. To what degree symphyseal fusion is influenced by size, given that all other selective factors are equal, is not known. The known tarsiiforms lack this characteristic, but it is not unlikely that some unknown taxa had symphyseal fusion. The causes of certain symphyseal configurations (in, e.g., *Washakius* or *Macrotarsius*) may be explained essentially by the same selective events which result in symphyseal synostosis.

C. Postcranial Morphology (Figs. 13–17)

On the first metatarsal of *Hemiacodon* (AMNH 12613, Fig. 17) the area of attachment for the peroneus longus is enormous compared with homologs on either known adapids or living strepsirhines. As it is known only in *Hemiacodon*, this keel-shaped process cannot be considered representative of the omomyids in general. For *Hemiacodon*, however, it suggests that unusually great forces, probably associated with grasping, were incurred at this area. Walker (1974) suggests this development in vertical clingers and leapers to be the result of greater components of gravitational forces acting on the process in a vertical postural position.

The advanced nail-bearing, as opposed to clawed, cheiridia and grasping feet of at least some omomyids, although perhaps not the extremes seen in *Hemiacodon*, would indicate terminal branch milieu ecology at least for the common ancestor with nails. Presence of nails clearly cannot be used beyond this rudimentary explanation as the ancestral nature of this feature does not necessarily set limits on its use for purposes other than those for which it orginally evolved.

Some of the omomyid, as well as adapid, postcranial evidence figured prominently in the hypothesis of Napier and Walker (1967a,b) that the common ancestor of known strepsirhine and haplorhine primates was a vertical clinger and leaper. Some views on their analyses have been published by Cartmill (1972), Szalay (1972), and Decker and

391

PHYLOGENY,
ADAPTATIONS,
AND DISPERSAL
OF TARSIIFORM
PRIMATES

Frederick S.
Szalay

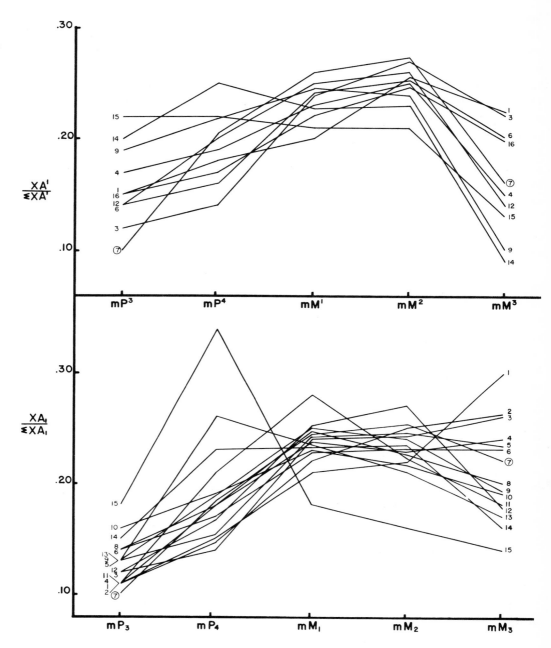

Fig. 18. Relative functional areas of omomyid cheek teeth (P_3–M_3). The dental modules (m) represent the mean "area" (xA) of the tooth (the mean product of the greatest buccolingual width and greatest mesiodistal length) divided by the sum of the mean individual dental areas for that species ($\sum xA$). The resulting graphs show the estimated relative importance of a given tooth within the tooth row of that taxon. This data is clearly incomplete partly because of lack of good samples and because of the usual absence of the anterior part of the dentition in the fossils. The functional importance of crown height and the extremely important qualitative differences of crowns are not reflected. Only when qualitative characters of the crowns and the relative size of the animal are considered

Szalay (1974). Stern and Oxnard (1973) recently presented a critical review of the reported locomotions of strepsirhines and *Tarsius* and a few of their morphological correlates, without treating any of the joints. Their assessment is built around an all-pervasive premise that the morphology studied in a given species is a reflection of characters acquired during the existence of that species as we know it, for locomotor behavior as we now recognize it.[*] The numerous methods suggested for the study of morphology, most helpful in taxonomic discrimination, do not stress the concurrent need to analyze phylogenetic aspects of form and mechanical function. Without clear ordering of morphocline polarity and assessment of morphotypes for taxa compared, and without hypotheses which envisage morphological stages in the studied lineages, such studies, irrespective of various sophisticated techniques employed, lack a necessary phylogenetic perspective.

Stern and Oxnard (1973, p. 34) pose direct questions about the locomotor mode, rather than the mechanical function, of Eocene primate remains, review the literature on the reported fossils, and comment in detail, mostly on aspects of the specimens discernible from the illustrations published. In a carefully measured manner they state (p. 38) that, "... vertical clinging of a type could have been an important locomotor behavior of primates ancestral to the Anthropoidea, without invoking the image of a tarsier or sifaka for the ancestral forms," a conclusion of some general significance but one which needs elaboration. The evidence that the Eocene adapid morphotype, a form more primitive than a haplorhine morphotype, was a vertical clinger and leaper is unlikely (see analysis presented in Decker and Szalay, 1974). This, therefore, makes the inference that the common ancestor of nonparomomyiform primates was a vertical clinger and leaper (as advocated by Napier and Walker, 1967*a*,*b*) also unlikely. Skeletal adaptations of known omomyids, however, although still poorly understood in the joint morphology, do indicate features that may be correlated with rapid locomotion and, at least in *Hemiacodon*, possible vertical posturing. Thus, there is some indication that the omomyid morphotype was an agile, rapid runner and perhaps a habitual leaper, but not necessarily a vertical clinger and leaper. To my knowledge, however, there is no

[*] In their analysis of the Mahalanobis-generalized distance for scapular measurements of a large number of primate species, Stern and Oxnard state that (p. 22), "although *Galago* and *Propithecus* are outlying to the main mass of forms [i.e., of the running–leaping quadrupedal primates], they do not exhibit any special relationships to one another, being on opposite sides of the central group." As a possible explanation they suggest (p. 22) "that there may be, morphologically at least, more than one kind of vertical clinging and leaping." The differences in the history in the two groups is perhaps purposefully sidestepped, but the point is made, namely that form and activity should be used synonymously, and that biological history as a source of explanation for differences in biological features, in as much as they affect mechanical function, is not necessary to consider. Both these assumptions must be rejected.

---

can the relative functional areas be fully appreciated. Thus, for *Uintanius ameghini* (15), for example, mP^3 and mP^4 do not adequately demonstrate the extremely divergent nature of these premolars. The numbers represent the following species: 1, *Washakius insignis*; 2, *Dyseolemur pacificus*; 3, *Ourayia uintensis*; 4, *Anemorphysis* spp.; 5, *Hemiacodon gracilis*; 6, *Omomys carteri*; 7, *Palaechthon alticuspis*, a paromomyid; 8, *Loveina zephyri*; 9, *Tetonius homunculus*; 10, *Chumashius balchi*; 11, *Ekgmowechashala philotau*; 12, *Teilhardina belgica*; 13, *Absarokius noctivagus*; 14, *Absarokius abotti*; 15, *Uintanius ameghini*; 16, *Shoshonius cooperi*.

PHYLOGENY, ADAPTATIONS, AND DISPERSAL OF TARSIIFORM PRIMATES evidence which would suggest the presence of mechanical functions required for vertical clinging and leaping in the morphotype of anthropoids.★

We may now ask the question: what was the natural history of the tarsiiform ancestor like? Considering the overall general similarities of most known fossil tarsiiforms (inasmuch as they were small arboreal species) to present day tarsiers, some small platyrrhines, cheirogaleids, and lorisids, and being fully aware of the specifics of phyletic relationships as discussed above, the general analogy that these fossils were not unlike the groups listed is a broadly valid one (see also Cartmill, 1972). Attempts to answer some questions, for example whether the ancestral tarsiiform (probably an omomyid) was diurnal or nocturnal, insectivorous or frugivorous, to what degree was it social (this is clearly tied to the first two questions, see especially Charles-Dominique, this volume), or what its characteristic posturing and locomotion were, are perhaps beyond the scope of this study or maybe of the available evidence. It seems reasonable to conclude, however, that the structural diversity displayed by the known omomyid fossils (assuming greater intrafamilial differences in the postcranial morphology than that known in the fossils) in the tarsiiform radiation of the early Tertiary produced species at least as diversified in their mode of life as any of the present-day primate families of similar size.

It is worthwhile to quote Gregory's (1951, p. 470) view of the primitive tarsiiformes:

> The advanced specializations of the early tarsioids for arboreal nocturnal habits, reduction of the olfactory sense, great enlargement of eyes, increasing ability to make huge leaps and grasp the tree in landing, all placed the premium of survival upon a rapidly increasing brain; this in turn may have been the focal point of selection in the post-tarsioid stages that are preserved in the New World monkeys.

We now recognize that such an assessment of the primitive specializations probably relies too heavily on the still poorly known mode of life of *Tarsius*. Many unexplained facts remain even in relation to the only living tarsiiform. *Tarsius* lacks a tapetum behind its retina and perhaps the extreme enlargement of its eyes is partly the result of this condition. If one suggests that the tarsiiform morphotype was nocturnal, it is difficult not to suggest that it probably evolved a tapetum. But in that case, one would be forced to state that *Tarsius* is a descendant of an advanced, diurnal tarsiiform which has lost its tapetum. It is not impossible, however, that the prototarsiiform was nocturnal, large-eyed, but without a tapetum. This question is clearly unresolved both for the ancestral tarsiiform or for species of *Tarsius*.

It is probably true that modern tarsiers, which are predatory, occasionally terrestrial crepuscular–nocturnal, silent, and not gregarious primates, are not likely to resemble very closely the tarsiiform or omomyid common ancestor. Nevertheless, an element of nocturnality along with insectivory appear to be likely behavioral components in the repertoire of the ancestral tarsiiform. Judged only from the generally small size of omomyids, their ancestor was also a very small form and perhaps more insectivorous or omnivorous than the putative ancestry of the adapids. As I have argued above on the subject of the enlargement of the promontory artery, this shift may have been a selective

---

★ *Microcebus murinus* has a considerably elongated calcaneum but does not vertical cling and leap. P. Charles-Dominique pointed out at the conference that this species uses the elongated distal segment of the foot primarily to increase its stride in rapid running among small branches and lianas.

response to increase blood supply to the brain and perhaps also to the eyes in order to supply an enlarged eyeball in an ancestral species that was nocturnal or predatory or probably both (see especially Charles-Dominique, this volume).

I cannot help noticing a very close parallel between the origin of the first tarsiiform from a primitive strepsirhine stock and the differentiation of the first cheirogaleid from lemurids, probably under very similar selection pressures from relatively unmodified lemuriforms (adapids and lemurids) during different geological times and geographical settings.

395

PHYLOGENY,
ADAPTATIONS,
AND DISPERSAL
OF TARSIIFORM
PRIMATES

IV. Dispersal

The tarsiiforms were an essentially Holarctic group of primates but their descendants have radiated into the southern continents (Fig. 19). Paleozoogeography is intimately interwoven with phylogeny, and discussion of tarsiiform distribution without the remaining primates would be somewhat out of context. Therefore, in order to place the postulated paleozoogeographic history of the tarsiiforms into perspective, I will briefly examine the dispersal history of the entire order.

The recent acceptance of continental drift, based on overwhelming evidence, carries with it a danger of "band-wagon" mentality for biogeographers. In spite of the advances in the study of geological earth history, phylogenetic analyses of the biological evidence should be assessed at first as independently as possible before attempts are made to decipher the dispersal history of a group. Only following the construction of phylogenies should one attempt to synthesize a sequence for the deployment of the known and postulated taxa.

No "final," comprehensive paleogeographical synthesis exists as yet which would allow us to pinpoint the position of the major land masses and their proximities in given geological moments (but see Cracraft, 1973; McKenna, 1973). Nevertheless attempting to reconstruct the adaptive radiation of any group of mammals without the existing pertinent geological and paleontological evidence is meaningless, even if the phylogeny appears clearly established. Therefore, I will first very briefly outline an approximate history of land and sea distribution from the late Cretaceous to the present which affected deployment of faunas, taking into account a large body of literature on plate motion and polar wander (see particularly results of the recent important symposium, Tarling and Runcorn, 1973).

Although a broad, midcontinental sea (reaching from north to south) divided North America in at least in the early part of the late Cretaceous, land connection existed between North America and Europe via Greenland and the Barents Shelf until probably the end of the early Eocene (McKenna, 1972; Szalay and McKenna, 1971). Since the Mesozoic separation fo North and South America, no land connection was reestablished between these until the end of the Miocene. Final separation of South America and Africa was in the Cretaceous and it appears that the distance between Africa and South America was always greater in the Cenozoic than that between the Americas.

Although the proximity of the African continent to Europe and Asia has not changed

FREDERICK S.
SZALAY

substantially during the Cenozoic, the role of the Tethys and Turgai seaways as effective barriers between Africa and Europe and between Europe and Asia, respectively, during the early Tertiary seems well established. The marine environments of the Turgai straits, during the early Cenozoic, and Beringia, intermittently throughout the Cenozoic, acted as filter routes. Sweepstake dispersal between Central and South America, and South America and Africa was possible throughout the Paleogene, although it has clearly become less and less likely during the latter half of the Cenozoic between South America and Africa as these plates had assumed their present day positions. The exact geographical and faunal relationships of Africa, Madagascar, and India represent an important and enigmatic problem. During the course of the gradual attainment of their present global positions, sweepstakes dispersal was clearly possible between Africa and Madagascar, and Madagascar and India. Just one of the many difficult problems centers on the time of successive positions of the Indian plate prior to its docking to Asia.

As it has been often pointed out in the past, primate dispersal was greatly influenced by the presence of warm forests with perennial food supply, and therefore indirectly by the climate and, of course, the available dispersal routes. Various summary accounts of Tertiary climates based on different approaches agree on a steady, if not regular, climatic deterioration in the northern hemisphere. Some of the causes that effect dispersal and presence or absence of vertebrate species (e.g., such factors as continental positions, orogenies, and subsequent climatic change) were recently reviewed by Walker (1972).

Two of the most recent syntheses of primate paleozoogeography are those of McKenna (1967) and Walker (1972). My interpretation of the evidence for deployment differs from these, partly because I postulate a lesser number of hitherto unknown stocks, but primarily because of my somewhat different understanding of primate phylogeny. Assumptions of phylogeny figured most significantly in interpreting primate dispersal, and this is evident in the syntheses both by McKenna and by Walker, and in my own advanced here. In the discussion to follow, I will invoke the useful model advocated by Kurtén (1967). In essence, this hypothesis suggests that higher categories are probably the result of adaptive radiation of a stock geographically isolated from its closest relatives. Invariably the invasion of the "same" or similar adaptive zones in different areas isolated by barriers, will result in new radiations, depending on space and time.

McKenna, in his view of primate dispersal, held that lemuroids were present only in Africa, whereas his omomyoids (he considered paromomyids also as a primitive family

FIG. 19. A reconstruction of the dispersal history of the Primates. The six background maps attempt to summarize some of the currently available plate tectonic information for the time periods depicted. Arrows indicate both genetic derivation and direction of dispersal. Question marks preceding abbreviations for the specific taxa represent suspected, postulated, or doubted presence (see text for discussion). Question marks affixed to the arrows indicate suspected, postulated, or doubtful dispersion in that direction, but do not signify statements on phylogenetic ties (see text for discussion). The following abbreviations are used: A, Paromoyiformes (including Paromomyidae, Picrodontidae, Plesiadapidae, and Carpolestidae); B, unknown primitive stock of Lemuriformes; B_1, Adapidae; B_2, tooth-combed Lemuroidea; B_3, Cheirogaleidae; C, unknown stock of Tarsiiformes; C_1, Omomyidae; C_2, Tarsiidae; D, Platyrrhini; E, unknown stock of Catarrhini; E_1, Cercopithecoidea; E_2, Hominoidea.

of that group) were present during the Paleocene in North America and, as he suggested, also in Asia. His omomyoids dispersed from Asia to Europe, and the plesiadapids followed the same route. He postulated the spread of lemuroids from Africa to Eurasia and subsequently to North America. Lorisoids were postulated to have evolved from lemuroids in Eurasia during the early medial Eocene. A late Eocene differentiation of precatarrhine and preplatyrrhine stocks in Eurasia and Central America, respectively, is shown by McKenna, prior to the dispersal of these into Africa and South America, respectively.

According to Walker, an ancestral primate stock was present throughout the world (except in Australasia and Antarctica) during the latest Cretaceous to medial Paleocene. Similarly, for the period spanning from late Paleocene to early Eocene, this author suggests the presence of lemuroids and omomyids throughout the entire New World and Old World, except for Madagascar which lacked the omomyids. From the late Eocene to early Oligocene, lemuroids are shown to be present in all places of the previous time period, except in South America where they have apparently become extinct. Omomyids at that time were restricted, according to Walker, to both of the Americas and Africa. During late Oligocene to early Miocene, in addition to the actual presence of omomyids in North America, platyrrhines in South America, and catarrhines and lorisoids in Africa, the tarsiids are shown to have been differentiated in Asia and lemurids to be present in Africa. I believe some of these last assumptions, unlike some of the previous ones, are well justified.

The following account represents a synthesis of my current views on primate phylogeny (see Szalay, this volume) fused with available paleogeographical information. A postulated ancestral stock, ancestral to and in general structurally probably very similar to the stem paromomyiforms of North America and Europe, was widely distributed throughout the northern continents during the late Cretaceous. These mammals, the earliest of which may already qualify to be called primates by basicranial and pedal criteria, were possibly distinct 80–70 million years ago, but the only Cretaceous record so far is the single tooth of *Purgatorius ceratops* (Van Valen and Sloan, 1965). The Paleocene primate fauna of western North America consists of only paromomyiforms, i.e., paromomyids, picrodontids, plesiadapids, and carpolestids. The known Paleocene sites of France and Germany share the plesiadapines, along with an endemic group of plesiadapids represented by *Saxonella*. That Paleocene paromomyids were also present in Europe is likely as judged by the probable phylogenetic ties of *Berruvius*, a form related to the North American *Navajovius*. Absence of picrodontids and carpolestids might indicate that these small forms were endemic to western North America, or simply the poor sampling of European Paleocene faunas may account for their absence.

The evidence from the Eocene primate faunas of western North America and Europe convincingly suggests that the paromomyiforms were superseded by the lemuriforms and tarsiiforms. This is so in spite of the fact that such genera as *Plesiadapis*, *Platychaerops*, *Phenacolemur*, and *Ignacius* linger on. There is no question in my mind that the stem lemuriform and tarsiiform radiations were already in existence during the Paleocene, but these arenas of evolution are largely unknown.

The lemuriform primates, restricted today to Madagascar, were once widely distributed on the Northern Hemisphere. They appear to be widespread during the entire span of the Eocene, with the greatest taxonomic diversity (known so far for that time) apparently in Europe. As I judge from the description and illustrations of the few known, tantalizing Asian fossils, they were probably even more varied east of the Turgai Straits in southern Asia. In North America, significantly, there are three known generic groups present: the early Eocene *Pelycodus*, the medial Eocene *Notharctus* and *Smilodectes*, and an undescribed late Eocene adapine (Wilson and Szalay, in press). *Pelycodus* is also present in Europe, and there, unlike in North America, the adapids are the dominant primates during the Eocene. The lemuriform paleozoogeographic record indicates a Holarctic Eocene distribution, yet the Cretaceous and Paleocene history is so far entirely lacking. Paleocene presence of adapids in Europe or during that time period in the tropical forests of southeastern North America certainly cannot be dismissed. Complete lack of knowledge of the first 20 million years of the Cenozoic of southeast Asia and of the land mammal faunas of the entire Paleocene and most of the Eocene of Africa make speculations difficult. In regard to Africa, the *three* successive primate faunas of the Oligocene from the Fayum deposits of Egypt, with the absence of adapids or noncatarrhine primates in general, might be significant. Considering factors of global tectonics, as a mere speculation, the possible Asian source for the colonizing tooth-combed lemurs of the Paleocene or Eocene might have been western Asia or the southern part of the continent as it appeared then without the Indian subcontinent. India could have acted as the place which received northern waifs and the subsequent source for the colonization of Madagascar or first the Somali Peninsula and subsequently Madagascar.

The lemuriforms, then, judged by the distribution of the Eocene taxa, may be originally an Asiatic group which gained entry into Europe and eastern North America. Dispersal to Madagascar from Asia or southern Europe probably occurred either in late Paleocene or as late as early Oligocene time, but clearly this question is unsettled. It is quite possible that Africa supported a varied lemuriform fauna during the late Cretaceous, Paleocene, and Eocene.

The lorisiform primates are enigmatic in terms of their place of origin. Either Africa or possibly southern Asia are likely places for the differentiation of the first lorisids from cheirogaleids (see Szalay and Katz, 1973; Cartmill, this volume), the latter derived from *bona fide* lemuroids. I suspect that the lorisids originated either in Africa or Asia from either (1) Malagasy cheirogaleids that reached Africa sometime during the late Oligocene, or from (2) endemic members of the Cheirogaleidae which were originally African, along with African tooth-combed lemuroids. At present, I am perhaps more inclined to favor lorisid evolution from Malagasy cheirogaleid waifs reaching Africa, or possibly Asia.

The interpretation of the deployment of haplorhine primates is no less problematical than that of the strepsirhines. In North America at the beginning of the Eocene an unprecedented radiation of the greatest known bulk of the tarsiiforms, the omomyids, is already under way, whereas, as noted, only one adapid genus, *Pelycodus*, is known there during that time. It is a pure guess that the tarsiiforms may have come north from their more southern and southeastern range of either continental North America or Asia as the climate warmed at the end of the Paleocene. In North America the last remnants of the midcontinental Paleocene Cannonball Sea disappeared at this time. The provenance

399

PHYLOGENY,
ADAPTATIONS,
AND DISPERSAL
OF TARSIIFORM
PRIMATES

of *Pelycodus* is likely to be from Europe where the genus was also present, but not during the Paleocene. Clearly, an Asiatic origin for this genus cannot be ruled out.

The omomyids persist as late as the early Miocene in North America, but their Mesoamerican fate, because there is no record, remains intriguingly unknown. These primitive tarsiiforms probably existed during the Paleocene of North America, perhaps in hitherto poorly sampled ecological facies of western North America, in the southern or eastern half of this continent, or in central and southern Asia. The North American diversity and abundance could be explained then either by local origins of the tarsiiforms from adapid ancestors, probably during the Paleocene in the southeastern part of North America, or by dispersal from Europe or central Asia. Deployment, if the origin was Asiatic, was either in the form of sweepstakes dispersal across the southern part of the Turgai Straits into Europe and from there to North America or, more likely, from Asia to North America through Beringia. In the case of a North American origin, the dispersal would have been in one of the opposite directions. The European microchoerines are somewhat specialized tarsiiforms that could be descendants of eastern North American or southern European anaptomorphine omomyids. The time of origin for this group may be the same as for the omomyid *Teilhardina*, to which the microchoerines might be specially related. Whether tarsiiforms were once diversified in Asia cannot be assessed, in spite of the presence of *Tarsius*, but it seems very likely that the Asiatic land mass, like North America, also contained a large tarsiiform primate fauna during the early and later Tertiary (Daszheveg and McKenna, 1975).

Perhaps as a direct result of the acceptance of a model of a tectonically mobile earth, the interesting question of the degree and kind of relationship between platyrrhines and catarrhines and caviomorph and phiomorph rodents was lately reexamined, mainly by Lavocat (1969) and Hoffstetter (1972). They have argued that the platyrrhines of South America are descendants of colonizers arriving from Africa. The geographical origins of hystricognathous rodents and anthropoid primates are probably tied together, as well argued both by Lavocat (1971) and Hoffstetter (1972). Late Eocene rodents with a clearly hystricognathous lower jaw from the late Eocene of Texas (Wood, 1972) represent some recently discovered, tantalizing evidence. Although both the Anthropoidea and Hystricognathi are probably monophyletic taxa, the African origin of these categories is not as convincing as Lavocat's and Hoffstetter's arguments suggest (see also Wood, 1973). As I see it, there are two underlying assumptions to their arguments, and there is no evidence which may be unequivocally interpreted to the exclusion of other assessments of the known facts. The assumptions underlying an Africa to South America dispersal hypothesis are (1) that current patterns of the Atlantic facilitated only a western dispersal between those two southern continents, and (2) that Africa received ancestral primate stocks prior to the radiation of the known Oligocene catarrhines.

In light of the widely separated paleopositions of North and South America during the Paleocene and Eocene (see particularly the faunal evidence presented by Simpson, 1948, 1967, and Patterson and Pascual, 1968) the confluent Pacific and Atlantic oceans and other factors, such as the then prevailing wind patterns, may have dictated oceanic currents between the southern continents different from those of today. Thus, the mechanism of dispersal via existing currents is questionable. The second assumption, the presence of undoubted primates in Africa during the early Tertiary, is clearly unresolved.* Admittedly if the introduction of the ancestral anthropoid stock into either South America or Africa is sought, it is perhaps more probable that some haplorhine stock found its way into Africa rather than South America during the late Eocene or early Oligocene, since the distance of dispersal from the north was probably less in the case of the former than in the latter. However, if a sweepstakes dispersal of primates and rodents from southern North America of the medial or late Eocene into South America is assumed, then the probability, at least as I see it at the present, that their descendants rafted to Africa is not less likely than westerly dispersal from Africa. For morphological reasons I favor a dispersal hypothesis in an easterly direction. What has actually happened in the course of anthropoid and hystricognath evolutionary history, accepting that these are monophyletic categories, is thus not at all resolved in spite of our knowledge that South America and Africa were substantially closer together during the Eocene than they are today. The meager evidence of some of the North American omomyid genera and of morphocline polarities of features shared by anthropoids suggest to me tarsiiform dispersal to South America, where the origins and radiation of the Platyrrhini occurred, and subsequently the dispersal of a platyrrhine into Africa, where the Catarrhini evolved from this colonizing stock.

V. Summary and Conclusion

In assessing the geneology within the Tarsiiformes, the most diverse and abundant evidence, the dentition, was primarily utilized. Judgments of the relationships of the group to other primates can be based also on basicranial and postcranial morphology, as well as on dental characters. Both the morphotypes of the basicranial organization and known postcranial remains of the Omomyidae point to an ancestral condition very similar to that found in adapid primates. The primitive omomyid character states of many known features are derived compared to the more primitive conditions of the adapid morphotype. A derivation from a species which would, on total morphological grounds, be considered as adapid rather than a paromomyiform is postulated for the stem tarsiiforms.

In examining the relationships within the Tarsiiformes, it is concluded that the most primitive group of this category is probably the Omomyidae, as exemplified by the Anaptomorphinae and Omomyinae. Specializations found in the basicranial morphology and dentitions of the Microchoerinae on the one hand and the Tarsiidae on the other are not shared between these two groups. Tibiofibular fusion of the Microchoerinae may or may not be a unique character of the subfamily, and it may be a characteristic of the Omomyidae. Possession of divergent specializations and sharing of only possibly primitive tarsiiform characters between the Microchoerinae and Tarsiidae strongly argue against special relationships and hence against the inclusion of the former group within the Tarsiidae. The specialized condition of the basicranium (see Szalay, this volume), as well as numerous unique, advanced soft anatomical attributes shared between

* Judged by the published figures, the Paleogene African *Azibius* recently described by Sudre (1975) appears to me to be a hyopsodontid-like condylarth rather than a paromomyiform primate.

the Tarsiiformes, Platyrrhini, and Catarrhini, warrant the monophyletic status of the suborder Haplorhini (see Luckett, this volume).

An attempt was made to envisage aspects of behavior and ecology of the ancestral tarsiiforms. The enlarged promontory artery of the tarsiiform ancestor, like the enlarged ascending pharyngeal artery of the nocturnal cheirogaleids, was probably a means of increasing the blood supply to the brain and possibly to the ophthalmic artery, and therefore to the eyes. This, in turn, very tentatively suggests (but only for the morphotype) that selection was for more richly vascularized eyes, possibly in relation to crepuscular and nocturnal feeding habits. These adaptations were accompanied by the apparent increase in the sophistication of rapid locomotion as evidenced by the tarsus. The swift locomoting habits of the tarsiiform morphotype, probably coupled with attributes for detecting insect prey, partly by vision and partly by hearing, supplied powerful selective forces for augmenting the centers in the brain responsible for an increase in visual and auditory memory and a more efficient processing and integration of visual and auditory stimuli. These conclusions are specific for the earliest tarsiiform adaptation and do not necessarily apply to the antecendent Adapidae, and therefore to primitive strepsirhines in general.

An inferred dispersal history for the primates is presented. As far as the meager fossil evidence suggests, paromomyiform primates were not present in South America and Africa during the Cretaceous and Paleocene, whereas on the northern continents they were well established during the early Tertiary. The main arena of strepsirhine evolution appears to have been Europe and Asia although presence of varied stocks of African strepsirhines prior to the Miocene cannot be ruled out. Dispersal of the original stock of tooth-combed lemurs to Madagascar or Africa was probably from Europe or Asia. Although the descendants of primitive haplorhines, the platyrrhines and catarrhines, have dispersed to the southern continents, the known record indicates the Tarsiiformes to have been a Holarctic group. The origin of the Anthropoidea is probably from southern North American tarsiiforms, and the direction of dispersal was to South America and subsequently across the Atlantic to Africa at low northern latitudes. Dispersal in the opposite direction cannot be ruled out, and if that is the case, origins of the anthropoids was more likely from Asiatic or southern European tarsiiforms.

VI. References

BOCK, W. J., and T. VON WAHLERT. 1965. Adaptation and the form–function complex. *Evolution* **19**:269–299.

CARTMILL, M. 1972. Arboreal adaptations and the origin of the order Primates, pp. 97–122. *In* R. Tuttle (ed.), *The Functional and Evolutionary Biology of Primates*. Aldine-Atherton, Chicago and New York.

CARTMILL, M. 1974. Pads and claws in arboreal locomotion, pp. 95–83. *In* F. A. Jenkins, Jr., ed., *Primate Locomotion*. Academic Press, New York.

CHARLES-DOMINIQUE, P. 1971. Eco-ethologie des prosimiens du Gabon. *Biol. Gabonica* **7**: 121–128.

CRACRAFT, J. 1973. Continental drift, paleoclimatology, and the evolution and biogeography of birds. *J. Zool. London* **169**:455–545.

CROMPTON, A. W., and K. HIIEMAE. 1970. Molar occlusion and mandibular movements during occlusion in the American opposum, *Didelphis marsupialis* L. *Zool. J. Linn. Soc.* **49**:21–47.

DECKER, R. L., and F. S. SZALAY. 1974. Origins and function of the pes in the Eocene Adapidae (Lemuriformes, Primates), pp. 261–291. *In* F. A. Jenkins, ed., *Primate Locomotion*. Academic Press, New York.

GAZIN, C. L. 1958. A review of the middle and upper Eocene primates of North America. *Smithson. Misc. Coll.* **136**:1–112.

403

PHYLOGENY,
ADAPTATIONS,
AND DISPERSAL
OF TARSIIFORM
PRIMATES

GREGORY, W. K. 1915. On the classification and phylogeny of the Lemuroidea. *Bull. Geol. Soc. Am.* **26**:426–446.

GREGORY, W. K. 1920. On the structure and relations of *Notharctus*, an American Eocene primate. *Mem. Am. Mus. Nat. Hist.* **3**:51–243.

GREGORY, W. K. 1951. Evolution emerging. *A Survey of Changing Patterns from Primeval Life to Man.* Vol. 1. Macmillan Co., New York.

HIIEMAE, K., and R. KAY. 1973. Evolutionary trends in the dynamics of primate mastication. *Symp. IVth Int. Cong. Primatol.* **3**:2–64.

HLADIK, A., and C. M. HLADIK. 1969. Rapports trophiques entre vegetation et primates dans la foret de Barro Colorado (Panama). *Terre Vie* **1**:25–117.

HLADIK, C. M., A. HLADIK, J. BOUESSET, P. VLADEBOUZE, G. VIROBENT, and J. DELORT-LAVAL. 1971. Le régime alimentaire des Primates dè l'île de Barro-Colorado (Panama). *Folia Primatol.* **16**:85–122.

HOFFSTETTER, M. R. 1972. Relationships, origins, and history of the ceboid monkeys and caviomorph rodents: a modern reinterpretation, pp. 323–347. *In* T. Dobzhansky, M. K. Hecht, and W. C. Steere, eds., *Evolutionary Biology*, Vol. 6. Appleton-Century-Crofts, New York.

HÜRZELER, J. 1946. Zur Charakteristik, systematischen Stellung, Phylogenes und Verbreitung der Necrolemuriden aus dem europaischen Eocaen. *Schweiz. Palaeontol. Gesell.* **10**:352–354.

HÜRZELER, J. 1948. Zur Stammesgeschichte der Necrolemuriden. *Schweiz. Palaeontol. Abh.* **66**:1–46.

JOLLY, A. 1972. *The Evolution of Primate Behavior.* Macmillan, New York.

KURTÉN, B. 1967. Continental drift and the palaeogeography of reptiles and mammals. *Comment. Biol.* **31**:1–8.

LAVOCAT, R. 1969. La systematique des rongeurs hystricomorphes et la derive des continents. *C.R. Acad. Sci. Paris* **5**:1496–1497.

LAVOCAT, R. 1971. Affinités systématiques des caviomorphes et des phiomorphes et origine africaine des caviomorphes. *Ann. Acad. Brasil. Cienc.* **43**:515–522.

LAY, D. M. 1972. The anatomy, physiology, functional significance and evolution of specialized hearing organs of gerbilline rodents. *J. Morphol.* **138**:41–120.

McKENNA, M. C. 1960. Fossil Mammalia from the early Wasatchian Four Mile Fauna, Eocene of northwest Colorado. *Univ. Calif. Publ. Geol. Sci.* **37**:1–130.

McKENNA, M. C. 1967. Classification, range, and deployment of the prosimian primates. *Coll. Int. Cent. Nat. Rech. Sci., Prob. Actuels Paleont.* **163**:603–610.

McKENNA, M. C. 1972. Was Europe connected directly to North America prior to the middle Eocene. *Ecol. Biol.* **6**:179–189.

McKENNA, M. C. 1973. Sweepstakes, filters, corridors, Noah's arks, and beached Viking funeral ships in palaeogeography. *In* D. H. Tarling, and S. K. Runcorn, eds., *Implication of Continental Drift to the Earth Sciences*, Vol. 1, Academic Press, London and New York.

NAPIER, J. R., and A. C. WALKER. 1967a. Vertical clinging and leaping: a newly recognized category of locomotor behaviour of primates. *Folia Primatol.* **6**:204–219.

NAPIER, J. R., and A. C. WALKER. 1967b. Vertical clinging and leaping in living and fossil primates, pp. 66–69. *In* D. Starck, R. Schneider, and J. H. Kuhn, eds., *Progress in Primatology*, Gustac Fischer Verlag, Stuttgart.

PATTERSON, B., and R. PASCUAL. 1968. Evolution of mammals on southern continents. V. The fossil mammal fauna of South America. *Rev. Biol.* **43**:409–451.

PETTER, J. J., and A. PEYRIERAS. 1970. Observations eco-ethologiques sur les lemuriens malgaches du genre *Hapalemur*. *Terre Vie.* **24** (3):356–382.

ROBINSON, P. 1968. The paleontology and geology of the Badwater Creek area, central Wyoming. Part 4. Late Eocene primates from Badwater, Wyoming, with a discussion of material from Utah. *Ann. Carnegie Mus.* **39**:307–326.

RUSSELL, D. E., P. LOUIS, and D. E. SAVAGE. 1967. Primates of the French early Eocene. *Univ. Calif. Publ. Geol. Sci.* **73**:1–46.

SCHLOSSER, M. 1887. Die Affen, Lemuren, Chiropteren, Insectivoren, Marsupialier, Creodonten, und Carnivoren des europaischen Tertiars und deren Beziehungen zu ihren lebenden und fossilen ausser europaischer Verwandten. *Beitr. Palaeont. Oesterreich-Ungarns und des Orients* **6**:1–227.

SELIGSOHN, D., and F. S. SZALAY. in press. Relationship between natural selection and dental morphology: tooth function and diet in *Lepilemur* and *Hapalemur*. *In* K. A. Joysey, ed., *IVth International Congress of Dental Morphology*, Academic Press, London.

SIMONS, E. L. 1961a. The dentition of *Ourayia*: its bearing on relationships of omomyid prosimians. *Postilla* **54**:1–20.

SIMONS, E. L. 1961b. Notes on Eocene tarsioids and a revision of some Necrolemurinae. *Bull. Brit. Mus. (Nat. Hist.). Geol.* **5**:43–69.

SIMONS, E. L. 1963. A critical reappraisal of Tertiary primates, pp. 65–129. *In* J. Buettner-Janusch, ed., *Evolutionary and Genetic Biology of Primates*, Vol. 1. Academic Press, New York and London.

SIMONS, E. L., and D. E. RUSSELL. 1960. Notes on the cranial anatomy of *Necrolemur*. *Breviora*. **127**.

SIMPSON, G. G. 1937. The Fort Union of the Crazy Mountain Field, Montana, and its mammalian faunas. *U.S. Nat. Mus. Bull.* **169**:1–287.

SIMPSON, G. G. 1940. Studies on earliest primates. *Bull. Am. Mus. Nat. Hist.* **77**:185–212.

SIMPSON, G. G. 1945. The principles of classification and a classification of mammals. *Bull. Am. Mus. Nat. Hist.* **85**:1–350.

SIMPSON, G. G. 1948. The beginning of the age of mammals in South America. Part 1. *Bull. Am. Mus. Nat. Hist.* **91**:5–239.

SIMPSON, G. G. 1955. The Phenacolemuridae, new family of early primates. *Bull. Am. Mus. Nat. Hist.* **105**:415–441.

SIMPSON, G. G. 1967. The Tertiary lorisiform primates of Africa. *Bull. Mus. Comp. Zool.* **136**:39–62.

STEHLIN, H. G. 1916. Die Saugetiere des schweizerischen Eocanes. Critscher Catalog der Materialen. *Abh. Schweiz. Paleontol. Ges.* **4**:1297–1552.

STERN, J. T., and C. E. OXNARD. 1973. Primate locomotion: some links with evolution and morphology. *Primatologia* **4**:1–93.

SUDRE, M. J. 1975. Un prosimien du Paléogène ancien du Sahara Nord-occidental: *Azibius trerki* n.g.n.sp. *C.R. Acad. Sc. Paris*, t.280, série Di. 1539–1542.

SZALAY, F. S. 1972. Paleobiology of the earliest primates, pp. 3–35. In R. Tuttle, ed., *The Functional and Evolutionary Biology of Primates*, Aldine-Atherton, Chicago and New York.

SZALAY, F. S. in press. Systematics of the Omomyidae (Tarsiiformes, Primates): taxonomy, phylogeny, and adaptations. *Bull. Amer. Mus. Nat. Hist.*

SZALAY, F. S., and R. L. DECKER. 1974. Origins, evolution, and function of the pes in the Eocene Adapidae (Lemuriformes, Primates), pp. 239–259. In F. A. Jenkins, Jr., ed., *Primate Locomotion*, Academic Press, New York.

SZALAY, F. S., and C. C. KATZ. 1973. Phylogeny of lemurs, galagos and lorises. *Folia Primatol.* **19**:88–103.

SZALAY, F. S., and M. C. McKENNA. 1971. Beginnings of the age of mammals in Asia. *Bull. Am. Mus. Nat. Hist.* **144**:269–318.

SZALAY, F. S., I. TATTERSALL, and R. L. DECKER. 1975. Phylogenetic relationships of *Plesiadapis*—postcranial evidence, pp. 136–166. In F. S. Szalay, ed., *Approaches to Primate-Paleobiology*, *Contributions to Primatology*, Vol. 5. Karger, Basel.

TARLING, D. H., and S. K. RUNCORN. 1973. *Implications of Continental Drift to the Earth Sciences*. Academic Press, London and New York.

TEILHARD DE CHARDIN, P. 1916–1921. Sur quelques primates des Phosphorites du Quercy. *Ann. Paleontol.*, **10**:1–20.

VAN VALEN, L., and R. E. SLOAN. 1965. The earliest primates. *Science* **150**:743–745.

WALKER, A. 1970. Post-cranial remains of the Miocene Lorisidae of East Africa. *Am. J. Phys. Anthropol.* **33**:249–262.

WALKER, A. 1972. The dissemination and segregation of early primates in relation to continental configuration, pp. 195–218. In W. W. Bishop and J. A. Miller, eds., *Calibration of Hominoid Evolution*. Scottish Academic Press.

WALKER. A. 1974. Locomotor adaptations in past and present prosimian primates, pp. 349–381. In F. A. Jenkins, Jr., ed., *Primate Locomotion*. Academic Press, New York.

WALKER, P., and P. MURRAY. 1975 (in press). An assessment of masticatory efficiency in a series of anthropoid primates with special reference to the Colobinae and Cercopithecinae. In R. Tuttle, ed., *Ninth International Congress of Anthropological and Ethnological Science. World Anthropology*, Vol. 1. Mouton Publishers, Hague, Netherlands.

WEIGELT, J. 1933. Neue Primaten aus der mitteleozanen (oberlutetischen) Braunkohle des Geiseltals. *Nova Acta Leopoldina*, N.S. **1**:97–156.

WERNER, C. F. 1960. Das Ohr. A. Mittel- und Innenohr. *Primatologica* **2**(5):1–40.

WILSON, J. A. 1966. A new primate from the earliest Oligocene, west Texas, preliminary report. *Folia Primatol.* **4**:227–248.

WILSON, J. A., and F. S. SZALAY. in press. New adapid primate of European affinities from Texas. *Folia Primatol.*

WOOD, A. E. 1972. An Eocene hystricognathous rodent from Texas: its significance in interpretations of continental drift. *Science* **175**:1250–1251.

WOOD, A. E. 1973. Eocene rodents, Pruett Formation, southwest Texas; their pertinence to the origin of South American Caviomorpha. *Pearce-Sellards Ser.* **20**:1–40.

WORTMAN, J. L. 1903. Studies of Eocene Mammalia in the Marsh Collection, Peabody Museum, Part 2. Primates. *Am. J. Sci.* **15**:163–176.

16

Evolution and Interrelationships of the Catarrhine Primates

ERIC DELSON AND PETER ANDREWS

I. Introduction

The object of this chapter is to present our assessment of phyletic relationships among Old World higher primates in the light of the "cladistic" methodology when possible, and to suggest possible phylogenetic and classification scheme(s) for this group consistent with this interpretation. In the first section of the paper, morphological data (mostly dental, cranial and postcranial skeletal) in a number of catarrhine groups will be presented. The typical morphologies which can be inferred within each group are taken to correspond to the hypothetical (ancestral) morphotype, that which would be expected in the latest common ancestor of the animals on which data have been provided. Fossils will be included with the groups to which they belong where such relationships are clear. Other extinct forms will be treated separately in this phase.

The second section attempts to provide a reconstruction of the morphology to be expected in the ancestral catarrhine based on a combination of the several subordinate morphotypes. This correctly implies that we do consider all animals treated here to have

ERIC DELSON · Department of Anthropology, Lehman College, City University of New York, Bronx, New York.
PETER ANDREWS · Sub-Department of Anthropology, British Museum (Natural History), Cromwell Road, London.

had a single common ancestor, of as yet uncertain age and distribution. With this catarrhine morphotype as a basis, each of the major taxa will be reanalyzed in terms of the relationships of its constituent genera, modern and extinct. The result of these analyses will be a series of possible cladogram/phylogenies, with the taxonomic problems discussed for each group synthesized in a final proposed classification.

The essence of cladistic interpretation of relationships as seen here is that only those linkages based on shared derived ("advanced," apomorphous) characters reflect true phyletic relationships. Those based on shared ancestral ("primitive," conservative, plesiomorphous) features merely reflect common ancestry at some previous point in time and are thus not useful in forming subgroups of a larger taxon. For a further discussion

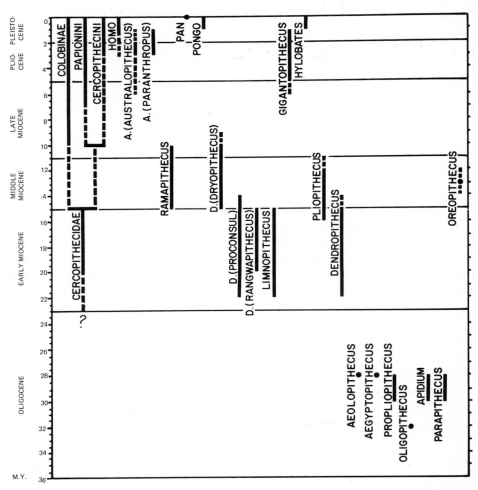

Fig. 1. Temporal ranges of cercopithecid higher taxa and all other genera (and some subgenera) of catarrhines. Dashed lines indicate uncertain ranges, large dots single occurrences.

of this subject reference may be made to other papers in this volume and also to Schaeffer *et al.* (1972), Eldredge and Tattersall (1975), and Delson (1976).

Fossils may be incorporated in the deduction of ancestral morphotypes in order to increase the range of character variation available for study, but their age is not considered. Theories of relationship may be tested by comparing reconstructed morphotypes to known early fossils in order to see if any do agree with the prediction (Delson, 1975c). There may be cases where polarity along the morphocline is unclear, and we take the view that early fossils are more likely to be conservative in most features, having changed less from the actual ancestral condition. Thus, the morphology present in one or more early fossil members of the group under consideration, if it is one of the two alternative choices, is tentatively accepted as the less-derived one. This may introduce some circularity, but it appears to us better than merely ignoring that line of evidence until (if ever) more data become available. Additional theoretical aspects of the application of cladistics to paleontology, and especially to classification, will be considered at relevant points in the general discussion.

II. Distribution of Morphological Characters in the Catarrhini

In this section the six main groups we can distinguish are treated together with their potential fossil relatives. Because temporal occurrence is not emphasized at this point, we include for reference a stratigraphic range chart (Fig. 1). Other than for the Old World monkeys, whose single-family status seems clear, taxonomic ranking of the groups is studiously avoided in order to preclude biasing the conclusions. This may make the discussion more verbose, but it should permit clearer analysis eventually.

The majority of our interpretations and conclusions are based on skeletal elements that can be studied in fossils as well as in modern species. Owing to the frequency with which dental remains of extinct forms are preserved, teeth have been subjected to intensive study. The major results are discussed by group and summarized in Table 1, which also includes a list of dental features of the catarrhine morphotype. Additional information drawn from cranial and postcranial morphology, soft tissues, and karyology is summarized in Table 2.

Many researchers have utilized data on the relative lengths of skeletal segments as indicators of taxonomic and/or locomotor affinity among the primates. Recently, however, Biegert and Maurer (1972) have shown that many such features may reflect allometric modifications, and their results have been confirmed by the work of Andrews and Groves (1975). Comparing relative lengths of individual long bones and especially of the arm or leg to an axial vertebral length, Biegert and Maurer found that almost all catarrhines fall on a single regression line for each measure. Arm length increases more rapidly than leg length as body size increases, and thus the rise in intermembral index (radius + humerus/femur + tibia) from small monkeys to gorillas is size dependent, not phyletically (nor fully locomotorily) significant. Thus the most interesting cases are those of nonalignment with this general trend. Several basic long-bone indexes are presented for reference in Table 2, along with indications of placement with respect to

TABLE I

DENTAL CHARACTERS OF THE CATARRHINE ANCESTRAL MORPHOTYPE AND THEIR DISTRIBUTION IN THE INFRAORDER[a]

Catarrhine ancestral morphotype[b]	Colobinae	Cercopithecini	Papionini	Hylobates	Pliopithecus and Dendropithecus	Homo (and Australopithecus)	Pongo and Pan	Gigantopithecus	Dryopithecus	Propliopithecus and Aegyptopithecus	Parapithecus and Apidium	Oreopithecus	Oligopithecus
I¹ spatulate, long mesiodistally	A	A	D?	A	A	A	A	O	A	A?	O	A	O
I² narrow conical crown	A	A	AD	D	A	D	AD	O	A	A?	O	A	O
C₁ tall and bilaterally compressed	A	A	A	A	A	D	A	D	A	A	A	A	A?
sexually dimorphic	A	A	A	D	A	D	A	A?	A	A	A	A	O
C¹ with prominent mesial groove not extending onto root	D	D	D	A	A	D	A	O	A	A	A	A	O
P⁻ only 2 (AA = P₂ retained— see text)	A	A	A	A	A	A	A	A	A	A	AA	A	A
P₃ single-cusped	A	A	A	A	A	D	A	D	A	A	★	D	A
"sectorial" (C¹-honing) and bilaterally compressed	A	A	A	A	A	D	D	D	D	A	★★	D	A
P₄ two subequal cusps	AD	A	A	A	A	A	A	A	A	A	AD	A	A
P⁻ two cusps, buccal more projecting	AD	A	A	A	A	D	A	A?	A	A	A	A	O
buccolingually broad	D	D	D	D	A	D	D	A	A	A	A	A	O
M_ six cusps—													
paraconid retained	D	D	D	D	D	D	D	D	AD	D	A	D(A?)	A
hypoconulid developed	D	D	D	A	A	A	A	A	A	A	A	A	A
buccal cingulum developed	D	D	D	AD	A	D	AD	D	A	A	D	A	A
talonid broad, lower than trigonid	D	D	D	A	A	A	A	A	A	A	A	A	A
M₃ larger than M₂ larger than M₁	A	A	A	D	A	D	AD	A	A	AD	AD	A	O(A?)
M₃ narrow with larger hypoconulid	A	D	A	A	A	D	D	A	A	A	A	A	O
M⁻ four cusps, hypocone small	D	D	D	A	A	D	D	D	A	A	A	D	O
lingual cingulum developed	D	D	D	AD	A	D	AD	D	A	A	A	A	O
M¹ (at least) with protoconule	D	D	D	D	A	D	AD	D?	A	A	A	A	O
crowns buccolingually wide	D	D	D	D	A	D	D	A	A	A	A	D	O
M² larger than M³ larger than M¹	AD	D	AD	A	A	D	D	O	A	A	A	A	O

[a] A = Retention of ancestral condition, D = Development of *a* derived condition, O = Character state unknown (not preserved), ? Indicates uncertainty, ★ P₃ of *Parapithecus* and *Apidium* bicuspid, but probably part of different morphocline, ★★ Retained P₂ of *Parapithecus* (and *Apidium*?) apparently adapted to honing C¹

[b] A dash indicates all teeth of the particular series (e.g., M⁻ refers to all upper molars in the series).

arm and leg allometries. Thoracic breadth is under similar allometric control among catarrhines (Andrews and Groves, 1975), but it is not yet certain how strongly such additional characters as vertebral number and lengths of hand, foot, clavicle, and tail are also dependent upon size (compare data in Table 2).

A. Old World Monkeys (Cercopithecidae)

The most characteristic feature of the cercopithecids is their dentition. As in most catarrhines, the dental formula is 2–1–2–3. Relative incisor size varies within the group. The canines are large stabbing weapons that show high sexual dimorphism; uppers, especially of males, present a deep, compressed cleft or sulcus on the mesial face that continues through the cervix onto the root. The P_3 is a unicuspid tooth with a sloping mesiobuccal flange for honing the C^1 and a distal fossa homoplastic, but probably not homologous, to the molariform-tooth talonid; in females the flange does not project far beyond the alveolar plane (if at all), but in males it sinks deeply below this level. All other cheek (postcanine) teeth are lophodont, with a single loph on premolars and two on molariform teeth (M and dP).

The molariform teeth of Cercopithecidae are all based on a single ground plan, consisting of a high ("hyposodont" of authors) crown with 4 marginal cusps linked by transverse ridges or loph(id)s, and 3 foveas separated by the 2 ridges. Upper teeth are, in general, mirror images of their mandibular isomer, with buccal and lingual feature reversed. The teeth widen or "flare" outward laterally from the cusp apexes to the cervix, especially on the buccal face of lowers and the lingual face of uppers, but there is no cingulum. Exceptions to this plan are found in M_3 and dP_3^3; in M_3 a hypoconulid is developed on the distal shelf in most forms; a paraconid is present on dP_3 mesial to the trigonid basin and is often joined to the protoconid by a paralophid. The P^4 consists of a small trigonid, subequal metaconid and protoconid linked by a molariform metalophid and a large talonid basin. The upper premolars are somewhat D-shaped, with straight buccal faces. On both P^4 and, especially, P^3 there is a prolongation of enamel onto the mesiobuccal root; this is apparently not functional and may represent an "overflow" effect from the canine–premolar honing field that induces the flange on P^3. Cercopithecid dental function involves shearing along upper buccal and lower lingual notches combined with crushing (and guidance control) by loph(id)s (R. Kay, personal communication).

Despite the arguments of Schultz (1970) that the Old World monkeys are remarkably uniform in their morphology, a more detailed examination of at least some character complexes denies this homogeneity, in fact, to the point where further characterization of the family as a whole becomes difficult. The cranium of monkeys is quite variable in both size and form, with two main types noted below. In general the brain is relatively smaller compared to body size than in modern apes. According to Radinsky (1973, 1974), cercopithecid brains are characterized by a V- or C-shaped arcuate sulcus anteriorly. The external auditory meatus is tubular.

Other characters common to all cercopithecids may depend on their generally small size among catarrhines. There is little variation in precausal vertebra number, with generally 18 or 19 thoracic + lumbar and 2–4 sacral, but caudal number ranges between

TABLE 2

SUMMARY OF THE DISTRIBUTION OF SELECTED NONDENTAL CHARACTERS IN MAJOR CATARRHINE GROUPS[a]

	Colobinae	Cercopithecini	Papionini	Hylobates	Pliopithecus	Dendropithecus	Homo (Australopithecus)	Pongo	Pan	Gigantopithecus	Dryopithecus (and Limnopithecus)	Propliopithecus and Aegyptopithecus	Parapithecus and Apidium	Oreopithecus	Oligopithecus
Relative depth of mandible — S, shallow; M, moderate; D, deep	D/M	D/M	M	S	S/M	S/M	M	M	M	D	M/D	M	M	M/D	M/S
Mandibular shape — C, constant; D, deepens; S, shallows mesially	C/S	D	D	D	D	D	C?	C	C	S	C/D?	C?	S?	C	C
Choanal shape — W, wide and low; N, narrow and high; V, variable	V	N	N	W	O	O	N	N	N	O	O	N	O	O	O
External auditory meatus — R, ring; T, tube	T	T	T	T	R	O	T	T	T	O	T	R	R	T	O
Inter-orbital distance — W, wide; I, intermediate; N, narrow	W	N	N	W	W	O	W	N	W	O	I/W	I/W	O	W	O
Length of ulnar olecranon — L, long; I, intermediate; S, short	L	L	L	S	L	L	S	S	S	O	L/I	L	O	S	O
Ulnar-carpal articulation — S, styloid; L, lunula; D, derived (several types)	S	S	S	L	S	O	D1	D2	D3	O	D4(2?)	O	O	O?	O
Astragalo-calcaneal joint — H, helical; R, rotational	R	H	H	H	H?	H?	R	H	H	O	H?	H?	O	R	O
Tail — P, present; A, absent	P	P	P	A	P?	O	A	A	A	O	O(P?)	P?	P?	O(A?)	O
Number of lumbar vertebrae	6–7	6–7	6–7	5	6–7	0	5	4	3–4	0	0	0	0	0	0
Number of sacral vertebrae	2–4	2–4	2–4	4–5	3	0	5–6	5–6	5–6	0	0	0	0	0	0
Intermembral index (range) (R + H/F + T) × 100	73–98 (105)★	74–90	81–100	120–155	94	96	64–79	135–150	100–125	0	?	0	0	119	0
Humero-femoral index (range) (H/F) × 100	71–90 (103)★	71–86	72–97 (103)★	115–125	85	90	68–74	128–147	98–122	0	?	0	0	117	0
Radio-humeral index (range) (R/H) × 100	90–110 (85–120)★	90–110	90–110+	105–135	105	104	65–85	92–109	73–100	0	86	0	0	95	0
Relation to allometric trend, arm: trunk — T, near trend; L, limb long; VL, very long	T	T	T	VL	T	O	T	L	T	O	O	O	O	VL	O
Relation to allometric trend, leg: trunk: — T, near trend; L, limb long; VL, very long	T	T	T	VL	T	O	VL	T	T	O	O	O	O	L	O
Appearance of ischial callosities — PO, postnatal; PR, prenatal	Early PR	Early PR	Early PR	Early PO	O	O	Never	Later PO	Later PO	O	O	O	O	O	O
Diploid chromosome number 2n=	44–48	48–72	42	44–52	O	O	46	48	48	O	O	O	O	O	O

[a] O, condition unknown; ?, condition uncertain; asterisks (★) indicate range increase by fossil.

3 and 30 (Schultz, 1970). Hand length is always less than foot length, and the upper limb longer than the lower. On the basis of data from several sources (sometimes measured by different methods), the intermembral index ranges from 73 to 100 (possibly to 105 in some fossils), the brachial (R/H) index from 90 (or 85?) to at least 116, and the humerus/femur index from 71 to 97 (100 or even 104 in some fossils). The trunk is relatively long and narrow, the clavicle short, and the scapula dorsolateral. The ulna possesses large olecranon and styloid processes with no carpal meniscus. The astragalo-calcaneal joint is typically not helical, as in most primates, but of a simple rotational form considered to be more derived (secondarily—see Szalay, 1975). Chromosome number is highly variable between groups as is the digestive system, but ischial callosities which appear early in prenatal development, are constant, although of varying separation across the midline.

The morphological diversity within the family permits the distinction of at least three main groups of genera. Most important are the subfamilies Colobinae and Cercopithecinae, but the latter may be further subdivided into tribes. The Colobinae, or leaf-eating cercopithecids, are characterized by facial and dental form and by their possession of sacculated, ruminant-like stomachs for processing of cellulose from the leaves that form the major part of their diet. The increased relief of their cheek teeth is similarly related to food processing (Walker and Murray, 1975). The molariform teeth are squarish, with low lateral flare, and the lowers present short trigonids and deep lingual (intercusp) notches. The protocone of P^3 is often reduced, especially in African colobines; on P_4 the metaconid may be subequal in size to the protoconid or may be strongly reduced or lacking. The incisors are relatively small, with I^2 typically conical and I_2 with a disto-lateral projection. Cranially the colobines are generally round skulled and especially short faced. The interorbital distance is high, as is total facial width. The nasals are short and wide, although they may lengthen in longer-faced forms such as *Nasalis*. The ethmoid bone contributes to the medial orbit wall; the lacrimal bone is also within the orbit, and the lacrimal fossa is formed partly by the maxilla. Choanal shape varies from wide and low to high and narrow, with no clear relationship to facial form. The mandibular body is either of relatively constant depth beneath the cheek teeth or is shallow mesially; the ramus is vertical. All colobines show some combination of expanded gonial region, "bulging" under the rear molars and possibly relatively deep corpus. Colobine long bones are usually gracile and the foot elongate, related to their principally arboreal locomotor pattern. The majority of colobines have 44 chromosomes; *Nasalis* has 48. Females do not have cyclic sexual swellings.

The Cercopithecinae have simple stomachs like those of other catarrhines (and most primates) and have cheek pouches for temporary storage of undigested food. For the most part their teeth are lower crowned (more "bunodont"), with little relief between cusp apexes and basin floors, long trigonids, and moderate to high lateral flare. The P^4 cusps either are subequal or the metaconid is wider and/or taller than the protoconid. The incisors are generally larger, with the uppers typically expanded and I^2 often tilted and not conical. Other features must be described for dentally distinct subgroups. Cranially the cercopithecines are long faced, with long and narrow nasals, low inter-orbital and facial width, and great facial height, especially in the zygoma. The ethmoid apparently grows forward in the midline and is covered by the frontal, the vomer expands to form part of the medial orbit wall, and the lacrimal expands mesially beyond the inferior orbit margin and envelops the lacrimal fossa; all of these features are clearly part of a single complex of facial lengthening (see Jones, 1972; Vogel, 1966; Verheyen, 1962—the latter two are especially important on catarrhine facial form). The choanae of cercopithecines are typically high and narrow, which may also reflect facial shape. The mandibular ramus is typically tilted back, but it may approach the vertical in shorter-faced forms (*Theropithecus, Cercopithecus*); a median mental foremen is present (it occurs rarely in other catarrhines; see Vogel, 1968); the body increases in depth mesially. The brain differs from that of colobines in several sulcal features reflecting rostral expansion of the occipital lobe and in positional changes of the sulcus rectus and the arcuate sulcus (Radinsky, 1974). The long bones of the often semiterrestrial cercopithecines are relatively robust. Chromosome numbers vary widely, but some patterns exist at lower taxonomic levels.

The Cercopithecini are mostly small arboreal forms, but the larger *Erythrocebus* is highly terrestrial, although gracile. They share strongly reduced third molars, with no hypoconulid on M_3 and a compressed distal loph on M^3. The teeth are generally elongate with low flare, but in *Allenopithecus* flare is as great as in *Cercocebus*. The ischial callosities are separated by an area of haired skin. Female sexual swelling occurs only in *C. (Miopithecus) talapoin*; otherwise this species is a craniodentally typical cercopithecin. Chromosome number is strongly variable within this group, even within species of *Cercopithecus*, ranging from 60 to 72, with most at 60, 66, or 72. This pattern of spacing by units of 6 is continued by *Erythrocebus* and *C. talapoin* at 54, and apparently by *Allenopithecus* at 48 (see, *inter alia*, Boer, 1971).

The Papionini include small or medium-sized semiarboreal species of *Macaca* and *Cercocebus*, along with larger and more terrestrial members of these genera, *Papio, sensu lato*, and *Theropithecus*. The chromosome number is 42 in all species, and female sexual swelling is pronounced. Dentally the group is most distinguished by the lack, or strong reduction, of enamel lingually on lower incisors, as well as relatively high lateral flare and often accessory cuspules on molars. *Theropithecus* departs from other papionins in converging on the high relief of colobines, while retaining large trigonids and flare; cranially this genus is also distinctive in its vertical ramus, anteriorly placed temporalis muscle, somewhat shortened face, and small incisors compared with large molars. These features are all part of a feeding complex described by Jolly (1970).

Determination of an ancestral morphotype for cercopithecids depends in part on characters present in other catarrhine taxa, but the ancestral state of some of the cercopithecid dental specializations can be discussed here. Such obviously derived conditions as the loss of M_3 hypoconulids in Cercopithecini and of lingual incisor enamel in Papionini can be discounted immediately, but other polarities are more difficult to assess. The mesial groove of the upper canine passing continuously onto the root, as well as the bilophodont molar pattern and the absence of dP_4–M_2 hypoconulids, are further shared derived features of all Cercopithecidae that are probably part of the ancestral cercopithecid morphotype, but are not to be expected in earlier ancestral catarrhines. For the lower premolars, it seems likely that the P_4 cusps were of subequal development in early cercopithecids (or perhaps the protoconid was slightly higher; see below), and we expect such forms to have possessed rather strong development of C^1/P_3 honing, with a P_3 flange extending below the alveolar plane. It would appear that a relatively large (macaque-like?) trigonid was present in the lower molars of ancestral cercopithecids, based on comparisons within the group and with other catarrhines, but it is more certain that the high relief and deep lingual notches on the lower molars of colobines and *Theropithecus* are derived within the Cercopithecidae. Finally, lateral flare may have been moderately present in the early Old World monkeys, a point to be discussed at greater length below, in connection with the reconstruction of a precercopithecid ancestor. In terms of other bodily systems, both types of digestive specialization within the Cercopithecidae (sacculated stomachs and cheek pouches) are just that, and are not to be expected in a common ancestor of the two subfamilies; Radinsky (1974) has suggested that the cercopithecine cerebral differences from colobines are derived, while the presence of an arcuate sulcus probably is a derived feature of monkeys among catarrhines; finally, Pocock (1925) considered that ischial callosities separated by haired skin (as in the Cercopithecini) were ancestral for the family, but he further thought that callosities were independently acquired in gibbons and monkeys, a point we contest.

B. Gibbons and Possible Extinct Relatives—Lesser Apes

An understanding of the relationships of the gibbons* to the other groups of catarrhine primates is fundamental to understanding the phylogeny of the Old World higher primates. The gibbons are usually linked with the Hominoidea (Simpson, 1945), but it has been suggested recently by Chiarelli (1968b) that gibbons belong to the Cercopithecoidea on the basis of their karyologic resemblances with the Colobinae. There is little support for this in other lines of evidence, especially when it is considered that within the gibbons themselves there is considerable variation, the concolor and siamang gibbons being phenetically closer to the great ape condition. Evidence from biochemistry (Goodman, this volume; Romero-Herrera et al., 1973), comparative anatomy (Remane, 1960; Le Gros Clark, 1971; Groves, 1972; Schultz, 1973; Tuttle, 1972, this volume) and paleontology (Le Gros Clark and Leakey, 1951; Zapfe, 1960; Andrews, 1973) is all consistent with the gibbons being a group of the "Hominoidea," or "anthropomorph" catarrhines (to use a term with less taxonomic implication).

Another line of evidence that might be taken to show the hominoid status of the gibbons is the structure of the teeth. Indeed gibbon teeth are remarkably similar to those of the great apes. They share elongate lower molars with 5 main cusps placed marginally, and squarish uppers with 4 cusps. On the lowers the hypoconulid is usually buccal to, or on, the midline, resulting in the appearance of 3 buccal and 2 lingual cusps. True crests crossing the lower molars are rare, but a mesial fovea (trigonid basin) is separated from the main talonid basin by a low ridge, and a distal fovea is similarly separated by a hypoconulid–entoconid crest. Other crests link neighboring cusps along the tooth

---

* Following Groves (1972), only one genus, *Hylobates*, is recognized to receive all gibbons, siamangs, "concolors," etc. The terms gibbon and lesser apes will here be used interchangeably with *Hylobates*, unless specifically noted otherwise.

margin. The cusps are separated internally by grooves defining a "y" or "+" pattern. On upper molars the mesial fovea is delimited by a crest from the paracone to the margin mesial to the protocone; a protoconule there is very rare in modern apes. The large talon area is separated from the trigon by the crista obliqua, linking protocone and metacone, sometimes via a metaconule. The hypocone is generally large and well separated, on the margin of the crown. Cingula may be present lingually on uppers and buccally on lowers, but not very strongly in modern forms. In gibbons it is quite rare on lower molars, and it varies among species on uppers from massive to merely a slight bulge (Frisch, 1965, 1973). The premolars of gibbons (and other "hominoids") are bicuspid (except for P_3), with the buccal cusp larger than the lingual; a cingulum may be present lingually on the uppers.

The dentition of gibbons is distinguished from that of other hominoids by such features as small hypocones, the small size of M_3^3, large canine with low sexual dimorphism and strong premolar honing, and incisor morphology. The third molars of modern gibbons are always shorter than the M_2^2, often shorter than M_1^1 (especially M^1), and there is often concomitant reduction of distal cusps. The teeth of males are on the average larger than those of females of the same taxon, but there is overlap even in canine height, with the female range often greater (Frisch, 1973); a similar instance among cercopithecids occurs in *Colobus guereza* (Delson, 1973). The canines of both sexes are high and sharp, and they participate in C^1/P_3 honing of the type seen in cercopithecids. The lower incisors of gibbons are moderately narrow, while the I^1 is rather broad; I^2 is conical in shape, as in colobines (Vogel, 1966).

The skull of gibbons is also strongly colobine-like: the face is even shorter; bone relationships in the orbit are as described above for colobines (Vogel, 1966); and the choanae are low and wide. The orbits are large and surrounded by a projecting marginal ring. The mandibular body is gracile, long and shallow, deeper anteriorly than posteriorly; the ascending ramus is vertical. The brain is relatively larger than it is in monkeys, but Radinsky (1974) implies it may be the most conservative in terms of sulcal geometry.

Gibbons present ischial callosities invariably, but they appear much later in ontogeny than they do in cercopithecids; Pocock (1925) reported that the area between the callosities was relatively hairy, as it is in *Cercopithecus*. No female sexual swellings occur. Most gibbons have 44 chromosomes, but siamangs have 50 and "concolors" 52. We follow Groves (1972) in recognizing these forms as the subgenera *H. (Symphalangus)* and *H. (Nomascus)*, respectively. In all but the latter form, specializations of the male sexual organs occur, with the testes para- or prepenial and the scrotum suppressed. It is finally worth noting that all gibbons so far observed live in parental family groups, do not build nests, are strongly territorial, and have displays in which the female plays a major part (Chivers, 1972). Groves (1972) has attempted to determine the polarity of evolutionary trends within the gibbons, and he has also discussed potential ancestral traits of hominoids, which will be considered below.

The major morphological features distinguishing modern gibbons from other higher primates are those related to their ricochetal, brachiating locomotion. It must be noted that we employ the term "brachiation" to imply habitual support of the body below extended arms during locomotion. Thus, the modern "great apes" are not brachiators, although many of the following morphological features that apparently relate to this type of locomotion are found also in modern larger apes (including man).

In all modern anthropomorphs the trunk is transversely broad and somewhat shortened, the scapula dorsal, the clavicle robust and long relative to trunk length (especially in gibbons and orangutans), and the humeral head large, dorsally facing, and medially directed. The tail has been lost, the lower back vertebrae reduced in number (mostly 17 or 18 in gibbons), and the sacrum lengthened (to 4 or 5 vertebrae in gibbons), which in turn leads to greater mobility of the shoulder and lessened flexibility of the trunk. The internal organs have been rearranged and supported craniocaudally, rather than dorsoventrally. The flexibility of the elbow joint has been increased by the reduction of the olecranon process, while the ulnar styloid has retreated somewhat from the wrist articulation; in gibbons there is a bony lunula within the small meniscus separating the ulna from the carpus (Lewis, 1972a). The lower ankle joint is helical.

Gibbons have carried locomotor specializations farther than most other catarrhines (save *Homo*, at least), with dominance of the upper limb, and especially its distal components. Thus the upper limb is much larger than the lower (intermembral index is 120 to 155), the humerus longer than the femur (index 114 to 136), the radius longer than the humerus (index 105 to 124), the hand very long compared to trunk size, and the thumb, although long, is short compared to the long palm and thus is out of the way of the hooklike grasping fingers (see Van Horn, 1972). The long bones generally are gracile with a minimum of muscle markings. Biegert and Maurer (1972) have clearly illustrated the appendicular lengthening compared to trunk height seen in gibbons. Among all catarrhines, *Hylobates* spp. fell farthest from the allometric trend lines of both arm and leg length on trunk length. The relative arm and leg lengths seen in gibbons are greater than in any other known form and would be expected only in a catarrhine whose body size greatly exceeded that of a gorilla. This specialization might be predicted for the forelimb, as seen also in *Pongo*, but not for the hindlimb as well. We are tempted to suggest that the ancestor of modern gibbons was much larger than any *Hylobates* species (also suggested by Dr. Colin Groves, personal communication).

Two rather well-known fossils have been long considered as potential ancestors or relatives of the gibbons, but their status in this regard is now increasingly in doubt.★ *Pliopithecus* is known as a number of supposed species from the middle Miocene of Europe. Its dentition has been considered in detail by Hürzeler (1954), and excellent cranial and postcranial remains of one variety have been described by Zapfe (1960). Andrews (1973) has shown that *Limnopithecus legetet* is closely related to *Dryopithecus* and not pertinent to gibbon evolution, but that the species known as *L. macinnesi* may well be gibbon-like. It was therefore necessary to transfer this species to the new monotypic genus *Dendropithecus* (Andrews, Pilbeam and Simons, 1976).† This species is known in two subspecific variants from the early (and probably middle) Miocene of East Africa (Andrews, 1973). *Dendropithecus* and *Pliopithecus* dentitions are well known and basically gibbon-like

---

★ Simons (1972) has recently reported a gibbon-like tooth from Miocene deposits in the Siwaliks.

† Owing to the lengthy delays beyond our control, this article has not yet appeared in print. The name Dendropithecus is therefore used as a *nomen nudum* in the present paper.

416

ERIC
DELSON
AND
PETER
ANDREWS

with narrow incisors (by comparison to contemporary *Dryopithecus*); canines sexually dimorphic, but large in both sexes and honing on a specialized high-crowned P$_3$; and a moderate degree of intercusp cresting. The elongate M$^3_3$, however, is unlike that of gibbons. No skull is known of *Dendropithecus*, but that of *Pliopithecus* is gibbon-like, with a short and broad face and nasals, gracile and mesially deepening mandibular body, and projecting orbital margins; the lack of an ossified tubular auditory meatus is distinctly unlike gibbons or other modern catarrhines. The skeleton of *Pliopithecus* is most gibbon-like in its gracility, but the limb proportions are more as in monkeys, although suggestive of a trend toward gibbons; the intermembral index is 94, the radiohumeral, 105, and the humerofemoral, 85. The ulna has large olecranon and styloid processes and lacks marking distally that would indicate the presence of a meniscus, as in gibbons, while the humerus retains an entepicondylar foramen. There are 6 or 7 lumbar vertebrae and 3 sacrals. Ankel (1965) suggested the presence of a long tail, but this is questionable and remains undocumented by direct fossil evidence. The data of Biegert and Maurer (1972) also show that *Pliopithecus* was typically catarrhine in relative limb proportions for its trunk size. *Dendropithecus* postcranial elements are more fragmentary, but ulnar morphology is similar between the two genera, as is that of the astragalus. The humerus and femur of the African form are morphologically similar to *Pliopithecus*, but more elongate and thus more gibbon-like; the distal humerus of *Dendropithecus* is also close in form to that of *Hylobates*, and there is no entepicondylar foramen; femora of the two fossils are of similar length, but the humerus of *Dendropithecus* is longer. Its estimated limb indexes are: intermembral, 96; radiohumeral, 104; humerofemoral, 90; all are close to *Pliopithecus*, as is overall size.

In the past opinions as to the relationship of these fossils and gibbons has varied widely. Ferembach (1958) suggested that *L. legetet* was a small "pongid," while "*L.*" *macinnesi* was morphologically conservative, much like early catarrhines; Groves (1972, 1974) has also denied these forms gibbon relationship, because of their lack of similarity to modern *Hylobates*. On the other hand, Tuttle (1972), Simons and Fleagle (1973), and others find these fossils to have morphology that is reasonable to expect in a gibbon ancestor, as the specializations of the modern gibbons must have developed mosaically. We consider *Dendropithecus* phenetically closer to gibbons than *Pliopithecus*, but evaluation of cladistic relationship depends on the comparison of these fossils to the ancestral catarrhine features to be determined below.

C. Larger "Anthropomorphs"

In this group especially, there is a problem with nomenclature, both formal and informal. Large apes will imply Miocene to modern nongibbon "hominoids" except *Oreopithecus* and *Homo*, unless otherwise qualified (e.g., modern, fossil, African, etc.). Of the modern forms, three genera are recognized: *Pongo*, *Homo*, and *Pan*. The latter is tentatively divided into the subgenera *P.* (*Pan*) with two species and *P.* (*Gorilla*). *Homo* and *Australopithecus*, or their ancestral morphotype as discussed by Eldredge and Tattersall (1975) and Delson (1976), are here considered men. For many characters the large apes are similar to gibbons, and these will not be further discussed. Otherwise, differences among the modern forms will be considered by organ system, rather than by taxon.

1. Modern Great Apes and Men. The dentition and trunk are most similar among larger apes and men. The shoulder girdle is generally as in gibbons, with clavicle longest in orangutans (as in gibbons), intermediate in *Homo*, and shortest in *Pan* (although still longer than in monkeys). The lower back is further shortened, with thoracic + lumbar vertebrae numbering 15 (or 16) in *Pongo*, 17 (of 16) in *Pan*, and 17 in *Homo*; sacral vertebrae number 5 or 6. Lewis (1972a, b) has shown that the wrist of larger apes is more derived than that of gibbons in the removal of the ulna from carpal articulation, through the increase in size of the meniscus and lack of the bony lunula therein. There appear to be three major types of wrist here, characterizing *Pongo*, *Pan*, and *Homo*, but questions remain as to the polarity of change and the functional value of the different morphologies. Orangutans seem most clearly adapted for suspensory locomotion, while *Pan* knuckle-walks habitually in the adult stage. *Homo* may have developed from a suspensory ancestral condition, but probably never had knuckle-walking or truly brachiating ancestors (see Tuttle, 1974, this volume). Further specializations of men include adaptations to bipedal striding, such as the short and broad ilium, inflexible foot, and related muscular changes. All large anthropomorphs present long upper limbs relative to trunk length, and in men the lower limb is further elongated. This is reflected also in Biegert and Maurer's (1972) analysis, in which the relative arm length of *Homo* is typical for catarrhines of its trunk length, while its leg is longer even than in *Hylobates*. The situation is reversed in *Pongo*, although its upper limb is not so extremely elongated as is the lower one in men. *Pan* species are "typical" catarrhines in limb length with the long forelimbs a result of the large size they share with orangutans and men. Unfortunately, there are no large catarrhines that are standard quadrupeds to serve as "controls" in order to separate the effects of allometry from those of locomotor pattern.

In dental morphology, the larger apes and men are close to the gibbons. Cingula may be more prominent on molars, and the third molars are not as reduced, although usually smaller than the second (Mahler, 1973). In *Pan* and *Pongo* sexual dimorphism is high, especially in the canines, with a marked decrease in canine–premolar honing. These teeth in men are even less dimorphic, metrically and morphologically, with the P_3 becoming bicuspid as an adaptation to increased grinding, while the canine takes on the functions of an incisor. In *P.* (*Pan*) and *Pongo* the central upper incisors are quite large, and the lowers are also often large compared to molar size; in gorillas the lowers are smaller, while I^2 is a more conical tooth than in other large apes (see Vogel, 1966). Incisor proportions vary greatly among taxa of men, but the ancestral condition was probably narrow and high, with a moderately large I^1. The mandibular body is of relatively constant depth in modern larger apes, neither very shallow nor very deep anywhere; the ascending ramus may be slightly back-tilted, and a chin is present in *H. s. sapiens*.

Cranial differences are more important, as has been shown especially by Vogel (1966). In *Pan* and *Pongo* the face is moderately long and high, with great depth in the maxilla and zygoma; the orbital construction and lacrimal fossa are as in colobines and gibbons; however, Schultz (1950) has noted that a frontomaxillary suture in the orbit wall is common only in *Pan*. The interorbital region is broad in *Pan*, narrower in *Pongo*, while the nasal bones of both are moderately elongated. In *Homo*, and less so in *Australo-*

418

ERIC
DELSON
AND
PETER
ANDREWS

pithecus, the face is shorter, the orbits widely spaced, and the nasals short and broad. The choanae are high and narrow in all large apes, including short-faced *Homo*. The brain is largest in these forms among primates, especially so in later men, but morphologically not different from gibbons in sulcal patterns at least (see Radinsky, 1974; Holloway, 1972). The chromosomes of larger apes are strongly similar, but *Homo* has 46, while *Pan* and *Pongo* have 48; Chiarelli (1968a) has suggested a simple behavioral model for the reduction. Ischial callosities are found in many individuals of *Pan* and *Pongo*, but they develop very late in ontogeny; none are present in *Homo* (Schultz, 1968; Rose, 1974).

2. *Extinct Large Apes*. A number of fossil forms have been considered as relatives of the modern larger apes. Three to be considered here are of Miocene–Pleistocene age: *Dryopithecus* (and *Limnopithecus*), *Ramapithecus*, and *Gigantopithecus*; three others are of Oligocene age, namely *Propliopithecus*, *Aegyptopithecus*, and *Aeolopithecus*. *Dryopithecus* is the best known of these, represented through the Miocene (and possibly latest Oligocene) in Africa, Europe, and Asia. Andrews (1973, 1974) recognized the seven species accepted by Pilbeam and Simons, adding the new subgenus *D. (Rangwapithecus)* to receive two new African species; the lack of clear distinction (other than zoogeographic) between *D. (Dryopithecus)* and *D. (Sivapithecus)* resulted in their synonymy. Other new species named since 1965 are considered synonyms of previous taxa, and material formerly placed as *Sivapithecus* (or *Kenyapithecus*) *africanus* is now assigned to species of *D. (Proconsul)*. *Dryopithecus* remains are mostly dental, with some cranial and postcranial portions represented. The dentition is basically similar to that in other apes, differing in smaller and more gracile incisors and canines; canines participating in honing with P$_3$ more than in modern larger apes; broader upper premolars; cingulum more common on cheek teeth, but variable among species; and lowers often elongate. The mandibular corpus is of moderate to great depth, either constant or deeper mesially; the simian shelf present on some *Pan* and *Pongo* is lacking. Parts of two skulls of *D. (P.) africanus* (Davis and Napier, 1963) indicate a relatively wide interorbital region and face of moderate length (perhaps slightly more than gibbons), no brow ridges (found in most other apes), and a brain of essentially modern ape form and size compared to body weight (Radinsky, 1974; Andrews, 1974). A frontal sinus is present in at least *D. (P.) major* and *D. (P.) africanus* and is found in modern larger apes only in *Pan* and *Homo*, not *Pongo*. Postcranial remains are known basically from *D. (P.) africanus* and *D. (D.) fontani* (= "*Austriacopithecus*" and "*Paidopithex*"). The ulnar olecranon of the former species is long and monkey-like (Preuschoft, 1973), while shorter and more apelike in the latter, younger taxon (Zapfe, 1960). The ulnar styloid of *D. africanus* has withdrawn from carpal articulation more than that of gibbons (Lewis, 1972b). Humeri and femora from Africa and Europe are morphologically similar to those of *P. (Pan)* species, but more gracile, while the finger bones are like those of *Pan* and some monkeys. The radiohumeral index of *D. africanus* is 86. Functional locomotor interpretations of *Dryopithecus* have varied among an agile, quadrupedal "probrachiator" similar to some modern colobines (Napier and Davis, 1959), a brachiator (Lewis, 1971, 1972b), a knuckle-walker (Pilbeam and Simons, 1971, Conroy and Fleagle, 1972), and an *Ateles*-like arboreal quadruped that may have suspended itself from its forelimbs, but that could not knuckle-walk (Schön and Ziemer, 1973; Preuschoft, 1973). In part, these differences are the re-

sult of concentration on the limited data from specific body systems, not always the total morphology available. On the one hand, Conroy and Fleagle (also Tuttle, 1974) argued that Lewis is incorrect in referring to the wrist of larger apes as that of a brachiator, as they in fact are primarily knuckle-walkers (or mixed, in the case of *Pongo*). On the other hand, Schön and Ziemer's investigation of the wrist bones suggests habitual dorsiflexion of the hand, as in *Ateles* and *Alouatta* when moving quadrupedally, but they do not deny the possibility of some arm-suspension as well. Further study and integration of data is obviously needed, but the similarity of *Ateles* to *Pliopithecus* and *Dendropithecus* in the forelimb also is most interesting.

Both *Gigantopithecus* and *Ramapithecus* have been suggested as possible human ancestors, although they differ greatly in their known gnathic morphology. The former genus is known from one mandible of probably late Miocene age in India and three mandibles and some thousand teeth of later Early to Middle Pleistocene age in South China; *Ramapithecus* is represented by a dozen or so teeth and fragmentary jaws of middle to late Miocene age (ca. 15–10 million years) in East Africa, India, China, and western Eurasia (see Simons, 1972, 1976). *Gigantopithecus* species were very large animals with heavy, robust mandibles; the body is deep, shallowing mesially, the symphysis heavily buttressed, ascending ramus possibly upright, and the face probably short (Pilbeam, 1970). The incisors are rather small and vertical; the canines are low crowned, robust, not honed on P_3, and probably function with like premolars than incisors; the P_3 is nearly bicuspid, but with a large trigonid and a small mesial flange. Other cheek teeth are strongly molarized, with thick enamel, low cusp relief, and high interproximal attrition; molar crowns may have increased in height from the Miocene to Pleistocene. *Ramapithecus* species were much smaller animals, in size like *D. africanus* or *D. nyanzae*, with dental arcades apparently diverging slightly posteriorly, much as in *Gigantopithecus*. The alveolar process of the maxilla is deep, the mandibular body shallow and robust, and the ramus may have been upright, leading again to suggestions of a broad- and short-faced primate adapted to heavy chewing. The incisors are quite small and slightly procumbent; the cheek teeth show low relief and thick enamel, much interproximal wear, and a steep wear gradient; the canine is low crowned; and the incipiently bicuspid P_3 has only a small honing flange. An interrelated assemblage of features indicates increased lateral chewing, without incorporation of the canine into the grinding apparatus (see Simons, 1964, 1972, 1975; Andrews, 1971; Walker and Andrews, 1973).

The mid-Oligocene Fayum beds of Egypt have yielded a number of fossil primate taxa that are here divided into three groups. Three are treated in this section, two others (*Apidium* and *Parapithecus*) considered next, and the last, *Oligopithecus*, is known so little that we hold it until last. These species have been discussed by Schlosser (1911), Kälin (1961), Simons (1965, 1972), Simons and Fleagle (1973), Szalay (1970, 1972) and Delson (1975c), among others. Of the three largest Fayum primates (all are smaller than modern apes and *Dryopithecus* spp.), *Aeolopithecus* is known from a single weathered mandible; *Propliopithecus* (including *Moeripithecus*) is represented by a number of mandible fragments; and *Aegyptopithecus* is the best known, with several mandibles and loose teeth, a nearly complete skull, and a number of undescribed postcranial elements.

420

ERIC
DELSON
AND
PETER
ANDREWS

The teeth of *Aeolopithecus* are too eroded to preserve crown details, but it is clear that the M_3 is short, the canines robust and tall, and the P_3 elongated and narrow, indicating well-developed honing. The type of *Propliopithecus* has no incisors (nor are any known *in situ*), but the canine is low and the single-cusped P_3 little adapted to honing; it is probably a female, however, and another specimen attributed to this genus has a well-developed honing facet on P_3. In *Aegyptopithecus* there are no known lower canines, but a C^1 is long and laterally compressed, while P_3 seems clearly a honing tooth. The P_4 of both genera has subequal trigonid cusps. The molar morphology of *Propliopithecus* and *Aegyptopithecus* is similar to that in *Dryopithecus*, and thus to other apes as well: there are 5 rounded cusps, the buccal 2 slightly larger and lower than the lingual; the hypoconulid is centrally placed on M_1 and more buccal on M_{2-3}; low ridges demarcate a raised trigonid, while the talonid is narrow and crowded by the 3 distal cusps; the hypoconulid is linked to the entoconid, rather than to the hypoconid, as in *Dryopithecus* species; buccal cingulum development is variable, from great to almost none; the third molar is slightly longer to much longer than the second, and always narrower. The upper premolars of *Aegyptopithecus* are much broader than long, while those of most modern forms (and even *Dryopithecus*) are more nearly square; lingual cingulum is present as a swelling, rather than as a shelf [as in *Dendropithecus* and *D. (Proconsul)*]. The upper molars have a cingulum that varies between a shelf and a swelling, the occlusal ridges vary in degree of development, and there is always a small distinct protoconule and a smallish hypocone, placed lingually. The mandible is of moderate and constant depth in most specimens.

The cranium of *Aegyptopithecus* has not been fully described, but a few important points may be noted. The interorbital pillar is rather wide, and although the face and nasal bones are long, a comparison of interorbital width/nasion–prosthion length results in an index almost identical with that in *D. africanus*, which is intermediate in value between *Hylobates* and *Pan*. There is marked postorbital constriction and moderate nuchal and sagittal crests, reflecting the large size of the face relative to the brain case. The chonanae are high and quite narrow. There is no ossified external auditory tube, merely a platyrrhine-like ring, as in *Pliopithecus*. Radinsky (1973) reported the brain to be essentially modern, with catarrhine-like large size and a central sulcus, increased visual cortex, smaller olfactory bulbs than in prosimians, but still a smaller frontal lobe than in modern anthropoids. A nearly complete *Alouatta*-like ulna, with long olecranon and robust, nearly straight shaft suggests an arboreal quadrupedal locomotor pattern for this genus (Fleagle, Simons, and Conroy, 1975; Conroy, 1976). The relationships of these early fossils will be considered later, but it may be noted here that although 4 species have previously been recognized, a reduction to as few as 2 is conceivable. The only major differences are in size, relative size of lower molars, and relative cingulum development.

D. *Parapithecus* and *Apidium*

These two genera, each apparently represented by a species in each of the upper two Fayum horizons, are, respectively, the second and first most abundant primate species. Both are small, about the size of marmosets or the smallest cercopithecids; are known mostly from teeth and jaws, with some fragmentary cranial and limb elements; and share a number of features that lead to their being discussed here together. One feature that sets these animals clearly apart from other catarrhines is their possession (retention) of 3 premolars, of which the most anterior (P_2) is strikingly caniniform. Delson (1973, 1975c) has suggested that this tooth might well be involved in honing of the C^1 in some *Parapithecus*, and the same may be true of *Apidium*. The lack of cingulum, especially on the lower molars, and some aspects of crown morphology also indicate a close link between these two taxa. Incisors are poorly known, but the lowers are quite small, narrow, vertical teeth, the laterals especially small. The canines are robust, but not very tall, and thus if honing was indeed present, it may not have been strongly developed, nor ancient. The P_2 is caniniform, with a single crown cusp and very little talonid development, if any (at least in the *P. fraasi* type); P_{3-4} are similar teeth, lower crowned, but more robust than P_2, with large protoconid, small metaconid and short talonid; P_4 is larger, with a higher metaconid. The lower molars, as in the three Fayum primates noted above, are moderately high crowned, but with low relief, having 4 main cusps and a small but distinct midline hypoconulid. The trigonid basin is moderately large, bounded by a rather distinct ridge linking metaconid and protoconid, with no paraconid clearly present. The upper teeth have not been formally described, but it can be noted that they are broad, with well-marked conules placed on the lines connecting the mesial and distal cusp pairs, not mesial to these lines, as in apes and *Oreopithecus*. *Apidium* is distinguished by its much smaller premolars compared to molar size; its tendency to polycuspidation, resulting in the well-known centroconid (mesoconid) on lower molars; an elongate M_3 but reduced M^3; and generally more bulbous cusps. *Parapithecus*, by contrast, is more crest oriented (but *P. fraasi* is bulbous), the molars wearing quite flat; M_3 is smallest of the molars, its talonid especially reduced; a paraconid may be present on dP_4; the lower molars are somewhat constricted by the in-turning median buccal notch.

The mandibular body is variable in both genera, usually of constant depth, but sometimes shallowing slightly mesial; depth is less in *P. fraasi* than in other species, but may vary intraspecifically; reconstructed rami are rather vertical. Cranial fragments referable to *Apidium* have been described: the frontal demonstrates postorbital closure and a fully fused metopic suture (Simons, 1959); the brain (Radinsky, 1974) *may* have had relatively larger olfactory bulbs than in *Aegyptopithecus*; there was again no ossified external auditory meatus, as in *Pliopithecus* and *Aegyptopithecus* (Gingerich, 1973). Humeri, ulnae, astragali, and calcanei described by Conroy (1976) indicate a generalized arboreal quadruped similar to *Saimiri* or *Cebus*.

E. *Oreopithecus*

This most enigmatic primate is known only from five lignite localities in Tuscany, which may be of late middle Miocene age (11–13 million years). Despite published reports (Leakey, 1968; Koenigswald, 1969) no close relatives are known from Africa. Numerous remains, including a crushed partial skeleton, have been incompletely

422

ERIC
DELSON
AND
PETER
ANDREWS

described (Hürzeler, 1958; Straus, 1963; Szalay and Berzi, 1973). After Hürzeler, the dentition has been discussed in greatest detail by Butler and Mills (1959), who considered it quite distinct from that of other catarrhines. The incisors are vertically implanted, with conical I^2 and large I^1 presenting a strong projecting lingual cusp. The canines are sexually dimorphic and robust, but not tall, although they do interlock; the upper canines show tip wear and did not hone strongly on P_3, nor do they have the cercopithecid continuation of the mesial sulcus onto the root; lower canines honed on the uppers, but not on the small I^2. The lower premolars are both bicuspid, with subequal protoconid and metaconid (sometimes smaller on P_3) and moderate talonid, especially on P_4; upper premolars have large paracone, lower protocone, and sometimes a small metacone. The lower molars are elongate, with the standard 4 catarrhine cusps marginally, a centroconid much as in *Apidium*, and a midline hypoconulid, small on M_{1-2}, larger on the M_3 talonid; the cusps are partly linked by crests, relief is high and lingual notches rather deep; a paraconid is variably present on M_1; there is little cingulum. The upper molars are elongate, as in modern catarrhines, not squarish, as in earlier Miocene forms; the 4 main cusps are joined by distinct protoconule and large metaconule, which are important in the formation of the ridge delimiting the small fovea anterior and the crista obliqua; the 4 main cusps are acute, the trigon deep but small in area. The crest delimiting the long, deep, and narrow fovea anterior ends at the protoconule, which restricts the fovea to the buccal part of the tooth. The strong crista obliqua passes via a well-defined metaconule and is further connected to the distolingual hypocone; this results in a large and open fovea posterior. With wear, the metaconule may merge into the crista obliqua. There is lingual cingulum on upper cheek teeth, but it is not continuous around the protocone.

The skull of *Oreopithecus* has previously been considered to closely approach hominids in morphology (Hürzeler, 1958), but Szalay and Berzi (1973) have recently shown that this was a misinterpretation of the crushed specimen. The face is somewhat gibbon-like, with broad interorbital region and sloping snout; there are large supra-orbital tori and clear sagittal and nuchal crests, the latter high on the vault; the mandible has a smoothly rounded symphysis, expanded gonial region, and vertical ramus; the body is of constant and rather great depth. The postcranial skeleton reveals some contrasting specializations by comparison with modern catarrhines. Straus (1963) has reported an intermembral index of 119, a radiohumeral index of 95, and a humero-femoral index of 117; all of these indexes are within the range of *Pan* spp. but perhaps not found together in a single species or individual of that genus. Straus considered that these indexes reflected short legs and not very long arms. Biegert and Maurer (1972), however, found that the relative elongation of the arm in *Oreopithecus* was greater than that in *Pongo*, although much less than in *Hylobates*. They apparently did not accept Straus' estimate of tibial length, but that value seems reasonable, and its inclusion yields a relative leg elongation greater than that of any catarrhine except *Homo* and *Hylobates*. Thus, both limbs are moderately elongated for body length, contrary to Straus' interpretation. There are 5 lumbar vertebrae, as is common in men and gibbons (fewer in other apes), and 6 in the narrow sacrum; the vertebrae also reveal a strong ventral keel, which otherwise occurs only in cercopithecids for attachment of a longitudinal ligament. The ulnar olecranon is reduced, as in modern apes, but this may in part reflect large size as well as suspensory postures. Straus considered the astragalus mobile and cercopithecid-like, while the calcaneus more closely resembled larger apes. Szalay (1975) has noted that the astragalocalcaneal joint is nonhelical, as found otherwise only in cercopithecines and *Homo. Oreopithecus* has often been considered a brachiator, especially as its remains are found in coal swamps where it would have been unlikely to walk terrestrially, but Zapfe (1958) has remarked that some Austrian *Pliopithecus* and *Dryopithecus* specimens also derive from lignite deposits, and he has suggested that they did in fact walk on the bog. Various cercopithecids also are known from such deposits, which may merely reflect forest habitats and/or taphonomic effects, not necessarily locomotor modes.

F. *Oligopithecus*

Oligopithecus savagei is represented by a single mandibular fragment with C_1–M_2, from the lowest Fayum horizon; it is thus the oldest known catarrhine (see Szalay, 1970, 1972; Simons, 1971, 1972). The canine is moderately tall, but broken below the tip. There are two premolars, of which the anterior (P_3) is unicuspid and adapted to hone the C^1, while P_4 is bicuspid as in other catarrhines. The fact that the honing flange on P_3 has been worn down through enamel to dentine, a condition not seen in any other catarrhines that hone, suggested to Delson (1975c) that such honing was a relatively recent acquisition in the ancestry of this animal. The P_4 metaconid is only slightly lower than the protoconid, a small paraconid may be discerned and the talonid is large but low. On the two preserved molars, cingulum is almost nonexistent; the 4 main cusps surround a large and wide talonid basin, while a short and somewhat higher trigonid may be bounded mesially by a small paraconid (M_1); a hypoconulid is present lingual to the midline, but its pattern of wear suggests incipient confluence with the entoconid; intercusp ridges are poorly developed. This specimen shows that *Oligopithecus* was clearly catarrhine in its possession of cuspate hypoconulid and broadened talonid, as well as the C^1–P_3 honing and deep mandibular body (compared to earlier primates).

III. Morphotype Deduction and Phylogeny Reconstruction

A. Postulated Ancestral Morphotype for Catarrhini

Based on the data presented above, a reconstruction has been attempted of the hypothetical ancestral morphotype of the catarrhines, essentially equivalent to the common ancestor of all these animals. Some comparisons have also been made with relevant conditions in other primate groups, especially the Paleogene forms thought to reflect conditions potentially ancestral for Anthropoidea/Haplorhini. A number of questions have arisen that will be discussed after descriptions of these features on which there is more certainty.

In the dentition the incisors of ancestral catarrhines would have been small relative to molar size, relatively narrow and high crowned; I^2 was probably conical or caniniform,

424

ERIC
DELSON
AND
PETER
ANDREWS

involved in honing the lower canines; I^1 was the widest (longest) tooth mesiodistally. The canines would have been tall and bilaterally compressed, showing sexual dimorphism in size and probably in form, with C^1 involved in honing on C_1 and P_3, and C_1 on I^2 as noted. A major problem involves the number of premolars; certainly 3 were present at one time, but for now it will be assumed that $P^2_{\overline{2}}$ was already lost and the time and manner of that loss will be discussed below. In that case, the anterior lower premolar, P_3, would have been bilaterally compressed, sexually dimorphic, and involved in C^1 honing, but without extreme extension of enamel onto the mesial root; the crown was of moderate height, not very high as in fossil "lesser apes," and only 1 main cusp was present. The P_4 would have had the two main cusps subequal in size, or the protoconid slightly higher than the metaconid; a small trigonid basin, possibly with a small marginal paraconid, was somewhat higher than a broad talonid basin. The upper premolars were homomorphic, bicuspid, with a broad oval shape, not elongate or trefoil-shaped as in some Paleogene forms; the paracone was higher than the protocone, especially on P^3. There may have been some lingual cingulum related to the broadening of these teeth, and it seems doubtful that a protoconule was present. The lower molars would have had 6 cusps, the 4 main ones as well as small paraconid and hypoconulid, the latter placed on the midline. The talonid was as broad as the trigonid, but slightly lower and longer mesiodistally. There may have been some buccal cingulum, and cusps probably did not rise far above the surrounding area—relief was low, although the crowns were moderately high. In length M_1 was smallest, with M_2 equal or slightly shorter than M_3; the latter was probably narrow, whatever its length, and may have had a larger hypoconulid. The upper molars would have had 4 cusps, the hypocone small, low, and cingulum-linked. As in other primates (e.g., adapids), with the increase in size and height of the hypocone, the paraconid would have been reduced in size even further, and the hypoconulid perhaps increased as well. Upper molars would have been rather wide, sub-square in shape, with lingual cingulum (probably related to the increase in width) present, but perhaps not continuous around the large protocone. A small protoconule would probably have been present on the tooth margin, mesial (and slightly buccal) to the protocone, not between paracone and protocone; a crest linking it to the paracone would have demarcated a distinct mesial fovea. No metaconule would have been present, but a strong crista obliqua set off the trigon from the fovea posterior. The size progression $M^2 > M^3 > M^1$ probably held; both large and small M^3 are considered derived characters. These features are summarized in Table 1.

In the skull, ancestral catarrhines are reconstructed as rather short-faced, with short, broad nasals, lacrimal fossa overlapping onto the maxilla, ethmoid participating in the orbit wall, a brain the relative size of that of *Aegyptopithecus* and with similar form, much like that of gibbons. The circumorbital bar would have been closed, with much filling in of the rear orbital wall; the symphysis and metopic suture fused early in life; the bulla was probably inflated and the whole region pneumatized, but there was no ossified external auditory tube, only a ring. Mandible depth was probably constant along the jaw and (absolutely) greater than that in Eocene primates, although these characters may have been variable; the ramus was likely upright. Choanal shape, among other features, is uncertain, but possibly low and wide. Basically, the face was most similar to gibbons and colobines among recent forms, as Vogel (1966) and Delson (1973), among others, have suggested.

Postcranial reconstructions are even chancier than the above, but for the sake of completeness, it is suggested that trunk form was much as in cercopithecids, perhaps more gracile; the tail was long. Small body size would have resulted in lower limbs only slightly longer than upper, radius slightly shorter than humerus, and chest narrow. Estimates of limb bone indexes (ranges) might be: radiohumeral, 90–100; humero-femoral, 80–90; intermembral, 90–95. The early catarrhine ulna would have had monkey-like olecranon and styloid processes, the humerus an entepicondylar foramen (present in prosimians), the lower ankle a helical joint (unlike the derived states in cercopithecids and *Oreopithecus*). It may finally be hypothesized that the ancestral catarrhine diploid chromosome number was 44 (as found in most gibbons and colobines) and that ischial callosities were present and separated by an area of haired, not specialized, skin.

B. Apes and Men

A substantial number of characters indicate that the current practice of linking the greater and lesser apes closely is justified, which in turn permits the reconstruction of the ancestral morphotype of the (modern) apes. The dental apparatus of the early apes appears to have been highly conservative. The only significant change in the dentition would have involved loss of honing (sharpening) effect of the P_3 on the upper C, with the correlated reduction of bilateral compression of these two teeth and the lower C. Cranial form was similarly conservative, except for a possible slight increase in relative brain size. Postcranial differences between early apes and the ancestral ape morphotype would have been mainly those related to the enormous size increases that are known to have taken place by the early Miocene (Le Gros Clark and Leakey, 1951; Pilbeam, 1969; Andrews, 1973). These include the broadening of the thorax, with all the changes associated with that: longer clavicle, dorsal position of scapula, elongation of the scapula, and medial rotation of the humeral head; and the lengthening of the forelimb relative to the hindlimb. Reduction of the lumbar and caudal regions and lengthening of the sacral articulation with the ilia may have been partly independent of the size increase, and so also might have been the changes in the elbow joint and the carpal articulation. According to this reconstructional hypothesis, all of these changes need not have occurred at one time, but all must at least have been well advanced before the major split among the apes occurred, between gibbons and others. Inferences based on this cladistic model result in specific interpretations of certain fossil apes, while a model permitting widespread parallelism in acquisition of these features is more flexible (and less precise), as will be seen especially with the lesser apes.

1. Evolution of the Larger Apes and Men. The shared derived features of the common ancestor of at least the modern forms would have included: reduction of canine honing; increase in size of incisors relative to molars (variable in modern *Pan*, however); development of facial lengthening, with associated elongation of nasal bones and probably of choanae; increase in relative brain size and complexity; continued decrease in number

426

ERIC
DELSON
AND
PETER
ANDREWS

of lower back vertebrae and increase in sacrals; loss of carpal lunula if present at first and increase in meniscus size; increase in overall body size and muscularity, leading to a longer forelimb and greater muscle markings on limb bones; reduction of ischial callosities; and probably increase in chromosome number to 48 (or 46).

Modern apes and men are all somewhat more derived than the *Dryopithecus* group in such features as very large I^1, small cingulum and slightly reduced M_3, more reduced canine honing, possibly more "advanced" carpal and especially elbow joint morphology (see below) and, according to some researchers, a more restricted locomotor pattern (size-related?). It thus appears that *Dryopithecus* (and the little-known *Limnopithecus legetet*) represents a form close to the ancestry of modern larger apes and men before the split between Asian and African lineages, possibly including that ancestor. Preuschoft (1973) has shown that the olecranon was long in some *D. (Proconsul)*, while it was quite short in *D. (Dryopithecus)*, suggesting greater diversity than expected. Moreover, Pilbeam (1969) has shown that at least *D. (P.) major* had a frontal sinus, not found in *Pongo* or "lower" catarrhines; if this feature were present in all species, it would imply a separation of *Pongo* before that of *Dryopithecus* and *Pan*/men. It has previously been suggested that *D. major* is ancestral to gorillas (and *D. africanus* to chimps), but this seems to carry ancestor-hunting too far; the similarities are more probably seen as ecological or size-related convergences. Pilbeam (1970) has suggested an origin for *Gigantopithecus* within *Dryopithecus*, and he (also Andrews, 1973) has gone so far as to posit the lineage *D. major*→(unpublished Turkish material)→*D. indicus*→"*D. giganteus*"→*Gigantopithecus* spp. Again, if one wishes to find "ancestors," this is most elegant, but in fact it is based on a morphocline involving too few characters and is not yet testable. Here, *Gigantopithecus* is considered the sister group of *Dryopithecus*/*Limnopithecus*; its supposed links to men seem unsubstantiated.

Recent cladistic studies of Plio–Pleistocene human relatives (Eldredge and Tattersall, 1975; Delson, 1976) have suggested that the more "gracile" type of *Australopithecus* represents a close approach to the morphotype expected in a late Miocene (5–7 million years) human ancestor. The paper by Pilbeam and Gould (1974), although problematical, does suggest an allometry within *Australopothecus* spp. supporting their placement in a single genus as well as a distinctive trend in cerebral increase within *Homo*, which potentially began with *A. africanus*. Delson (1976) has accepted two subgenera of *Australopithecus*, with perhaps a third for the actual common ancestor, while realizing that this is in itself a stretching of cladistic methods; Robinson's view of *Homo* vs. *Paranthropus* is more cladistic superficially, but neglects both the strong shared features of *Australopithecus* species and the derived nature of *A. (P.) robustus*. Looking farther back in time, it appears that the status of *Ramapithecus* as a human ancestor is not as clear as some have thought, but known parts do agree well with the postulated morphotype, differing only in the more honing type of P_3 and U-shaped dental arcades (both conservative). The wide face and very narrow incisor region demonstrated by Walker and Andrews (1973) do not clash with this reconstruction, but require further study. *Ramapithecus* is clearly distinct from contemporary *Dryopithecus* species and must be considered a sister group either of *Dryopithecus* or of *Australopithecus* plus *Homo* (Fig. 2).

Little more can be said here about the three most apelike of the Fayum primates,

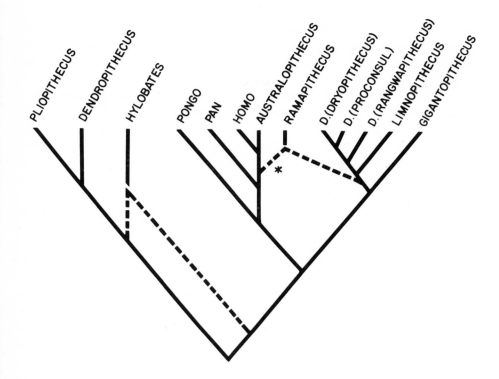

FIG. 2. Relationships of the Miocene to modern apes and men. In Figs. 2–5, there is no true time axis, but lineage branchings closer to the top of the cladogram are later than those below, along any given clade/lineage. Dashed lines indicate possible alternatives; the preferred choice among these, if any, is marked with an asterisk.

other than emphasizing that they are strongly conservative. This is to be expected of early forms, and is not (altogether) circular reasoning, as their age was not considered when the ancestral morphotype was inferred. Until these animals are fully studied, it seems best to group them closely and to consider them either the potential sister group of all later apes or as part of the *Dryopithecus* (or *Pliopithecus*) group (see Fig. 3 and pp. 429–430). We argue below for one of the two latter views, but the first has merit in that *Aegyptopithecus* essentially conforms to an ape ancestral morphotype. If gibbons diverged before this time, the retention by *Aegyptopithecus* of a long olecranon and an auditory tube would imply convergence in these complexes between greater and lesser apes. Szalay (1975 and this volume, chapter 5) has tentatively proposed that an auditory tube may have developed independently in monkeys and apes, but his reasoning is incomplete, and he further has denied the chance of convergence in this region among other primates (to the degree of similarity seen in catarrhines). Even more confusingly, the teeth of *Aegyptopithecus* show close similarities both to those of *Dendropithecus* (in the uppers) and to *Limnopithecus/Dryopithecus* (in lowers, see also Andrews, 1970), while retaining "primitively" large cingula. In part resolution of this potential dilemma depends on the interpretation of the lesser apes.

428

ERIC
DELSON
AND
PETER
ANDREWS

2. *Phylogeny of Gibbons and Relatives*. In many respects *Hylobates* is the most conservative of all modern catarrhines. Vogel (1966) expressed this view in detail as regards the facial skeleton and incisors, arguing that the group of features shared by colobines, gibbons, and men (and *Oreopithecus*) in the facial region were to be regarded as ancestral. The difficulty with this view is that those characters in *Homo* are not conservative retentions, but surely secondarily derived character states. Koenigswald (1968, 1969) has also remarked upon similarities between gibbons and monkeys, indeed implying a special phyletic relationship between them; in fact, such similarities as do exist are all ancestral retentions, the rare presence of semibilophodont molars in gibbons not having much phyletic weight other than as an indication of how early catarrhine teeth might have been altered in the ancestry of cercopithecids (see below). Compared to the ancestral ape, *Hylobates* has continued to retain basic catarrhine dental and cranial form, as well (possibly) as chromosome number, ischial callosities,* and intermediate development of sacral dominance and rigidity. Thus, the derived characters of modern gibbons are limited to: the shallow mandible, deeper posteriorly; reduced sexual dimorphism; projecting orbits; bony lunula in the carpal meniscus (if not ancestral); extreme elongation of hind- and forelimb (especially the radius) and the ricochetal brachiating form of locomotion; and the paired-family social organization. It is therefore most difficult to reconstruct earlier stages in the evolutionary history of *Hylobates*. Tuttle's (1972) scenario is most intriguing but suffers because the stages are not as far along as he suggested. Instead of early and late prehylobatines, his earlier phases (depending on the models discussed below) would correspond more closely to our concept of early catarrhines and ancestral apes, and even his later stages may require some reordering of the mosaic sequence of evolutionary events.

These problems become all too apparent when the focus shifts to the fossil forms *Pliopithecus* and *Dendropithecus*. These two are extremely similar, with the lesser-known *Dendropithecus* slightly more derived toward gibbons in its loss of the entepicondylar foramen, generally less musclemarked limb bones and longer humerus with respect to the femur (although still within the estimated ancestral catarrhine range). Recalling the characters presented above, *Pliopithecus* is linked specially to *Hylobates* only by its projecting (if somewhat lateral) orbits, somewhat shallow mandible (not as extreme), molars with low cusps and simple crowns, gracile and long humerus, and especially quite long radius (see also Simons, 1972). On the other hand, as Groves (1972) has argued, *Pliopithecus* is in most other characters among the most conservative of all known catarrhines: its basic cranial and dental patterns are ancestral, as are those of gibbons; it lacks an ossified external auditory meatus and any indication of a carpal meniscus; and it retains typical catarrhine body proportions, the ancestral number of lumbar and sacral vertebrae, humeral entepicondylar foramen, a long ulnar olecranon, and perhaps a femoral third

---

* Groves (1972, p. 81) has attempted to reconstruct the course of morphological evolution within *Hylobates*, including the suggestion that the ischial callosities (and perhaps 44 chromosomes) of the otherwise derived *H. (H.) lar* group might be secondary features. No strong supporting evidence for this theory is presented, and it would seem that Groves was overly impressed with *H. lar* as a derived lineage. In fact, no one group would be expected (empirically or theoretically) to have only derived characters, and the concept of mosaic evolution suggests that unrelated derived character states would be well dispersed within a group.

trochanter and a tail (possibly in the process of reduction) (see also Simons and Fleagle, 1973). Moreover, although the radius of these forms is long relative to the humerus, the data of Biegert and Maurer (1972) reveal that neither bone is expecially long compared to trunk length, certainly not comparable to *Hylobates*.

There are two possible interpretations of the phyletic position of these smaller Miocene apes. First, *Pliopithecus* (and *Dendropithecus*?) may have been part of an ape side branch which converged on gibbons in the above-listed features (probably before the ancestors of modern gibbons became so specialized), while still retaining a number of conservative character states; this view has been put forward by Groves (1972, 1974), less strongly by Simons and Fleagle (1973) and essentially also by Ferembach (1958), on fewer data. The presence of a long olecranon in *D.* (*Proconsul*) spp. (Preuschoft, 1973) implies that this genus also retained ancestral features, and thus the split between gibbons and other apes might have been earlier than generally thought. The demonstration in *Pliopithecus* (and/or *Dendropithecus*) of derived characters not shared with *Hylobates* would be a strong argument for this model, indicating that relationship between these "lesser apes" was only phenetic, not phyletic.

In the second hypothesis, *Pliopithecus* (and especially *Dendropithecus*) might represent forms close to the actual ancestry of *Hylobates*, in which case the carpal, elbow joint, and auditory region modifications considered above as shared derived features of all apes were in fact developed independently in gibbons and larger apes; most authors, including Andrews (1973, 1974) and Simons (1972), have accepted this view, while ignoring the convergence question. There is no problem in explaining the disparity between relative age and "primitiveness" between the two fossil genera. *Pliopithecus* (of the middle Miocene) may represent a more conservative lineage that exited Africa still retaining features already lost in the ancestry of forms we know as early Miocene *Dendropithecus*. This explanation would be especially meaningful if the African form proves to have auditory and carpal modifications similar to those of *Dryopithecus* spp. A phyletic relationship between *Hylobates* and these fossil genera might be supported directly if it were shown that the two groups shared additional derived features apart from other apes not likely to be convergent, such as might be found by a detailed study of the morphocline polarity of the cheek teeth, joint surfaces or ear region. Further and more intensive analyses of these functional complexes would certainly permit more explicit assessments of convergence (see also Szalay, chapter 5, this volume; and Tuttle, this volume).

At present, it is not possible to decide unequivocally whether greater weight should be given to those characters linking gibbons to other apes or to *Pliopithecus* (with *Dendropithecus* potentially intermediate). In the latter case, larger and lesser apes would be the major sister groups, probably best ranked at family level; *Propliopithecus*, *Aegyptopithecus*, and *Aeolopithecus* might then best be considered as the sister group of some or all *Dryopithecus*-like taxa, perhaps as a tribe of Dryopithecinae (this view is currently held by Andrews). On the other hand (as tentatively preferred by Delson), it may be better to consider that *Pliopithecus* and *Dendropithecus* were persistently conservative holdovers that had in some few ways converged upon the gibbons. In this case, following Groves (1972, 1974), and less precisely, Zapfe (1960), a family Pliopithecidae could be recognized

430

Eric
Delson
and
Peter
Andrews

with one subfamily for *Pliopithecus* and *Dendropithecus*, another for the three Oligocene taxa; gibbons would then be granted subfamily status within a single family including all apes. More specific taxonomic rankings are presented in the final classification, but in any case, it does not seem meaningful to keep men and other large apes apart at the family level any longer. Simpson (1963) has presented the most persuasive arguments for such separation, based on the different adaptive zone occupied by what are here called "men," yet he used almost the same set of intergeneric relationships as we have. The fact that gibbons and other apes are phyletically sister groups must certainly be indicated in a classification now, with the level dependent on the model followed for gibbons. The question then devolves to the placement of *Pan* among modern genera—with *Homo* (plus *Australopithecus*, etc.) on grounds of clade, or with *Pongo* by Simpson's grade argument, which is tentatively followed here; the *Dryopithecus* group would appear to be the sister group of all three modern large apes, as shown in both versions of Fig. 3.

C. Evolutionary History of Monkeys

Current knowledge of the interrelationships of the cercopithecids may be greater than that for the apes. As noted above (p. 413), the mesially grooved C^1, molar bilophodonty, and absence of hypoconulids on dP_4–M_2 are considered to be features of the

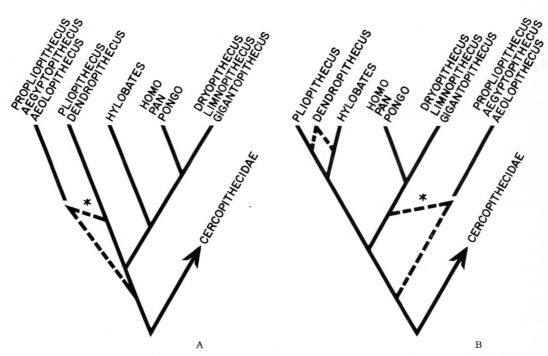

Fig. 3. Alternatives in the higher-level phylogeny of apes and men; notes as in Fig. 2. A: *Hylobates* closer to other modern apes, *Pliopithecus* persistently conservative. B: *Hylobates* and *Pliopithecus* closely related, forming the sister group of other apes.

earliest cercopithecid—in a sense, the presence of these characters defines their holder as the first "true" cercopithecid. It was also suggested that such an animal would have P^4 cusps and canine honing similar to that postulated for the ancestral catarrhine morphotype, except perhaps with a stronger P_3 honing flange. By comparison with the ancestral catarrhine morphotype and variability among known cercopithecids, it may be expected that early members of the family had incisors of moderate to small size relative to molars, conical I^2, variably a crista obliqua (see Koenigswald, 1969; Delson, 1975c, and below), and long M_3; it is not certain how chewing would have progressed in an animal with loph(id)s and also a crista obliqua, but one such (*Victoriapithecus*) is now under study by Delson. The cranial and truncal patterns of the ancestral catarrhine were probably retained by ancestral cercopithecids, which appear conservative in all but dental features. Such forms would thus have had the skull, chromosomes, and incisors of a colobine, ischial callosities and perhaps body form of a *Cercopithecus*, and cheek teeth (with low, rounded cusps, large and shallow trigonid, moderate flare, low relief, and large M_3) of a macaque. As discussed by Napier (1970) and Delson (1973, 1975b), these animals may have been early Miocene arboreal or semiarboreal inhabitants of seasonal deciduous forests. In this environment, fruit might not have been abundant throughout the year, and a dental adaptation that allowed early monkeys to supplement their diet with leaves would have been of great value.

The earliest known fossil cercopithecids are of early Miocene age, from Napak (Uganda) and Wadi Moghara (Egypt). *Prohylobates* from Moghara (see Simons, 1969) is apparently less than fully bilophodont, but incomplete material makes further analysis difficult. The Napak molar (Pilbeam and Walker, 1968) is clearly cercopithecine-like (ancestral), while the frontal fragment may be colobine-like (also ancestral); Radinsky (1974) has suggested that its endocranial cast is possibly gibbon-like, but a conservative early monkey would also have that morphology. The first evidence for the separation of the two modern subfamilies is from the middle Miocene of Maboko Island (Kenya), whence come specimens assigned to two species or morphs of *Victoriapithecus* (see Koenigswald, 1969). One type has long molars with large trigonids and low relief and may be associated with larger and more terrestrially-adapted limb bones. A second and more common set of molars are squarer, with the same low relief (shallow lingual notches), but short, colobine-like trigonids; they may be associated with a few more arboreal-seeming limb fragments. Delson (1973, 1975a) has argued that these fossils suggest an early phase in a split between colobines, that began to concentrate on folivory (with concomitant changes in teeth and later stomach) in an arboreal environment, and cercopithecines, that retained or enlarged an eclectic diet but experimented with semi-terrestrial locomotion, leading in turn to allometric facial lengthening (and related reorganization of facial bones) and to cheek pouches. It is possible that a reduction to 42 chromosomes also occurred in the ancestral cercopithecine at this (mid-Miocene) time.

By the late Miocene (11 million years?), colobines may have divided into more arboreal and more semiterrestrially adapted forms. The former seem to have remained in Africa, splitting into at least three units: modern *Colobus* subgenera; latest Miocene *Libypithecus* (Egypt); and Plio–Pleistocene species of the large *Paracolobus* and *Cerco-*

432

ERIC
DELSON
AND
PETER
ANDREWS

pithecoides in eastern and southern Africa. On the other hand, some early colobines left Africa by way of a semiopen corridor to Eurasia, whence *Mesopithecus* and its possible descendant *Dolichopithecus* became increasingly terrestrial in Europe. Colobines first appear in Asia by the latest Miocene, but relationships of the modern genera (*Presbytis*, *Pygathrix*, and *Nasalis*) to the extinct European lineage are uncertain.

A major split among cercopithecines also may have occurred around 10–12 million years ago, leading to the differentiation of cercopithecins and papionins. The former group may have reinvaded the high forest to compete with colobines, converging on their dental patterns by decreasing flare while also reducing M_3^3 distally. Diversification of chromosome numbers may have aided in rapid adaptive radiation into new niches—the pattern of difference between species in multiples of 6 is interesting, as is the fact that the most distinctive forms have the lower numbers (42 in papionins, 48 in *Allenopithecus*, 54 in *Erythrocebus* and *C. talapoin*, 60–72 in others). The Papionini share the derived dental feature of reduced lower incisor lingual enamel (and often increased flare), which must be of late Miocene age at least, as climatic changes allowed the tribe to divide into its three major components by the early Pliocene: Eurasian conservative macaques and their larger and more terrestrial descendants, *Paradolichopithecus* and *Procynocephalus*; African conservative *Parapapio* and *Cercocebus* and more derived *Papio* and *Dinopithecus* species; and dentally derived *Theropithecus*, represented today by a relict species, but far more widespread and of large size during the Plio–Pleistocene. This pattern of evolutionary relationship is reflected in Fig. 4.

The large amount of data now available on the evolution of cercopithecids has been summarized quite briefly here because of its recent analysis elsewhere (Delson, 1973, 1975a), including more complete treatment of fossil taxon ranges, generic relationships, and classification. More relevant to this paper is the attempt to work backward from the reconstructed morphotype of the ancestral monkey to *its* predecessor, a form that would not be called a monkey, on dental evidence at least. As discussed by Delson (1975c), such an animal would not (yet) have possessed the bilophodont dental pattern of later cercopithecids; that is, the rearrangement of cusps and development of loph(id)s in response to functional adaptive pressures (for increased shearing?) had not begun. The results of that analysis have been included in the deduction of the ancestral catarrhine morphotype, and as presented above, such an early catarrhine would (by definition) be potentially ancestral to both apes and monkeys.

The upper molars of precercopithecids might be expected to have 4 cusps and a crista obliqua, and the lowers to have 5 cusps (no paraconid, moderate hypoconulid) without clear talonid crests. The metaconid ridge is simple to envisage. Other features of known cercopithecids that were probably lacking in their early ancestors would include very high crowns, high relief, and lateral flare on molars. Because crown height and especially relief are greater in cercopithecids than in most other primate families, they would have been less developed in the early catarrhine ancestor of monkeys. The distribution of molar flare, most pronounced lingually on uppers and buccally on lowers, agrees with the distribution of cingulum in other catarrhines and most other primates as well. It may be suggested that whereas cingulum itself was increased in the ancestors of catarrhines in order to widen cheek teeth, it was then lost in the later history of this

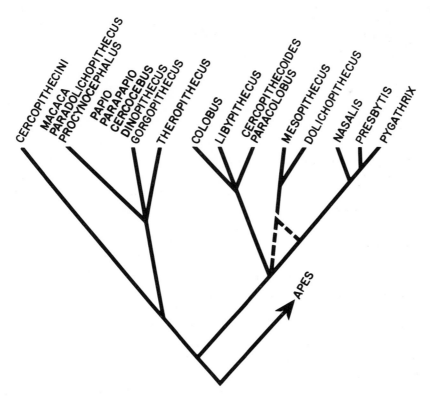

FIG. 4. Relationships of the Cercopithecidae; notes as in Fig. 2.

group. While apes seem to have lost it by simple reduction of cingular shelves as the tooth crown lengthened [except in *D. (Proconsul)* and *D. (Rangwapithecus)*], in monkeys the cingulum was incorporated directly into the external tooth wall, producing a smooth surface slightly bulged out near the cervix-flare. It is therefore expected that some cingulum would have been present in precercopithecids, much as in generalized early catarrhines. One of the major benefits of cladistics as applied to paleontology is that no early fossils have been included in this assessment of the precercopithecid ancestor, and such fossils can now be tested against the reconstruction without fear of circularity.

D. *Parapithecus* and *Apidium*

Simons (1972, and earlier articles summarized in Delson, 1975c) has argued strongly that new specimens of *Parapithecus* show strong morphological similarities to certain monkeys and thus represent the ancestor of Cercopithecidae. Taxonomically, he has suggested that *Parapithecus* (and its close ally *Apidium*) should be considered as a subfamily of monkeys, rather than as the family Parapithecidae, as is more common. Morphologically, Simons' view rests upon high molar crowns, a general shape (and

434

Eric
Delson
and
Peter
Andrews

size) similarity between *Parapithecus* and *Cercopithecus* (*Miopithecus*) *talapoin* and the cranial evidence of symphyseal and frontal fusion and postorbital closure in *Apidium* at least. It will be recalled that relatively high crowns (by comparison to those of earlier primates) are common to all Fayum catarrhines and also that *C. talapoin* is a rather distinctive member of a highly derived tribe of cercopithecids. Therefore, it has little to recommend it for comparison to potential cercopithecid relatives (much less ancestors) except size; comparison with either dentally conservative macaques, or preferably with the postulated precercopithecid morphotype, is of greater interest. Such comparison (Delson, 1975c) reveals that *Parapithecus* has a very low P_4 metaconid, short M_3, and lack of cingulum, none of which conform to expectation. The presence of a dP_4 paraconid and especially of 3 premolars are probably ancestral retentions, but the development of C^1 honing on P_2 would be a major derived feature. It seems quite unlikely that such a specialization would be lost and then redeveloped on P_3 if P_2 were dropped; Simons has admitted that if *Parapithecus* were ancestral to cercopithecids, they would have developed C_1–P_3 honing convergently with apes, but there seems little to support this hypothesis. Nor do the alternatives of losing either P_3 or P_4 seem likely: such occurrences have purportedly been documented among other potential lineages of early primates (J. Schwartz, personal communication), but the morphologies involved were quite distinct from those here considered. Another possible character of importance is the strong development of conules on the upper molars of both *Parapithecus* and *Apidium*, which might be incorporated into functional lophs to increase shear; the metaconule of *Cebus* spp. may be involved in a phenetically similar situation (A. Rosenberger and W. Kinzey, personal communications). Contrary to their reasoning, however, it seems that such conules would not be expected in an intermediate stage of loph building in early cercopithecids, because the highest point on a "conule–loph" between the main cusps would be at the conule apex; in known cercopithecids, on the other hand, the midpoint of a loph is a deep notch.

Overall, it appears that *Parapithecus* and *Apidium* are not easily regarded as phyletically linked to Old World monkeys. They both are rather conservative forms in many ways (cusp pattern, retention of 3 premolars, and an auditory ring), but they also share a number of derived features not found in contemporary or later catarrhines (P_2 honing, little cingulum, extra cusps). Simons' argument that they have reached anthropoid or "monkey" grade is well taken, but there is a marked difference between grade (or phenetic) similarity and phyletic relationship. The phenetic similarity does suggest that the Parapithecidae (as we here rank them) may have been the ecological vicars of Old World monkeys in the Fayum region at least, but such morphological resemblances as do exist are seen as convergent and are not the result of a common ancestry with monkeys.

The actual place of parapithecids in primate phylogeny is then open to question, with three other potential answers. First, they could be specially related to the apes, but this seems as little supported by shared derived features as does linkage with monkeys. Second, they could be equally related to monkeys and apes, forming a "triple point" of the type strict cladists abhor but working systematists must always consider. This view is interesting, but still assumes that P_2 was lost independently in apes and monkeys after the time when the three groups separated from a common ancestor; evidence for such independent loss would support this view. The last hypothesis is the most extreme, but at present seems the only acceptable one: parapithecids are essentially the sister group of all other catarrhines so far discussed. From a common ancestor with 3 premolars (none honing) and perhaps some cingulum, the parapithecids diverged to lose cingulum and develop P_2 honing, strong conules, and other characters, while another catarrhine branch lost P_2, developed P_3 honing, and emerged as the ancestor described above (p. 423). Delson (1975c) arrived at this view with less substantiating evidence and then offered the suggestion that in some ways parapithecids were well suited geographically, temporally, and morphologically to be the long sought sister group of the platyrrhines. Hoffstetter (1974) has also considered this point in a commentary on the study by Gingerich (1973), but more detailed analysis of known morphology is urgently needed. One other possible relationship of the parapithecids has previously been suggested: *Oreopithecus*.

E. *Oreopithecus*

The summary of the known morphology of *Oreopithecus bambolii* presented above intentionally did not consider any of the many theories regarding its phylogenetic position. As reviewed by Hürzeler (1958) and Straus (1963), these have ranged among the same alternatives noted for parapithecids. Some early authors viewed *Oreopithecus* as monkey-like, and Szalay (1975) has recently returned to a suggestion of this affinity, based on shared derived features of the lower ankle joint. As the rotational, rather than the helical, type of astragalocalcaneal surface is found in *Homo* as well as cercopithecines, it seems likely to be a convergent pattern related to locomotion, rather than a shared feature of common ancestry in all animals that present it. The potential implication of at least some terrestriality in *Oreopithecus* is unexpected, to say the least (but see also Riesenfeld, 1975). Most modern authors have accepted *Oreopithecus* as an ape of some type, based on such (shared derived) characters as the reduced ulnar olecranon, 5 lumbar vertebrae, high intermembral and brachial indexes, and general shape of the trunk and axial skeleton. Once again, it is likely that most, or even all, of these features are under allometric control and thus the corollary of the nondiagnostic criterion of size. Nonetheless, *Oreopithecus* does appear to have significantly long arms and (to a lesser degree) legs for its trunk length, thus suggesting some potential links with apes. Hürzeler (1958) has argued that *Oreopithecus*, among apes, is specially related to men, because of its short face, bicuspid P_3, small canines, and some other characters, and Straus (1963) more or less accepted this view. Szalay and Berzi's (1973) study of the crushed skull has shown that some of these features are not as Hürzeler had implied, while others are seen to be merely conservative among catarrhines, and the short legs are definitely not manlike; but the small canine and bicuspid P_3 still remain. A fourth alternative has been to consider *Oreopithecus* as a distinctive type of ape or "hominoid," separated at a high taxonomic level from the others. It combines conservative facial features with small canines, bicuspid P_3, large internal cusp on I^1, deep jaw with large gonion, and especially the large conules and centroconid, all derived features isolating *Oreopithecus* from other modern types of catarrhines.

436

ERIC
DELSON
AND
PETER
ANDREWS

It is just those features, however, which combine to link it to *Apidium*, as suggested by Gregory and later by Simons (1960). Since then, Simons has refrained from pressing his point, but his 1972 review of both genera adds other features in common. The two forms are among the few catarrhines that present clear protoconule and metaconule, the latter perhaps as part of a distal upper molar loph. They are nearly unique among primates in possession of a centroconid, whose relationships to other cusps are quite similar in the two genera, although *Orepithecus* is much more cristodont than is *Apidium*. Furthermore, it is most suggestive to note that if indeed an *Apidium*-like form were to have lost a honing P_2, the two remaining premolars would be bicuspid and nonhoning, and the canine probably small, as seen in *Oreopithecus*—of course the actual reduction would be under selectional control and not so simplistic. This relationship is more worthy of consideration now than ever before, but it in turn brings up the difficulties mentioned above and more: the shared derived features of apes and *Oreopithecus* would be convergences if *Apidium* is far from the modern catarrhines; and parapithecids must be closer to the last if *Oreopithecus* is. Even a relationship among parapithecids, cercopithecids, and *Oreopithecus* would be conceivable, but much less so than others have thought. Further detailed study is needed of both *Oreopithecus* and all Fayum primates—either clear examples of nonconvergent shared derived features or morphologically intermediate fossils to show the morphocline are needed to permit choice among the alternatives. Figure 5 summarizes the major options available in terms of the phylogeny of *Oreopithecus*, parapithecids and other main catarrhine groups.

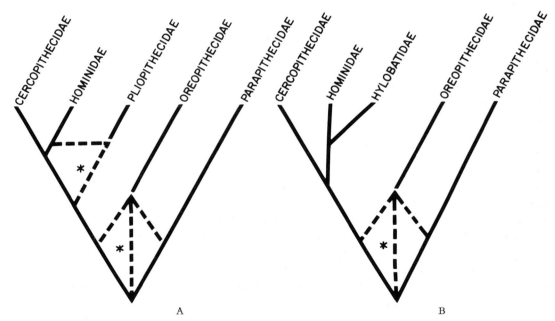

FIG. 5. Alternatives in the higher-level phylogeny of catarrhines; notes as in Fig. 2. A: as in Fig. 3A, modern families roughly equally distant from one another. B: as in Fig. 3B, hylobatids and hominids phyletically linked, classified as Hominoidea.

F. *Oligopithecus*

437

EVOLUTION AND
RELATIONSHIPS
OF THE
CATARRHINES

None of the above has yet answered the question of Old World monkey ancestry asked previously. In his original description of *Oligopithecus*, Simons considered a relationship to monkeys possible, but he has since withdrawn from that view. Delson (1973, 1975c) was unable to find any known fossil primate that matched the ancestral precercopithecid morphotype in all respects, but he did suggest that a series of such features were found variably in species of *Oligopithecus* and *Propliopithecus*: 2 premolars, the anterior with well-developed C^1 honing, the posterior with subequal metaconid and protoconid and 2 large foveas; lower molars that have lost the paraconid and moderately enlarged the trigonid, linking protoconid and metaconid by a ridge (only in *Propliopithecus*); high tooth crowns with low relief; moderate-sized hypoconulid with evidence of appression to the entoconid (this is clearest in *Oligopithecus*, but also indicated in some *Propliopithecus*); a cingulum, especially buccally, around the lower molars (in *P. haeckeli*, very little in *O. savagei*, and none (?) in *P. markgrafi*, which instead shows lateral sloping of the lower crown, almost as in monkey flare). *Propliopithecus* is clearly similar to *Aegyptopithecus* and *Aeolopithecus*, while *Oligopithecus* is quite distinctive in its lack of cingulum combined with conservative molar form and "new" honing, but the known variation in these Fayûm catarrhines suggests that in some unknown animal, probably in another environment, this mosaic of features was combined in a single species that began the lineage leading eventually to cercopithecids. *Oligopithecus* itself may be considered as a basically conservative remnant of the ancestral (modern-type) catarrhines. As such, it may be classed either as the sister group of all nonparapithecids, or of all catarrhines, or be placed with the relatively conservative *Aegyptopithecus* group. Because of the lack of knowledge about its anterior and upper dentition, much less the skull and postcranium, we here take the easy way out and consider it as Catarrhini, *incertae sedis*.

IV. Conclusions—Phylogeny and Classification

A. The Relationship between Classification and Phylogeny

Two of the most important aspects of evolutionary studies are the determination of phylogeny and the erection of a classification of the group under study; the relative importance accorded to these two types of result differs widely among evolutionists. In the preceding section, we have attempted to present sufficient argument validating the phylogenies that we have determined—at this point, it becomes possible to formalize some of these relationships in the Linnean hierarchy. The relationships between phylogeny and classification have been specifically discussed in recent years, with the resulting delineation of two major opposing views.

The cladists (e.g., Hennig, 1966; Nelson, 1972, 1973) strongly assert that a classification should precisely reflect the phylogeny on which it is based, the only open questions being which branch points (nodes) on the cladogram need be formally named. Some authors have suggested that taxonomic rank should be determined by the age of origin

438

Eric
Delson
and
Peter
Andrews

of a group or taxon, but this is considered superfluous by most practitioners—thus, at least one rank must be determined *a priori*, as is true as "evolutionary" systematics. Only one cladistic classification could thus be erected from a phylogeny in which all nodes are understood, and in this classification all taxa would be maximally monophyletic (holophyletic), that is, all descendants of an ancestral taxon would be grouped together at all ranks. Nelson (1974) has recently cited some of Darwin's (1871) writings on the classification of catarrhines to show that Darwin accepted the above tenets, at least in theory.

The second major school of taxonomic philosophy has been termed "evolutionary" by its practitioners (e.g., Simpson, 1961, 1963, 1971, this volume; Mayr, 1969, 1974; Bock, 1973), but a more neutral term would be preferable. Supporters of their views consider that a classification should be more than just a verbalized phylogeny, namely, an "information-retrieval system." Thus, in selected cases, one member of a sister-group pair would have to be raised in taxonomic rank by comparison to the other in order to reflect great divergence from the ancestral condition and/or numerical diversity. The basic logic of cladistic analysis and the cladists' desire to classify or group related taxa together are now fully accepted by the "evolutionary" school (see Mayr, 1974), but they wish to add further information to their classifications than can be found in a phylogeny or cladogram. In retort, the cladists have argued that classification cannot serve this function without introducing the element of "art" acknowledged by Mayr and others, thus decreasing reproducibility given the same data; further, it has been asked whether the information put into such a classification can in fact be "retrieved" without additional textual material (e.g., Cracraft, 1973). Mayr (1974) and others have in turn attempted to show that Darwin was in fact an "evolutionary" taxonomist, who aimed to produce "cladistic" classifications in theory, but who was pragmatic enough to bend his rules in certain cases.

As will be obvious from the body of this study, we are fully aware of the power and importance of cladistic analysis in the determination of phylogenetic relationships. We further accept that a classification must be based on a specific phylogeny, with which it should agree and not contradict. But we argue that there may in fact be cases in which one of two sister groups has diverged so strongly from the ancestral condition that to accord it equal taxonomic rank with its sister would be a distortion of the Linnaean system. This situation might occur especially when dealing with fossil organisms, given an ancient split from which a single low-rank taxon persisted conservatively for a short time and a large, diverse, and increasingly derived group of taxa proliferated over a much longer time (even if all are today extinct), it seems wasteful to rank both "sisters" at elevated levels in the hierarchy. Mayr (1974, p. 121) has commented on this point with examples from Darwin's *Origin* (1895, pp. 420–421) and from Darwin's classification of the barnacles. It may be useful here to cite a passage from the *Descent of Man* (1871, pp. 187–188) that follows immediately after one of the lines cited by Nelson (1974, p. 456), considering the classification of man. Darwin has suggested that man is clearly an anthropoid primate and thus might be ranked as

> merely a Family, or possibly even only a Sub-family. If we imagine three lines of descent
> proceeding from a common stock, it is quite conceivable that two of them might after the

lapse of ages be so slightly changed as still to remain as species of the same genus, whilst the third line might become so greatly modified as to deserve rank as a distinct Sub-family, or even Order. But in this case, it is almost certain that the third line would still retain through inheritance numerous small points of resemblance with the other two. Here, then, would occur the difficulty, *at present insoluble*, how much weight we ought to assign in our classifications to strongly marked differences in some few points,—that is, to the amount of modification undergone; and how much to close resemblance in numerous unimportant points, as indicating the lines of descent or genealogy. To attach much weight to the few but strong differences is the most obvious and perhaps the safest course, though it appears more correct to pay great attention to the many small resemblances, as giving a truly natural classification [emphasis ours].

The remainder of that section is devoted to proving that man is a catarrhine, indeed a "hominoid," and that our ancestors would have also been so, but Darwin makes no further decision concerning the relative ranking of *Homo* (the only "hominid" known to Darwin). As usual, Darwin's views are penetrating, not only for his time but for ours: his reference to three forms from a common stock fits precisely with our interpretation of man, chimpanzee, and gorilla, and his inability to choose between the cladistic and "evolutionary" philosophy is exactly our dilemma. In fact, although this is almost by chance, we opt for the subfamilial rank he mentions, as a potential compromise between tradition and strict cladism.

B. Application to the Catarrhines

To return to the question of catarrhine classification, and to begin at the beginning, it is necessary in either the cladistic or the "evolutionary" system to assign at least one supraspecific rank *a priori*. Considering the results of this symposium and the general concensus among primatologists (and also Simpson's 1945 analysis of all mammalian families), it appears reasonable to assign the rank of family to the Old World monkeys and that of infraorder to the catarrhines as a whole. From this base, Simpson (1963) has argued that the diversity among apes and men requires two families, of which he allocated one for men alone. On an "evolutionary" basis, this may be too high a ranking if monkeys are allocated a single family; on a cladistic basis, the important question is the geometry of the phylogeny involved. Our two alternative interpretations of the latter are given in Fig. 3, from which we infer somewhat different classifications, in each case accepting the lowest major node as a split between two families.

Following Fig. 3A, it appears that the apes can be divided into a "modern" versus a conservative, archaic group. Pending evidence to the contrary, the three Fayum genera are allied most probably with the conservative Miocene forms in a single family, while the latter genera (*Pliopithecus* and *Dendropithecus*) may be set off as a derived subfamily. No name has been seriously proposed for a family–group taxon encompassing or based on the three Fayum genera, but Straus (1961), in a paper that lacked some understanding of the Rules of Nomenclature, did propose a family Propliopithecidae; this nomen may be employed here. A second family, Hominidae, would then encompass the modern apes and the *Dryopithecus* group, which seems phyletically closer to the great apes. Gibbons could be accorded a subfamily of their own, while the modern apes and all men would rank as one tribe, the dryopiths as another, in a second subfamily. Within

440

ERIC
DELSON
AND
PETER
ANDREWS

the Dryopithecini, *Gigantopithecus* might be given subtribal rank, but this may be pushing our fragmentary knowledge to the limits; similarly, *Pongo* might be set off from *Pan* and men on a cladistic argument as a subtribe. At this rate, in order to distinguish Pan from *Homo* and *Australopithecus* (and perhaps *Ramapithecus*), a new rank lower than subtribe would have to be employed. It is at this point that we would fall back on Darwin's reasoning and raise the rank of the taxon for men, leaving Hominidae with Hylobatinae, Ponginae (Pongini and Dryopithecini) and Homininae. For those wishing to employ a fully cladistic classification, the subtribe Hominina would be available for *Pan, Homo*, and the fossils noted.

If, on the other hand, the arrangement illustrated in Fig. 3B were substantiated, the major split between anthropomorph families would separate gibbons from great apes. Here, a family Hylobatidae might be subdivided into Hylobatinae and Pliopithecinae if necessary. The placement of *Dendropithecus* would depend on clear indication of its relationship, now in doubt. A second family, Hominidae, might be cladistically divided into Homininae vs. either Dryopithecinae or Propliopithecinae, depending upon whether the Fayum forms were considered the sister group of Neogene large apes or of the dryopiths—they might even be considered a sister group of all hominids, but this seems least likely to us at present. Once again, if the rank of *Homo* and its closest relatives were raised to subfamily level, the Hominidae would include Homininae, Ponginae (Pongini, Dryopithecini), and Propliopithecinae; or alternatively, Homininae, Ponginae, and Dryopithecinae (Dryopithecini, Propliopithecini). The last of these, favored here, is in many ways closest to the presently most accepted classification, save for the high-level separation of the gibbons.

Four other fossil genera are less well known. *Apidium* and *Parapithecus* appear to share a number of traits, although many of these may be ancestral for the catarrhines in the widest sense, as may be seen from Table 1. They do seem to share the derived conditions of P_2 honing and loss of cingulum on lowers, however, and it is our opinion that they are best grouped as the family Parapithecidae. *Oreopithecus* is highly distinctive, being conservative in many characters and strongly derived in a number of others (Tables 1 and 2). It may also be given family rank, although one could also argue that it be ranked as Catarrhini *incertae sedis*. We take the latter action with regard to *Oligopithecus*, although it most probably is closer to modern forms than to the parapiths, because our knowledge of it is so meager.

Grouping at the suprafamilial level depends again on which of the models of catarrhine phylogeny (Fig. 5) is followed. It appears rather clear to us that in either case, the Parapithecidae are to be regarded as the sister group of at least the modern families, if not of all other catarrhines. Thus, one may question whether there is a need to separate modern apes and monkeys any longer at the superfamily level. One potential (cladistic) division of the infraorder might then be Parapithecoidea vs. Cercopithecoidea (a name antedating Hominoidea by 4 years). Oreopithecidae would be *incertae sedis*, unless it were shown clearly to fit into one of the two superfamilies. This would certainly work well if the phylogeny of Fig. 5A were accepted, but under that of Fig. 5B, the hylobatids and hominids are surely closer to one another phyletically than either is to the cercopithecids. Thus, here, it might be useful to retain the common Hominoidea vs. Cerco-

TABLE 3. CLASSIFICATIONS OF THE INFRAORDER CATARRHINI

"Evolutionary" classifications—reflecting divergence of "men"		Strict cladistic classification
I	II	III
After Figs. 3A and 5A (preferred by Delson)	After Figs. 3B and 5B (preferred by Andrews)	
(Cercopithecoidea)	(Cercopithecoidea)	Cercopithercoidea
Cercopithecidae	Cercopithecidae	Cercopithecidae
Cercopithecinae	Cercopithecinae	Cercopithecinae
Cercopithecini	Cercopithecini	Cercopithecini
Papionini	Papionini	Papionini
Colobinae	Colobinae	Colobinae
Cercopithecidae *incertae sedis*	Cercopithecidae *incertae sedis*	Cercopithecidae *incertae sedis*
†*Victoriapithecus*	†*Victoriapithecus*	†*Victoriapithecus*
†*Prohylobates*	†*Prohylobates*	†*Prohylobates*
	(Hominoidea)	Hominidae
Hominidae	Hominidae	Homininae
Homininae	Honininae	Hominini
Homo	*Homo*	Hominina
†*Australopithecus*	†*Australopithecus*	(infratribe unnamed)
†(*Ramapithecus*)	†(*Ramapithecus*)	*Homo*
Ponginae	Ponginae	†*Australopithecus*
Pongini	*Pongo*	†(*Ramapithecus*)
Pongo	*Pan*	(infratribe unnamed)
Pan	†Dryopithecinae	*Pan*
†Dryopithecini	†Dryopithecini	Pongina
†(Dryopithecina)	†(Dryopithecina)	*Pongo*
†*Dryopithecus*	†*Dryopithecus*	†Dryopithecini
†*Limnopithecus*	†*Limnopithecus*	†Dryopithecina
†(Gigantopithecina)	†(Gigantopithecina)	†*Dryopithecus*
†*Gigantopithecus*	†*Gigantopithecus*	†*Lmnopithecus*
(Ponginae *incertae sedis*)	(Dryopithecini *incertae sedis*)	†Gigantopithecina
†(*Ramapithecus*)	†(*Ramapithecus*)	†*Gigantopithecus*
Hylobatinae	†Propliopithecini	(Dryopithecini *incertae sedis*)
Hylobates	†*Propliopithecus*	†(*Ramapithecus*)
†Pliopithecidae	†*Aegyptopithecus*	Hylobatinae
†Pliopithecinae	†*Aeolopithecus*	*Hylobates*
†*Pliopithecus*	Hylobatidae	Hominidae *incertae sedis*
†*Dendropithecus*	(Hylobatinae)	†*Pliopithecus*
†Propliopithecinae	*Hylobates*	†*Propliopithecus*
†*Propliopithecus*	†(Pliopithecinae)	†*Aegyptopithecus*
†*Aegyptopithecus*	†*Pliopithecus*	†*Aeolopithecus*
†*Aeolopithecus*	Hylobatidae *incertae sedis*	†*Dendropithecus*
	†*Dendropithecus*	
†(Parapithecoidea)	†(Parapithecoidea)	†Parapithecoidea
†Parapithecidae	†Parapithecidae	†Parapithecidae
†*Parapithecus*	†*Parapithecus*	†*Parapithecus*
†*Apidium*	†*Apidium*	†*Apidium*
(Catarrhini *incertae sedis*)	(Catarrhini *incertae sedis*)	Catarrhini *incertae sedis*
†Oreopithecidae	†Oreopithecidae	†Oreopithecidea
†*Oreopithecus*	†*Oreopithecus*	†*Oreopithecus*
Family *incertae sedis*	Family *incertae sedis*	Family *incertae sedis*
†*Oligopithecus*	†*Oligopithecus*	†*Oligopithecus*

† Indicates wholly extinct taxon.

442

Eric
Delson
and
Peter
Andrews

pithecoidea division, with Parapithecidae also given superfamily rank or potentially a new, higher rank (between infraorder and superfamily). The most conservative arrangement, which we here favor tentatively, is to accept 5 families within the Catarrhini, with no further subdivision.

In order to demonstrate both our uncertainty and the major variants conformable with our phyletic inferences, we have provided three parallel catarrhine classifications (Table 3). The first reflects the phylogeny of Fig. 3A and 5A, the second 3B and 5B. In both, "men" have been differentially raised to subfamilial rank. Family–group taxa listed parenthetically are optionally named or alternatives, while genera in parentheses have more than one possible placement. A third classification is strictly cladistic, accurately reflecting the sequence of cladogram branching points and nothing else, as well as indicating the uncertainties involved in the relationships of *Pliopithecus*, *Aegyptopithecus*, etc. Nelson (1972) has argued that fossils should be ranked *incertae sedis* if there is doubt as to their placement, but denies this procedure in most cases for recent organisms. We disagree, holding that all taxa must be treated in the same manner, with those least understood (empirically most often fossils) being relegated when necessary to the *incertae sedis* category. The order of generic taxa within family–group higher taxa is determined by neither of the criteria discussed by Nelson (1973), but instead merely by date of authorship, save that nominate genera are listed first.

V. Summary

The distribution of morphological characters is presented for the major groups of catarrhine primates. Concentration is on dental characters (summarized in Table 1), with the addition of cranial, postcranial, and some other features (Table 2). Utilizing the methods of cladistic analysis, and without considering the geological age of any taxa (see Fig. 1), ancestral morphotypes are reconstructed for each higher taxon. The phyletic relationships of *Hylobates* are seen as central to understanding the phylogeny of the catarrhines—it may be closest to the modern apes and men or it may be linked to *Pliopithecus*, which retains many ancestral features and may have converged partially on gibbons postcranially. It appears that the living apes and man (including *Australopithecus*) form a natural group apart from the more conservative dryopiths; the position of the three Fayum "apes" is equivocal. Phylogeny cladograms reflecting the major alternatives are presented in Figs. 2 and 3. *Apidium* and *Parapithecus* on the one hand, and *Oreopithecus* on the other, are conservative in many respects but have derived characters setting them off from other catarrhines.

A hypothetical morphotype of the ancestral catarrhine is reconstructed from those of the lower taxa. None of the oldest fossils fit this morphotype completely, but it does appear that *Parapithecus* and *Apidium* separated from other catarrhines (unless *Oreopithecus* is indeed related to them) at a stage before the loss of P_2, which may have already developed into a canine honing tooth in these forms. The phylogeny of apes, Old World monkeys (Fig. 4), and catarrhines in general (Fig. 5) are reviewed and interpreted. Classifications must be based on such phylogenies, but need not reflect them perfectly. In the catarrhine classification(s) presented, Hominidae is expanded to include men and apes (excluding either *Hylobates* and *Pliopithecus* or *Pliopithecus* and the three Fayum "apes"), but the human lineage is raised in rank to a subfamily. Neither Parapithecidae nor Oreopithecidae appear to be phyletically close to Cercopithecidae, and *Oligopithecus* must for now be ranked as Catarrhini, *incertae sedis*.

ACKNOWLEDGMENTS

We thank Drs. Niles Eldredge, Colin Groves, and Frederick Szalay for discussions relevant to the reasoning in this paper, and the curators of numerous paleontological and mammalogical collections for permission to study specimens in their care. We further acknowledge the financial support of the Wenner–Gren Foundation for Anthropological Research and the National Science Foundation (Grant GS–42628 to ED). The figures were ably prepared by Miss Lorraine Meeker, and Miss Marlyn Mangus provided assistance in the preparation of the manuscript.

VII. References

ANDREWS, Peter. 1970. Two new fossil primates from the lower Miocene of Kenya. *Nature* **228**:537–540.

ANDREWS, Peter. 1971. *Ramapithecus wickeri* Mandible from Fort Ternan, Kenya. *Nature* **231**:192–194.

ANDREWS, Peter. 1973. Miocene Primates (Pongidae, Hylobatidae) of East Africa. Thesis, Cambridge University.

ANDREWS, Peter. 1974. New species of *Dryopithecus* from Kenya. *Nature* **249**:188–190.

ANDREWS, PETER, and COLIN GROVES. 1975. Gibbons and brachiation. *Gibbon Siamang* **4**:167–218.

ANDREWS, Peter, David PILBEAM and E. SIMONS. 1976 (in press). A review of the African Cenozoic apes (Hominoidea, Primates). *In* V. J. Maglio, ed. *Mammalian Evolution in Africa*. Princeton University Press.

ANKEL, Friderun. 1965. Der Canalis Sacralis als Indikator fur die Länge der Caudalregion der Primaten. *Folia Primatol.* **3**:263–276.

BIEGERT, Josef, and R. MAURER. 1972. Rumpfskclettlänge, Allometrien und Körperproportionen bei catarrhinen Primaten. *Folia Primatol.* **17**:142–56.

BOCK, Walter J. 1973. Philosophical foundations of classical evolutionary classification. *Syst. Zool.* **22**:375–392.

BOER, L. E. M. de. 1971. Cytology of the Cercopithecidae with special reference to the karyotype of *Cercopithecus hamlyni* Pocock. *Proc. 3rd Int. Congr. Primatol.* (*Zurich, 1970*). **2**:121–126.

BUTLER, Percy M., and J. R. E. MILLS. 1959. A contribution to the odontology of *Oreopithecus*. *Bull. Brit. Mus.* (*Nat. Hist.*), *Geol.* **4**:1–30.

CHIARELLI, Brunetto. 1968a. From the karyotype of the apes to the human karyotype. *S. A. J. Sci. Spec. No.* **64**(2):72–80.

CHIARELLI. 1968b. Caryological and hybridological data for the taxonomy and phylogeny of the Old World primates, pp. 151–186. *In* B. Chiarelli, ed., *Taxonomy and Phylogeny of Old World Prinates with References to the Origin of Man*. Rosenberg and Sellier, Torino.

CHIVERS, D. 1972. The siamang and the gibbon in the Malay Peninsula. *Gibbon Siamang* **1**:103–135.

CONROY, GLENN C. 1976 (in press). The earliest anthropoid postcranial remains from the Oligocene of Egypt. *J. Hum. Evol.*

CONROY, Glenn C., and John G. FLEAGLE. 1972. Locomotor behavior in living and fossil pongids. *Nature* **237**:103–104.

CRACRAFT, Joel. 1973. Comments in discussion of symposium papers on contemporary systematic philosophies. *Syst. Zool.* **22**:398–399.

DARWIN, Charles, 1859. On the Origin of Species . . . Murray, London.

DARWIN, Charles. 1871. The Descent of Man. . . Murray, London.

DAVIS, P. R., and John NAPIER. 1963. A reconstruction of the skull of *Proconsul africanus* (R.S.51). *Folia Primatol.* **1**:20–28.

444

ERIC
DELSON
AND
PETER
ANDREWS

DELSON, Eric. 1973. Fossil Colobine Monkeys of the Circum-Mediterranean Region and the Evolutionary History of the Cercopithecidae (Primates, Mammalia). Thesis, Columbia University.

DELSON, Eric. 1975a. Evolutionary history of the Cercopithecidae. *Contrib. Primatol.* **5**:167–217.

DELSON, Eric. 1975b (in press). Paleoecology and zoogeography of the Old World monkeys. *In* R. Tuttle, ed. *Primate Functional Morphology and Evolution*, Mouton, Hague. pp. 1–27.

DELSON, Eric. 1975c (in press). Toward the origin of the Old World monkeys. *Actes C.N.R.S. Coll. Int. No. 218, Evol. Verts.*, pp. 689–698.

DELSON, Eric. 1976d (in press). Models of early hominid phylogeny. *In* C. J. Jolly, ed., *African Hominids of the Plio–Pleistocene*. Duckworth, London.

ELDREDGE, NILES, and Ian TATTERSALL. 1975. Evolutionary models, phylogenetic reconstruction and another look at hominid phylogeny. *Contrib. Primatol.* **5**:218–242.

FEREMBACH, Denise. 1958. Les limnopithèques du Kenya. *Ann. Paléontol.* **44**:149–249.

FLEAGLE, J. G., E. L. SIMONS and G. C. CONROY. 1975. Ape limb bone from the Oligocene of Egypt. *Science* **189**:135–137.

FRISCH, J. E. 1965. Trends in the evolution of the hominoid dentition. *Bibl. Primatol.* **3**:1–130.

FRISCH, J. E. 1973. The hylobatid dentition. *Gibbon Siamang* **2**:55–95.

GINGERICH, P. D. 1973. Anatomy of the temporal bone in the Oligocene anthropoid *Apidium* and the origin of Anthropoidea. *Folia Primatol.* **19**:329–337.

GROVES, Colin P. 1972. Systematics and phylogeny of gibbons. *Gibbon Siamang* **1**:1–89.

GROVES, Colin P. 1974. New evidence on the evolution of the apes and man. *Vestn. Ustred. Ustavu geol.* **49**:53–56.

HENNIG, Willi. 1966. *Phylogenetic Systematics*. University of Illinois Press, Chicago.

HOFFSTETTER, Robert. 1974. *Apidium* et l'origine des Simiformes (= Anthropoidea). *C.R. Acad. Sci., Paris* **278D**:1715–1717.

HOLLOWAY, Ralph L. 1972. Australopithecine endocasts, brain evolution in the Hominoidea, and a model of hominid evolution, pp. 185–203. *In* R. Tuttle, ed., *The Functional and Evolutionary Biology of Primates*. Aldine, Chicago.

HÜRZELER, Johannes. 1954. Contribution à l'odontologie et à la phylogenèse du genre *Pliopithecus* Gervais. *Ann. Paléontol.* **40**:5–63.

HÜRZELER, JOHANNES. 1958. *Oreopithecus bambolii* Gervais, a preliminary report. *Ver. Nat. Ges. Basel* **69**:1–47.

JOLLY, Clifford J. 1970. The large African monkeys as an adaptive array, pp. 141–174. *In* John R. Napier and P. H. Napier, eds., *Old World Monkeys*. Academic Press, New York.

JONES, R. Trevor. 1972. The ethmoid, the vomer and the palatine bones from the baboon skull. *S. Afr. J. Sci.* **1972**:156–161.

KÄLIN, J. 1961. Sur les primates de l'Oligocène Inférieur d'Egypte. *Ann. Paléontol.* **47**:3–48.

KOENIGSWALD, G. H. R. von 1968. The phylogenetical position of the Hylobatinae, pp. 271–276. *In* B. Chiarelli ed., *Taxonomy and Phylogeny of Old World Primates with References to the Origin of Man*. Rosenberg and Sellier, Torino.

KOENIGSWALD, G. H. R. von 1969. Miocene Cercopithecoidea and Oreopithecoidea from the Miocene of East Africa. *Fossil Vertebrates of Africa*. Academic Press, London, **1**:39–51.

LEAKEY, Louis S. B. 1968. Upper Miocene primates from Kenya. *Nature* **218**:527–530.

LE GROS CLARK, W. E. 1971. *The Antecedents of Man: An Introduction to the Evolution of the Primates*, 3rd ed. University Press, Edinburgh.

LE GROS CLARK, Wilfred E., and Louis S. B. LEAKEY. 1951. The Miocene Hominoidea of East Africa. *Fossil Mamm. Afr. (Br. Mus. Nat. Hist., London)* **1**:1–117.

LEWIS, Owen J. 1971. Brachiation and the early evolution of the Hominoidea. *Nature* **230**:577–579.

LEWIS, Owen J. 1972a. Evolution of the hominoid wrist, pp. 207–222. *In* R. Tuttle, ed., *The Functional and Evolutionary Biology of Primates*. Aldine, Chicago.

LEWIS, Owen J. 1972b. Osteological features characterizing the wrists of monkeys and apes, with a reconsideration of this region in *Dryopithecus (Proconsul) africanus*. *Am. J. Phys. Anthropol.* **36**:45–58.

MAHLER, P. 1973. Metric Variation in the Pongid Dentition. Thesis, University of Michigan.

MAYR, Ernst. 1969. *Principles of Systematic Zoology*. McGraw-Hill, New York.

MAYR, Ernst. 1974. Cladistic analysis or cladistic classification? *Z. Zool. Syst. Evol.* **22**:344–359.

NAPIER, John R. 1970. Paleoecology and catarrhine evolution, pp. 55–95. *In* J. R. Napier and P. H. Napier, eds. *Old World Monkeys*. Academic Press, New York.

NAPIER, J. R. and P. R. DAVIS. 1959. The forelimb skeleton and associated remains of *Proconsuls africanus*. *Fossil Mamm. Afr. (Br. Mus. Nat. Hist., London)* **16**:1–69.

NELSON, Gareth. 1972. Phylogenetic relationship and classification. *Syst. Zool.* **21**:227–230.

NELSON, Gareth. 1973. Classification as an expression of phylogenetic relationships. *Syst. Zool.* **22**:344–359.

NELSON, Gareth. 1974. Darwin–Hennig classification: a reply to Ernst Mayr. *Syst. Zoo.* **23**:452–458.

PILBEAM, David. 1969. Tertiary Pongidae of East Africa: evolutionary relationships and taxonomy. *Yale Peabody Mus. Bull.* **31**:1–185.

PILBEAM, David. 1970. *Gigantopithecus* and the origins of Hominidae. *Nature* **225**:516–519.

PILBEAM, David. 1972. *The Ascent of Man.* Macmillan, New York.

PILBEAM, David, and Stephen Jay GOULD. 1974. Size and scaling in human evolution. *Science* **186**:892–901.

PILBEAM, David, and E. L. SIMONS. 1971. Humerus of *Dryopithecus* from Saint-Gaudens, France. *Nature* **229**:408–409.

PILBEAM, David, and Alan WALKER. 1968. Fossil monkeys from the Micoene of Napak, North-East Uganda. *Nature* **220**:657–660.

POCOCK, R. I. 1925. The external characters of the catarrhine monkeys and apes. *Proc. Zool. Soc., Part 4* **1925**: 1479–1579.

PREUSCHOFT, H. 1973. Body posture and locomotion in some East African Miocene Dryopithecinae, pp. 13–46. *In* M. Day, ed., *Human Evolution*, Vol. 11, Symp. Soc. Study Human Biol.

RADINSKY, Leonard. 1973. *Aegyptopithecus* Endocasts: Oldest record of a pongid Brain. *Am. J. Phys. Anthropol.* **39**:239–248.

RADINSKY, Leonard. 1974. The fossil evidence of anthropoid brain evolution. *Am. J. Phys. Anthropol.* **41**:15–27.

REMANE, Adolf. 1960. Zahne und Gebiss, pp. 637–845. *In* H. Hofer, D. Starck, and A. Schultz, eds., *Primatologia*, Vol. 3, Pt. 2. Karger, Basel.

RIESENFELD, ALPHONSE. 1975. Volumetric determination of metatarsal robusticity in a few living primates and in the foot of *Oreopithecus*. *Primates* **16**:9–15.

ROMERO-HERRERA, A. E., H. LEHMANN, K. A. JOYSEY, and A. E. FRIDAY. 1973. Molecular evolution of myoglobin and the fossil record: a phylogenetic synthesis. *Nature.* **246**:389–395.

ROSE, M. D. 1974. Ischial tuberosities and ischial callosities. *Am. J. Phys. Anthropol.* **40**:375–383.

SCHAEFFER, Bobb, Max K. HECHT, and NILES ELDREDGE. 1972. Phylogeny and paleontology. *Evol. Biol.* **6**:31–46.

SCHLOSSER, Max. 1911. Beiträge zur Kentniss der Oligozänen Landsäugetiere aus dem Fayum: Aegyptem. *Beitr. Paläeont. Geol. Öst.-Ung,* **24**:51–167.

SCHÖN, M. A., and Linda K. ZIEMER. 1973. Wrist mechanism and locomotor behavior of *Dryopithecus (Proconsul) africanus*. *Folia Primatol.* **20**:1–11.

SCHULTZ, Adolph H. 1950. The physical distinctions of Man. *Proc. Am. Philos. Soc.* **94**:428–449.

SCHULTZ, Adolph H. 1968. The Recent hominoid primates, pp. 122–195. *In* S. L. Washburn and Phyllis C. Jay, eds., *Perspectives on Human Evolution*, Vol. 1. Holt, Rinehart and Winston, New York.

SCHULTZ, Adolph H. 1970. The comparative uniformity of the Cercopithecoidea, pp. 41–51. *In* John R. Napier and P. H. Napier, eds., *Old World Monkeys*. Academic Press, New York.

SCHULTZ, Adolph H. 1973. The skeleton of the Hylobatidae and other observations on their morphology. *Gibbon Siamang* **2**:1–54.

SIMONS, Elwyn L. 1959. An anthropoid frontal bone from the Fayum Oligocene of Egypt: the oldest skull fragment of a higher primate. *Am. Mus. Novit.* **1976**:1–16.

SIMONS, Elwyn L. 1960. *Apidium* and *Oreopithecus*. *Nature.* **186**:824–826.

SIMONS, Elwyn L. 1964. On the mandible of *Ramapithecus*. *Proc. Natl. Acad. Sci., U.S.A.* **51**:528–636.

SIMONS, Elwyn L. 1965. New fossil apes from Egypt and the initial differentiation of Hominoidea. *Nature* **205**:135–139.

SIMONS, Elwyn L. 1969. Miocene monkey (*Prohylobates*) from northern Egypt. *Nature* **223**:687–689.

SIMONS, Elwyn L. 1971. Relationships of *Amphipithecus* and *Oligopithecus*. *Nature* **232**:489–491.

SIMONS, Elwyn L. 1972. *Primate Evolution: An Introduction to Man's Place in Nature.* Macmillan, New York.

SIMONS, Elwyn L. 1976 (in press). Diversity among the early hominids: a vertebrate paleontologist's viewpoint. *In* C. J. Jolly, ed., *Early African Hominidae*. Duckworth, London.

SIMONS, Elwyn L., and J. FLEAGLE. 1973. The history of extinct gibbon-like primates. *Gibbon Siamang* **2**:121–148.

SIMPSON, George Gaylord. 1945. The principles of classification and a classification of mammals. *Bull. Am. Mus. Nat. Hist.* **85**:1–350.

SIMPSON, George Gaylord. 1961. *Principles of Animal Taxonomy.* Columbia University Press, New York.

SIMPSON, George Gaylord. 1963. The meaning of taxonomic statements, pp. 1–31. *In* S. L. Washburn, ed., *Classification and Human Evolution*. Aldine, Chicago.

SIMPSON, George Gaylord. 1971. Remarks on immunology and catarrhine classification. *Syst. Zool.* **20**:369–370.

ERIC
DELSON
AND
PETER
ANDREWS

STRAUS, William L., Jr. 1961. Primate taxonomy and *Oreopithecus*. *Science* **133**:760–761.

STRAUS, William L., Jr. 1963. The Classification of *Oreopithecus*, pp. 146–177. *In* S. L. Washburn, ed., *Classification and Human Evolution*. Aldine, Chicago.

SZALAY, Frederick S. 1970. Late Eocene *Amphipithecus* and the origins of catarrhine primates. *Nature* **227**:355–357.

SZALAY, Frederick S. 1972. *Amphipithecus* revisited. *Nature* **236**:179–180.

SZALAY, Frederick S. 1975 (in press). On haplorhine phylogeny and the status of the Anthropoidea. *In* R. Tuttle, ed., *Primate Functional Morphology and Evolution*. Mouton, Hague.

SZALAY, Frederick S., and Annalisa BERZI. 1973. Cranial anatomy of *Oreopithecus*. *Science* **180**:183–185.

TUTTLE, Russell. 1972. Functional and evolutionary biology of hylobatid hands and feet. *Gibbon Siamang* **1**:136–206.

TUTTLE, Russell. 1974. Darwin's apes, dental apes, and the descent of man: Normal science in evolutionary anthropology. *Curr. Anthropol.* **15**:389–398.

VAN HORN, Richard N. 1972. Structural adaptations to climbing in the gibbon hand. *Am. Anthropol.* **74**:326–334.

VERHEYEN, Walter N. 1962. Contribution à la craniologie comparée des primates. *Mus. R. Afr. Cent., Tervuren, Belg. Ann. Sér. 8VO, Sci. Zool.* **105**:1–246.

VOGEL, Christian. 1966. Morphologische Studien am Gesichtsschädel Catarrhiner Primaten. *Bibl. Primatol.* **4**:1–226.

VOGEL, Christian. 1968. The phylogenetic evaluations of some characters and some morphological trends in the evolution of the skull in catarrhine primates, pp. 21–55. *In* B. Chiarelli, ed., *Taxonomy and Phylogeny of Old World Primates with References to the Origin of Man*. Rosenberg and Sellier, Torino.

WALKER, Alan, and Peter ANDREWS. 1973. Reconstruction of the dental arcades of *Ramapithecus wickeri*. *Nature* **244**:313–314.

WALKER, Phillip, and Peter MURRAY. 1975 (in press). An assessment of masticatory efficiency in a series of anthropoid primates with special reference to the Colobinae and Cercopithecinae. *In* R. Tuttle, ed., *Primate Functional Morphology and Evolution*. Mouton, Hague.

ZAPFE, Helmuth. 1958. The skeleton of *Pliopithecus* (*Epipliopithecus*) *vindobonensis* Zapfe and Hürzeler. *Am. J. Phys. Anthropol.* **16**:441–457.

ZAPFE, Helmuth. 1960. Die Primatenfunde aus der miozänen Spaltenfüllung von Neudorf an der March (Devinska Nova Ves), Tschechoslowakei. *Mem. Suisse Paleontal.* **78**:5–293.

17

Parallelism, Brachiation, and Hominoid Phylogeny

RUSSELL TUTTLE

I. Introduction

A renaissance of comparative primatological studies occurred during the past 15 years. It flourishes unabated. It is characterized by intensive focuses on function and inferential phylogenies and a dazzling battery of new and refurbished gadgetry and methods ranging from electromyography and cineradiography to computers and multivariate statistics for many seasons. This, coupled with an outpouring of information on the behavior of captive and free-ranging primates, their biomolecular particularities, and a remarkable number of new fossils, promises to elucidate aspects of some primate careers in a refreshingly solid note. Hopefully man will be among this newly illuminated lot.

Control over questions of homology and the closely related phenomenon, parallelism, is fundamental to proper theoretic evolutionary applications of results from all empirical primatological studies. To the question on homology put by the organizers of this symposium, unfortunately one must initially respond that unequivocal evidence that particular osseous, ligamentous, and muscular features are indeed homologous must await breakthrough in biochemical genetics and other developmental sciences on the heritability and operative mechanisms of structural genes. During the prospectively long

Georgia.

interim, evolutionary morphologists should execute controlled experiments and systematic comparisons and utilize careful logic in the face of Murphy's Laws and Mother Nature's ill-bred pups.

This enterprise is now facilitated by indications of basic genetic relationships among primates through studies of their serum proteins, enzymes, chromosomal morphology, and structure of DNA. But we need fossils to document phylogeny and especially for time considerations and the discernment of relative divergences. The vacillation by advocates of biomolecular dating regarding exact times of furcation in most lineages is at least as telling on the uncertainties of their evolutionary applications, as, for instance, the changes of mind by Simons and Pilbeam over which dryopithecine species represents the protochimpanzee (Tuttle and Basmajian, 1974a, p. 343).

Discussions, or more commonly passing comments, on parallelism are numerous in the evolutionary primatological and anthropological literature. Here I will discuss questions of parallelism in anthropoid primates with special emphasis on certain features of the forelimb that traditionally have been associated with brachiation. I will begin with a brief discussion of definitions of parallelism, convergence, and homology by several authoritative evolutionary biologists. This is a necessary preliminary because the omission of stated operational definitions of parallelism has led to ambiguities and contributed to controversy in some anthropological literature.

I will conclude with a speculative phylogeny of the Hominoidea based chiefly on evidence from the postcranial anatomy of extant forms.

II. Parallelism

A. Review of Concepts

In *The Meaning of Evolution* (1949, pp. 181–183), Simpson compared parallelism with convergence and commented that there is no really fundamental difference between these two evolutionary phenomena. He presented no vertebrate examples of parallelism, but he cited the wings of pterodactyls, birds, and bats as convergent structures premised on homologous forelimbs. The wings of insects and birds exemplify convergence that is negligibly based on homology (p. 183). Simpson further noted that convergence on a grand scale is seen in the comparison of South and North American mammals (p. 184). This is illustrated by a suite of comparisons among marsupials, insectivores, rodents, and carnivores; liptoterns, camels, and horses; and notoungulates, rhinoceroses, and chalicotheres (pp. 178–179).

In *The Major Features of Evolution* (1953, pp. 170–171, 176–177) Simpson mentioned parallelism and convergence *passim* in discussions of adaptation but he did not specifically define them. In *Principles of Animal Taxonomy* (1961, pp. 78–79) he defined parallelism, convergence, and homology as follows:

> Parallelism is the development of similar characters separately in two or more lineages of common ancestry and on the basis of, or channeled, by characteristics of that ancestry.
> Convergence is the development of similar characters separately in two or more lineages

without a common ancestry pertinent to the similarity but involving adaptation to similar ecological status.

Homology is resemblance due to inheritance from a common ancestry.

Simpson (1961, p. 103) remarked that parallelism may be difficult or practically impossible to distinguish from homology on the one hand and convergence on the other. But he added that the distinction between homology and parallelism can frequently be made from the fact that homologous characters are primitive for the group but parallel characters are not (1961, p. 105). He noted that the distinction of parallelism from convergence is usually fairly easy if the convergence has not also been affected by community of ancestry, that is to say, if it does not really have an element of parallelism (1961, p. 106).

Applications of concepts of parallelism and convergence pervade Simpson's essays on the Mammalia, collected in *The Geography of Evolution* (1965). Generally, he considered aspects of the adaptive radiations in South America, Australia, and North America–Eurasia–Africa to exemplify convergent evolution (Simpson, 1965, pp. 84–85). He cited the predaceous marsupial carnivores of South America and Australia as examples of convergence in the parallel radiations of the two continents (Simpson, 1965, p. 117). Although Simpson (1965, pp. 117–118) once considered the porcupines of the Western Hemisphere to have evolved convergently with the African porcupines, in a later essay he modified this view. Therein he stated that

> The close resemblance of South American caviomorph rodents . . . with certain African rodents . . . is probably as much parallelism as convergence: they had different immediate ancestors in North America and Eurasia, respectively, but those ancestors were almost certainly themselves closely related. [Simpson, 1965, p. 232.]

Similarly, he concluded that the ceboid monkeys and certain Old World primates (presumably the cercopithecoid monkeys) evidenced parallelism as much as convergence (Simpson, 1965, p. 232).

Romer (1949, p. 115; 1966, pp. 3, 136) considered parallelism to be a very common evolutionary phenomenon. Unlike Simpson, he sharply distinguished parallelism from convergence (Romer, 1949, p. 115; 1966, p. 3). Thus he stated,

> In convergence, two forms with similarities in directly adaptive structures (as wolf and marsupial "wolf") have come from radically different ancestors with basically different patterns of organization; in Gregory's terminology, the habitus is similar, the heritage is different. In true parallelism, both habitus and heritage are similar; the ancestral types were closely related, and evolutionary progress, stage by stage, has been closely comparable in the two or more lines concerned. [Romer, 1949, p. 115.]

Bigelow (1958, p. 50) adopted Romer's definitions of parallel evolution and convergence. But he disagreed with Romer's sharp distinction between them. Hennig (1966, p. 117) also accepted Romer's definitions. But, unlike Bigelow, he did not contest the view that they are quite different phenomena. Hennig (1966, p. 117) remarked, "To parallelism in the broader sense belong above all those cases in which characters certainly absent in the stem species of a monophyletic group occur independently in the successor species."

Rensch (1960, p. 192) distinguished three forms of "parallel evolution":

(1) parallelisms due to similar hereditary factors, including cases of parallel mutations;
(2) parallelisms resulting from parallel selection acting on homologous structures or organs; and
(3) parallelisms caused by parallel selection affecting analogous structures and organs.

While stating that there are numerous examples of parallel evolution in which two or all three of these types are mixed, Rensch (1960, pp. 192, 198, 200) nevertheless drew a rather sharp contrast between parallelisms due to heredity and those due to selection. Rensch exemplified hereditary parallelism almost exclusively with invertebrate forms. In his discussion of parallelisms in consequence of parallel selection he relied more heavily upon examples from the vertebrates, including mammals.

Mayr (1963, p. 609) devoted only one paragraph to parallelism and convergence in his landmark treatise on *Animal Species and Evolution*. He discussed these phenomena more comprehensively in *Principles of Systematic Zoology* (1969). Like Romer, he considered parallelism and convergence to be very common phenomena (1969, p. 243). Mayr (1969, p. 202) stated that the similarity between two taxa may be composed of the following:

1. Similarities resulting from joint possession of characteristics shared with a common ancestor
   a. Ancestral characters (symplesiomorphs of Hennig) shared with a remote ancestor
   b. Derived characters (synapomorphs of Hennig) shared with a more recent ancestor
2. Similarities resulting from joint possession of independently acquired phenotypic characteristics produced by a shared genotype inherited from a common ancestor (similarity through parallel evolution)
3. Similarities resulting from joint possession of independently acquired phenotypic characteristics that are not produced by a genotype inherited from a common ancestor (similarity through convergence).

In summary, the concensus of authors surveyed here is that parallelism is the independent development of similar features in closely related taxa and is usually attributable to similar selective pressures acting on a common genetic base. In evolutionary biological literature, discussions generally focus on the morphological expression of parallelism. Passing comments and implicitness characterize most treatments of similarity in the selective complexes that are assumed to be causatively related to parallel features. Yet one of the vital components in the conceptualization of parallelism is selection. Thus, as we subsequently consider several primate lessons on parallelism, I will endeavor to give this aspect of the problem detailed attention.

B. An Operational Definition

Ideally, in order for parallelism to be ascribed, two or more natural, reproductively isolated, closely related groups should evidence basically similar, though independently derived, *adaptive complexes*. An adaptive complex is composed of (1) the morphological complex—a limited set of coadapted features which act together to produce one or more functions characteristic of the species*; (2) the genetic complex—structural genes upon which the morphological complex is premised and the

* If the terminology of Bock and von Wahlert (1965) were employed here, "features" would read *faculties* and "functions" would read *biological roles*.

mechanism whereby its components are developmentally coordinated; and (3) the selective complex—environmental features that acted with the genetic complex to produce the morphological complex. Basic similarities in all three constituents of the adaptive complexes of anthropoid forms should be evidenced if they are to be attributed to parallelism. As with convergence, primate species exhibiting true parallelism presumably would occupy fundamentally similar niches or at least those aspects of niche pertaining to the adaptive complexes in question. Herewith I employ the functional definition of niche advocated by Whittaker *et al.* (1973). Accordingly, presumed parallel adaptive complexes in two or more species should underlie their functioning in basically similar intracommunity roles, especially at the time that the adaptive complexes would be optimally effective.

III. Brachiation

The problem of brachiation, i.e., the extent to which selection for arm-swinging locomotion or other manual suspensory behaviors contributed to the human morphological condition, has nagged the anthropological community for three quarters of a century (Tuttle, 1974). Illumination and resolution of the problem are hampered by ambiguous conceptualizations of the activity or activities subsumed under "brachiation." This ambiguity is premised on inadequate naturalistic and experimental studies of primates designated brachiators. Changes of mind by established authors on the subject (e.g., Washburn, 1973; Ashton *et al.*, 1971), and arguments by other researchers have created a renewed sense of urgency regarding precise concepts of brachiation and other positional behaviors and concerning which primates are to be considered true brachiators versus adjectivally suspended or prefixed forms. During debates over the problem of brachiation, gibbons, siamangs, orangutans, chimpanzees, gorillas, man, ateline monkeys, alouattine monkeys, and colobine monkeys were designated "brachiators," "modified brachiators," "semibrachiators," or "brachiating primates" by one or more authors.

Despite the terminological flux, several modern authors have ascribed certain features in the locomotive systems of anthropoid primates to parallel evolution of brachiation. Some, exemplified hereinafter (p. 455) by Simons (1962, 1967) provided operational definitions of brachiation. Others, like Mayr, did not. Although Mayr (1969, p. 243) attempted to illustrate the "extremely widespread" nature of parallel evolution with a statement that *Pongo* and *Hylobates* acquired certain structural similarities in connection with a brachiating mode of life, he did not define the brachiating mode that presumably was implicated in the process.

A. Arboreal Positional Behavior

A survey of literature and personal observations evidence that anthropoid primates may variously employ their forelimbs in upraised positions in the following arboreal situations:

1. As sole propellant organs to traverse horizontal distances in free flight between a superstrate and a landing site.
2. As sole or principal propellant organs for swinging along beneath superstrates.
3. During transfers, to test and to secure positions in the terminal branches of adjacent canopy while retaining grips on the base structure with feet or tail (if prehensile) or both.
4. Moving quadrupedally while suspended beneath superstrates.
5. As sole supporting organs prior to vertical drops.
6. In combination with posterior appendages for suspension beneath a superstrate prior to vertical drops.
7. As principal or sole supporting organs upon first contact in landing sites.
8. As prominent supporting organs in combination with the hindlimbs upon initial contact with landing sites.
9. As prominent propellant organs to haul themselves upwards on vertically or obliquely disposed branches or from a base branch directly on to a higher one.
10. As prominent supporting or propellant organs while climbing tree trunks, vines, and other vertical supports.
11. As sole supporting organs while feeding and foraging beneath superstrates.
12. In combination with posterior appendages while feeding and foraging beneath superstrates.
13. While reaching for food or nesting materials from sitting or bipedal postures atop branches.
14. To steady themselves as they move bipedally atop a branch.
15. As sole or prominent suspensory supporting organs during play, display, and other nonmaintenance activities.

Different combinations of these activities characterize the locomotive and postural repertoires of each extant anthropoid species. I will now briefly sketch the extent to which these positional behaviors characterize hylobatid apes, orangutans, chimpanzees, gorillas, spider monkeys, and black-and-white colobus monkeys. This will serve as a background against which we may discern whether forelimb elongation, pollical reduction, and wrist structure in the Hominoidea and certain monkeys evidence parallel evolution in response to similar selective complexes for brachiation. It will also highlight some of the gaps in our knowledge of positional behavior in nonhuman primates.

Gibbons, chimpanzees and spider monkeys have been reported by field observers to traverse notable horizontal distances in free flight by arm-swinging from a superstrate to a landing site. The mechanism commonly employed by hylobatid apes to effect powerful forelimb propulsion, which I term ricochetal arm-swinging (Tuttle, 1969a), appears to be significantly different from those of other arm-swinging primates. Gibbons and siamangs propel themselves forwards and upwards by simultaneous retraction of the shoulder and flexion of the elbow joints (Tuttle, 1969a, 1972a) while other arm-swinging primates appear simply to swing beneath a branch and then release it at the upper end of a forward trajectory. Some elbow flexion may be employed by chimpanzees and perhaps also by ateline monkeys to gain momentum, but only the hylobatid apes possess a highly complex multiple joint muscle mechanism that may be associated with ricochetal arm-swinging (Tuttle, 1969a, 1972a).

Arm-swinging along branches or between juxtaposed networks of limbs and vines or both have been observed in gibbons, siamangs, chimpanzees, spider monkeys, orangutans, and young gorillas. A principal difference between modes of arm-swinging beneath branches in *Ateles* by contrast with the apes is the regular employment of the prehensile tail by *Ateles* (Carpenter and Durham, 1969). The consensus of observations is that arm-swinging is an uncommon or rare mode of locomotion among orangutans and gorillas. It is employed somewhat more frequently by chimpanzees though they apparently prefer to knuckle-walk on the ground or large branches. Spider monkeys may arm-swing more frequently than chimpanzees; but the impressionistic nature of available accounts of their naturalistic locomotion (Carpenter, 1935; Erikson, 1963; Eisenberg and Kuehn, 1966) makes this comparison problematic. Richard's (1970) quantitative study of *Ateles geoffroyi* on Barro Colorado Island is virtually useless for our present purpose because she only presents data on brachiation [defined strictly after Napier (1963) as the subject "propelling itself along using its arms only" (p. 252)] and a category termed "swing and grasp." Her scheme does not distinguish between tail-assisted arm-swinging (which in fact characterizes ateline monkeys more than brachiation of a hominoid sort does) and their bridging and other transfer behaviors.

Reports in secondary sources of arm-swinging by black-and-white colobus monkeys remain unconfirmed by my survey of primary literature and augmentary observations of the species in Tanzania. Further, Morbeck (1975) did not observe arm-swinging by free-ranging guerezas in Kenya. Her study, like mine, was focused on the locomotive and postural behavior of the monkeys.

Transfers between adjacent areas of small flexible branches, wherein the posterior appendages grasp base supports and a forelimb reaches ahead carefully to test and to secure new vantage points, have been reported for orangutans, chimpanzees, and spider monkeys. Among apes, transferring is most common as a mode of progression in the orangutan. Gibbons, spider monkeys, and orangutans have been observed to bridge gaps in the canopy with their bodies while young animals climb over them.

Although quadrupedal suspensory posturing beneath superstrates probably occurs sporadically in other species, it has been reported specifically only in the orangutan.

Hanging by the forelimbs alone during descents to nearby lower branches or to the ground has been reported in chimpanzees, gorillas, and black-and-white colobus monkeys. Orangutans, hylobatid apes, and spider monkeys probably also sporadically descend from branch to branch in this manner.

Vertical drops over extensive distances have been reported in ateline monkeys but the exact nature of their appendicular contacts prior to release of the base support have not been described. However, it is likely that the forelimbs, in various combinations with the tail, may be predominant supporting appendages prior to some drops.

Momentary suspension solely or principally from the forelimbs immediately after landing from a swing, leap, or drop has been described for chimpanzees and black-and-white colobus monkeys. It probably also occurs in hylobatid apes and ateline monkeys.

I noted that hauling actions of the forelimbs are conspicuous constituents of the locomotive repertoire of black-and-white colobus monkeys, especially when they progress in the periphery of trees. Gibbons, chimpanzees, orangutans, and spider monkeys may also engage in hauling actions subsequent to leaps or transfers into new supports but this behavior has not been specifically ascribed to them.

Pulling actions by the forelimbs are employed by gibbons, orangutans, chimpanzees, gorillas, guerezas, and spider monkeys when climbing tree trunks or vines. Although some field researchers have attributed special prominence to forelimb pulling actions in orangutans and spider monkeys when climbing vertical substrates, available reports do not permit clear assessments and comparisons of the relative importance of forelimbs versus hindlimbs as propellant organs during climbing in any species.

Gibbons, siamangs, orangutans, and chimpanzees have been observed hanging by one forelimb and drawing foods to mouth with the free hand. This is a frequent, if not the predominant, feeding posture in the hylobatid apes, but it is much less common in chimpanzees and orangutans. During feeding while suspended from one forelimb, the apes commonly grasp nearby branches with their feet though these actions often appear to provide little substantial augmentary support except perhaps to inhibit rotation about the supportive limb. Gibbons and orangutans also hang from a synergistic combination of one forelimb and one or more hindlimbs while bringing foods to mouth with the free hand. Spider monkeys feed while hanging from several synergistic combinations of forelimbs, hindlimbs, and tail or by the tail alone. I never saw black-and-white colobus monkeys pluck food while suspended by the forelimbs. Instead they assumed a wide variety of sitting, prone, and occasionally bipedal postures atop limbs as they drew in and plucked sprigs of leaves.

While feeding in sitting or bipedally standing positions, hylobatid apes, chimpanzees, orangutans, and geurezas employ their hands in positions above the head either to reach distant food-bearing branches or to steady themselves with grips on overhanging limbs.

During bipedal progression along branches, chimpanzees sometimes grasp overhanging branches. This action probably occurs also in hylobatid apes and orangutans, but such employment has not been described specifically in reports on the Asian apes.

Suspensory play postures that implicate the forelimbs have been described in gibbons, orangutans, chimpanzees, and gorillas and undoubtedly occur also in spider monkeys.

Until recently, one category of suspensory forelimb posturing, display swinging, had not received the attention that its prospective role in the adaptive complex of certain arm-swinging species might merit. Spider monkeys have been observed to hang by synergistic combinations of cheiridial and tail holds as they shake branches and otherwise engage in noisy reactions to people. But the most dramatic development of display swinging occurs in gibbons and siamangs. They frequently engage in ricochetal arm-swinging during altercations with conspecifics (Tuttle, 1972a).

In brief, extensive, rapid, bimanual suspensory progression is only characteristic on one lineage, the hylobatid apes. The "rapid brachiation" that has been attributed to chimpanzees and spider monkeys is probably vertical dropping and leaping between supports instead of true ricochetal arm-swinging of the hylobatid sort. Suspensory employment of the forelimbs, other than in rapid brachiation, commonly occurs in orangutans, chimpanzees, hylobatid apes, and spider monkeys. Suspensory postures are quite rare in guerezas and gorillas.

B. Forelimb Elongation

Elongate forelimbs extend the reach of animals that possess them. This is particularly advantageous during arboreal foraging and feeding (Grand, 1972), bridging or other transfers in the canopy, catching landing sites after leaps and drops, and grasping large tree trunks during vertical climbs. Further, depending upon the relative lengths of arm to forearm and the development and attachments of the forelimb muscles, mechanical advantage may be gained by selective elongation of forelimb segments (Keith, 1926; Tuttle and Basmajian, 1974b). This would be useful in the employment of forelimbs as propellant organs during ricochetal and other rapid arm-swinging, and perhaps also when hauling movements are employed in vertical climbs and when hoisting to positions atop branches, especially subsequent to landings in springy foliage.

Simons (1962, p. 292) proposed that forelimb elongation developed independently "at least 3, and possibly as many as 6 times" among the Anthropoidea. He later revised the lower limit of his expectance to "at least five times" (Simons, 1967, p. 241). He considered the lineages culminating in the gibbons (including siamang), African apes (chimpanzee and gorilla), orangutan, spider monkey, and *Oreopithecus bambolii* to exemplify true parallel evolution of forelimb elongation as part of their adaptations for brachiation. He defined brachiation as arm-swinging or bimanualism (Simons, 1962, p. 291) and later expressed that brachiation is one of the few available solutions to the problem of rapid arboreal progression in a relatively large animal (Simons, 1967, p. 241).

Now, in regard to the criterion noted previously (p. 450), we may query whether the adaptive complexes of these 5 taxa evidence sufficient basic similarity to merit the ascription "true parallelisms" (Simons, 1967, p. 241). At the outset, Simons' case is premised on demonstration that forelimb elongation characterizes the 5 taxa in his scheme and that it was independently evolved in each of them. He cited the relatively high values (greater than 100) of the intermembral indices as evidence for forelimb elongations (Simons, 1962). He did not present figures or references to support this factual point; however, that the living forms in his sample possess high intermembral indices and other evidence of elongate forelimbs has been documented abundantly by Mollison (1910), Schultz (1930, 1937, 1953, 1956, 1973a), and Erikson (1963). Straus (1963) determined the intermembral index of the 1958 skeleton of *Oreopithecus* to be approximately 119, a value within the range of variation of the gorilla alone among hominoid primates.

We may reasonably assume that the long bones implicated in forelimb elongation are homologous [*sensu* Romer (1970, p. 9), Rensch (1960, p. 192), and Mayr (1969, p. 85)] between all forms in Simons' example. If biomolecular features provide an adequate basis for inferences about structural genes (but compare views of Goodman and Kohne, both this volume), we may assume that the genetic complexes related to forelimb elongation are similar in the extant apes. *Ateles* and *Oreopithecus* are somewhat problematic in this regard; the former because of its clear-cut genetic distance from the apes in all molecular features examined to date, and the latter because it is an enigmatic fossil (Straus, 1963; Delson and Andrews, this volume). Further, the extent to which the humerus vs. bones of the forearm (radius and ulna) contribute to overall forelimb elongation differs among the various apes, *Ateles*, and *Oreopithecus*. This variation, expressed in the brachial index, does not coincide exactly with the distribution of variations in the intermembral indices of the same forms.

In the extant species, values of the brachial index are highest in the most frequent arm-swingers—gibbon and siamang—followed by spider monkey, orangutan, chimpanzee, and gorilla [based on Schultz (1937, p. 300) and Erikson (1963, p. 145]. Straus (1963) determined the brachial index of *Oreopithecus* to be approximately 95, a value similar to those of chimpanzee and orangutan (particularly the mean value of the former). *Per contra*, siamang and orangutan exhibit the highest intermembral indices, followed by gibbon, gorilla (and *Oreopithecus*), chimpanzees, and spider monkey [based on Schultz (1937, p. 298) and Erikson (1963, p. 145)]. Erikson (1963, p. 149) attempted to eliminate the obfuscating factor of intertaxonal variations of hindlimb length in the intermembral index by constructing proportion diagrams based on constant trunk length. He then established the following sequence of relative forelimb length: gibbon (184),* orangutan (148), spider monkey (128), gorilla (121), and chimpanzee (119).

Fossil evidence is scant although suggestive regarding the question of independent acquisition of forelimb elongation in spider monkeys, *Oreopithecus*, gibbons, orangutans, and the African apes. Students of possible Paleogene precursors of the ceboid monkeys have not discerned that any of them possessed elongate forelimbs or other indications of "brachiation." Thus, the ateline monkeys probably developed this feature *de novo* in the New World. *Oreopithecus* represents the earliest clear indication of forelimb elongation among anthropoid primates. Its dentition is remarkably unique. It represents the culmination of an ancient distinct lineage (Hürzeler, 1968; Simons, 1960, 1972; Straus, 1963; Delson and Andrews, this volume).

The most likely candidates for precursors of hylobatid apes, *Limnopithecus macinnesi* and *Pliopithecus vindobonensis*, do not evidence forelimb elongation (Clark and Thomas, 1951; Zapfe, 1960; Tuttle, 1972a). They are dentally distinct from presumed contemporary precursors of the pongid apes (Andrews, 1974; Delson and Andrews, this volume). Because of the absence of hindlimb and vertebral remains, relative forelimb length cannot be determined in *Dryopithecus* (*Proconsul*) *africanus*, the only species among possible antecedents of the extant pongid apes for which relatively complete forelimb remains are known. However, its brachial index is low (86) by comparison with many other anthropoid primates (Napier and Davis, 1959, p. 38).

No Neogene postcranial fossils attributable to protoorangutans are available. Thus the question of their independent acquisition of forelimb elongation remains open. Some biomolecular studies evidence that orangutans are rather more distinctive from the African apes than some morphologists had maintained they were. But the time and extent of their divergence have not been established. A distinct possibility remains that

---

* Figures in parentheses are mean values.

elongate forelimbs characterized the common pongid ancestor of orangutans and the African apes, especially if the furcation occurred sometime between 10 and 20 million years ago.

In sum, Simons' case for parallel evolution of forelimb elongation in the apes appears to rest on relatively solid foundations regarding basic similarity in the morphological complex and probably also, although somewhat more equivocally, in the genetic complex. The morphological criterion is most likely met also by *Ateles* and *Oreopithecus*. But basic similarity between their genetic complexes and those of apes is disputable, at least on logical grounds. However, one may surmise that if a case could be made for sufficient basic genetic similarity between *Ateles* and the apes, it probably follows that *Oreopithecus* must have possessed even closer similarity to the apes in the genetic mechanisms related to forelimb elongation.

That *Ateles*, *Oreopithecus*, and, to a lesser extent, the hylobatid apes acquired forelimb elongation independently from each other and from the pongid apes is fairly assured. The case for independent acquisition of this feature by orangutan and the African apes after their divergence from common ancestors is much more problematic and doubtful but not at all implausible.

Discerning degrees of resemblance among the selective complexes of the 5 taxa is nearly as equivocal as determining degrees of similarity in their genetic complexes and the number of independent acquisitions. But available information on the naturalistic behavior of the apes and spider monkeys is sufficient to deny the inference that rapid brachiating progression was an important component in the selective complexes of the African apes and, to a lesser extent, orangutans and spider monkeys.

What indeed might underlie the elongate forelimbs of all 5 taxa in Simons' model is a heritage of feeding on fruits, new leaves, buds, and flowers in the periphery of trees. Except for gorilla, this is evidenced by current practice in all extant forms that he considered. Whether seated on a branch or suspended beneath it, a long reach would be advantageous. Further, climbing about in the less stable regions of the canopy would be facilitated by a capacity to reach somewhat distant supports, especially in relatively large-bodied animals. Once elongate forelimbs has been acquired in response to these selective influences, the more eye-catching suspensory behaviors could develop in some species. I have sketched one such hypothetical sequence for the hylobatid apes (Tuttle, 1972*a*).

Thus, Simons' model would probably gain greater credibility if we substituted a concept of terminal-branch foraging, feeding, and transferring for rapid brachiation. But there still remain many unresolved tensions in this scheme concerning whether the African apes evolved elongate forelimbs independently from orangutans and the latter also from the hylobatid apes.

C. Pollical Reduction

Controversy over the phylogenetic significance of pollical reduction in anthropoid primates has occurred *pari passu* with debates on the problem of brachiation. In order to deny a notable heritage of brachiation in the hominid lineage some nonbrachiationists dichotomized pollical development in the Hominoidea, stating that thumbs are reduced in all apes but are well developed in man. This led to misinterpretations of the hylobatid thumb which is in fact remarkably well developed (Tuttle, 1969a, 1972a, 1974; Van Horn, 1972). Further, once pollical reduction was linked to brachiation it contributed to the premature characterization of guerezas and other colobine monkeys and ateline monkeys as brachiators.

Among primates, the most extreme reduction of the pollex occurs in guerezas (*Colobus, sensu lato*) and *Ateles*. External morphological expression of this feature is remarkably similar in the two groups. They commonly lack an external thumb although some *Colobus* and *Ateles* evidence a small nailless nubbin (Straus, 1942). Internally the pollices of guerezas and spider monkeys are generally characterized by rudimentariness or complete loss of terminal and sometimes also proximal phalanges. The pollical metacarpal bones are always present although small (Straus, 1942).

Tendons of the extrinsic pollical muscles (flexor digitorum profundus and extensor pollicis longus) are usually absent or extremely vestigial. Development of the intrinsic pollical muscles varies from clearly differentiated in some *Ateles* to quite undifferentiated in other *Ateles* and *Colobus*. Straus (1942) found that a specimen of *Ateles* with some external development of the thumb also possessed the most clearly identifiable thenar musculature. Hill (1962) noted the intrinsic thenar muscles to be undifferentiated in *Ateles*. The relative mass of all intrinsic thumb muscles is strikingly small in guerezas by comparison with other catarrhine primates (Fig. 1) (Tuttle, 1972b). Data are not available on relative mass of the pollical muscles in *Ateles*.

The great apes (*Pongo pygmaeus, Pan troglodytes, Pan paniscus,* and *Pan gorilla*) normally possess a nailed pollex with a full complement of bones. While their external thumbs might appear to be short, they are not in fact inordinately undersized by comparison with most other anthropoid primates (Schultz, 1936, 1956; Straus, 1942). That reduction has occurred is most clearly evidenced in development of the pollical long flexor tendon. It is nearly always totally absent in orangutans and is characteristically absent or vestigial in chimpanzees and gorillas, especially by comparison with its development in man, hylobatid apes, and many cercopithecine monkeys (Straus, 1942; Tuttle, 1970).

The pollical long extensor tendon is normally present in all great apes. Furthermore, some gorillas possess an extensor pollicis brevis tendon which attaches to the base of the proximal phalanx thus mimicking the human condition. Although the intrinsic pollical muscles are remarkably differentiated in the Pongidae and are even characterized by the development of additional fasciculi, especially in *Pongo* (Day and Napier, 1963; Tuttle and Rogers, 1966; Tuttle, 1969b, 1970), their relative mass is significantly smaller than that of man, hylobatid apes, and many cercopithecine monkeys (Fig. 1) (Tuttle, 1972b, p. 280). This is probably chiefly attributable to smallness of muscles in the thenar eminence; the adductor pollicus muscle is well developed in the Pongidae, especially in *Pan* (Tuttle, 1972b).

In brief, available evidence indicates that the morphological expression of reduced pollices is strikingly similar in *Colobus* and *Ateles. Per contra*, pollical reduction in the Pongidae only resembles that of *Colobus* and *Ateles* in rudimentariness or absence of the

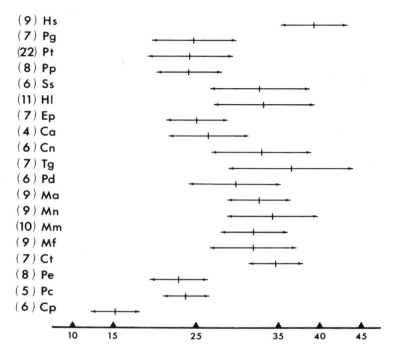

FIG. 1. Relative weight of total intrinsic thumb muscles expressed as a percentage of total intrinsic hand muscles in catarrhine primates. Vertical bars represent mean values and double-tipped arrows the 90% limits of populations. Numbers of specimens are indicated between parentheses beside each species. Hs, *Homo sapiens*; Pg, *Pan gorilla*; Pt, *Pan troglodytes*; Pp, *Pongo pygmaeus*; Ss, *Symphalangus syndactylus*; Hl, *Hylobates lar*; Ep, *Erythrocebus patas*; Ca, *Cercopithecus aethiops*; Cn, *Cercopithecus nictitans*; Tg, *Theropithecus gelada*; Pd, *Papio doguera*; Ma, *Macaca arctoides*; Mn, *Macaca nemestrina*; Mm, *Macaca mulatta*; Mf, *Macaca fascicularis*; Ct, *Cercocebus torquatus*; Pe, *Presbytis entellus*; Pc, *Presbytis cristatus*; Cp, *Colobus polykomos*.

long flexor tendon. Degrees of similarity in the genetic complexes underlying pollical reduction in the three groups cannot be determined now. The genetic mechanism(s) underlying pollical reduction could be quite simple, like certain deformities of the human hand (lobster claw and polydactyly) which are controlled by single genes with variable expression in affected individuals (Stern, 1973, p. 385). On the other hand, more complex mechanisms like multiple allelism or polygenetic inheritance could be involved as might be true of hallucal reduction in *Pongo* (Tuttle and Rogers, 1966).

We may fairly assuredly postulate that reduction of pollical structures occurred independently at least three times, that is to say, in *Ateles, Colobus* (*sensu lato*), and the Pongidae. But whether *Pongo* and *Pan* evolved from common ancestors with reduced long flexor tendons or developed this feature in parallel is debatable.

The selective complexes related to pollical reduction are probably quite dissimilar between *Colobus* and the Pongidae and perhaps also at least in part between *Ateles* and the other two taxa.★

★ I originally made this suggestion at Burg Wartenstein Symposium No. 48. It is recorded as part of the proceedings (July 19, 1970). The tape is the property of the Wenner–Gren Foundation for Anthropological Research, Inc., New York.

Pollical reduction in guerezas cannot be attributed to brachiation because, notwithstanding statements by Pocock (1925:1514), Straus (1942) and others to the contrary, they do not engage in arm-swinging locomotion or suspensory feeding to a notable extent. Instead I have suggested that a hooklike grasp, employing manual digits II–V may be of special importance to colobus monkeys while running on contiguous leafy substrates and while regaining vantage points atop limbs subsequent to leaps onto pliable substrates. Prominent thumbs could be a hindrance during these hauling actions unless they were equipped with special mechanisms for quick release, in line with digits II–V, and the avoidance of snags (Tuttle, 1972b, p. 290).

In the montane forest of Arusha National Park, guerezas move rather heavily over areas of the canopy where leafy or denuded springy branches are contiguous. The flexibility of their shoulder joints and vertebral columns seem to be particularly important in adjusting to springy substrates. During rapid movement along a stout horizontal branch or in either direction atop obliquely disposed stout branches, they bound forward with flexion and extension of the vertebral column more notable than that observed in Sykes' monkeys and baboons in similar contexts. Running guerezas propel themselves forward by extension of the hindlimbs together, while concurrently reaching ahead with both extended forelimbs. When a guereza moves up an inclined branch, the forelimbs exert propellant forces by flexing together subsequent to contact with the substrate.

Guerezas use their forelimbs to haul themselves over horizontal springy surfaces and upwards along moderately inclined gradients of flexible branches. On several occasions when they moved rapidly upwards on modestly inclined networks of small branches, their forelimbs seemed to be especially prominent (sometimes nearly equal to the hindlimbs) in effecting propulsive strokes. But the forelimbs never appeared to supersede the hindlimbs as primary propellant organs in that context.

Stern and Oxnard (1973) also believe that pollical reduction is a response to different selective complexes acting on apes, ateline monkeys, and colobine monkeys. But they stress that the thumb would be a hindrance during "the most distinctive rapid locomotion practiced by all colobines, i.e., tremendous leaps from tree to tree," and especially "in grasping branches at the end of a giant leap" (Stern and Oxnard, 1973, p. 61). My field observations of gray langurs (Presbytis entellus) in Ceylon and guerezas in Tanzania indicate that these statements are gross overgeneralizations of the locomotor modes of colobine monkeys.

Even when they fled from me at a rapid pace, black-and-white colobus monkeys never executed dramatic leaps of the sort that I have observed in langurs. Out of 78 leaps downwards or along approximately horizontal trajectories, 51 traversed 10 feet or less, 26 between 10 and 15 feet, and only 1 (executed by a juvenile animal) between 25 and 30 feet. Generally, the longer the leap, the greater the tendency for footholds to be concurrent with or to precede handholds, although the manner in which the monkeys struck landing sites made it difficult to discern the existence of subtle differences between hand and foot contacts. Morbeck (1975) also found that the amount of leaping displayed by guerezas in Kenya is insufficient to confirm the Stern–Oxnard explanation of pollical reduction.

It is also noteworthy that pollical reduction is less advanced in the most dramatic colobine leapers, i.e., the langurs, than in guerezas. Furthermore other cercopithecoid monkeys, e.g., vervets and patas monkeys, evidence rather diminutive intrinsic thenar musculature (Tuttle, 1972b, pp. 280–282), yet they are not exceptional arboreal acrobats in comparison with close congeneric species and the langurs.

D. The Wrist

Although it has been generally recognized that hominoid wrist joints are distinct from those of other catarrhine primates, the nature of the distinction was detailed only recently by Lewis (1965 et seq.). He discovered that a new synovial inferior radioulnar joint has developed in the Hominoidea concomitant with retreat of the ulna from the proximal row of carpal bones. This new joint is characterized by a neomorphic head of the ulna and a triangular articular disk against which the ulnar head articulates inferiorly. The primitive ulnar head is retained as the ulnar styloid process. It articulates with a meniscus in the apes and with a homolog of the meniscus in man (Lewis, 1965). Lewis originally explained that ontogenetically mere regression of the ulna leads to the appearance of a meniscus; thus it developed quite simply.

He concluded that his observations "demonstrate that some progress, at least, towards a diarthroidal inferior radioulnar joint had occurred more than once among the primates, yielding yet another example of the frequency of parallel evolution" (Lewis, 1965, p. 284). He did not specify which of the three hominoid forms in his study (Hylobates lar, Pan troglodytes, Homo sapiens) had developed this complex of features in parallel.

Subsequently, Lewis (1969) studied a more comprehensive sample of hominoid wrists, including an orangutan and mountain and lowland gorillas. He also discovered that the human intraarticular meniscus has a more complex developmental history than that which he had postulated in 1965 (Lewis, 1972b, p. 216). Noting that hominoid wrists possess enhanced capacities for pronation and supination and adopting Avis' (1962) definition of brachiation as a particular form of arm-swinging locomotion, he attributed the evolution of the hominoid wrist to selection for brachiation. He now rejected the hypothesis on parallelism and suggested instead that "Since an attempted brachiating habit in monkeys (viz., Ateles★) has apparently not effected wrist joint modifications similar to those seen in the Hominoidea, it seems likely that the similarities between the living members of the latter are the result of true genetic affinity rather than of parallelism." (Lewis, 1969, p. 264). He proposed that the extant Hominoidea evidence a graduated series ranging from Hylobates, which possesses the least advanced exclusion of the ulna from the wrist joint, through Pan, Homo, and Gorilla to Pongo, which possesses the most advanced condition. Finally, he commented that "the published description of Proconsul africanus . . . leaves little doubt that this species had not acquired the morphological features characteristic of the hominoid wrist; this would seem to preclude it from the place in hominid ancestry suggested for it . . ." (1969, pp. 265–266).

★ Author's note.

But shortly Lewis experienced another change of mind. Upon viewing casts of the wrist bones of *Dryopithecus* (*Proconsul*) *africanus*, he began to champion it as the common advanced brachiating ancestor of *Pan*, *Gorilla*, and *Homo* (Lewis, 1971*a*, 1972*a*,*b*). He determined that the meniscus-containing wrist joint of *Dryopithecus africanus* was more advanced than that of the extant Hylobatidae because in the lesser apes the ulnar carpal bones (triquetral and pisiform) are quite monkey-like (1971*a*). Further, Lewis (1972*a*) described certain features of the midcarpal joint (relating to shapes and relationships of the hamate, triquetral, capitate, lunate, scaphoid and centrale bones) which clearly distinguish extant hominoid primates and presumably also *Dryopithecus africanus* from monkeys. He related some of these features to a special locking mechanism during extension of the wrist as it is employed for brachiation. However, he noted that gibbons lack the special midcarpal locking mechanism. Since *Dryopithecus africanus* evidenced it Lewis' (1972*a*) contention that the Hylobatidae diverged prior to achievement of the *Dryopithecus* state was further supported.

Lewis' (1972*a*,*b*) statements on the position of *Pongo* in his model are rather puzzling. He stated that "the osteological evidence points to the origin of the orangutan line from a structural grade approximating more closely to, but still inferior to, that of *D. africanus*" (1972*a*, p. 56), but he then remarked that the special midcarpal joint mechanism is quite well developed and is perhaps even more effective in *Pongo* than it was in *Dryopithecus africanus* (1972*a*, p. 56). He favored development of the midcarpal joint mechanism *de novo* in *Pongo* because "there appears little doubt . . . that the Asiatic apes diverged early from the brachiating lineage leading to *Pan* and *Gorilla*, and that their locomotor behavior developed its own special characteristics" (1974, pp. 159–161).

Per contra, Lewis (1972*a*) employed another feature—early ontogenetic fusion of the os centrale to the scaphoid bone—to deny parallelism in *Pan*, *Gorilla*, and *Homo*. He explained the union of the two bones as an adaptation to the "tensional forces" of brachiation, particularly as part of the special midcarpal locking mechanism (1974). Fusion had to occur at a time when members of the lineage shared more common locomotive patterns than are evidenced by the three extant forms. Since *Dryopithecus africanus* probably possessed an unfused os centrale, the common ancestors of *Pan*, *Gorilla*, and *Homo* must have been more advanced brachiators than *Dryopithecus africanus* was (Lewis, 1972*a*, 1974).

In brief, Lewis' current model states that the Hominoidea are a monophyletic group in which retreat of the ulna and concomitant changes of the ulnar carpal bones in response to selection for brachiation constituted a novel base for other changes in the wrist pertaining to suspensory behavior and exploitation of resources in the small flexible branches of the forest canopy (1972*b*, 1974). The Asian apes diverged early. Even though the Hylobatidae are the brachiators *par excellence*, they retained many primitive carpal features because ricochetal arm-swinging probably requires some limitation on rotation of the wrist (Lewis, 1972*b*, p. 211). Some or most of the similarities between orangutans, on the one hand, and the African apes and man, on the other, in advanced features of the wrist are perhaps attributable to parallelism. Advanced features of the wrist unique to *Pongo* or the Hylobatidae are to be considered novel adaptations for their distinctive repertoires of suspensory behavior. The extant African pongids and the hominids, inclusive of *Australopithecus* (Lewis, 1973), are closely linked by a common history of advanced brachiation. *Dryopithecus africanus* represents a somewhat less derived wrist morphology than later members of the African pongid–hominid stem. But *D. africanus* was somewhat more advanced in wrist morphology than the lineage of *Pongo* and even more advanced than the stem Hylobatidae.

Several aspects of Lewis' model have been cogently challenged by researchers with divers theoretic inclinations (Conroy and Fleagle, 1972; Schön and Ziemer, 1973; Zwell and Conroy, 1973; Tuttle and Basmajian, 1974a, Tuttle, 1975; Jenkins and Fleagle, 1975). In particular, those who conducted studies on additional collections of primate wrist bones and performed experiments discovered that some of Lewis' morphological features are not always distributed among the anthropoid primates as he stated and that his functional interpretations of certain features are equivocal or spurious: For instance, Schön and Ziemer (1973) observed a notable spiral facet where the triquetral bone articulates with the hamate bone in *Alouatta* and *Ateles*. They concluded that notwithstanding Lewis' (1972a) claim to the contrary the spiral facet in these ceboid monkeys more closely resembles that of *Dryopithecus africanus* than those of the four extant apes do. Similarly, Schön and Ziemer (1973) were unable to confirm that the extant apes more closely resemble *Dryopithecus africanus* than *Alouatta* and especially *Ateles* do in the configuration of the midcarpal capitular facet and "waisting" of the capitate bone. Lewis (1972a, 1973, 1974) related these features to strengthening the midcarpal joint during brachiation. Jenkins and Fleagle (1975) also reported the presence of the spiral hamate facet, "waisted" capitate, and dorsal orientation of the centrale facet on the capitate in *Macaca*.

In addition, Lewis has persistently slighted evidence (Tuttle, 1967, 1969a,b,c, 1970) that the Asian apes possess remarkable capacities for dorsiflexion (also termed extension) of the wrist and that among apes, only the chimpanzee and gorilla consistently exhibit marked limitations of carpal dorsiflexion. Since suspensory behaviors constitute more fundamental and frequent aspects of the maintenance activities in the Asian apes than in the African apes it is requisite that we seek alternative explanations for the functional and adaptive significance of limited wrist extension in the African apes (Tuttle, 1967 *et seq.*; Conroy and Fleagle, 1972; Jenkins and Fleagle, 1975).

Upon overview, I must unfortunately register skepticism that Lewis has presented adequate evidence to evaluate questions of common heritage versus parallelism in the wrists of hylobatid apes, orangutans, African apes, and man. His most compelling argument pertains to retreat of the ulna. But this feature occurs also in lorisine primates (Cartmill and Milton, 1974). Thus, much more detailed embryological studies would seem to be needed to document common vs. parallel developments of this and related carpal features.

Further, more comprehensive studies on the entire carpus in large samples of primates, including the quantification of presumed functionally diagnostic features and radiographic studies of living subjects, must be conducted before we can assuredly resolve carpal conundrums in important fossils like *Dryopithecus africanus* and *Pliopithecus vindobonensis*.

Finally, Lewis' (1969, 1971a) initial functional and evolutionary interpretations were greatly hampered by his adherence to a narrow concept of "brachiation." We may surely applaud his subsequent modifications of this concept to encompass feeding and other suspensory behavior (Lewis, 1971*b*, 1972*a,b*, 1974). But this should induce a very thorough reconsideration of the entire model, including both its morphological and its phylogenetic aspects. Indeed we might discover that retreat of the ulna was basic to the stem Hominoidea and that it evolved primarily in response to selection for suspensory feeding and vertical climbing behavior that required flexible wrists and especially capacities for adduction and extension as well as pronation and supination. I also surmise that the lineage leading to orangutan might be found to have had a more intimate common heritage of suspensory behavior with the stem leading to the African apes and man than Lewis postulated.

E. Overview and Conclusions

Invocations of brachiation with narrow connotations of rapid arm-swinging loco-motion, particularly of the hylobatid sort, do not properly explain many of the presumed or documented morphological similarities between extant apes, monkeys, and fossil primates that have been designated brachiators. When my criterion for basic similarity of the adaptive complexes of respective species is applied within the Anthropoidea, we may fairly conclude that forelimb elongation probably developed, to a large extent, by parallel evolution for suspensory foraging, feeding, and transferring behavior in proto-

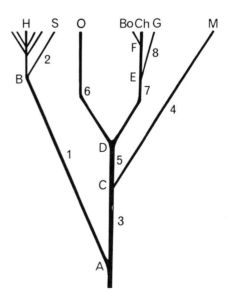

FIG. 2. Phylogeny of extant hominoid primates. H, gibbons excluding siamang; S, siamang; O, orangutan; Bo, bonobo; Ch, common chimpanzee; G, gorilla; M, man. A, furcation of the hylobatid apes; B, furcation of the siamang; C, furcation of the hominid lineage; D, bifurcation of the orangutan and troglodytian lineages; E, furcation of gorilla; F, furcation of bonobo; 1, hylobatid evolution; 2, evolution of siamang; 3, prehominid evolution; 4, hominid evolution; 5, protopongid evolution; 6, evolution of orangutan; 7, troglodytian evolution; 8, evolution of gorilla. (See text for details.)

hylobatid and protopongid apes. Some degree of ulnar retreat from the carpal bones may have occurred prior to divergence of the hylobatid lineage and the pongid–hominid lineages. Further changes in carpal structure related to adduction, extension, pronation, and supination probably occurred in each of the two lineages *pari passu* with forelimb elongation. Reduction of the long pollical flexor tendon in the protopongids probably occurred as they achieved large body size. Parsimoniously one could interpret this to have occurred prior to divergence of the protoorangutan and prototroglodytian stems. One also might postulate either that protohominids diverged from small-bodied apes that retained fully developed pollices, or alternatively, that the protoorangutans and prototroglodytians achieved large body size and diminution of the long pollical flexor tendon by parallel evolution while the protohominids independently developed (or redeveloped) a more hominid condition of the pollex.

Together with the apes, *Ateles* probably illustrates parallel or convergent evolution in elongation of the forelimbs for suspensory foraging, feeding, and transferring behavior. And together with *Colobus*, *Ateles* may illustrate parallel or convergent evolution of pollical reduction as part of manual adaptation for rapid hauling movements by the forelimb. Until the phylogenetic history and genotypes of catarrhine and platyrrhine primates are better understood, I prefer conservatively and undogmatically to ascribe these two character complexes to convergence.

IV. Hominoid Phylogeny

The hypothetical phylogeny that I present here (Fig. 2) is based upon postcranial evidence from extant forms. It is generally compatible with modern cranial evidence and available fossil remains. Aspects of it are contrary to the views of many biomolecular evolutionary biologists, especially those who are cladistically disposed (see Goodman, this volume). I will speculate on hypothetical forms at each point of furcation (A–F in Fig. 2) and the major adaptive trends (1–8 in Fig. 2) that ultimately led to the extant forms. Current information is insufficient for conclusive inferences on times of divergence. However, I believe that this scheme would fit comfortably within the period extending from early Miocene to late Pleistocene times.

A. Hylobatid Emergence

At the furcation of the ancestors of modern hylobatid apes (Fig. 2A), the stem hominoids were small by comparison with most modern hominoids; adults weighed between 8 and 15 lb and had slight builds. Many features of the trunk which characterize modern Hominoidea were incipiently or more notably developed in response to orthograde positional behavior, including some arboreal bipedalism, vertical climbing, and forelimb suspension. The transverse diameter of the thorax was greater than the sagittal diameter; the ribs were angled; and the thoracic vertebral bodies protruded into the thoracic cavity. The diaphragm was dome shaped; its muscle fibers tended to converge toward a central area. Parts of the dorsal mesentery were foreshortened so that the liver, certain segments of the small and large intestines, and other abdominal organs were either affixed or closely juxtaposed to the diaphragm or posterior abdominal wall or both. There was a tendency for fusion of the sternebrae in late adolescence or adulthood. Individuals normally possessed 5 or 6 lumbar vertebrae. The lumbar region was still rather flexible but the mass and fasciculation of the lumbar back muscles were diminished in comparison with those of fully pronograde arboreal animals. The ischial callosities were small. The tail was short or absent and there was incipient development of a pelvic diaphragm from the rudimentary and otherwise transformed pelvocaudal muscles. Some of the proximal caudal (or coccygeal) vertebrae were incorporated into the sacrum so that it was composed of more than three segments. The pelvis was relatively wide especially in the ilia. An incipient sacrotuberous ligament was distinguishable. The gluteal muscles were arranged much like the pattern that characterizes hylobatid and modern African apes (Sigmon, 1974, 1975).

Both forelimbs and hindlimbs were relatively long but the forelimbs did not approximate modern hylobatid dimenions. The clavicle was relatively long. The scapula had begun to assume the shape of a "brachiator's" (Erikson, 1963). It was positioned toward the posterior wall of the thorax. The shoulder joint could be positioned so that it faced laterally. The shoulder joint was quite mobile and powered by well developed arm-raising muscles. The latissimus dorsi muscle did not attach to the iliac crest. The elbow joint was quite extensible although the olecranon process probably had not reduced markedly. The wrist joint was mobile, permitting extensive flexion, extension, adduction, and rotation. The proximal wrist articulation had begun to assume modern hominoid form. The long digital flexor muscles were very well developed. The pollices and halluces were large, supplied with powerful extrinsic and intrinsic muscles, and capable of wide divergence.

The protohylobatids (termed "late pre-hylobatines" in Tuttle, 1972a) probably climbed on vertical supports such as branches, small tree trunks, and vines more frequently than their predecessors did. Further, they sometimes stood bipedally on firm horizontal and springy supports in order to reach overhanging food-bearing branches. In large trees they opportunistically ascended directly to higher branches by hoisting instead of returning to the core of the tree and climbing up the trunk. Similarly, some short descents to lower supports were executed directly by forelimb suspension beneath the base branch. Because they often foraged in the periphery of trees they began to employ bridging behaviors to transcend short gaps in the canopy. Larger gaps were sometimes crossed by leaping and hauling actions if alternate routes were not available to them.

B. Later Hylobatid Trends

The hylobatid (Fig. 2, branch 1) trends related to suspensory foraging and feeding activities probably were firmly established without major changes in body size or the general form of the torso and proximal segments of the limbs. The frequency of forelimb suspensory postures surpassed the frequency of other postures during feeding in the periphery of trees. Arm-swinging over short distances along branches and between closely juxtaposed superstrates became a regular mode for securing new vantage points for feeding. Bridging behavior as a means to transfer from tree to tree also became more frequent. Remarkable forelimb elongation and reduction of the olecranon process of the ulna probably advanced with the increased incidence of suspensory behavior. Trends toward a ball-and-socket configuration of the proximal carpal joint, for the proximal capitate and hamate bones to form a knoblike process (Jenkins and Fleagle, 1975), and for structural limitations of overextension in metacarpophalangeal joints II–V also advanced. Culminant transformations of the thorax, shape of the scapulas, position of the scapulas on the posterior thoracic wall, diaphragm, and visceral attachments; reduction of the tail; development of the pelvic diaphragm; and other complexes of features in modern hylobatid apes probably also occurred as suspensory behaviors became chronic.

Continued selection for vertical climbing on small trunks, branches, and vines resulted in development of a ball-and-socket configuration and special ligamentous mechanisms of the pollical carpometacarpal joint (Van Horn, 1972), the deep cleft between manual digits I and II, and for particular arrangements of the pollical musculature (Tuttle, 1969a, 1972a). Much the same selective complex, with the important addition of bipedal running atop relatively stable branches and small flexible substrates, resulted in the long, widely divergent, heavily muscled hallux of modern Hylobatidae (Tuttle, 1972a).

As the forelimb suspensory complex was established in the hylobatid apes, the center of gravity shifted cranially into the thoracic region. This presented special challenges to them during bipedal activities. Consequently, some further reductions in length and flexibility of the lumbar region of the vertebral column and elongation of the ilia probably occurred. Long ilia provide better leverage for the gluteus medius muscles which may act as powerful extensors of the hip joints during bipedal behavior and vertical climbing.

Rather late in hylobatid evolution the ricochetal arm-swinging complex, particularly the multiple-joint–muscle complex of the forelimb (Tuttle, 1969a, 1972a), evolved from an advanced stage of the suspensory foraging and feeding complex. This enabled the lesser apes to transcend gaps in the canopy during routine locomotive activities and it was incorporated as a fundamental component in the dramatic displays that characterize modern hylobatid intergroup altercations.

C. Furcation and Trends of the Siamang

The Pleistocene (Hooijer, 1960) and Recent siamangs (*Symphalangus* or *Hylobates syndactylus*) probably represent the upper limit of body size that could be achieved in a lineage characterized by the ricochetal arm-swinging adaptive complex. Modern siamangs are nearly twice the size of other modern gibbons. They possess more elongate forelimbs than all species of *Hylobates* except *Hylobates concolor* (Groves, 1972) and other postcranial features which Schultz (1933, 1973b) reasonably postulated are advanced over the basal hylobatid condition. Siamangs probably evolved after the ricochetal arm-swinging complex had been fully established in the Hylobatidae, perhaps because they began to exploit a new fare in which leaves and young parts of vines were common (Fig. 2B and branch 2). Chivers (1972) noted that lar gibbons (which are sympatric with siamangs) appear to be much more frugivorous than siamangs. He designated the siamang a "folivore." Generally in primates the species most disposed to folivory are among the largest forms in their respective groups, for instance, indriids among lemurs, howlers and perhaps woolly spider monkeys (Zingeser, 1973) among ceboid monkeys, certain colobines among cercopithecoid monkeys, and gorillas among the hominoids.

Size increase led to certain modifications in siamang lifeways, most notably a greater predispostion for bridging vs. ricochetal arm-swinging locomotion. Nevertheless, they are still quite capable of dramatic "leaps" by forelimb propulsion and employ them routinely in their displays (Chivers, 1972).

Groves' (1972) theory that the pristine hylobatid apes were siamang-like brachiators and that other gibbons evolved from them by size reduction (and other transformations) is not convincing. Ricochetal arm-swinging probably was not characteristic of the common ancestor of great apes and the lesser apes. If the ricochetal arm-swinging morphological complex is correctly attributable to dramatic locomotion in trees, the fact that modern siamangs engage in this behavior less frequently than other gibbons do indicates that their size limits them from arm-swinging in circumstances in which their smaller predecessors and contemporaries could ricochetally brachiate.

D. Hominid Emergence and Trends

Subsequent to furcation of the protohylobatids, body weight increased in the stem hominoids and they became more stocky (Fig. 2, branch 3). Vertical climbing and bipedal behavior were somewhat more conspicuous components in their positional repertoires. Some species were less inclined to forage and feed in suspensory postures than the proto-hylobatids were.

The protohominid furcation (Fig. 2C) probably occurred while body size was still relatively small, perhaps no more than 20 or 30 lb. Climbing on vines and tree trunks and bipedal standing and reaching were notable components of their foraging behavior. Hoisting, suspensory descents to nearby lower substrates, and transfers between springy regions of the canopy were common. Occasionally gaps in the canopy were traversed by leaping with hindlimb propulsion. The hands were prominent in catching springy landing sites. Bipedal walking and running were common when suitable substrates were available.

The pollices and halluces were relatively long, divergent, and muscular. But the hands and feet lacked deep clefts of a hylobatid sort. The thorax was further advanced in the broad, square-shouldered configuration that characterized the stem hominoids. As the scapulas came to lie fully on the posterior wall of the thorax, the clavicles were further elongated and the acromial processes of the scapulas enlarged. Arm-raising as part of bipedal reaching overhead, climbing on vines and tree trunks, and, to a lesser extent, suspensory foraging and short-distance locomotion advanced development of the arm-raising and scapular rotatory muscles. The scapulas did not acquire elongate vertebral borders nor were they otherwise transformed to the form of advanced brachiators (Erikson, 1963). Mobility of the shoulder joints and extension of the elbow joints (which was facilitated by reduction of the olecranon process of the ulna) permitted a wide range of reaching and grasping movements with hands above the head. The wrist and finger joints also remained quite mobile. Early ontogenetic fusion of the os centrale to the scaphoid bone probably occurred sometime after palmigrade postures were no longer employed for arboreal locomotion. The fusion may be related to stabilization of the midcarpal region during object manipulation and thus might have occurred rather late in hominid evolution.

In the stem Hominoidea suspensory behavior such as hoisting, transferring, hauling, and occasional arm-swinging along branches has selected for relatively well-developed flexor digitorum superficialis muscles which attach to the middle phalanges of the fingers. Climbing, foraging, and suspensory behavior and increasing weight continued this trend early in the hominid lineage (Tuttle, 1972*b*).

Increased emphasis on the hindlimbs for climbing, leaping, and bipedal foraging, feeding, and locomotion led to emphasis on the quadriceps femoris and triceps surae muscles. The center of gravity was located at a relatively low level in the lumbar region of the abdominal cavity. Hence the lumbar region did not shorten and become inflexible to the extent characteristic of the great apes. The ilia did not become exceptionally elongate. They continued to widen. Although powerfully developed as part of the vertical climbing complex, the latissimus dorsi muscles did not achieve extensive muscular attachment to the ilia; instead their iliac attachments were confined to the posterior segments of the iliac crests.

Bipedal locomotion was performed with knee joints flexed and femora flexed, abducted, and laterally rotated at the hip joints. This lowered the center of gravity during branch walking and running and selected for well developed gluteal abductor–extensor muscles. But the hip and knee joints retained the capacity to fully extend and to over-extend as part of bipedal foraging and vertical climbing behavior.

Although transferring and leaping continued to be important modes of moving from tree to tree, the protohominids also began to run bipedally between trees as they sporadically foraged on or near the ground for fallen fruits, the produce of herbs and shrubs, and insects. Increased seasonal exploitation of resources in the shrub layer and on the ground may have led some populations to expand into relatively open habitats near riverine forests somewhat in the manner of modern chimpanzees in western Tanzania (Kano, 1972).

Dramatic size increase beyond that of the stem protohominids (and perhaps the radiation of several bipedal species) probably did not occur until terrestrial bipedal locomotion and foraging had become commonplace. At some point in the terrestrial career of the early hominids, small vertebrate prey were sought to supplement insect and vegetable resources. This may have occurred on the forest floor, in forest fringe areas, or still more open woodlands. The critical factors which predisposed the protohominids to bipedalism were the relatively low position of the center of gravity (Morton, 1924, Tuttle, 1969*b*, 1974), substantial development of the lower limbs, and a notable apprenticeship of arboreal bipedalism and other orthograde positional behavior.

While the pelvic tilt mechanism may have developed incipiently in arboreal protohominids (Tuttle, 1974), habitual employment of the extended lower limb as a "propulsive strut" (Haxton, 1947); cranial retreat of the lower portion of the gluteus maximus muscle (Tuttle *et al.*, 1975; Tuttle and Basmajian, 1975); other transformations of the gluteal complex toward the hominid pattern (Sigmon, 1974, 1975); permanent convergence of the hallux; and the singular hominid pattern of foot contact and propulsion (Kondo, 1960) occurred with habitual diurnal terrestrial foraging and locomotion in open areas.

Bipedalism permitted the terrestrial protohominids to survey their surroundings while moving or alarmed and to carry objects over relatively long distances. They could climb trees to escape most predators. If caught in open areas they might have brandished or thrown objects and otherwise performed noisy displays. At night the protohominids slept in trees or on rocky ledges. The earliest hominid hunters probably worked from ambush and by stalking. Mobbing and body contact with prey also might have been common prior to the innovation of projectile weapons and puncture poisoning. Sustained long-distance running probably developed late in hominid evolution as part of particular hunting practices.

E. Protopongid Evolution

The evolving protopongid stock (Fig. 2, branch 5) was chiefly characterized by a steady increase of body size and development of a suspensory adaptive complex that was more advanced than that of the collaterally evolving protohominid lineage. By the time protopongids had achieved weights between 40 and 60 lb. they evidenced several complexes of features that are shared by modern orangutans and African apes.

The forelimbs became elongate, especially in the forearm and manual digits II–V. The proximal and middle phalanges of manual digits II–V were convexly curved ventrally and had strong flexor sheaths and bands attached to them. These structures increased the mechanical advantage of the powerful long digital flexor muscles. The flexor digitorum superficialis muscle was remarkably well developed. Large body size and forelimb suspensory behavior selected for total devotion of the radial component of the flexor digitorum profundus muscle to its indicial tendon. Concordantly, the long pollical flexor tendon diminished. But counterselection for fine manipulation advanced development of the intrinsic pollical musculature (Tuttle and Rogers, 1966; Tuttle, 1969a, 1970).

The capacity to overextend metacarpophalangeal joints II–V was limited. *Per contra*, mobility of the wrist was retained and probably further advanced. This was requisite in order that the stocky protopongid apes could travel and forage versatilely in the canopy and on vertical vines and tree trunks.

Marked reduction of the olecranon process of the ulna and other changes in the elbow complex permitted full extension of the humeroulnar joint and facilitated pronation and supination of the hand. Reduction of the olecranon process might be related to development of a special close-packed positioning mechanism that allows forelimb suspension with a minimum of transarticular muscle force (Tuttle and Basmajian, 1974b).

The body of the scapula was transformed to the basic shape of "brachiators" (Erikson, 1963), the glenoid cavity faced craniad, the spine of the scapula was oriented obliquely on the body, and the acromion process was robust. The clavicle was elongate, robust, and oriented craniolaterally. The arm-raising and scapular rotatory muscles were powerfully developed. The sternum was broad. Unlike the hylobatids and hominids, the protopongids developed a funnel-shaped thorax; but like other hominoids, the transverse diameter of the thorax exceeded the sagittal diameter and the scapulas lay on its posterior aspect.

The lower region of the vertebral column was remarkably transformed. The lumbar region shortened; the lumbar vertebrae became robust; intervertebral mobility decreased; and the lower back muscles completed the change to modern pongid configuration. This to some extent offset the trend for the center of gravity to be high because of massiveness of the forelimbs.

The hindlimbs continued to be important propellant organs, especially for vertical climbing and occasional hand-assisted bipedalism atop branches. The sacrum was strengthened by sacralization of additional coccygeal segments; the pelvis continued to broaden; and the ilia became remarkably elongate. Like the Hylobatidae, the Pongidae developed especially elongate ilia to provide better leverage for the gluteal extensor muscles. Because the center of gravity of the body was located in the thoracic region this mechanism maintained balance as the protopongids climbed and were bipedal in trees.

The powerfully developed latissimus dorsi muscle attached extensively on the iliac crest in advanced protopongids. This was facilitated by elongation and wideness in the ilium and shortness of the lumbar spine. The latissimus dorsi muscle is a primary retractor of the humerus, especially from positions wherein the forelimb is raised overhead. The iliac attachment may have increased its mechanical advantage during vertical climbing and bridging behavior and may have transferred weight of the lower body directly to the supportive limb as protopongids enagaged in certain suspensory activities.

The sacrotuberous ligament was distinct and the gluteus maximus muscle resembled the pattern that Sigmon (1975) reported to be common to extant hylobatid and African apes. The hallux was robust, heavily muscled, and divergent. The heel was not markedly protuberant though it was powered by a robust triceps surae complex.

The protopongids probably foraged widely in the forest canopy. Arm-swinging along branches and between closely juxtaposed supports was common. They did not engage in ricochetal arm-swinging. Although they may have executed vertical drops or leaped across gaps in the canopy by hindlimb propulsion when suddenly alarmed, they generally moved between trees by arm-swinging along familiar routes or transferring (cf. bridging) where the canopy was sparse. Suspensory feeding was more common than in modern chimpanzees but perhaps less frequent than in hylobatid apes. They may have ventured to the ground sporadically in order to sample foods and to cross small open spaces in the forest. Except when carrying objects, they moved quadrupedally with hands in fist-walking or palmigrade postures and feet plantigrade.

F. Furcation and Trends of Orangutan

Divergence between protoorangutans and prototroglodytians (Fig. 2D) probably occurred as the former became fully specialized for canopy dwelling and the latter for terrestrial knuckle-walking. The earliest representatives of the two lineages were allopatric and perhaps quite remote from each other geographically. Modern orangutans represent a culminant adaptation in large-bodied pongid apes to evergreen dipterocarp (Richards, 1952, pp. 329–330) and swamp forests.

The protoorangutans continued to increase in size at least until mid-Pleistocene times (Fig. 2, branch 6). Their hands, and particularly their feet and hip complexes, were specially transformed so that they could distribute their considerable bulk and climb versatilely through the canopy. Ground-level resources were generally scarce. In many areas the ground was permanently or seasonally soggy or submerged.

Their forelimbs became extremely elongate thereby permitting them to reach new supports from secure positions during bridging transfers, to forage widely from secure vantage points, and to climb large tree trunks. Further proximal positioning of the distal end of the ulna and, like hylobatid apes, trends toward a ball-and-socket configuration in the proximal carpal articulation and a knoblike process of the capitate and hamate bones in the midcarpal region (Jenkins and Fleagle, 1975) enhanced mobility of their wrists. The capacity to adduct the wrist widely permitted orangutans to hang by manual grips from vines and springy branches. General mobility of the wrist enabled them to reach in many directions and thus to grasp new supports safely despite precarious positions among base supports.

The ·singular adaptive complex represented by the hindlimb of *Pongo* is chiefly related to their bridging and pedal suspensory foraging and feeding behavior. The feet of protoorangutans were transformed to powerful prehensile appendages wherein pedal digits II–V were elongate and convexly curved and the extrinsic pedal digital flexor musculature was remarkably developed and devoted entirely to digits II–V (Tuttle, 1969a, 1970, 1972a). The hallux reduced dramatically, although, as in the pollex, intrinsic musculature elaborated to preserve some prehensile capacity (Tuttle and Rogers, 1966; Tuttle, 1970). The feet acquired a permanently inverted (supinated) set. The plantar flexor muscles and the posterior protuberance of the heel were quite modestly developed in comparison with structures related to powerful grasping with pedal digits II–V (Tuttle, 1970, 1972a).

The hip musculature and the hip joint became uniquely adapted (among Hominoidea) for an extremely wide range of movement. The tensor fasciae lata muscle and iliotibial tract were lost. The gluteus maximus and gluteus minimus muscle complexes of *Pongo* are quite distinctive from those of other extant apes and man (Sigmon, 1974, 1975).

It is now faddish to entertain the idea of pristine terrestrial origins for the modern great apes including the orangutan. Although it is certainly reasonable to propose that protoorangutans occasionally ventured to the ground, I doubt that these sporadic excursions left any noticeable imprint on their postcranial morphology (Tuttle, 1975; Tuttle and Basmajian, 1974a). MacKinnon (1974a,b) argued that because modern male orangutans grow too large for easy arboreal locomotion and must travel on the ground, past races that were larger must have been ever more terrestrial in their locomotive habits although still highly dependent on trees for food and nesting. It might be instructive to recall the empirical premise upon which this hypothesis is based, especially as MacKinnon cast aside the caution exhibited in his monograph (1974a, pp. 68–69) and proposed that in prehistoric southern China large bands of orangutans ("twice as bulky as their descendants") ranged on the ground like modern gorillas and were protected from predators by enormous males (1974b, p. 212).

Hooijer (1948) compared isolated teeth from Holocene cave deposits in the Padang Highlands of central Sumatra with teeth in the skulls of 80 modern orangutans and isolated teeth from Pleistocene deposits in Java, southern China, and Indo-China. He concluded that the Sumatran subfossil form had teeth 16% more robust than those of modern orangutans and that the Chinese and Indo-Chinese orangutans were even larger than the subfossil form. But Hooijer did not systematically examine whether modern Sumatran and Bornean orangutans exhibit significant differences in tooth size nor how large individual teeth relate to the size of the whole dentition and supporting cranial structures. In Hooijer's modern sample, of the 66 specimens with permanant teeth, 54% were Bornean, 26% were Sumatran, and 20% were undetermined. Further, the 4 age–sex classes possessing permanent teeth were distributed geographically as follows: juvenile I: 77% Bornean, 8% Sumatran, 15% unknown; juvenile II: 50% Bornean, 38% Sumatran, 12% unknown; adult female: 62% Bornean, 19% Sumatran, and 19% unknown. Thus, if modern Sumatran orangutans possess larger teeth or longer tooth rows than their Bornean neighbors this would to some extent obscure the difference between Hooijer's Sumatran subfossil form and modern *Pongo*. There appears to have been a low (33% or 91 specimens) proportion of adult males in the Sumatran caves (Marcus, 1969). Therefore, it is conceivable that a systematic comparison between a large sample of extant Sumatran orangutans, especially adult females, and those of the subfossil form might reveal less dramatic differences than the results of Hooijer's original study. Finally, we really need articulated teeth and postcranial remains of the fossil and subfossil forms before advocating troops of pongid behemoths in the post-Hipparion Sino-Malayan fauna.

G. Prototroglodytian Trends and Evolution of African Apes

The prototroglodytian apes (Fig. 2, branch 7) probably evolved in evergreen and mosaic forest that contained open spaces and opportunities to feed on the ground. Some populations may have lived in or ventured seasonally into mixed deciduous forests (Richards, 1952, p. 339), savanna woodland, and occasionally still more open areas for produce of certain trees, shrubs, and herbaceous plants.

The knuckle-walking complex of the forelimb and plantigrade foot with prominent plantar flexing mechanism evolved as terrestrial locomotion replaced arm-swinging and bridging as the commonplace mode for moving between trees. But because body weight was not inordinately great and because they retained many advanced features of the proto-pongid suspensory foraging–feeding adaptive complex, the prototroglodytians continued to exploit trees for their daily fare and nesting.

Knuckle-walking became morphologically feasible by transformation of the flexible long-fingered hands inherited from arboreally adapted protopongid apes into efficient terrestrial supportive and propulsive organs (Tuttle, 1969a,b, 1970; Tuttle and Basmajian, 1974a). The initial transitional period was especially facilitated by the remarkably well-developed flexor digitorum superficialis muscle and proper flexor muscles of the wrist.

The wrist flexors maintained alignment of the palm with the forearm. The flexor digitorum superficialis muscle guarded against excessive stress on the metacarpophalangeal joints of weight bearing digits and also secondarily assisted the proper wrist flexors to support the wrist. Osseoligamentous supportive mechanisms and knuckle pads (Tuttle, 1969a, p. 349; Tuttle and Basmajian, 1974a, pp. 311, 322) evolved rapidly as knuckle-walking became chronic. The characteristic configuration of weight-bearing metacarpal heads (Tuttle, 1967, pp. 192–193; 1969b, 1970, pp. 201, 204) that induces overextension of the metacarpophalangeal joints is part of a mechanism to disperse potentially harmful stresses upon initial contact of the knuckled hand with the ground and to provide better leverage for propulsive employments of the fingers.

The wrist of the prototroglodytians lost some of the mobility that characterized the protopongid wrist. Extension (or dorsiflexion) and abduction (or radial deviation) were especially restricted in full-fledged knuckle-walkers. Bony features related to limited mobility in the wrists of modern African apes include (1) the deeply concave distal articular surface of the radius, (2) volar and ulnar inclination of the distal radius relative to its shaft, (3) the configuration of the proximal articular surface of the scaphoid and adjacent nonarticular bony ridge, (4) the relationship of the dorsal edge of the scaphoid bone and the concave part of the capitate facet in the midcarpal joint, (5) the radioulnar dimension of the radial articular surface of the proximal carpal row being greater than its dorsoventral dimension, and (6) fusion of the os centrale with the scaphoid bone (Tuttle, 1967, pp. 189–190; Jenkins and Fleagle, 1975). Features 1 and 2 are both extension-limiting and abduction-limiting. Features 3 and 4 are primarily extension-limiting. Features 5 and 6 are chiefly associated with weight transmission during knuckle-walking, but they also limit mobility in the proximal carpal and midcarpal joints, respectively.

As extension-limiting osseoligamentous mechanisms developed in the wrist, the long digital flexor muscles shortened and became "tendonized." This permitted the flexor digitorum superficialis muscle to act synergically with intrinsic osseoligamentous mechanisms to safeguard the metacarpophalangeal joints during overextension (Tuttle et al., 1972; Tuttle and Basmajian, 1974a).

In advanced knuckle-walkers the elbow complex may have been modified somewhat beyond the condition of their fully arboreal predecessors in order to accommodate overextended positions during quadrupedalism. Overextension is facilitated by deep olecranon fossae and reduced olecranon processes. Experimental studies on human subjects evidence that overextended elbows are more effective than less extended elbows to resist the potentially damaging torque forces incurred when falling onto the hands (Carlsöö and Johansson, 1962; Tuttle and Basmajian, 1974b). Diminution and perhaps other restructuring of the olecranon process of the ulna (Tuttle and Basmajian, 1974b) and the development of steepness on the lateral aspect of the olecranon fossa (McHenry, 1975) stabilize the humeroulnar joint of African apes. During knuckle-walking the flexion–extension axis and lateral buttress of the olecranon fossa are more or less perpendicular to the line of travel. Thus, the elbow region is advantageously positioned to resist shear stresses.

Although the hip and thigh of the troglodytians were not altered dramatically from the protopongid condition, their feet were rather remarkably transformed. The heel became a prominent supportive structure during terrestrial positional behaviors like squatting and quadrupedal stance and locomotion. The foot of troglodytians, like hominids, is plantigrade; the heel strikes the ground early in the contact phase of the locomotive cycle of the hindlimb. Plantar flexion is an important component of the propulsive thrust of the hindlimb during knuckle-walking progression and especially when climbing up trees. The relative length of the "power arm" of the foot in African apes is rivaled only by man among extant primates (Schultz, 1963, pp. 156–157; Tuttle, 1970, p. 215). The triceps surae complex is very well developed in modern troglodytians (Tuttle, 1970, pp. 215, 229).

The troglodytians increased in body size as their adaptations for terrestrial locomotion were perfected and as they supplemented traditional arboreal fares with plants near the ground. Different populations invaded new habitats and their successors adapted in new niches.

Modern gorillas (Fig. 2E and branch 8) are a culminant result of adaptation by quite large troglodytians to ground dwelling and heavy reliance on the produce of herbaceous plants, like *Aframomum*, and outsized woody grasses like bamboo. This adaptive complex admirably equipped them for permanent habitation in forested montane regions. In protogorillas the foot was further transformed for terrestrial plantigrade functions with some sacrifice of prehensile ability (Tuttle, 1970, pp. 222–224, 236–237; 1972b, p. 229). In addition to foot structure and body size, gorillas differ strikingly from chimpanzees in the alimentary tract and especially in the external genitalia.

Bonobos (*Pan paniscus*) might represent the general morphological pattern of the stem prototroglodytian apes more closely than common chimpanzees (*Pan troglodytes*) do (Fig. 2F). But it is also plausible that ancestral bonobos arrived relatively late in the true tropical rain forest (Richards, 1952, p. 338) of the Zaire Basin. According to this theory their predecessors, like other troglodytians, would have developed the knuckle-walking and plantigrade adaptive complexes in evergreen or mosaic forests that were marginal to the dense wet forests. Some reduction of body size may have occurred in proto-bonobos to assuage the limitations on versatile arboreal positional behavior caused by their heritage of advanced adaptations for terrestrial locomotion. Neoteny might constitute a prime mechanism whereby size reduction and greater locomotive versatility were achieved by *Pan paniscus*. The skulls of bonobos are strikingly reminiscent of juvenile common chimpanzee skulls. Sexual dimorphism is less in the skulls of *Pan paniscus* than *Pan troglodytes* (Fenart and Deblock, 1973, 1974). Coolidge (1933, p. 32) remarked that the postcranial skeleton of an adult female bonobo recalls the skeleton of juvenile common chimpanzees. Some adult bonobos retain the white perianal tuft of hair that characterizes infant common chimpanzees (Coolidge, 1933, p. 54).

V. Summary

Traditionally parallelism has been invoked to explain many similarities among anthropoid primates. And primates are often employed to exemplify the manifest nature of parallelism in the animal kingdom. Herein I recalled concepts of parallelism by several notable evolutionary biologists, proposed an operational definition that emphasizes the selective complex, and applied it in several lessons pertaining to the role that brachiation may have played in the evolution of selected postcranial complexes in the Hominoidea, *Ateles*, and *Colobus*. When consideration is given to what each species actually does in natural contexts it becomes quite problematic to explain features like forelimb elongation and pollical reduction in all anthropoid primates that posesss them on the basis of parallel selection for a single form of forelimb positional behavior, like rapid brachiation.

Comparisons of postcranial features in extant species led me to surmise that the stem Hominoidea were small creatures in which the trunk was specially adapted to a variety of orthograde positional behaviors. The lesser apes evolved their characteristic features during a period in which forelimb suspension for feeding and locomotion were prominent components of the positional repertoire. The richocetal arm-swinging adaptive complex developed for traversing gaps in the canopy and display-swinging. Siamangs evolved after the ricochetal arm-swinging complex was well advanced. Siamangs probably represent the upper limit of body weight that could be achieved by animals possessing this adaptive complex.

The hominid lineage diverged from the stem hominoids while they were still relatively small-bodied. Emergent hominids were predisposed to terrestrial bipedalism by a heritage of bipedal and other orthograde positional behavior in trees. The hominids retained some symplesiomorphic hominoid features in the trunk. They did not possess many of the highly advanced features that are associated with arm-swinging in the great apes and gibbons. Transformations of the hindlimb to the modern hominid condition occurred as terrestrial habitation became chronic.

Subsequent to furcation of the hominid lineage, protopongid apes evolved as relatively large-bodied creatures in which forelimb suspensory behavior was a conspicuous component of the foraging and feeding repertoire. Protopongids traversed the canopy by transferring and arm-swinging on customary pathways. Vertical climbing was common and they occasionally made brief excursions to the ground. The protopongids did not engage in ricochetal arm-swinging. Many of the features in trunk and forelimb that characterize both *Pongo* and *Pan* developed during the protopongid phase of hominoid evolution.

Orangutans evolved in tropical forests that were perennially or seasonally wet and that provided few opportunities for terrestrial travel and feeding. Transferring, climbing, and arm-swinging were common modes of traversing the canopy and reaching foods. The forelimbs advanced in features that are traditionally ascribed to "brachiators," and the hindlimbs became remarkably specialized for powerful pedal digital grasping and versatile positioning in springy regions of the canopy.

Troglodytian apes evolved in forests that offered opportunities for terrestrial feeding and that otherwise induced excursions on the ground. After the knuckle-walking and plantigrade pedal complexes were firmly established in the troglodytian lineage, populations of protogorillas became especially adapted to certain herbaceous plants that characterize African evergreen forests. Gorillas spread into montane forests. Bonobos may have evolved by neoteny from advanced troglodytian apes that deployed into the dense wet forests of central Africa.

ACKNOWLEDGMENTS

477

PARALLELISM,
BRACHIATION,
AND HOMINOID
PHYLOGENY

Research upon which this paper is based was supported by NSF Grant No. GS-3209 and previous grants from the National Science Foundation and Public Health Service (Tuttle, 1967 *et seq.*). I am especially grateful to Mrs. Lita Osmundsen and the Wenner–Gren Foundation for sponsoring the field studies in Ceylon and Tanzania that enabled me to observe primates in natural habitats and for the opportunity to learn once more at Burg Wartenstein. I thank Patrick Luckett, Fred Szalay, and participants in the conference for their good company and provocative comments on ideas presented on this paper. Perspectives on the habitats of early troglodytian apes and hominids were stimulated by discussions at Kyoto University with Dr. J. Itani and at the University of Tokyo with Dr. T. Nishida. I thank Drs. S. Kondo, J. Ikeda, H. Watanabe, and H. Ishida and the Japan Society for the Promotion of Science for this opportunity to exchange ideas.

VI. References

ANDREWS, P. 1974. New species of *Dryopithecus* from Kenya. *Nature* **249**:188–190.

ASHTON, E. H., R. M. FLINN, C. E. OXNARD, and T. F. SPENCE. 1971. The functional and classificatory significance of combined metrical features of the primate shoulder girdle. *J. Zool., Lond.* **163**:319–350.

AVIS, V. 1962. Brachiation: the crucial issue for man's ancestry. *Southwest. J. Anthropol.* **18**:119–148.

BIGELOW, R. S. 1958. Classification and phylogeny. *Syst. Zool.* **7**:49–59.

BOCK, W. J., and G. VON WAHLERT. 1965. Adaptation and the form–function complex. *Evolution* **19**:269–299.

CARLSÖÖ, S., and O. JOHANSSON. 1962. Stabilization of and load on the elbow joint in some protective movements. *Acta Anat.* **48**:224–231.

CARPENTER, C. R. 1935. Behavior of red spider monkeys in Panama. *J. Mammal.* **18**:171–180.

CARPENTER, C. R., and N. M. DURHAM. 1969. A preliminary description of suspensory behavior in nonhuman primates, pp. 147–154. *Proc. 2nd Int. Congr. Primatol., Atlanta, Georgia 1968*, Vol. 2. Karger, Basel and New York.

CARTMILL, M., and K. MILTON. 1974. The lorisiform wrist joint. *Am. J. Phys. Anthrop.* **41**:471.

CHIVERS, D. J. 1972. The siamang and gibbon in the Malay Peninsula, pp. 103–135. *In* D. M. Rumbaugh, ed., *Gibbon and Siamang*, Vol. 1. Karger, Basel and New York.

CLARK, W. E. LeGros, and D. P. THOMAS. 1951. Associated jaws and limb bones of *Limnopithecus macinnesi*. *Fossil Mammals of Africa, No. 3*, British Museum (Natural History), London.

CONROY, G. C., and J. G. FLEAGLE. 1972. Locomotor behaviour in living and fossil pongids. *Nature* **237**:103–104.

COOLIDGE, H. J., Jr. 1933. *Pan paniscus*. Pygmy chimpanzee from south of the Congo River. *Am. J. Phys. Anthropol.* **18**:1–59.

DAY, M. H., and J. R. NAPIER. 1963. The functional significance of the deep head of flexor pollicis brevis in primates. *Folia Primatol.* **1**:122–134.

EISENBERG, J., and R. E. KUEHN. 1966. The behavior of *Ateles geoffroyi* and related species. *Smithson. Misc. Coll.* **151**:1–63.

ERIKSON, G. E. 1963. Brachiation in New World monkeys and in anthropoid apes. *Symp. Zool. Soc., London* **10**:135–164.

FENART, R., and R. DEBLOCK. 1973. *Pan paniscus–Pan troglodytes*. Crâniométrie. Etude comparative et ontogénique selon les méthodes classique et vestibulaire. *Ann. Mus. Afr. Cent. Tervuren, Belg.* **1**:1–593.

FENART, R., and R. DEBLOCK. 1974. Sexual differences in adult skulls of *Pan troglodytes*. *J. Hum. Evol.* **3**:123–133.

GOLDMAN, E. A. 1920. Mammals of Panama. *Smithson. Misc. Coll.* **69**:1–309.

GRAND, T. I. 1972. A mechanical interpretation of terminal branch feeding. *J. Mammal.* **53**:198–201.

GROVES, C. P. 1972. Systematics and phylogeny of gibbons, pp. 1–89. *In* D. M. Rumbaugh, ed., *Gibbon and Siamang*, Vol. 1. Karger, Basel and New York.

HAXTON, H. A. 1947. Muscles of the pelvic limb, a study of the differences between bipeds and quadrupeds. *Anat. Rec.* **98**:337–346.

HENNIG, W. 1966. *Phylogenetic Systematics*. University of Illinois Press, Urbana. 263 pp.

HILL, W. C. O. 1962. *Primates, Comparative Anatomy and Taxonomy, Vl Cebidae, Part B.* Edinburgh University Press, Edinburgh. 537 pp.

HOOIJER, D. A. 1948. Prehistoric teeth of man and of the orang-utan from central Sumatra, with notes on the fossil orang-utan from Java and southern China. *Zool. Med. Mus. Leyden* **29**:175–301.

HOOIJER, D. A. 1960. The orang-utan in Niah Cave pre-history. *Sarawak Mus. J.* **9**:408–421.

HÜRZELER, J. 1968. Questions et reflexions sur l'histoire des anthropomorphes. *Ann. Paleontol.* **54**:195–233.

JENKINS, F. A., Jr., and J. G. FLEAGLE. 1975. Knuckle-walking and the functional anatomy of the wrist in living apes. *In* R. Tuttle, ed., *Primate Functional Morphology and Evolution*. Mouton, Hague.

KANO, T. 1972. Distribution and adaptation of the chimpanzee on the eastern shore of Lake Tanganyika. *Kyoto Univ. Afr. Stud.* **7**:37–129.

KEITH, A. 1926. *The Engines of the Human Body*. J. B. Lippincott Co., Philadelphia. 256 pp.

KONDO, S. 1960. Anthropological study on human posture and locomotion mainly from the viewpoint of electromyography. *J. Fac. Sci. Univ. Tokyo, Sect. V.* **2**:189–260.

LEWIS, O. J. 1965. Evolutionary change in the primate wrist and inferior radio-ulnar joints. *Anat. Rec.* **151**:275–286.

LEWIS, O. J. 1969. The hominoid wrist joint. *Am. J. Phys. Anthropol.* **30**:251–268.

LEWIS, O. J. 1971*a*. Brachiation and the early evolution of the Hominoidea. *Nature* **230**:577–578.

LEWIS, O. J. 1971*b*. The contrasting morphology found in the wrist joints of semibrachiating monkeys and brachiating apes. *Folia Primatol.* **16**:248–256.

LEWIS, O. J. 1972*a*. Osteological features characterizing the wrists of monkeys and apes, with a reconsideration of this region in *Dryopithecus (Proconsul) africanus*. *Am. J. Phys. Anthropol.* **36**:45–58.

LEWIS, O. J. 1972*b*. Evolution of the hominoid wrist, pp. 207–222. *In* R. Tuttle, ed., *The Functional and Evolutionary Biology of Primates*. Aldine-Atherton, Chicago.

LEWIS, O. J. 1973. The hominoid os capitatum, with special reference to the fossil bones from Sterkfontein and Olduvai Gorge. *J. Hum. Evol.* **2**:1–11.

LEWIS, O. J. 1974. The wrist articulations of the Anthropoidea, pp. 143–169. *In* F. A. Jenkins, Jr., ed., *Primate Locomotion*. Academic Press, New York.

MACKINNON, J. 1974*a*. The behavior and ecology of wild orang-utans (*Pongo pygmaeus*). *Anim. Behav.* **22**:3–74.

MACKINNON, J. 1974*b*. *In Search of the Red Apes*. Collins, London. 222 pp.

MARCUS, L. F. 1969. Measurement of selection using distance statistics in the prehistoric orang-utan *Pongo pygmaeus palaeosumatrensis*. *Evolution* **23**:301–307.

MAYR, E. 1963. *Animal Species and Evolution*. Belknap Press, Cambridge. 797 pp.

MAYR, E. 1969. *Principles of Systematic Zoology*. McGraw-Hill, New York. 427 pp.

McHENRY, H. 1975. Multivariate analysis of early hominid humeri. *In* G. Giles and J. S. Friedlaender, eds., *Measures of Man*. Shenkman, Cambridge, Massachusetts.

MILLER, L. E. 1916. Field notes, in mammals collected on the Roosevelt Brazilian expedition, with field notes by Leo E. Miller. By J. A. Allen. *Bull. Am. Mus. Nat. Hist.* **35**:559–610.

MOLLISON, Th. 1910. Die Korperproportionen der Primaten. *Morphol. Jahrb.* **42**:79–304.

MORBECK, M. E. 1975. Positional behavior in *Colobus guereza*: a preliminary quantitative analysis. *In* S. Kondo, M. Kawai, A. Ehara, and S. Kawamura (eds.), *Proceedings from the Symposia of the Fifth Congress of the International Primatological Society*. Japan Science Press Co., Ltd., Tokyo.

MORTON, D. J. 1924. Evolution of the human foot. II. *Am. J. Phys. Anthropol.* **7**:1–52.

NAPIER, J. R. 1963. Brachiation and brachiators. *Symp. Zool. Soc., London* **10**:183–195.

NAPIER, J. R., and P. R. DAVIS. 1959. The fore-limb skeleton and associated remains of *Proconsul africanus*. *Fossil Mammals of Africa, No. 16*, British Museum (Natural History), London.

POCOCK, R. I. 1925. The external characters of the catarrhine monkeys and apes. *Proc. Zool. Soc., London* **1925**: 1479–1579.

RENSCH, B. 1960. *Evolution Above the Species Level*. Columbia University Press, New York. 419 pp.

RICHARD, A. 1970. A comparative study of the activity patterns and behavior of *Alouatta villosa* and *Ateles geoffroyi*. *Folia Primatol.* **12**:241–263.

RICHARDS, P. W. 1952. *The Tropical Rain Forest*. Cambridge University Press, Cambridge. 450 pp.

ROMER, A. L. 1949. Time series and trends in animal evolution, pp. 103–120. *In* G. L. Jepsen, E. Mayr, and G. G. Simpson, eds., *Genetics, Paleontology and Evolution*. Princeton University Press, Princeton, New Jersey.

ROMER, A. L. 1966. *Vertebrate Paleontology*. 3rd ed. University of Chicago Press, Chicago. 468 pp.

ROMER, A. L. 1970. *The Vertebrate Body*. 4th ed. W. B. Saunders Co., Philadelphia. 601 pp.

SCHÖN, M. A., and L. K. ZIEMER. 1973. Wrist mechanism and locomotor behavior of *Dryopithecus* (*Proconsul*) *africanus*. *Folia Primatol.* **20**:1–11.

SCHULTZ, A. H. 1930. The skeleton of the trunk and limbs of higher primates. *Hum. Biol.* **2**:203–438.

SCHULTZ, A. H. 1933. Observations on the growth, classification and evolutionary specialization of gibbons and siamangs. *Hum. Biol.* **5**:212–255, 385–428.

SCHULTZ, A. H. 1936. Characters common to higher primates and characters specific for man. *Q. Rev. Biol.* **2**:259–283, 425–455.

SCHULTZ, A. H. 1937. Proportions, variability and asymmetries of the long bones of the limbs and the clavicles in man and apes. *Hum. Biol.* **9**:281–328.

SCHULTZ, A. H. 1953. The relative thickness of the long bones and the vertebrate in primates. *Am. J. Phys. Anthropol.* **11**:277–311.

SCHULTZ, A. H. 1956. Postembryonic age changes. *Primatologia* **1**:887–964.

SCHULTZ, A. H. 1963. Relations between the lengths of the main parts of the foot skeleton in primates. *Folia Primatol.* **1**:150–171.

SCHULTZ, A. H. 1973a. Age changes, variability and generic differences in body proportions of recent hominoids. *Folia Primatol.* **19**:338–359.

SCHULTZ, A. H. 1973b. The skeleton of the Hylobatidae and other observations on their morphology, pp. 1–54. *In* D. M. Rumbaugh, ed., *Gibbon and Siamang*, Vol. 2. Karger, Basel and New York.

SIGMON, B. A. 1974. A functional analysis of pongid hip and thigh musculature. *J. Hum. Evol.* **3**:161–185.

SIGMON, B. A. 1975. Functions and evolution of hominoid hip and thigh musculature. *In* R. Tuttle, ed., *Primate Functional Morphology and Evolution*. Mouton, Hague.

SIMONS, E. L. 1960. *Apidium* and *Oreopithecus*. *Nature* **186**:824–826.

SIMONS, E. L. 1962. Fossil evidence relating to the early evolution of primate behavior. *Ann. N.Y. Acad. Sci.* **102**:282–295.

SIMONS, E. L. 1967. Fossil primates and the evolution of some primate locomotor systems. *Am. J. Phys. Anthropol.* **26**:241–254.

SIMONS, E. L. 1972. *Primate Evolution, An Introduction to Man's Place in Nature*. The Macmillan Co., New York. 322 pp.

SIMPSON, G. G. 1949. *The Meaning of Evolution*. Yale University Press, New Haven, Connecticut. 364 pp.

SIMPSON, G. G. 1953. *The Major Features of Evolution*. Columbia University Press, New York. 434 pp.

SIMPSON, G. G. 1961. *Principles of Animal Taxonomy*. Columbia University Press, New York. 247 pp.

SIMPSON, G. G. 1965. *The geography of Evolution*. Chilton Co., Philadelphia. 249 pp.

STERN, C. 1973. *Principles of Human Genetics*, 3rd ed. W. H. Freeman and Co., San Francisco. 889 pp.

STERN, J. T., and C. E. OXNARD. 1973. Primate locomotion: Some links with evolution and morphology. *Primatologia* **4**(11):1–93.

STRAUS, W. L., Jr. 1942. Rudimentary digits in primates. *Q. Rev. Biol.* **17**:228–243.

STRAUS, W. L., Jr. 1963. The classification of *Oreopithecus*, pp. 146–177. *In* S. L. Washburn, ed., *Classification and Human Evolution*. Aldine-Atherton, Chicago.

TUTTLE, R. H. 1967. Knuckle-walking and the evolution of hominoid hands. *Am. J. Phys. Anthropol.* **26**:171–206.

TUTTLE, R. H. 1969a. Quantitative and functional studies on the hands of the Anthropoidea, I. The Hominoidea. *J. Morphol.* **128**:309–364.

TUTTLE, R. H. 1969b. Knuckle-walking and the problem of human origins. *Science* **166**:953–961.

TUTTLE, R. H. 1969c. Terrestrial trends in the hands of the Anthropoidea, a preliminary report, pp. 192–200. *Proc. 2nd Int. Congr. Primat., Atlanta Georgia 1968*, Vol. 2. Karger, Basel and New York.

TUTTLE, R. H. 1970. Postural, propulsive, and prehensile capabilities in the cheiridia of chimpanzees and other great apes, pp. 167–253. *In* G. H. Bourne, ed., *The Chimpanzee*, Vol. 2. Karger, Basel and New York.

TUTTLE, R. H. 1972a. Functional and evolutionary biology of hylobatid hands and feet, pp. 136–206. *In* D. M. Rumbaugh, ed., *Gibbon and Siamang*, Vol. 1. Karger, Basel and New York.

TUTTLE, R. H. 1972b. Relative mass of cheiridial muscles in catarrhine primates, pp. 262–291. *In* R. Tuttle, ed., *The Functional and Evolutionary Biology of Primates*. Aldine-Atherton, Chicago.

TUTTLE, R. H. 1974. Darwin's apes, dental apes, and the descent of man: normal science in evolutionary anthropology. *Curr. Anthropol.* **15**:389–426

TUTTLE, R. H. 1975. Knuckle-walking and knuckle-walkers: a commentary on some recent perspectives on hominoid evolution. *In* R. Tuttle, ed., *Primate Functional Morphology and Evolution*. Mouton, Hague.

TUTTLE, R. H., and BASMAJIAN, J. V. 1974a. Electromyography of forearm musculature in gorilla and problems related to knuckle-walking, pp. 293–347. *In* F. A. Jenkins, Jr., ed., *Primate locomotion*. Academic Press, New York.

TUTTLE, R. H., and BASMAJIAN, J. V. 1974b. Electromyography of brachial muscles in *Pan gorilla* and hominoid evolution. *Am. J. Phys. Anthropol.* **41**:71–90.

TUTTLE, R. H., and BASMAJIAN, J. V. 1975. Electromyography of *Pan gorilla*: an experimental approach to the problem of hominization. *In* S. Kondo, M. Kawai, A. Ehara, and S. Kawamura (eds.), *Proceedings from the Symposia of the Fifth Congress of the International Primatological Society.* Japan Science Press Co., Ltd., Tokyo.

TUTTLE, R. H., and C. M. ROGERS. 1966. Genetic and selective factors in reduction of the hallux in *Pongo pygmaeus.* *Am. J. Phys. Anthropol.* **24**:191–198.

TUTTLE, R. H., J. V. BASMAJIAN, E. REGENOS, and G. SHINE. 1972. Electromyography of knuckle-walking: results of four experiments of the forearm of *Pan gorilla. Am. J. Phys. Anthropol.* **7**:255–266.

TUTTLE, R. H., J. V. BASMAJIAN, and H. ISHIDA. 1975. Electromyography of the gluteus maximus muscle in gorilla and the evolution of hominid bipedalism. *In* R. Tuttle, ed., *Primate Functional Morphology and Evolution.* Mouton, Hague.

VAN HORN, R. N. 1972. Structural adaptations to climbing in the gibbon hand. *Am. Anthropol.* **74**:326–334.

WASHBURN, S. L. 1973. Primate studies, pp. 467–485. *In* G. H. Bourne, ed., *Nonhuman Primates and Medical Research.* Academic Press, New York.

WHITTAKER, R. H., S. A. LEVIN, and R. B. ROOT. 1973. Niche, habitat, and ecotope. *Am. Nat.* **107**:321–338.

ZAPFE, H. 1960. Die Primatenfunde aus der miozanen Spaltenfüllung von Neudorf an der March (Devinska Nova Ves), Tschechoslowakei. *Schweiz. Palaontol. Abh.* **78**:1–269.

ZINGESER, M. R. 1973. Dentition of *Brachyteles arachnoides* with reference to alouattine and ateline affinities. *Folia Primatol.* **20**:351–390.

ZWELL, M., and G. C. CONROY. 1973. Multivariate analysis of the *Dryopithecus africanus* forelimb. *Nature* **244**: 373–375.

Index